SYMPOSIA OF THE
SOCIETY FOR EXPERIMENTAL BIOLOGY

NUMBER XXXVI

SYMPOSIA OF THE
SOCIETY FOR EXPERIMENTAL BIOLOGY

The Journal of Experimental Botany
is published by the Oxford University Press
for the Society for Experimental Biology

(SYMPOSIA OF THE)
SOCIETY FOR EXPERIMENTAL BIOLOGY
(Gt. Britain)

NUMBER XXXVI

THE BIOLOGY OF PHOTORECEPTION

Published for the Society for Experimental Biology

CAMBRIDGE UNIVERSITY PRESS

CAMBRIDGE

LONDON   NEW YORK   NEW ROCHELLE

MELBOURNE   SYDNEY

Published by the Press Syndicate of the University of Cambridge
The Pitt Building, Trumpington Street, Cambridge CB2 1RP
32 East 57th Street, New York, NY 10022, USA
296 Beaconsfield Parade, Middle Park, Melbourne 3206, Australia

First published 1983

Printed in Great Britain at The Pitman Press, Bath

Library of Congress catalogue card number: 82–22032

British Library Cataloguing in Publication Data
The biology of photoreception. – (Society for
Experimental Biology symposium; 36)
1. Photosynthesis
I. Cosens, D. J. II. Vince-Prue, D.
III. Series
581.1'3342   QK882

ISBN 0–521–25152–4

CONTENTS

PREFACE

The thirty-sixth Symposium of the Society entitled 'The Biology of Photoreceptors', was conceived in London, nurtured in Edinburgh and Littlehampton, and held in Norwich. It was run in association with the British Photobiology Society. Taking a cue from a predecessor: 'Receptor Mechanisms', the Symposium set out to embrace an exchange of information and ideas between parallel problems in plant and animal photoreceptor systems. The aim was to seek from present knowledge, by juxtaposing a botanical and a zoological viewpoint in each session, common mechanisms or principles, or to define differences.

The sessions were organised to consider, firstly, photoreceptor pigments and their physiological properties, comparing situations where light is an energy source and an environmental cue; secondly, the mechanisms for transducing light energy into either chemical or electrical energy; thirdly, particular properties of light: the ways in which plants and animals perceive the direction, wavelength and colour of light, and its pattern of polarisation in the sky. Finally some selected topics were chosen to illustrate current ideas and research.

By its nature such a Symposium cannot be comprehensive, indeed its constituent parts are often dictated by factors that defy any organisation. Yet it is worthwhile to convey the fascination and excitement of the topic; and the contributors, diverse botanists, zoologists and others, were successful in this, and a cohesion did develop during the meeting which is reflected in the present volume. The arrangement of chapters follows the original sequence of the meeting but have not been grouped formally into sections, and there is an additional chapter by Professor Galston who unfortunately was unable to attend.

The Symposium was held at the University of East Anglia from the 8th to 10th September, 1981 and enjoyed many facilities made available by that University. Dr John Noble-Nesbitt undertook the duties of Local Secretary splendidly. My co-convenor/editor Dr D. Vince-Prue and I are also indebted to the contributors and to colleagues for their help during the planning of the Symposium. It is a pleasure to record the help and co-operation of colleagues and the Cambridge University Press during the final editing and preparation of this volume.

Derek Cosens

THE NATURAL RADIATION ENVIRONMENT: LIMITATIONS ON THE BIOLOGY OF PHOTORECEPTORS. PHYTOCHROME AS A CASE STUDY

HARRY SMITH

Department of Botany, University of Leicester, Leicester LE1 7RH, UK

Introduction

Almost all of the contributions to this volume are concerned with mechanisms in photobiology. The elucidation of mechanisms is currently the dominant drive throughout the whole of modern biology, and it is only to be expected that identification of photoreceptor action should appear as an important objective at the present time. There is a sense, however, in which the pursuit of mechanistic solutions, involving as it often does the application of exceptionally artificial conditions, can lead us away from an appreciation of the role of photoreceptors in the whole life of the organism. We need to know not only how a photoreceptor acts but also what function it serves. This latter function may indeed be very obvious as, for example, in vision and photosynthesis. But in a number of other cases it is not always readily apparent what adaptive value the possession of a particular photoreceptor confers upon the organism. This question is particularly important for signal-transducing photoreceptors: those that serve to provide the organism with information on the characteristics of the radiation environment to which the organism is exposed. For these purposes the term *function* may indeed be constructively defined as being the capacity of a photoreceptor to acquire information about the radiation environment (Smith, 1978, 1981*a*, 1982). It is logical to conclude that each signal-transducing photoreceptor will only have a single specific function as defined above. Furthermore, the existence of more than one signal-transducing photoreceptor in any organism presumably implies that each photoreceptor would have a *different* fundamental function. This requirement, that each photoreceptor should have only one function, does not preclude the possibility of any photoreceptor being involved in a wide range of different biological phenomena. Indeed, the overt metabolic and developmental responses to the same environmental information would be expected to be very different in different cells, at different stages of development, and in different organisms. Finally, in this context, and perhaps most importantly, we should expect the mechanism of action of any signal-transducing photoreceptor to be a molecular expression of the photoreceptor's fundamental perceptual function; in other words, models of photoreceptor action should be consistent with what can be determined about the operation of the photoreceptor in natural environment conditions. These considerations taken together mean that our search for the mechanism of action of signal-transducing photoreceptors is likely to be hampered unless we take due account of the ecological and evolutionary aspects of the photoreceptors' overall role in the life of the organism. In order to exemplify this point of view I shall use as a case study the signal-transducing photoreceptor found in all green plants: phytochrome.

Acquisition of information from the natural light environment

Before dealing with phytochrome in particular it will be useful to cover more generally the question of what information can be derived from the natural light environment. In this consideration I have arbitrarily excluded visual perception. Firstly, it is worth a superficial glance at the very wide range of biological photophenomena that exist (Table 1). The majority of these phenomena are, of course, due to signal transduction and thus involve acquisition of information from the environment. These phenomena have been studied in detail under closely-controlled laboratory conditions but each must have a role in the whole life of the organism. It is also possible to construct quite an extensive list of biological photoreceptors (Table 1). This table has been constructed principally from Presti & Delbrück (1978) with some additions. Although the list looks quite long the number of different chemical entities is really remarkably small when one considers the range of light-absorbing molecules present in the biosphere. In terms of the consideration in the Introduction, those photoreceptors which act to transduce environmental signals, rather than to acquire light energy, should exhibit a specific function definable in terms of the information acquired about the natural light environment. The available information can be quite readily categorised as shown in Table 2; thus it is possible to acquire information on the amount of light, the spectral distribution of the light, the direction of the light, the duration of exposure to light, and the plane of polarisation. In theory, the acquisition of this information

Table 1. *Major biological photophenomena and photoreceptors*

Photophenomena	Photoreceptors
Photoreactivation	UV reactivating enzyme
Photosynthesis	Provitamins D
Photokinesis	Rhodopsin
Phototaxis	Bacteriorhodopsin
Phototropism	Chlorophyll
Polarotropism	Bacteriochlorophyll
Photonasty	Protochlorophyll
Photomorphogenesis	Haem pigments
Photoperiodism	Flavoproteins
Vision	Carotenoids
	Phytochrome

Table 2. *Information that may be derived from the light environment*[a]

Information	Perception mechanism
Light quantity	Photon-counting
Light quality	Photon ratios
Direction	Photon gradients
Duration	Timing of light-dark transitions
Polarisation	Dichroic photoreceptor arrangement

[a] Excluding visual perception.

requires only a relatively limited range of perception mechanisms. Thus the perception of light quantity must involve photon-counting. Similarly, the perception of the spectral distribution of radiation must involve the estimation of ratios of photons in two or more wavelength bands; this of course depends upon photon-counting in the two or more bands plus some device for comparing the quantities thus measured. Perception of the direction of the actinic beam must depend upon the detection of photon gradients in space – again a matter of photon-counting at different points in the organism. The perception of the duration of exposure to light depends on timing of light–dark transitions; this may involve simple photon-counting, but also could involve other events that occur at the natural light–dark transitions, such as changes in the spectral distribution of radiation. Finally, the perception of the plane of polarisation of natural radiation must clearly depend upon dichroic arrangements of photoreceptors.

Table 3. *Some physiological effects of light*[a]

Photophenomena	Perception
Photosynthetic unicells	
Motility – rate	Photon-counting
Motility – direction	Photon-counting
Division rate	Photon-counting
Filamentous algae	
Chloroplast orientation	Photon-counting + photon ratios
Growth orientation	Photon gradient + dichroism
Fungi	
Growth orientation	Photon gradients
Sporulation	Photon-counting
Plants	
Germination	Photon-counting + photon ratios
De-etiolation	Photon-counting + photon ratios
Growth orientation	Photon gradient
Pigment synthesis	Photon-counting
Stem elongation	Photon ratios
Flowering and dormancy	Photon-timing
Insects	
Diapause	Photon-timing
Activity	Photon-counting
Orientation of flight	Photon-counting (gradients?)
Fish	
Vertical migration	Photon-counting
Birds	
Migration	Photon-timing
Homing	Dichroism
Breeding	Photon-timing
Mammals	
Breeding	Photon-timing
Implantation	Photon-timing
Moulting	Photon-timing

[a] Excluding energy transduction.

When these theoretical aspects are considered in relation to some of the actual photophenomena exhibited by different organisms it is possible to identify, at least in theory, the perception mechanisms involved. Table 3 lists a few selected, well-studied, biological photophenomena with suggestions for the perception mechanisms that must be involved. The list is not comprehensive and is not intended to be absolutely rigorous. The principal intention of the table is to show that perception mechanisms for photon-counting and for photon ratios can account for the majority of biological photophenomena if we assume that the organisms have evolved additional mechanisms to allow the distribution of photon counts and of photon ratios in either space or time to be determined. Keeping in mind these very general ideas – which I readily admit to being vulnerable to considerable criticism – I would now like to move to a consideration of phytochrome as being a superbly-designed device for the perception of photon ratios.

Properties of phytochrome

The properties of phytochrome have been well reviewed (Pratt, 1979, 1982) and also are covered in some detail in other chapters in this volume, consequently it is not necessary for me to do more than give a brief account of those properties that I regard as being essential for an understanding of the fundamental perceptual function of phytochrome. The most crucial property of phytochrome as far as its role in the whole life of the organism is concerned is its photoreversible photochromicity. The molecule exists in two forms, Pr, which absorbs maximally at around 660 nm and is converted by light into the other form, Pfr, which absorbs maximally at around 730 nm and is photoconverted back to Pr. Both Pr and Pfr are relatively stable components, although in etiolated tissues Pfr is subject to rather rapid degradation. In etiolated tissues phytochrome is only synthesised in the Pr form and the synthesis is a zero order reaction; Pfr breakdown (known as 'destruction') is a first order reaction, except in certain cereals, and Pfr can also be converted to Pr in darkness in some dicotyledonous plants (dark reversion). The interconversions can be described by the following scheme:

$$\underset{\text{Synthesis}}{\xrightarrow{\hspace{1.5cm}}} \; Pr \; \underset{hv;\,(\lambda_{max}\,730\,nm)}{\overset{hv;\,(\lambda_{max}\,660\,nm)}{\rightleftharpoons}} \; Pfr \; \underset{\text{Destruction}}{\xrightarrow{\hspace{1.5cm}}}$$
Reversion

The absorption spectra of the Pr and Pfr forms of phytochrome overlap below 730 nm, and thus a photoequilibrium is established. Thus in broadband irradiation, to which the plant is naturally exposed, the proportions of Pr and Pfr present will be determined very largely by the proportions of red and far-red light present in the incident radiation. Consequently it has become common practice to use the ratio R:FR (that is, the ratio of red to far-red radiation) to describe natural radiation spectra in terms relevant to the operation of phytochrome

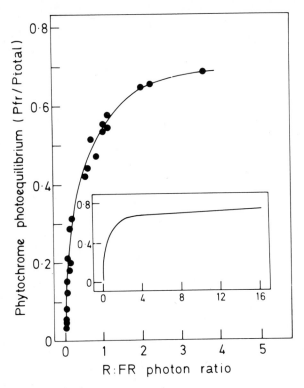

Fig. 1. The relationship between R:FR and phytochrome photoequilibrium (from Smith & Holmes, 1977).

within the plant (Holmes & Smith, 1975; Smith & Holmes, 1977). In the work described here R:FR* is defined as follows:

$$R:FR = \frac{\text{Photon fluence rate between 655 and 665 nm}}{\text{Photon fluence rate between 725 and 735 nm}}$$

When R:FR is determined by spectroradiometry for a large number of natural and artificial broadband radiation sources and compared with the measured Pfr/Ptotal established in etiolated tissue by the same sources, the relationship between the two parameters approximates a rectangular hyperbole (Fig. 1). This relationship allows one to convert R:FR into estimated Pfr/Ptotal values. As will be seen this is a very useful technique. I wish to stress, however, that this estimate of Pfr/Ptotal does not give a numerical estimate of the average photoequilibrium present within green tissues. Since chlorophyll absorbs red

* In many publications, R:FR is symbolised as ζ.

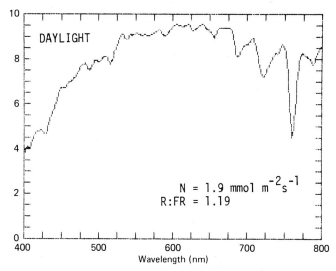

Fig. 2. A typical spectroradiometric scan of global radiation received by a horizontal cosine-corrected detector exposed to natural light (solar angle >10 deg.). Spectral photon fluence rates: 10 on ordinate scale = $6.0\,\mu$mol m^{-2} s^{-1} nm^{-1}.

light but does not absorb far-red light, any phytochrome within green tissues will be present at a lower Pfr/Ptotal than that estimated from Fig. 1; the estimate, however, serves to convert R:FR data into a physiologically meaningful parameter.

The natural radiation environment in relation to the function of phytochrome

It is clear from the above that any changes in the natural radiation environment which result in big differences in R:FR may also result in large differences in Pfr/Ptotal within the exposed tissues. From Fig. 1 it is obvious that for R:FR values between 0 and 2.5, any change in R:FR will cause a big change in Pfr/Ptotal. On the other hand for R:FR values above 2.5, even very large changes in R:FR will only have a very small effect on Pfr/Ptotal. With this in mind it is useful to analyse the natural radiation environment, concentrating particularly on the spectrum of natural radiation in the red and far-red and looking for significant changes which might provide information of ecological value to the green plant.

Figure 2 shows the typical spectroradiometric scan of natural daylight using a horizontal cosine-corrected detector (that is, global radiation). Assuming the sun is more than 10 deg above the horizon, the global

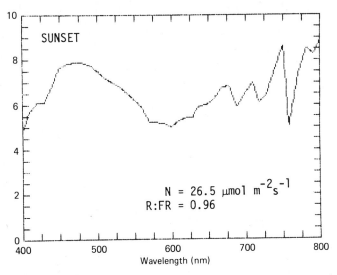

Fig. 3. Typical spectroradiometric scan of global radiation at sunset. Spectral photon fluence rates: 10 on ordinate scale = $0.1\,\mu\text{mol m}^{-2}\,\text{s}^{-1}\,\text{nm}^{-1}$.

daylight spectrum is remarkably constant. In particular R:FR varies very little indeed during the day and is hardly affected by even very overcast conditions which might reduce the total fluence rate by a factor of ten. Using the definition of R:FR as outlined above, we have observed a remarkable constancy in its value at 1.15 ± 0.02 (Holmes & Smith, 1977a). This value of R:FR would establish in etiolated tissue a Pfr/Ptotal of approximately 0.55 (see Fig. 1). A comprehensive survey of natural radiation shows that there are three ecologically significant conditions which yield substantial changes in R:FR as compared with normal daylight (see Morgan & Smith, 1981; Smith, 1982). These three conditions are: (a) twilight; (b) underwater, and (c) under vegetation canopies.

Twilight

Twilight is here defined operationally as that period during which the sun subtends an angle of between +10 deg and −10 deg to the horizon, and clearly extends for varying periods of the 24 h cycle, depending upon latitude and time of year. Figure 3 shows a spectroradiometric scan taken at sunset in early spring in a temperate location (Leicester, UK). R:FR progressively decreases as the sun's angle to the horizon declines. This aspect of the spectrum at twilight is due to the direct solar beam being refracted and scattered by the earth's atmosphere; consequently, the degree to which R:FR is reduced at twilight is dependent upon the relative contribution of the direct beam to the global radiation. Thus, the decline in R:FR at the end of day is very variable, since clouds on the horizon substantially affect the proportion of the direct solar beam present in the global radiation. Nevertheless it has been suggested (Wagner, 1976; Holmes & Wagner, 1980) that the drop in R:FR – and the consequent drop in Pfr/Ptotal – is involved in the initiation of dark-time measurement in photoperiodism. There is very little direct evidence for this, and such a conclusion seems quite unwarranted. The most important argument against phytochrome detecting the progressive reduction in R:FR at the end of day is that the spectral changes are least noticeable at high latitudes where photoperiodic phenomena are seen at their most dramatic. A full spectroradiometric analysis of natural radiation at high latitudes has not yet been made but Goldberg & Klein (1977) have shown with broadband filter measurements that the ratio of red to far-red radiation shows only small diurnal changes. This, of course, is to be expected since the angle which the sun subtends to the horizon at high latitudes is never very great. Consequently, and in agreement with the conclusion of Salisbury (1981), it seems most unlikely that end-of-day reduction in R:FR acts as the initiating signal for dark-timing in photoperiodism.

Underwater

The spectrum of radiation underwater is determined (*a*) by the spectrum reaching the surface, (*b*) by the absorptive properties of the water and of any dissolved or particulate matter in that water, and (*c*) by the scattering of light by molecules and particles within the water. There has been a considerable study of underwater spectra and it is quite clear that the topic is very complex. Jerlov (1966, 1975) recognises three classes of coastal waters and nine classes of oceanic waters, each characterised by the spectrum of radiation found at different depths. Spence (1975, 1981) recognises three categories of lake waters, again characterised by their light attenuation properties. It is not possible to go into the necessary detail in this article (see Morgan & Smith, 1981; Smith, 1982) and I will concentrate solely on the R:FR ratio. Water itself absorbs quite strongly in the far-red, a factor which contributes the large absorption trough seen in the spectrum of natural daylight (Fig. 2). With increasing depth therefore, and ignoring any contribution from dissolved or particulate matter, there is a gradual increase in R:FR underwater; indeed this is the *only* natural situation in which R:FR is found significantly to increase above the value for natural daylight. It is

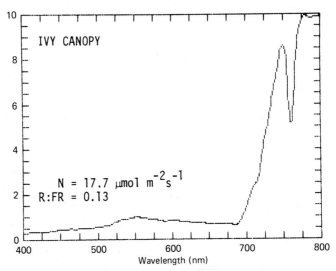

Fig. 4. Typical spectroradiometric scan of global radiation beneath a dense vegetative canopy. Spectral photon fluence rates: 10 on ordinate scale $= 0.2\,\mu$mol m^{-2} s^{-1} nm^{-1}.

possible, in theory, therefore, that phytochrome could act in underwater plants as a depth sensor (Spence, 1981). It can be seen from Fig. 1 however, that increasing R:FR beyond about 2.5 has very little effect on Pfr/Ptotal, and thus phytochrome would appear to make a very poor detector of R:FR underwater. A further complication is that the presence of any dissolved or particulate matter that absorbs in the red – for example, phytoplankton – will have a counteracting effect on the increase in R:FR with increasing depth. Morgan & Smith (1981) have calculated that for normally turbid waters the presence of organic particulate matter would make R:FR a highly unpredictable parameter if it were to be used as a depth sensor.

Canopies

Under vegetation canopies there is a very marked change in R:FR due to the absorption of the red and the blue radiation, and a substantial part of the green radiation, by the vegetation. Leaves are virtually transparent to far-red, and thus under-canopy spectra are typically similar to that shown in Fig. 4. A wide survey of reported R:FR values under canopies (see Morgan & Smith, 1981) shows that R:FR can range from the value found in daylight (around 1.15) down to values as low as 0.05. A glance at Fig. 1 will show that this range of R:FR encompasses about 70% of the theoretically possible range of Pfr/Ptotal; consequent-

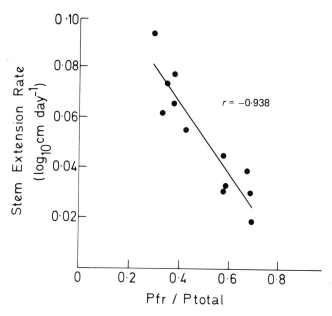

Fig. 5. The linear relationship between stem growth rate and estimated phytochrome photoequilibrium for *Chenopodium album* seedlings (from Morgan & Smith, 1976, 1978*a*).

ly a small drop in R:FR due to canopy shading will have a big effect on Pfr/Ptotal. This means that phytochrome could act as a very sensitive sensor of vegetation shade. This possibility has been exhaustively tested by growing plants under broadband white light in which R:FR has been systematically varied whilst keeping photosynthetically active radiation (PAR) (= 400–700 nm) constant and uniform. In these simulation experiments it was found that low R:FR treatments, even at fluence rates much higher than those found under natural canopies, caused extreme increases in stem extension growth in a range of shade-avoiding herbaceous species (for example, *Chenopodium album, Senecio jacobea, Chamaenerion angustifolium, Sinapis alba*) (Morgan & Smith, 1979). Furthermore, if the R:FR values of the light sources used in these experiments were converted to Pfr/Ptotal by the use of the curve in Fig. 1, a direct linear relationship was revealed between stem extension growth and Pfr/Ptotal (Fig. 5; Morgan & Smith, 1976, 1978*a*). When shade-avoiding species were compared with woodland herbs in these simulation experiments an interesting ecological relationship emerged; shade-avoiding species reacted very strongly, in terms of their extension growth, to reduction in R:FR at the same total PAR, but the shade-tolerant species, although exhibiting small increases in growth, did not show a marked response (Morgan & Smith, 1979). There is, therefore, persuasive evidence that phytochrome acts as a sensor of canopy shading and, in the case of shade-avoiding species, this information is used to bring about very large changes in the rate of stem extension.

The fundamental function of phytochrome

From the above it can be seen that only by comparing the properties of phytochrome with the observed changes in the natural radiation environment can we say with any degree of certainty what the ecological significance of phytochrome is. Evidence supports the view that the fundamental perceptual function of phytochrome is to detect the relative amounts of red and far-red light in the incident radiation; at the present state of higher plant evolution this perceptual capacity seems to be most important in allowing a plant to detect shading by other vegetation. It should not be presumed, however, that the fundamental function of phytochrome is to detect shading. It may be that at earlier stages of evolution the capacity to detect depth underwater was conferred upon plants by a phytochrome-like photoreceptor whose physical properties were sufficiently different from the present molecule that the R:FR ratios found under clear water would produce changes in the photoreceptor sufficiently large to bring about biological responses. It is further conceivable that selection pressures could later have brought about the evolution of the phytochrome molecule as we know it today. It is certainly true that shading by other vegetation is very significant in the competition for light and thus would be expected to represent a powerful selection pressure. Such questions represent vain speculation, of course, but they serve to underline the main point of this article: that we should seek the fundamental function of signal-transducing photoreceptors not in terms of overt biological responses, such as shade-avoidance or photoperiodism, but in terms of the acquisition and interpretation of environmental information.

Implications for the biology of phytochrome

The above considerations imply that phytochrome operates through the establishment of photoequilibria. Unfortunately, this view runs counter to established thought amongst 'phytochromologists'. The currently most popular theory of phytochrome action is that Pr is an inactive form which is converted by light to Pfr, which is active and which initiates the biological responses. This view, derived from the brilliantly conceived original work of Borthwick & Hendricks (Borthwick, Hendricks, Parker, Toole & Toole, 1952), is so well-entrenched in the phytochrome literature that it is hardly ever questioned, and yet its support rests entirely upon correlations between phytochrome measured by *in-vivo* spectrophotometry in etiolated tissues, and responses to radiation treatments, usually of a narrow band of wavelengths, almost always given over a short period of time. Even these correlations are incomplete, with a number of rather celebrated exceptions being given the term 'paradoxes' (see Hillman, 1967). At first sight, this theory does not conflict with the conclusion derived above that phytochrome must operate through photoequilibria since, under continuous irradiation, the concentration of Pfr would seem to be related directly to the photoequilibrium between Pr and Pfr established by the radiation. Unfortunately, however, on a longer term basis, it can be shown that due to the differential stability of Pr and Pfr, the concentration of Pfr becomes independent of R : FR and of fluence rate (Schäfer & Mohr, 1974). It is predicted that under low R : FR conditions Ptotal will increase, and under high R : FR conditions Ptotal will decrease, such that in both cases the concentration of Pfr becomes the same at steady-state. That this is certainly so in some cases has been shown by measuring the Ptotal in light-grown norflurazon-bleached leaves of *Zea mays* grown under broadband radiation at different R : FR values (Smith, 1981*b*). These results show that whilst extension growth of the leaves is directly related to Pfr/Ptotal it is not correlated at all with Pfr concentration. It is therefore difficult to account for these data – and, by implication, for the role of phytochrome in detecting canopy shade – on the basis of the simple 'Pfr-as-active-form hypothesis'.

When the responses of light-grown plants to changes in R : FR are studied using linear-displacement transducers it can be seen that plants react very quickly indeed to canopy shade (Morgan & Smith, 1978*b*; Morgan, O'Brien & Smith, 1981; Morgan, Child & Smith, 1981). With shade-avoiding species such as *Chenopodium album* and *Sinapis alba*, the addition of far-red to broadband white light (reducing R : FR) leads to a substantial increase in stem extension rate within 7–15 min. By applying the added far-red light through fibre-optic probes it has been possible to show that both the growing internode itself and the developed leaves can act as sites of photoreception for these responses; when the light is applied directly to the internode the growth response occurs after the short lag mentioned above but, if the leaf is the site of perception, growth responses do not begin for at least three hours. These data indicate that the growth rate changes may be mediated via a transportable 'elongation factor'; furthermore, they go some way towards accounting for the ability of the plant to 'average-out' the R:FR signals received by various parts of the plant body. Another advantage of the fibre-optic application of narrowband light to plants growing in broadband background white light is that the antagonistic effects of red and far-red radiation can be demonstrated. Morgan, O'Brien & Smith (1981) were able to prove conclusively that phytochrome was the photoreceptor for these responses by demonstrating red–far-red photoreversibility of stem extension growth as measured by the transducer methods. Similarly it was possible to show that the increment in growth rate was related to the Pfr/Ptotal established by the incident radiation as shown in Fig. 6. A remarkable aspect of this figure is the very dramatic rise in growth rate brought about by a very small decrease in Pfr/Ptotal. On the basis of work with etiolated tissues it had become automatic to assume that small changes in Pfr/Ptotal at relatively high levels of Pfr/Ptotal would have very little effect indeed; again these data are difficult to accommodate within the present central dogma of phytochrome action.

The amount of light in the natural environment in relation to the function of phytochrome

The simulation experiments referred to above (Figs. 5 and 6) show quite clearly a linear relationship between estimated Pfr/Ptotal and extension growth. Clearly, if phytochrome is to operate through the establishment of photoequilibria, then it is necessary to ascertain to what extent there is sufficient light in the natural environment for photoequilibria to be established. There are two problems here. Firstly, where the amount of light absorbable by Pr and Pfr is so low that the rates of the photoconversions are of a similar order of magnitude to the rates of Pfr destruction and Pr synthesis, then a true photoequilibrium is not established. Under these conditions a steady-state is achieved but the proportions of Pr and Pfr are not a simple function of the spectral distribution of photons; they are determined also by the rates of synthesis and degradation of phytochrome. Secondly, at high fluence rates the true photoequilibrium may not be established because of the accumulation of photoconversion intermediates between Pr and Pfr and between Pfr and Pr. The accumulation of intermediates is due to their conversion into the final products (Pr and Pfr) being thermochemical,

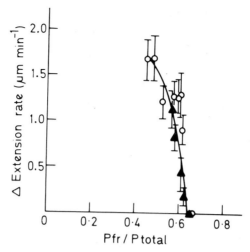

Fig. 6. The relationship between the growth rate increment (ΔER) brought about by the addition of far-red light via a fibre-optic probe direct to the growing internode, and the phytochrome photoequilibrium (Pfr/Ptotal) established by the light sources. *Sinapis alba* seedlings were used grown under background white light with growth rate monitored by linear displacement transducers (from Morgan, O'Brien & Smith, 1981). Data are from different experiments.

rather than photochemical, reactions. Thus the range over which the Pfr/Ptotal will be determined solely by the spectral distribution of photons in the incident radiation will be subject to low and high limits; and unfortunately it is not easy to estimate these limits, as they might apply in Nature, with any degree of accuracy. It seems likely however, that photon fluence rates (400–800 nm) below $10\,\mu$mol m^{-2}s^{-1} will not allow the achievement of photoequilibrium whereas those above $300\,\mu$mol m^{-2}s^{-1} may allow the accumulation of intermediates – particularly at lower temperatures. The eco-physiological significance of these limitations has not yet been grappled with, but its importance is obvious. Within the upper and lower limits, R:FR detection is effectively fluence-rate compensated. In other words, the perception of canopy-shade can occur independently of fluctuations in total light quantity due to climatic or seasonal factors. This fluence-rate compensation is shown clearly by the transducer experiments of Morgan, Child & Smith (1981) who demonstrated that the rapid effects of R:FR on *Sinapis alba* extension growth were independent of total fluence rate.

Recently, Holmes & Wagner (1980) produced a table in which they claimed to show that a very wide range of photomorphogenetic phenomena could be induced by the light present in the natural environ-

Table 4. *Photon fluence requirements for saturation of some phytochrome responses in relation to the natural light environment.* Saturation requirements (μmol m^{-2}) are given in parentheses

Natural light environment[a]	Time to reach saturation				
	Bean hook opening[b] (0.55)	Lettuce germination[c] (55)	*Xanthium* light break[d] (165)	Oat mesocotyl[e] (1650)	*Pharbitis* light break[f] (11 000)
Daylight	0.002 s	0.28 s	0.8 s	8 s	55 s
Below Canopy	0.25 s	28 s	80 s	13 min	1.5 h
Sunset	1 s	1.8 min	5.5 min	55 min	6.1 h
Moonlight	1 h	4.7 d	13.4 d	19 wk	2.4 yr

[a] The daytime data are taken from a typical spring day, cloudless sky at midday (daylight and canopy), and sunset at Leicester, UK. The moonlight data are calculated from the spectroradiometric scan published by Salisbury (1981). *Sources:* [b] Klein, Withrow and Elstad (1956); [c] Withrow (1959); [d] Toole *et al.* (1955); [e] Loercher (1966); [f] Takimoto & Hamner (1965).

ment, even including the very small amounts of phytochrome-absorbable radiation present in moonlight. In this survey they related the total fluence of red light calculated for an 8 h period, under a range of different environmental conditions, to the total fluence required for a number of photomorphogenetic phenomena. These calculations require reciprocity throughout the whole of the 8 h period, neglect the counteracting effect of the far-red radiation present in the natural radiation conditions used, and ignore the problem of attainment of photoequilibrium at low fluence rates. An alternative, but hardly better, way of comparing the photon fluence requirements of phytochrome-mediated phenomena with the available light in the natural environment is to estimate how long it would take for individual photophenomena to achieve saturation under selected natural conditions. This has been attempted in Table 4 using a range of photomorphogenetic phenomena whose saturation characteristics were reported many years ago; as with Holmes & Wagner (1980) only the red component of the natural radiation conditions has been taken into account and no attempt has been made to account for non-attainment of photoequilibrium. It will be seen from Table 4 that some of the calculations are quite absurd, but they do give a graphic illustration of the dangers of determining the light requirements of photophenomena which presumably have some role under natural conditions solely from experiments in the physiological dark room.

Final comment

The implications of the above story are clear but uncomfortable. Much of what we know about the biology of phytochrome, deduced as it has been from experiments carried out with etiolated seedlings grown in darkness and exposed, usually, to brief periods of radiation of a narrow band of wavelengths, seems quite irrelevant to the function of phytochrome in plants growing in the natural environment. On the other hand only by such carefully-controlled experiments has it been possible to achieve the knowledge and understanding we now have of the physical properties of phytochrome. Perhaps the most realistic approach is to assume that the phenomena observed in the dark room and laboratory are manifestations, albeit often markedly distorted by the unnatural growth and radiation conditions, of the responses of plants to changes in R:FR in the natural environment. What is required, therefore, is a model of phytochrome action that will simultaneously provide a sound basis for the photoreceptor's fundamental R:FR perceptual function, and account for the often bizarre responses observed in the laboratory.

References

BORTHWICK, H. A., HENDRICKS, S. B., PARKER, M. W., TOOLE, E. H. & TOOLE, V. K. (1952). A reversible photoreaction controlling seed germination. *Proc. Natl. Acad. Sci., USA*, **38**, 662–6.

GOLDBERG, B. & KLEIN, W. H. (1977). Variations in the spectral distribution of daylight at various geographical locations on the earth's surface. *Solar Energy*, **19**, 3–13.

HILLMAN, W. S. (1967). The physiology of phytochrome. *Ann. Rev. Plant Physiol.*, **18**, 301–24.

HOLMES, M. G. & SMITH, H. (1975). The function of phytochrome in plants growing in the natural environment. *Nature, London*, **254**, 512–14.

HOLMES, M. G. & SMITH, H. (1977a). The function of phytochrome in the natural environment. I. Characterisation of daylight for studies in photomorphogenesis and photoperiodism. *Photochem. Photobiol.*, **25**, 533–8.

HOLMES, M. G. & WAGNER, E. (1980). A re-evaluation of phytochrome involvement in time measurement in plants. *J. Theor. Biol.*, **83**, 255–65.

JERLOV, N. G. (1966). Aspects of light measurement in the sea. In: *Light as an Ecological Factor*, ed. R. Bainbridge, G. C. Evans & O. Rackham, pp. 91–8. Oxford: Blackwell.

JERLOV, N. G. (1975). Light measurements in the sea in terms of quanta. In: *Light as an Ecological Factor*, vol. 2, ed. G. C. Evans, R. Bainbridge & O. Rackham, pp. 521–4. Oxford: Blackwell.

KLEIN, W. H., WITHROW, R. B. & ELSTAD, V. B. (1956). Response of the hypocotyl hook of bean seedlings to radiant energy and other factors. *Plant Physiol.*, **31**, 289–94.

LOERCHER, L. (1966). Phytochrome changes correlated to mesocotyl inhibition in etiolated *Avena* seedlings. *Plant Physiol.*, **41**, 932–6.

MORGAN, D. C., CHILD, R. & SMITH, H. (1981). Absence of fluence rate dependency of phytochrome modulation of stem extension in light-grown *Sinapis alba* L. *Planta*, **151**, 497–8.

MORGAN, D. C., O'BRIEN, T. & SMITH, H. (1981). Rapid photomodulation of stem extension in *Sinapis alba* L. Studies on kinetics, site of perception and photoreceptor. *Planta*, **150**, 95–101.

MORGAN, D. C. & SMITH, H. (1976). Linear relationship between phytochrome photo-equilibrium and growth in plants under simulated natural radiation. *Nature, London*, **262**, 210–12.

MORGAN, D. C. & SMITH, H. (1978*a*). The relationship between phytochrome photo-equilibrium and development in light grown *Chenopodium album* L. *Planta*, **142**, 187–93.

MORGAN, D. C. & SMITH, H. (1978*b*). Simulated sunflecks have large, rapid effects on plant stem extension. *Nature, London*, **273**, 534–6.

MORGAN, D. C. & SMITH, H. (1979). A systematic relationship between phytochrome-controlled development and species habitat, for plants grown in simulated natural radiation. *Planta*, **145**, 253–8.

MORGAN, D. C. & SMITH, H. (1981). Non-photosynthetic responses to light quality. In: *Encyclopedia of Plant Physiology*, Vol. 12a, ed. P. Nobel, pp. 109–34. Berlin: Springer-Verlag.

PRATT, L. H. (1979). Phytochrome: function and properties. In: *Photochemical and Photobiological Reviews*, ed. K. C. Smith, pp. 59–124. New York: Plenum.

PRATT, L. H. (1982). Phytochrome. *Ann. Rev. Plant Physiol.* (In press).

PRESTI, D. & DELBRUCK, M. (1978). Photoreceptors for biosynthesis, energy storage and vision. *Plant, Cell Environ.*, **1**, 81–100.

SALISBURY, F. B. (1981). Twilight effect: initiating dark time measurement in photoperiod-ism of *Xanthium*. *Plant Physiol.*, **67**, 1230–8.

SCHAFER, E. & MOHR, H. (1974). Irradiance dependency of the phytochrome system in cotyledons of mustard (*Sinapis alba* L.). *J. Math. Biol.*, **1**, 9–15.

SMITH, H. (1978). Light quality as an ecological factor. In: *Plants and their Atmospheric Environment*. ed. J. Grace, E. D. Ford & P. G. Jarvis, pp. 93–110. Oxford: Blackwell.

SMITH, H. (1981*a*). Function and evolution of plant photoreceptors. In: *Plants and the Daylight Spectrum*, ed. H. Smith, pp. 499–508. London: Academic.

SMITH, H. (1981*b*). Evidence that Pfr is not the active form of phytochrome in light-grown maize. *Nature, London*, **293**, 163–5.

SMITH, H. (1982). Light quality, photoperception and plant strategy. *Ann. Rev. Plant Physiol.*, **33** (In press).

SMITH, H. & HOLMES, M. G. (1977). The function of phytochrome in the natural environment. III. Measurement and calculation of phytochrome photoequilibrium. *Photochem. Photobiol.*, **25**, 547–50.

SPENCE, D. H. N. (1975). Light and plant response in fresh water. In: *Light as an Ecological Factor* II. ed. G. C. Evans, R. Bainbridge & O. Rackham, pp. 93–133. Oxford: Blackwell.

SPENCE, D. H. N. (1981). Light quality and plant response underwater. In: *Plants and the Daylight Spectrum*, ed. H. Smith, pp. 245–76. London: Academic.

TAKIMOTO, A. & HAMNER, K. C. (1965). Studies on red-light interruption in relation to timing mechanisms involved in the photoperiodic response to *Pharbitis nil*. *Plant Physiol.*, **40**, 852–4.

TOOLE, E. H., TOOLE, V. K., BORTHWICK, H. A. & HENDRICKS, S. B. (1955). Interaction of temperature and light in germination of seeds. *Plant Physiol.*, **30**, 473–8.

WAGNER, E. (1976). The nature of periodic time measurement: energy transduction and phytochrome action in seedlings of *Chenopodium rubrum*. In: *Light and Plant Development*, ed. H. Smith, pp. 419–43. London: Butterworth.

WITHROW, R. B. (1959). A kinetic analysis of photoperiodism. In: *Photoperiodism and Related Phenomena in Plants and Animals*, ed. R. B. Withrow, pp. 439–71. Washington: American Association for the Advancement of Science.

PROPERTIES AND ORGANISATION OF PHOTOSYNTHETIC PIGMENTS

J. BARBER

Department of Pure and Applied Biology, Imperial College, Prince Consort Road, London SW7 2BB, UK

Introduction

The starting point for this chapter must be the recognition that the majority of the photosynthetic pigments act as a light-capturing antenna organised in such a way that the energy of an absorbed photon is efficiently transferred to a specific photochemical reaction centre where energy conversion occurs. This concept is shown diagrammatically in Fig. 1 where it can be seen that the initial energy conversion process is the creation of a metastable charge-transfer complex. These processes take place in picoseconds and are termed the primary light reactions. Subsequent electron-transfer events then occur within the time scale from picoseconds to milliseconds, and these stabilise the energy initially stored in the reaction centre (Barber, 1978). As discussed by Mathis in this volume, a number of redox components are involved in these slower secondary electron-transfer processes, the nature of which depends on the type of organism. In oxygen-evolving plants and algae, two different primary light reactions occur and act in series, so that there is sufficient redox separation to bring about the oxidation of water and reduction of carbon dioxide. However, in the case of photosynthetic bacteria, the oxidising potential created in the reaction centre is less positive than in oxygen-evolving systems, so that these organisms cannot extract electrons from water but use substances like hydrogen, hydrogen sulphide and a range of organic compounds as reducing agents for carbon dioxide. Consequently, photosynthetic bacteria contain only one type of reaction centre having properties that depend on the particular species.

The idea that each reaction centre is served by many light-harvesting pigment molecules stems back to the elegant experiments of Emerson & Arnold reported in 1932. They showed, using *Chlorella* as the experimental organism, that a short flash of light of saturating intensity gave rise to the evolution of one molecule of oxygen per 2400 chlorophyll *a* molecules. If we accept that two photosystems must act together and that the evolution of one molecule of oxygen involves the excitation of four electrons, this means that about 300 chlorophyll *a*

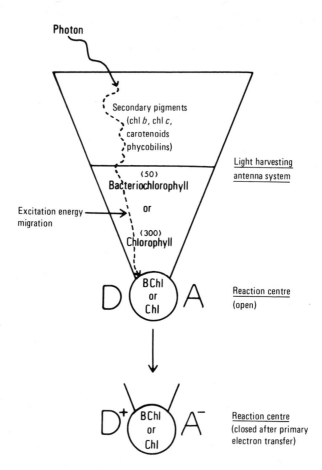

Fig. 1. A diagrammatic representation of a photosynthetic unit consisting of a light-collecting pigment system and a reaction centre. An excitation induced by photon absorption is efficiently transferred through the pigment system and is channelled to the reaction centre where stabilisation of the excitation is accomplished by a charge-transfer process.

molecules are associated with each photoactivated reaction centre. Since these light-harvesting chlorophyll *a* molecules have a complement of secondary light-harvesting pigments (see below), the complete antenna system is able to capture light energy over a wide spectral range, and allows photosynthetic electron flow to occur at high rates even when the rate of photon interception by a given antenna complex is low. The number of light-harvesting chlorophylls per reaction centre is often lower in photosynthetic bacteria than in O_2-evolving organisms, being in the region of 60 to 100, although there are exceptions – notably the green bacteria.

In this chapter I will only briefly summarise the main properties of the photosynthetic pigments because they have been dealt with extensively in other books and reviews (Goodwin, 1965; Vernon & Seely, 1966; Anderson, 1975; Seely, 1977; Song, 1978; Clayton & Sistrom, 1978; Boardman, Anderson & Goodchild, 1978; Sauer, 1979; Thornber & Barber, 1979; Cogdell & Thornber, 1980; Glazer, 1981; Hiller & Goodchild, 1981). Instead, I will place emphasis on the current ideas concerning their organisation as macroscopic protein complexes and discuss some of the factors that may control their inter-relationships as intrinsic and extrinsic membrane components.

Photosynthetic pigments

All O_2-evolving organisms contain chlorophyll *a*, while all photosynthetic bacteria contain bacteriochlorophyll. The remaining pigments are termed secondary or accessory pigments, the nature of which varies widely among the different types of photosynthetic organism. These secondary pigments fall into three general classes: various forms of chlorophyll other than chlorophyll *a* and bacteriochlorophyll, carotenoids, and phycobilins. Details of the distribution of the different types of light-harvesting pigments among the various classes of eukaryotic and prokaryotic photosynthetic organisms are given in Tables 1 and 2, while the chemical structures and spectral properties of some of the more common pigments are presented in Figs. 2–5. Although organisms may grow photosynthetically, even when the level of secondary pigments is lowered or depleted (by varying the growth conditions or by mutation), there is an absolute requirement for the presence of either chlorophyll *a* or bacteriochlorophyll. This is probably because these two types of chlorophyll are not only necessary as antenna pigments but also because they form an essential part of the photochemical reaction centre. In these reaction centres they act as the primary electron donors and seem to exist as a 'special pair' (Mathis & Paillotin, 1981, but also contrast with the recent report of Wasielewski, Norris, Shipman, Lin & Svec, 1981).

All chlorophylls have two main absorption bands in the 'visible' portion of the spectrum, in the blue and in the red or near infra-red (see Fig. 2). The carotenoids (Fig. 5) absorb in the blue while the phycobilins absorb in the green and orange-red regions of the spectrum (see Fig. 3). When compared with the *in-vitro* spectra, the *in-vivo* absorption bands are broadened with their long wavelength maxima shifted to the red. An extreme example of this is bacteriochlorophyll *b* which absorbs at about

Table 1. *Occurrence of pigments in photosynthetic organisms: distributions of chlorophylls and biliproteins in plants and bacteria*[a]

	Higher plants	Chlorophyceae	Conjugatophyceae	Charophyceae	Prasinophyceae	Euglenophyceae	Xanthophyceae	Chloromonadophyceae	Eustigmatophyceae	Haptophyceae	Chrysophyceae	Phaeophyceae	Bacillariophyceae	Dinophyceae	Cryptophyceae	Rhodophyceae	Cyanophyceae	Chlorobacteriaceae	Thiorhodaceae	Athiorhodaceae
Chlorophylls																				
Chlorophyll a	+	+	+	+	+	+	+	+	+	+	+	+	+	+	+	+	+			
Chlorophyll b	+	+	+	+	+	+														
Chlorophyll c_1							+			+	+	+	+							
Chlorophyll c_2							+	(+)		+	+	+	+	+	+					
Bacteriochlorophyll a																		+	+	+
Bacteriochlorophyll b																				+
Bacteriochlorophyll c																		+		
Bacteriochlorophyll d																		+		
Bacteriochlorophyll e																		[b]		
Biliproteins																				
C-phycocyanin																	+			
R-phycocyanin																+				
Allo-phycocyanin																+	+			
Allo-phycocyanin B																+	+			
B-phycoerythrin																+				
C-phycoerythrin																	+			
R-phycoerythrin																+				
Others[c]															+					

[a] Taken from Thornber & Barber (1979).

[b] In 'brown-coloured' *Chlorobium phaeobacterioides* and *C. phaeovibrioides*.

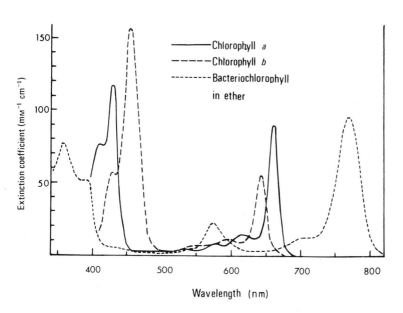

Structure of chlorophyll *a*   Structure of bacteriochlorophyll

Fig. 2. The structures of chlorophyll *a* and bacteriochlorophyll together with their absorption spectra (and that of chlorophyll *b*) measured with the pigments dissolved in ether. The group R shown in the structure for bacteriochlorophyll has not been identified with certainty but is probably a phytyl similar to that in chlorophyll *a*.

Table 2. Occurence of pigments in photosynthetic organisms: distribution of carotenoids in plants and bacteria[a]

	Higher plants	Chlorophyceae	Conjugatophyceae	Charophyceae	Prasinophyceae	Euglenophyceae	Xanthophyceae	Chloromonadophyceae	Eustigmatophyceae	Haptophyceae	Chrysophyceae	Phaeophyceae	Bacillariophyceae	Dinophyceae	Cryptophyceae	Rhodophyceae	Cyanophyceae	Chlorobacteriaceae	Thiorhodaceae	Athiorhadaceae
Olefins																				
Bicyclic hydrocarbons (e.g. α, β, γ-carotene)	+	+	+	+	+	+	+	+	+	+	+	+	+	+	+	+	+			
Bicyclic xanthophylls (e.g. lutein, echinonone)	+	+	+	+	+	+			+		+	+				+	+			
Bicyclic xanthophyll epides (e.g. violoxanthin)	+	+	+	+	+	+			+		+	+				+ (occasionally)				
Bicyclic xanthophylls (C-8 keto) (e.g. siphomoxanthin)		+			+															
Monocyclic xanthophyll glycosides (e.g. myxoxanthophyll)																	+			
Acyclic hydrocarbon and xanthophylls (e.g. spirilloxanthin series and spheroidenone series)																			+	+
Acyclic xanthophyll glycosides (e.g. oscilloxanthin)																	+			

Aryls
Monocyclic 1,2,5-trimethyl hydro-
caron (e.g. chlorobactene)   +

Monocyclic 1,2,3-trimethyl xantho-
phyll (e.g. okenone, warmingone)   +

Allenes
Bicyclic xanthophyll epides
(e.g. neoxanthin, vaucheraxanthin,   + + + + + + + + + +
fucoxanthin, peridinin)

Acetylenes
Bicyclic monoacetylenes   + + + +
(e.g. diatoxanthin)

Bicyclic monoacetylene epoxides   + + + + +
(e.g. diadinoxanthin)

Bicyclic diacetylenes   +
(e.g. alloxanthin)

---

[a] Biosynthetic intermediates which may occur in traces in organisms are not listed; the carotenoid examples given are not necessarily present in all species.

790 nm when dissolved in organic solvents, whereas the absorption maximum *in vivo* is at 1020 nm! The reason for the red shift, particularly for chlorophyll *a* and bacteriochlorophyll, has been the subject of much research (Katz & Norris, 1973). It has been argued that the displacement of the long wavelength peak is due to pigment–pigment, pigment–lipid and pigment–protein interactions (Katz, Norris & Shipman, 1976). Since it seems that all photosynthetic pigments are complexed with protein, it is likely that the red shift involves distortion of the electronic state of the chromophores by the influence of the associated protein (Thornber, 1975). Moreover, because there are several different pigment–protein complexes within a particular type of organism, it is not surprising that the overall *in-vivo* spectrum is complex and indicative of several spectral forms of the same pigment. For example, isolated monomeric bacteriochlorophyll *a* dissolved in organic solvents has a red absorption maximum at about 770 nm, but *in vivo* there exist forms of this pigment absorbing at 800, 850 and 870–890 nm; these are often termed B800–850 and B870, respectively (Cogdell & Thornber, 1980). In the case of isolated monomeric chlorophyll *a*, the red absorption maximum is in the region of 660–665 nm, depending on the organic solvent used for solubilisation. *In vivo*, however, a number of different environments exist for chlorophyll *a*, giving rise to several spectral forms having long wavelength absorption Q_y transition bands ranging from 662 to 704 nm. Because the individual bands overlap each other it is difficult to know precisely how many forms of chlorophyll *a* exist, but resolution of the red absorption band *in vivo* into its Gaussian components led French, Brown & Lawrence (1972) to suggest that there were four major forms of this pigment having peaks at 662, 670, 677 and 684 nm and two minor bands at 693 and 704 nm. These forms are symbolised by the notation Ca-662, Ca-670, etc. Chlorophyll *b* is readily detected *in vivo* by its maximum at 650 nm.

The presence of various forms of chlorophyll *a* and bacteriochlorophyll, coupled with the existence of the secondary light-harvesting pigments, allows photosynthetic organisms to maximise absorption of the energy available in the solar radiation incident on the Earth's surface (after transmittance through the atmosphere). The captured energy within the excited pigment molecules must be transferred to the reaction centre, and it is therefore not surprising that the absorption maxima of the reaction centre chlorophylls are usually at longer wavelengths than those of the antenna pigments. Also, the various pigments which make up a particular antenna system are arranged so that the longer wavelength-absorbing forms are physically closer to the

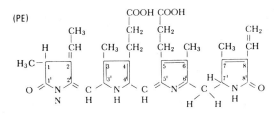

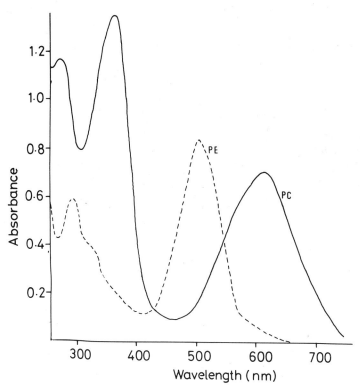

Fig. 3. The structures of phycocyanobilin (PC) and phycoerythrobilin (PE) and their absorption spectra measured in chloroform.

reaction centre (see section dealing with energy transfer, p. 38). In this way the migration of energy to the reaction centre is energetically 'downhill' so that the 'special pair?'* acts as a sink allowing photochemical trapping to occur.

Because O_2-evolving organisms have two different types of reaction centre they have two different antenna systems. The two complete systems are known as photosystem one (PS1) and photosystem two (PS2) (see Thornber & Barber, 1979). The reaction centre of PS1 is called P700 since the chlorophyll involved in the photochemical trapping absorbs light maximally at 700 nm, while the chlorophyll in the PS2 reaction centre absorbs maximally at about 680 nm and is thus signified by P680. Both PS1 and PS2 have chlorophyll *a* in their respective antenna systems, but for PS2 the red absorption maximum is about 670 nm while for PS1 it is 680 nm. Moreover, beyond 700 nm only the chlorophylls associated with PS1 absorb light. Although in photosynthetic bacteria there is only one type of reaction centre, the long wavelength absorption maximum of the associated 'special pair?' does vary depending on the organism. For organisms containing only bacteriochlorophyll *a*, the maximum absorption of the reaction centre chlorophyll is at about 870 nm (P870), while for those which contain bacteriochlorophyll *b* it is at 960 nm (P960).

The ability of photosynthetic organisms to synthesise a range of different secondary pigments seems to be a reflection of their ecological niche. For example, red algae are frequently found growing at considerable depths in the oceans so that they need to capture light that has not been absorbed by surface organisms containing chlorophyll and carotenoids. To do this they synthesise the phycobiliproteins: phycoerythrin, phycocyanin and allophycocyanin, which absorb maximally in the green and orange-red regions of the spectrum (phycoerythrin 490–576 nm, phycocyanin 615 nm, allophycocyanin 654 nm) and so make maximum use of the light transmitted by the window of the chlorophyll/carotenoid absorption spectra.

All photosynthetic organisms contain carotenoids (carotenes and xanthophylls) which help to capture energy in the blue region of the spectrum (for example, β-carotene dissolved in hexane has absorption bands which peak at 430, 450 and 480 nm). However, the carotenoids not only act in a light-harvesting capacity but also serve to protect chlorophyll against photooxidation (Cogdell, 1978). In some cases the absorption spectrum of carotenoid can be shifted into the green portion

---

* The query sign has been introduced since there is debate as to whether the reaction centre chlorophylls form a special pair.

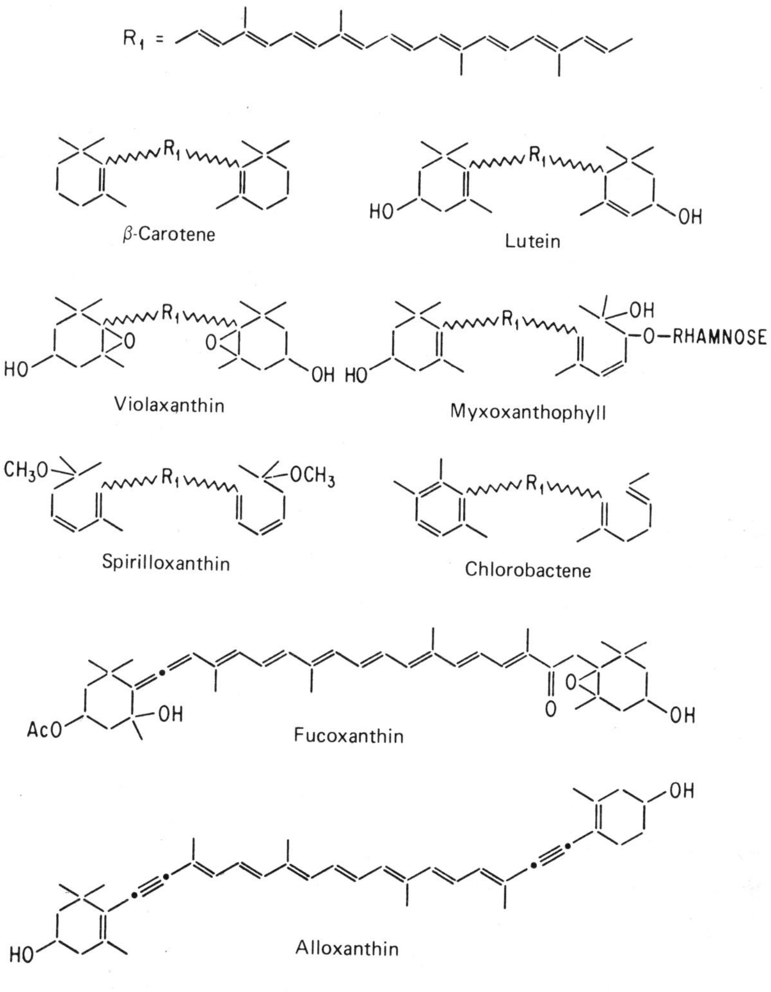

Fig. 4. Structures of some carotenoids commonly found in plants and bacteria.

of the spectrum so as to function in the same way as the phycobilins. For example, the marine alga *Codium* contains the carotenoid siphonoxanthin, which normally absorbs at 450 nm, but *in vivo* in complexes with protein in such a way as to shift its maximum to 540 nm (J. M. Anderson, private communication).

Pigment–protein complexes

It seems very likely that all the *in-vivo* pigments are conjugated with protein. The phycobilins are incorporated into proteins (biliproteins)

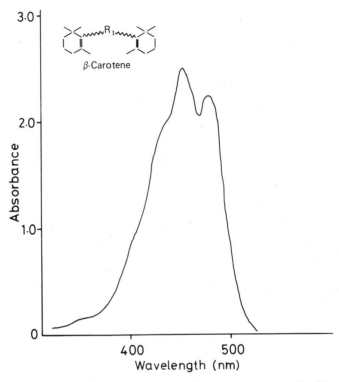

Fig. 5. Structure of β-carotene and its absorption spectrum measured in chloroform.

some of which are water-soluble and can be extracted with ease. But the majority of the other pigment–proteins are not water-soluble and have to be isolated by using various detergents. This is because these pigment–proteins exist as hydrophobic complexes embedded in the lipid matrix of a membrane. Moreover, their extraction is hampered by the fact that the pigment is usually not covalently bonded to the protein and thus is easily lost from the complex during the isolation unless great care is taken. Nevertheless, during the past few years considerable progress has been made in characterising the pigment–protein complexes of both eukaryotic and prokaryotic organisms.

Chlorophyll–protein complexes

For higher plants and algae, non-ionic (Triton X-100 and digitonin) and anionic (sodium and lithium dodecyl sulphates) detergents have been used for solubilisation of the pigment–protein complexes, followed by either polyacrylamide gel electrophoresis (PAGE) (carried out in the presence of the detergent), chromatography, or density-gradient centrifugation. With gel electrophoresis, the numbers of pigmented bands which appear depend on the conditions used, and their origins are being actively studied (Thornber & Markwell, 1981). Even so, it has been known for some years that a large proportion of the chlorophyll can be attributed, on SDS gels, to two major bands. One of these has a molecular weight (M_r) of 100–130 × 10³ and is called CPI; the other band has a M_r of about 30 000, and is often referred to as CPII (Thornber, 1975). CPI contains mainly chlorophyll a and the reaction centre of PS1 (P700), and seems to be ubiquitous to all O_2-evolving eukaryotic and prokaryotic organisms. The chlorophyll:P700 ratio is variable but is often around 40 even when isolated by other procedures. The complex seems to be composed of a major polypeptide of M_r 60–71 × 10³, plus a number of smaller minor polypeptides. However, according to Thornber, Alberte, Hunter, Shiozawa & Kan (1976) there may be two polypeptides of about the same M_r in the region of 48 000 and based on this assumption these authors have suggested a model for the CPI complex which has about 40 chlorophylls per P700. But because the chlorophyll:P700 ratio can vary considerably from 40 (it is often more) and because of the uncertainty of the nature of the apoprotein, no widely accepted model for CPI is yet available. However, a recent study by Mullet, Burke & Arntzen (1980) has given a working picture of the PS1 macroscopic complex involving about 120 chlorophyll a molecules arranged as an antenna system around one P700.

The CPII band contains both chlorophyll a and chlorophyll b but, unlike CPI, has no reaction centre. Its function is to act as an antenna (mainly for PS2) and it is often called the light-harvesting chlorophyll a/chlorophyll b pigment–protein complex (LHC). It has a M_r of about 30 000 and seems to contain six or seven chlorophyll molecules (three or four chlorophyll a and three chlorophyll b), and it is a major component of the thylakoid membrane, usually representing about 50% of the total chlorophyll and protein present. Because of the size of the *in-vivo* light-harvesting system per reaction centre, the monomeric LHC must undergo oligomerisation to form larger complexes (Mullet & Arntzen, 1980). Like CPI, this chlorophyll b-containing protein has a high proportion of hydrophobic amino acids. When completely denatured it is thought to consist of three polypeptides, two having an M_r of about 25 000 while the third is smaller (Thornber & Barber, 1979). However, it remains to be shown whether one or more of the constituent polypeptides carry pigments. From fluorescence polarisation and circular dichroism measurements, Van Metter (1977) has proposed a structural model for the monomeric LHC in which the three chlorophyll b

molecules are closely associated with each other in the centre of the protein and are encircled by the three or four chlorophyll *a* molecules which are closer to the periphery of the complex.

On SDS gels, the CPI and CPII bands can correspond to about 75–85% of the total chlorophyll. It has been assumed that the majority of the remaining chlorophyll, which is very easily solubilised during the detergent treatment (giving rise to a free pigment-band on PAGE), is the light-harvesting chlorophyll *a* that is closely associated with the reaction centre of PS2 (P680). Recently, the use of new detergents, such as the zwitterionic detergent Deriphat 160, and new gel and chromographic techniques have enabled new bands to be detected (Markwell, Thornber & Boggs, 1979; Anderson, 1980); and in particular the existence of a chlorophyll protein of M_r 50 000, containing P680 has attracted attention (Delepelaire & Chuo, 1979). A picture is emerging that the chlorophyll of higher plants and green algae is associated with three main oligomeric macromolecular complexes. These are: (i) a PS1 chlorophyll–protein complex containing P700 and 120 antenna chlorophyll *a* molecules, (ii) a PS2 complex containing P680 and about 60 antenna chlorophyll *a* molecules, and (iii) the light-harvesting chlorophyll *a/b* complex (LHC). In the past these complexes have been viewed as being in physical contact (Butler, 1978), but there is increasing evidence that they should be viewed as spatially separated entities (Anderson, 1980) embedded in the lipid matrix of the thylakoid membrane as shown in Fig. 6(A). The location of the carotenoids is not yet clear. To date no pure caroteno-protein has been isolated from plants or green algae and such complexes probably do not exist. This is because the carotenoid molecules must be situated very close to the chlorophyll molecules in order to allow excitonic interactions to occur. Thus, the *in-vivo* chlorophyll complexes must almost certainly contain carotenoid which in many cases is lost during detergent treatments – although not in all cases, as shown by Anderson & Barrett (1979).

Detergent treatment of photosynthetic bacterial membranes has also been rewarding in terms of isolating pigment–protein complexes. With the weakly zwitterionic detergent lauryl dimethylamine oxide (LDAO), it has been possible to isolate reaction centres from several bacterial species and use them as convenient experimental material to investigate the mechanisms giving rise to primary charge separation (Blankenship & Parson, 1979). Isolated reaction centres usually have M_rs of about 100 000, and contain four bacteriochlorophyll and two bacteriopheophytin molecules. Two of the bacteriochlorophylls form the 'special pair?' and are termed P870 and P960 for organisms containing

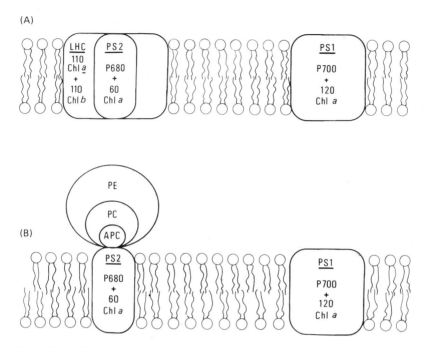

Fig. 6. (A) A diagrammatic model showing the relationships and composition of the PS1 and PS2 light-harvesting chlorophyll complexes within the higher plant chloroplast thylakoid membrane, based on the model given by Anderson (1980) and Mullet, Burke & Arntzen (1980). The LHC–PS2 complex is located predominantly in the appressed membrane of grana while the PS1 complex is found mainly in the non-appressed membranes (see Fig. 8). The PS1 and the PS2 core complexes are probably common to O_2-evolving organisms. (B) A diagrammatic model of the light-harvesting system of a phycobilin-containing organism (in this case a member of Rhodophyceae, e.g. *Porphyridium cruentum*). Here the LHC complex shown in (A) has been replaced by a phycobilisome containing phycoerythrin (PE) phycocyanin (PC) and allophycocyanin (APC) in the proportions 84%, 11% and 5% respectively. The model is based on that given by Gantt, Lipschultz & Zilinskas (1976) and is consistent with energy transfer studies (see Fig. 6).

bacteriochlorophyll *a* and *b* respectively. The reaction centre protein of the purple bacterium, *Rhodopseudomonas sphaeroides*, consists of three polypeptides with M_rs of 21-, 24- and 27×10^3 respectively, occurring with a stoichiometry of $1:1:1$ (Dutton, Leigh, Prince & Tiede, 1979). As yet no reaction centre complex equivalent to the bacterial systems has been isolated from O_2-evolving organisms. With regard to the light-harvesting pigments in purple bacteria there appear to be two main caroteno-bacteriochlorophyll–protein complexes, one corresponding to B870 and the other containing B800–850; in some organisms only the B870 protein occurs (Cogdell & Thornber, 1979).

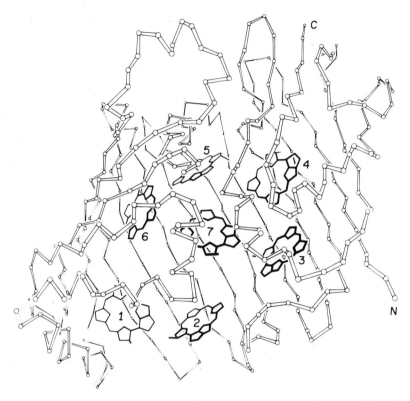

Fig. 7. Schematic diagram showing the polypeptide backbone and the seven chlorophyll molecules in one subunit of the trimeric bacteriochlorophyll *a*–protein from a green bacterium, *Prosthecochloris aestuarii* (from Fenna & Matthews, 1976).

The B800–850 complex is small, containing two polypeptides, each of M_r 10 000, in a 1:1 ratio. In this complex, one of these polypeptides has two bacteriochlorophyll molecules associated with it and the two chromophores are sufficiently close to result in exciton interaction, thus giving rise to the 850 nm absorption maximum. The other polypeptide contains one bacteriochlorophyll molecule, absorbing at 800 nm, and a carotenoid molecule. The B870 protein has an apparent M_r of about 20 000, consisting of one polypeptide, two bacteriochlorophylls and two carotenoid molecules (Cogdell & Thornber, 1980). However, the best characterised of all chlorophyll-containing protein complexes is the water-soluble bacteriochlorophyll *a*–protein complex of the 'green' photosynthetic bacteria (see Olsen, 1980 for review). This complex, unlike the hydrophobic complexes discussed above, has lent itself to detailed structural analysis. In an isolated crystalline state, high resolution X-ray analysis (Fenna & Matthews, 1976) has shown that the pigment–protein forms a trimer. Each monomer, which has a M_r of 45 000, has an overall shape resembling a flattened hollow cylinder and contains seven chlorophyll molecules (see Fig. 7). In the trimer all the chlorophyll molecules are completely surrounded by protein, the exposed surface of which is formed mainly by fifteen strands of β-pleated sheets. Within the monomer, the average centre-to-centre distance of the porphyrin rings of the chlorophylls is about 1.2 nm and the phytyl chains lie close together within the centre. The water solubility of the complex is probably accounted for by a high percentage of polar amino acid residues exposed at the surface.

Phycobiliproteins

These water soluble pigment-proteins are readily extracted from cyanobacteria* (Cyanophyceae) and red algae (Rhodophyceae) as well as from the cryptomonads and are ideal candidates for detailed structural studies (see Glazer, 1981). Unlike chlorophyll-proteins, the biliproteins have chromophores which are covalently bound to protein. They can be categorised into three general classes: phycoerythrin, phycocyanin and allophycocyanin. These phycobiliproteins consist of two polypeptide chains, α and β, of similar amino acid composition and size (the M_r of the α-chain ranges from 10 000 to 20 000 and that of the β-chain from 14 000 to 22 000), and have one chromophore per subunit (MacColl & Berns, 1979). In the cyanobacteria (blue-green algae) and red algae, the phycobiliproteins complex together to form macroscopic structures of 32–35 nm in diameter known as phycobilisomes consisting of about 100 chromophore molecules (Gantt, Lipschultz & Zilinskas, 1976). These phycobilisomes occur as extrinsic membrane proteins, and like LHC preferentially transfer energy to the chlorophyll antenna of PS2 as shown in Fig. 6B.

Peridinin-chlorophyll protein

A water-soluble caroteno-chlorophyll pigment–protein has been extracted and characterised from several dinoflagellates (Siegelman, Kycia & Haxo, 1976; Song, 1978). The pigment-protein obtained from *Glenodinium* sp. can contain most of the carotenoid, peridinin, and 20–40% of the chlorophyll *a*. This protein monomer has a M_r of about 32 000 and contains four peridinin molecules and one chlorophyll *a*

---

* Cyanobacteria are non-typical photosynthetic bacteria since they have two photosystems as do higher green plants and algae; they are, therefore sometimes referred to as algae.

molecule and has maximum absorptions at 476 nm and 667 nm. A detailed molecular model has been postulated based on spectral analysis. It is suggested that a centrally located chlorophyll molecule separates the carotenoid molecules into two pairs, each of which is subjected to exciton interactions due to their closeness (Song, Koka, Prezelin & Haxo, 1976).

Comment

Although there is still much work to be done to isolate and characterise the wide variety of pigment-proteins which seem to occur in photosynthetic organisms (see Barrett & Anderson, 1977; Thornber, Markwell & Reinman, 1979; Murata, 1981), it can be concluded with confidence that the pigments are always closely associated with protein and that their organisation is such as to give rise to efficient energy transfer among them.

Membrane fragments

When photosynthetic bacteria are subjected to mechanical disruption, small membrane fragments in the form of vesicles can be obtained. These are known as chromatophores and are formed by 'pinching off' portions of the cell membrane, especially where invaginations occur. Chromatophores have photosynthetic properties very similar to those of the cell membrane system from which they were derived, except for the fact that they are 'inside-out' – that is, the membrane surface which normally faces the inside of the cell becomes the exterior surface of chromatophores. Since these isolated vesicles maintain full photosynthetic activity but scatter light far less than intact bacterial cells, they have been an extremely valuable material for spectroscopic investigations (Clayton & Sistrom, 1978).

In contrast to bacteria, fragmentation of the chloroplast thylakoid membrane yields fractions that are usually not fully functional in terms of the overall electron-transport processes, and which do not contain a full complement of photosynthetic pigments. This is because the thylakoid membrane normally exists as a mixture of granal (stacked) and stromal (unstacked) lamellae (Coombs & Greenwood, 1976). Cleavage of the stromal lamellae from the grana-stacks can be achieved by a number of techniques, including mild detergent treatment (particularly the use of digitonin) and by mechanical treatments using the French press (Boardman *et al.*, 1978).

The isolated stromal lamellae have been found to contain predomi-

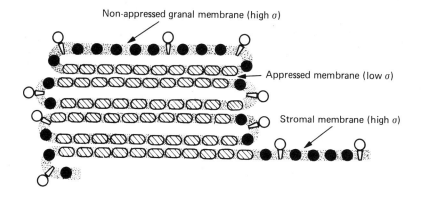

Fig. 8. A model for the possible distribution of functionally active protein components and of surface electrical charge (σ) in, and on, the appressed (partition) and non-appressed thylakoid membranes of higher plant chloroplasts. CF_1 is the extrinsic component of the coupling factor, while CF_0 is the intrinsic hydrophobic component.

nantly the pigments and electron-transfer components belonging to PS1. That is, isolated stromal lamellae vesicles have a low chlorophyll fluorescence yield at room temperature but a high yield of 735 nm emission at 77 K, a red absorption maximum at about 680 nm, a high chlorophyll *a* to chlorophyll *b* ratio, a high concentration of P700, and are able to support PS1-mediated reactions. On the other hand, the granal membrane fractions are enriched in PS2 since they have a high fluorescence yield at room temperature, a low chlorophyll *a* to chlorophyll *b* ratio, and are able to support PS2 reactions including O_2-evolution. However, all granal preparations maintained some PS1 activity, the extent of which depended on the method of fragmentation. Recently Akerlund, Andersson, Persson & Albertsson (1979) have extended the separation of mechanically-derived fragments to the use of phase partition techniques. They were able to obtain a vesicle fraction which was topographically 'inside-out' and apparently derived from the partition membranes of the grana where the lamellae are appressed (see Fig. 8). These 'inside-out' vesicles seem to contain only PS2 and thus a picture is emerging (see Fig. 8) to suggest that PS2 is mainly located in the appressed thylakoids which constitute the partition region, while PS1 is positioned in the non-appressed lamellae which includes the membranes exposed at the surfaces of the grana stacks as well as in the stroma (Andersson & Anderson, 1980; Anderson, 1981). As a consequence of this there will be a lateral heterogeneity in the distribution of PS2 and PS1 pigments within the membrane (Barber, 1980*a*, 1982). Since the LHC is closely associated with PS2, this protein would be expected to occur in the appressed regions of the grana and, indeed, its presence is required for normal membrane stacking to occur (Arntzen, 1978; Steinbeck, Burke & Arntzen, 1979). As mentioned earlier, the LHC exists *in vivo* as an oligomeric form of the monomer (M_r 30 000). This macroscopic complex has been isolated using both digitonin (Wessels, van Alphen-van Waveren & Voorn, 1973) and Triton X-100 (Burke, Ditton & Arntzen, 1978), and is the subject of much investigation especially in terms of reconstitution studies (Ryrie, Anderson & Goodchild, 1980; Mullet & Arntzen, 1980).

The cyanobacteria and red algae do not contain chlorophyll *b* and therefore have no LHC. However, it seems that, as shown in Fig. 6(B), the phycobilisomes can be directly compared with the LHC because both types of pigment–protein transfer energy preferentially to PS2 and their concentrations are variable depending on growth conditions (Thornber, 1975). Mechanical disruption of cyanobacteria and red algae under appropriate salt conditions yields isolated intact phycobilisomes (Gantt *et al.*, 1976) which are ideal candidates for energy transfer studies (Searle, Barber, Porter & Tredwell, 1978). As will be discussed later, the chloroplast LHC becomes phosphorylated under certain light conditions, but it has yet to be determined whether phycobilisomes undergo the same process.

Energy transfer

Mechanism

The function of the light-harvesting pigments is to capture photons and transfer the resulting excitation energy to a specific reaction centre. Several mechanisms of energy migration are possible (Barber, 1978) but the inductive resonance mechanism of Förster (see Knox, 1975) seems most likely for long range transfer. However, within the subunits of the various macroscopic pigment–protein complexes, very fast exciton exchanges may occur (Knox, 1977) which are detectable spectroscopically. According to the Förster theory, the rate of energy transfer is proportional to r^{-6} (where r is the distance between the donor and acceptor molecules) and to the degree of overlap of the emission and

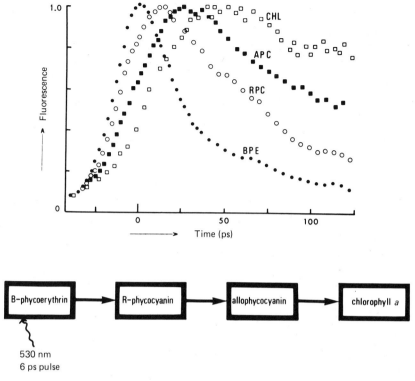

Fig. 9. Time-resolved energy transfer sequence to chlorophyll *a* via the accessory pigments of the red alga *Porphyridium cruentum*. The measurements were made by selectively exciting B-phycoerythrin using a 6 ps pulse generated by a mode-locked Nd-glass laser. The emission from the pigments was selected out by means of interference filters and time-resolved by using a streak camera (from Porter *et al.*, 1978).

absorption spectra of the donor and acceptor molecules respectively. For chlorophyll, the extremely good overlap of the fluorescence and absorption bands allows high rates of energy transfer even when the molecules are spaced several nanometres apart. For example, when the molecules are spaced by 1.5 nm, the pairwise transfer rate can be as high as $10^{12}\,\mathrm{s^{-1}}$ (also depends on the relative orientation of the two molecules to each other) which is 5000 times faster than the deactivation rate leading to fluorescence (Mathis & Paillotin, 1981). If two different types of pigments are involved, the preferred direction of transfer is to the longer wavelength-absorbing molecule because of the requirement for good overlap between the emission and absorption bands. Therefore, chlorophyll *b*, carotenoids and phycobilins transfer their energy to chlorophyll *a*, which in turn transfers its excitations to the long wavelength-absorbing chlorophylls within the reaction centres. This transfer process competes efficiently with other deactivation processes, such as fluorescence. Although energy transfer can be inferred from the excitation spectra of photosynthetic processes, such as O_2-evolution (Blinks, 1954) and chlorophyll fluorescence (Duysens, 1952), it has recently been possible to monitor directly the kinetics of the transfer using picosecond laser fluorimetry. This technique has been effectively used to detect energy migration among the phycobili-proteins of *Porphyridium cruentum* (Porter, Tredwell, Searle & Barber, 1978) and of *Anacystis nidulans* (Brody, Tredwell & Barber, 1981). Figure 9 shows that energy transfer from phycoerythrin → R-phycocyanin → allophycocyanin → chlorophyll *a* is detected when a 6 ps pulse of 530 nm light is absorbed by phycoerythrin. Such a transfer sequence is consistent with the model given in Fig. 6(B).

Chlorophyll fluorescence

In the case of PS2, the level of chlorophyll fluorescence varies at room temperature, depending on the state of its reaction centre. When the reaction centre is open and able to initiate charge separation, the fluorescence is low corresponding to a quantum yield of the PS2 chlorophyll *a* antenna of about 3%. On closing the reaction centre the yield rises to about 10%. However, for PS1 the room temperature fluorescence is very low (less than 0.5%) and usually does not change even when the reaction centre is closed. Presumably, even when the PS1 reaction centre is not functioning it remains a quencher of fluorescence (but see Telfer, Barber, Heathcote & Evans, 1978). Thus at room temperature most of the fluorescence (which peaks at 685 nm) is from the chlorophyll antenna of PS2 and, because of the existence of the other weakly fluorescent antenna chlorophylls, the 'apparent' fluor-escence yield of the complete system is reduced to 1% and 3% for open and closed PS2 reaction centres respectively.

At liquid N_2 temperature ($-196\,°C$) the *in-vivo* chlorophyll fluor-escence is more complex than that at room temperature. Three emission bands appear, peaking at 685, 695 and above 730–735 nm. The 695 nm band seems to originate from the chlorophyll *a* antenna of the PS2 complex while the 685 nm band is assumed to originate from LHC (Butler, 1978). The 730–735 nm emission comes from PS1 and can be used, like the 695 nm band for PS2, to monitor excitation density in the vicinity of the PS1 reaction centres (Butler, 1978). The change in fluorescence yield on going from a fully open reaction centre condition (often designated as the F_0 level) to the closed reaction centre (F_m

condition) allows detailed investigations (Malkin & Kok, 1966; Lavorel & Etienne, 1977) of the mechanism of charge separation within the PS2 reaction centre and also of the transfer of energy between the antennae of adjacent PS2 light-harvesting complexes (intrasystem transfer) and between the light-harvesting systems of PS2 and PS1 (intersystem transfer or spillover). The degree of transfer between adjacent PS2 units has been investigated by a number of workers and stems from the original studies of A. Joliot & P. Joliot (1964). From the various approaches it appears that, under the most favourable conditions (when the membranes are stacked in the way shown in Fig. 5), the size of the domain in which an exciton can migrate contains at least four PS2 reaction centres (Paillotin, Swenberg, Breton & Geacintov, 1979; Geacintov, Breton, Swenberg & Paillotin, 1977). According to the model of the thylakoid membrane in Fig. 5, the PS1 complexes are distributed in the non-appressed membranes and therefore, like PS2, energy transfer between PS1 units may also be possible. Evidence for this has come from studies by several workers (Fork & Amesz, 1967; Borisov & Ilina, 1973; Delepelaire & Bennoun, 1978), with the conclusion that migration of an exciton may occur among three separate PS1 light-harvesting complexes under conditions when their reaction centres are closed.

Spillover

When thylakoid membranes are isolated and suspended in a medium containing cations at a sufficient concentration to 'screen' the negative electrical charges on their surfaces (for example: 5 mM $MgCl_2$ or 100 mM KCl), the membranes retain their normal granal (stacked) and stromal (unstacked) condition (Barber & Chow, 1979; Barber, 1980a). In this state these preparations show a high degree of intrasystem energy transfer, but a low degree of intersystem transfer (low spillover from PS2 to PS1). However, if the isolated membranes are placed in a low salt solution with poor electrostatic screening conditions, there is a dramatic alteration in their organisation. The membranes unstack and there is a change in the spatial distribution of the chlorophyll–protein complexes (Staehelin, 1976; Arntzen, Armond, Briantais, Burke & Novitzky, 1976). In this condition there seems to be an intermixing of the LHC-PS2 and PS1 pigment–protein complexes and as a consequence, there is a decrease in the intrasystem transfer (Briantais, Vernotte & Moya, 1973; Marsho & Kok, 1974; Hipkins, 1978) and an increase in energy transfer from PS2 to PS1 (Barber, 1976; Arntzen, 1978). Because at room temperature the PS1 fluorescence is very weak, the increase in spillover from PS2 to PS1 can be recorded as a decrease in the F_m fluorescence level. There have been many investigations into the mechanisms controlling spillover (Barber, 1976; Williams, 1977) and the concept that it may involve the lateral movements of pigment–protein complexes has emerged only in the last few years (Barber, 1979; Barber, Chow, Scoufflaire & Lannoye, 1980). Recently we have carefully analysed the kinetics of the rise of the F_m level which occurs when cations are added to the medium so as to cause a transition from the maximum spillover (unstacked membranes) to the minimum spillover (normal granal and stromal membranes) state. The kinetics of the rise have been correlated with aggregation theory, and diffusion coefficients have been calculated for different temperatures (Rubin, Barber, Paillotin, Chow & Yamamoto, 1981). The way in which changes in electrostatic screening of surface charges can bring about both lateral diffusion of the pigment-proteins and the creation of appressed (grana partitions) and non-appressed membrane regions has been discussed in detail in recent reviews (Barber, 1980*a, b*, 1982).

State 1–State 2 changes

It has been known for many years that the quantum efficiency of photosynthesis is remarkably constant over a wide spectral range (Myers, 1971). Such a constancy would at first sight not seem possible bearing in mind that the absorption spectra of the two photosystems are not identical. It was the work of Bonaventura & Myers (1969) and of Murata (1969) which first clearly indicated that plants and algae are able to regulate energy transfer from PS2 to PS1 in such a way as to give rise to the most efficient use of incident light of a particular spectral quality. The phenomenon of spillover seems to have physiological significance, at least for photosynthesis occurring under light-limiting conditions (for example: for leaves deep in a canopy or algae growing at depths in oceans and lakes). Essentially it can be demonstrated that when an organism is illuminated with light of wavelengths less than 700 nm, which is preferentially absorbed by the light-harvesting antenna of PS2, the photosynthetic apparatus responds to the imbalance and increases energy transfer from the PS2 to the PS1 light-harvesting system to optimise electron flow through the two reaction centres (P680 and P700). This maximum spillover state is known as State 2. On the other hand if PS1 is over-excited, for example, with excess light above 700 nm, the spillover of energy from PS2 to PS1 is minimised and the organism is said to be in State 1 (see Barber, 1976; Williams, 1977). In 1957,

Emerson demonstrated the 'enhancement effect' which was simply a measure of the imbalance of the flow of quanta to PS2 and PS1 (see Myers, 1971; Barber, 1976; Williams, 1977). Therefore, 'enhancement' is minimum in State 2 (good balance between PS2 and PS1) and a maximum in State 1 (poor balance between the two photosystems). These changes in quantum distribution have been detected by measuring the quantum efficiency of photosynthesis (Bonaventura & Myers, 1969) and by measuring the relative yields of chlorophyll fluorescence at 685 nm (from PS2) and 735 nm (from PS1) at liquid N_2 temperatures (Murata, 1969). Using optical fibre techniques and monitoring chlorophyll fluorescence at room temperature, State 1–State 2 transitions can be monitored indirectly in intact leaves still attached to the parent plant (Barber, Horler & Chapman, 1981; Chow, Telfer, Chapman & Barber, 1981). Such a measurement is shown in Fig. 10. Here the fluorescence signal has been induced by modulated blue light, designated Light 2 because it is preferentially absorbed by the light-harvesting pigments of PS2. On initiating the excitation the fluorescence shows a complex induction curve which is partly due to the closing and opening of PS2 reaction centres (due to the lag in establishing a steady-state rate of CO_2 fixation) and is termed the Kautsky transient (see Barber, 1976). However after a few minutes of illumination with Light 2 the modulated fluorescence reaches a constant level. The question arises as to the balance of the excitation of the two photosystems at this steady state. If Light 2 is over-exciting PS2 relative to PS1, then superimposing additional PS1 light should alleviate the imbalance and increase the electron-transport rate – there should be enhancement. An increase in electron transport can be seen indirectly by a lowering of the fluorescence level, because more of the PS2 reaction centres will be open. However, if excitation of PS1 is not limiting, because Light 2 is equally distributed between both photosystems, the introduction of the additional PS1 light will have no effect. So as not to detect any additional fluorescence due to the PS1 light (called Light 1) this excitation is non-modulated and a lock-in amplifier is used to record only the modulated emission. As can be seen in Fig. 10, superimposing Light 1 (710 nm) on the modulated Light 2 at the steady-state level had very little effect. Under these conditions, Light 2 seems to be exciting each photosystem equally, and the leaf is apparently in a high spillover, low enhancement state (State 2). However, if Light 1 is left on for some time the photosynthetic apparatus reponds to the over-excitation of PS1 by minimising spillover so that when the Light 1 is turned off, Light 2 over-excites PS2 and there is a significant increase in the fluorescence

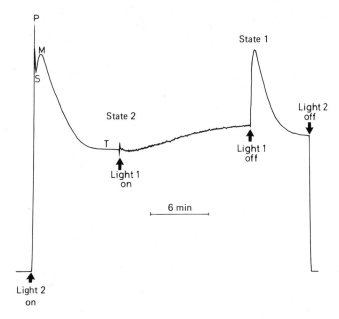

Fig. 10. State 1–State 2 changes in a leaf attached to an intact pea plant monitored via changes in modulated chlorophyll fluorescence by utilising branched optical fibres. Before turning on the modulated light 2 (blue-green light, $1.3\,\mathrm{Wm^{-2}}$ transmitted by Schott BG18 (4 mm) + BG38 (2 mm) filters) the leaf was exposed to non-modulated far-red PS1 light ($710 \pm 12\,\mathrm{nm}$) at $7.5\,\mathrm{Wm^{-2}}$ (Light 1) for 10 min. The experiment was conducted at room temperature. The notation PSMT is used to describe chlorophyll fluorescence transients as discussed by Barber (1976).

yield as shown in Fig. 10. Thus at this stage the leaf has been driven into a maximum enhancement state with poor spillover (State 1). These light-induced changes between State 1 and State 2 can be continuously repeated and seem to occur in all normal higher plants. However, for mutants lacking chlorophyll *b*, and therefore the LHC, no State 1–State 2 change is seen (Chow *et al.*, 1981). After long dark periods leaves are found to be in State 1. The way in which the PS2 and PS1 light-harvesting complexes can regulate energy transfer seems to hinge on the phosphorylation of exposed segments of the LHC (Allen, Bennett, Steinback & Arntzen, 1981).

Phosphorylation of LHC and State 1–State 2 changes

The fact that the LHC can become phosphorylated has been emphasised by the work of Bennett (1977, 1979*a*,*b*). He has shown that some of the threonine residues of this pigment–protein complex become phosphory-

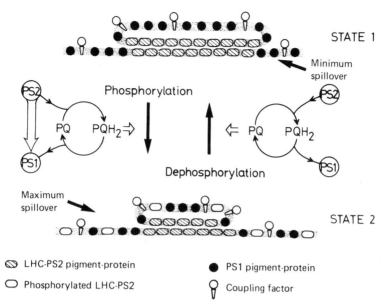

STATE 1

Minimum spillover

Phosphorylation

Dephosphorylation

Maximum spillover

STATE 2

⊘ LHC-PS2 pigment-protein   ● PS1 pigment-protein

○ Phosphorylated LHC-PS2   ♀ Coupling factor

Fig. 11. A model to relate changes in the phosphorylation of LHC–PS2 to changes in the degree of thylakoid stacking and energy transfer from PS2 to PS1 during State 1–State 2 transitions. The membrane-bound kinase responsible for the phosphorylation is activated by reduced plastoquinone (PQH_2) while the dephosphorylation by a membrane-bound phosphatase occurs when the plastoquinone pool is predominantly in an oxidised state.

lated under certain light conditions (1979*c*). Although Bennett (1979*b*) had suggested earlier that the phosphorylation process may be the basis of the State 1–State 2 regulation, it was not until recently that the link was firmly established (Bennett, Steinback & Arntzen, 1980). Even more recently it has been shown that the membrane-bound kinase, which induces the phosphorylation, is activated when PS2 is over-excited and the plastoquinone pool (which serves to transfer electrons from PS2 to PS1) is in a reduced state (Horton, Allen, Black & Bennett, 1981). The kinase activity is, however, inhibited if the plastoquinone pool is in an oxidised state (Horton & Black, 1980, 1981). Such a condition is created in the dark or by introducing excess PS1 light. Thus PS2 and PS1 light act in an antagonistic manner on the kinase activity and thus on the phosphorylation of LHC. Dephosphorylation of LHC is brought about by a membrane-bound phosphatase which is insensitive to the redox state of plastoquinone and competes with the kinase. The control of kinase activity by the redox state of plastoquinone has been nicely summarised by Allen *et al.* (1981) and is shown diagrammatically in Fig. 11. Of course the question arises as to how the phosphorylation of the exposed surface of the LHC can bring about changes in energy transfer from PS2 to PS1. Figure 11 summarises a possible mechanism based on the current understanding of the distribution of the pigment–protein complexes of PS2 and PS1 in the thylakoid membrane (see Fig. 8) and takes into account knowledge gained from investigating changes in spillover due to subjecting thylakoid membranes to different cation levels (Barber, 1980, 1980a). It is envisaged that the phosphorylation of some of the LHC changes the surface electrical properties of exposed segments of the LHC–PS2 complex (making them electrically more negative). It is suggested (Barber, 1982) that the increase in net negative charge, due to the introduction of the phosphate groups, brings about coulombic repulsion between the LHC–PS2 complexes within the membrane, and, in particular, between adjacent membrane surfaces in the granal partition regions. As a consequence, some unstacking will occur and the phosphorylated components will mix freely with the components in the non-appressed membranes, assuming that they are sufficiently 'fluid' and free diffusion can occur. In this way more intimate contact will be made between some of the LHC–PS2 and the PS1 complexes and energy transfer (spillover) will occur (State 2). Thus, State 1 is a condition in which there is a low level of phosphorylation, maximum segregation of PS2 and PS1 light-harvesting complexes and maximum stacking. There is already some evidence for this model (Barber, 1982), but new experiments have to be conducted to test it thoroughly.

Conclusion

It now seems certain that all photosynthetic pigments are associated with proteins. The monomeric forms of the pigment-proteins aggregate, forming macroscopic structures which are usually hydrophobic and embedded in the lipid matrix of a membrane (intrinsic proteins). In some cases, however, these structures are polar and attached to the surface of the membrane (extrinsic proteins) and therefore can be readily isolated without the need for detergents. The various pigments within the subunits and the macroscopic complexes are arranged in such a way as to allow efficient energy transfer to a 'primary light-harvesting complex' which directly serves the reaction centres. In all cases the reaction centre complex and its most closely associated 'primary light-harvesting array' are embedded in the hydrophobic environment of the membrane. These light-harvesting arrays plus reaction centre complexes are probably present, for both PS1 and PS2, in all O_2-evolving organisms, with the variations being restricted to the complexes that contain the secondary pigments. In higher plants, there are two main intrinsic macroscopic pigment–protein complexes, the LHC–PS2 complex and the PS1 complex, which seem to be spatially separated from each other in the plane of the membrane. However, this spatial relationship can be varied depending on certain conditions (ionic levels and phosphorylation of LHC), emphasising the dynamic nature of the membrane structure and the fluidity of the lipid matrix. It seems to be the ability of the pigment proteins of higher plant chloroplasts to vary their distances apart which gives rise to changes in both intra- and intersystem energy-transfer and underlies the State 1–State 2 transitions. Whether similar concepts can be applied to those organisms which lack LHC, but contain phycobiliproteins, has yet to be determined. The importance of changes in spatial relationships between the various pigment-proteins of photosynthetic bacteria is not clear, but these organisms contain only one type of reaction centre and thus there is no need for any mechanism comparable with the State 1–State 2 transitions observed in O_2-evolving organisms.

I would like to thank the Science Research Council and the Agricultural Research Council for financial support and Dr David Chapman, Suzanne Cheston and Lyn Barber for their help in the preparation of this manuscript.

References

AKERLUND, H. E., ANDERSSON, B., PERSSON, A. & ALBERTSSON, P. A. (1979). Isoelectric points of spinach thylakoid membrane surfaces as determined by cross partition. *Biochim. Biophys. Acta*, **552**, 238–46.

ALLEN, J. F., BENNETT, J., STEINBACK, K. E. & ARNTZEN, C. J. (1981). Chloroplast protein phosphorylation couples plastoquinone redox state to distribution of excitation energy between photosystems. *Nature, London*, **291**, 1–5.

ANDERSON, J. M. (1975). The molecular organisation of chloroplast thylakoids. *Biochim. Biophys. Acta*, **416**, 191–235.

ANDERSON, J. M. (1980). Chlorophyll–protein complexes of higher plant thylakoids: distribution, stoichiometry and organisation in the photosynthetic unit. *FEBS Lett.*, **117**, 327–31.

ANDERSON, J. M. (1981). Consequences of spatial separation of photosystem 1 and 2 in thylakoid membranes of higher plant chloroplasts. *FEBS Lett.*, **124**, 1–10.

ANDERSSON, B. & ANDERSON, J. M. (1980). Lateral heterogeneity in the distribution of chlorophyll–protein complexes of the thylakoid membranes of spinach chloroplasts. *Biochim. Biophys. Acta*, **416**, 191–235.

ANDERSON, J. M. & BARRETT, J. (1979). Chlorophyll–protein complexes of brown algae: P700 reaction centre and light-harvesting complexes. In: *Chlorophyll Organisation and Energy Transfer in Photosynthesis*, pp. 81–104. Ciba Foundation Symposium 61 (new series). Amsterdam: Excerpta Medica.

ARNTZEN, C. J. (1978). Dynamic structural features of chloroplast lamellae. In: *Current*

Topics in Bioenergetics, vol. 7, ed. D. R. Sanadi & L. P. Vernon, pp. 111–60. New York: Academic Press.

ARNTZEN, C. J., ARMOND, P. A., BRIANTAIS, J. M., BURKE, J. J. & NOVITZKY, W. P. (1976). Dynamic interactions among structural components of the chloroplast membrane. *Brookhaven Symp. Biol.*, **28**, 316–37.

BARBER, J. (1976). Ionic regulation in intact chloroplasts and its effect on primary photosynthetic processes. In: *The Intact Chloroplast*, ed. J. Barber, pp. 89–134. Topics in Photosynthesis, vol. 1. Amsterdam: Elsevier/North-Holland.

BARBER, J. (1978). Biophysics of photosynthesis. *Rep. Prog. Phys.*, **41**, 1158–99.

BARBER, J. (1979). Energy transfer and its dependence on membrane properties. In: *Chlorophyll Organisation and Energy Transfer in Photosynthesis*, pp. 283–304. Ciba Foundation Symposium 61 (new series). Amsterdam: Excerpta Medica.

BARBER, J. (1980a). An explanation for the relationship between salt-induced thylakoid stacking and the chlorophyll fluorescence changes associated with changes in spillover of energy from photosystem II to photosystem I. *FEBS Lett.*, **118**, 1–10.

BARBER, J. (1980b). Membrane surface charges and potentials in relation to photosynthesis. *Biochim. Biophys. Acta*, **594**, 253–308.

BARBER, J. (1982). Influence of surface charges on thylakoid structure and function. *Ann. Rev. Plant Physiol.*, **33**, 261–95.

BARBER, J. & CHOW, W. S. (1979). A mechanism for controlling and stacking and unstacking of chloroplast thylakoid membranes. *FEBS Lett.*, **105**, 5–10.

BARBER, J., CHOW, W. S., SCOUFFLAIRE, C. & LANNOYE, R. (1980). The relationship between thylakoid stacking and salt-induced chlorophyll fluorescence changes. *Biochim. Biophys. Acta*, **591**, 92–103.

BARBER, J., HORLER, D. N. H. & CHAPMAN, D. J. (1981). Photosynthetic pigments and efficiency in relation to the spectral quality of absorbed light. In: *Plants and the Daylight Spectrum*, ed. H. Smith, pp. 341–54. London and New York: Academic Press.

BARRETT, J. & ANDERSON, J. M. (1977). Plant chlorophyll–protein complexes. *Plant Sci. Lett.*, **9**, 275–83.

BENNETT, J. (1977). Phosphorylation of chloroplast membrane polypeptides. *Nature, London*, **269**, 344–6.

BENNETT, J. (1979a). Chloroplast phosphoproteins, phosphorylation of polypeptides of the light-harvesting chlorophyll–protein complex. *Eur. J. Biochem.*, **99**, 133–7.

BENNETT, J. (1979b). The protein that harvests sunlight. *Trends Biochem. Sci.*, **4**, 268–71.

BENNETT, J. (1979c). Chloroplast phosphoproteins. The protein kinase of thylakoid membranes is light-dependent. *FEBS Lett.*, **103**, 342–4.

BENNETT, J., STEINBACK, K. E. & ARNTZEN, C. J. (1980). Chloroplast phosphoproteins: regulation of excitation energy transfer by phosphorylation of thylakoid membranes. *Proc. Natl. Acad. Sci., USA*, **77**, 5253–7.

BLANKENSHIP, R. E. & PARSON, W. W. (1979). Kinetics and thermodynamics of electron transfer in bacterial reaction centres. In: *Photosynthesis in Relation to Model Systems*, ed. J. Barber, pp. 71–114. Topics in Photosynthesis, vol. 3. Amsterdam: Elsevier/North Holland.

BLINKS, L. R. (1954). The photosynthetic function of pigments other than chlorophyll. *Annu. Rev. Plant Physiol.*, **5**, 93–114.

BOARDMAN, N. K., ANDERSON, J. M. & GOODCHILD, D. J. (1978). Chlorophyll–protein complexes and structure of mature and developing chloroplasts. In: *Current Topics in Bioenergetics*, vol. 7, ed. D. R. Sanadi & L. P. Vernon, pp. 35–109. New York: Academic Press.

BONAVENTURA, C. & MYERS, J. (1969). Fluorescence and oxygen evolution from *Chlorella pyrenoidosa. Biochim. Biophys. Acta*, **189**, 366–83.

BORISOV, A. Y. & ILINA, M. D. (1973). The fluorescence lifetime and energy migration mechanism in photosystem 1 of plants. *Biochim. Biophys. Acta*, **305**, 364–71.

BRIANTAIS, J. M., VERNOTTE, C. & MOYA, I. (1973). Intersystem exciton transfer in isolated chloroplasts. *Biochim. Biophys. Acta*, **325**, 530–8.

BRODY, S. S., TREDWELL, C. & BARBER, J. (1981). Picosecond energy transfer in *Phorphyridium cruentum* and *Anacystis nidulans*. *Biophys. J.*, **34**, 439–49.

BURKE, J. J., DITTON, C. J. & ARNTZEN, C. J. (1978). Involvement of the light-harvesting complex in cation regulation of excitation energy distribution in chloroplasts. *Arch. Biochem. Biophys.*, **187**, 252–63.

BUTLER, W. L. (1978). Energy distribution in the photochemical apparatus of photosynthesis. *Annu. Rev. Plant Physiol.*, **29**, 345–78.

CHOW, W. S., TELFER, A., CHAPMAN, D. J. & BARBER, J. (1981). State 1–State 2 transition in leaves and its association with ATP-induced chlorophyll fluorescence quenching. *Biochim. Biophys. Acta*, **638**, 60–8.

CLAYTON, R. K. & SISTROM, W. R. (eds.) (1978). *The Photosynthetic Bacteria*. New York: Plenum Press.

COGDELL, R. J. (1978). Carotenoids in photosynthesis. *Philosoph. Trans. Roy. Soc. London*, **B284**, 569–79.

COGDELL, R. J. & THORNBER, J. P. (1979). The preparation and characterization of different types of light-harvesting pigment–protein complexes from some purple bacteria. In: *Chlorophyll Organisation and Energy Transfer in Photosynthesis*, pp. 61–79. Ciba Foundation Symposium 61 (new series). Amsterdam: Excerpta Medica.

COGDELL, R. J. & THORNBER, J. P. (1980). Light-harvesting pigment–protein complexes of purple photosynthetic bacteria. *FEBS Lett.*, **122**, 1–8.

COOMBS, J. & GREENWOOD, A. D. (1976). Compartmentation of the photosynthetic apparatus. In: *The Intact Chloroplast*, ed. J. Barber, pp. 1–51. Topics in Photosynthesis, vol. 1. Amsterdam: Elsevier/North-Holland.

DELEPELAIRE, P. & BENNOUN, P. (1978). Energy transfer and site of energy trapping in photosystem 1. *Biochim. Biophys. Acta*, **502**, 183–7.

DELEPELAIRE, P. & CHUA, N. H. (1979). Lithium dodecyl sulphate/polyacrylamide gel electrophoresis of thylakoid membranes at 4 °C: characterizations of two additional chlorophyll *a*-protein complexes. *Proc. Natl Acad. Sci., USA*, **76**, 111–15.

DUTTON, P. L., LEIGH, J. S., PRINCE, R. C. & TIEDE, D. M. (1979). The photochemical reaction centre of photosynthetic bacteria as a model for studying biological charge separation and electron transfer. In: *Light Induced Charge Separation in Biology and Chemistry*, ed. H. Gerirscher & J. J. Katz, pp. 411–48. Berlin: Dahlem Konferenzen.

DUYSENS, L. N. M. (1952). Transfer of excitation energy in photosynthesis. Thesis, Utrecht.

EMERSON, R. & ARNOLD, W. (1932). The photochemical reaction in photosynthesis. *J. Gen. Physiol.*, **16**, 191–205.

FENNA, R. E. & MATHEWS, B. W. (1976). Structure of a bacteriochlorophyll *a*–protein from *Prosthecochloris aertuarii*. *Brookhaven Symp. Biol.*, **26**, 170–82.

FORK, D. C. & AMESZ, J. (1967). Energy transfer between units of system 1 in algae. *Biochim. Biophys. Acta*, **143**, 266–8.

FRENCH, C. S., BROWN, J. S. & LAWRENCE, M. C. (1972). Four universal forms of chlorophyll *a*. *Plant Physiol.*, **49**, 421–9.

GANTT, E., LIPSCHULTZ, C. A. & ZILINSKAS, B. A. (1976). Phycobilisomes in relation to the thylakoid membranes. *Brookhaven Symp. Biol.*, **28**, 347–57.

GEOCINTOV, N. E., BRETON, J., SWENBERG, C. E. & PAILLOTIN, G. (1977). A single pulse picosecond laser study of exciton dynamics in chloroplasts. *Photochem. Photobiol.*, **26**, 629–38.

GLAZER, A. N. (1981). Photosynthetic accessory proteins with bilin prosthetic groups. In: *The Biochemistry of Plants*, vol. 8, ed. M. D. Hatch & N. K. Boardman, pp. 51–96. London and New York: Academic Press.

GOODWIN, T. W. (ed.) (1965). *Chemistry and biochemistry of plant pigments.* London and New York: Academic Press.

HILLER, R. G. & GOODCHILD, D. J. (1981). In: *The Biochemistry of Plants*, vol. 8, ed. M. D. Hatch & N. K. Boardman, pp. 1–49. London and New York: Academic Press.

HIPKINS, M. F. (1978). Kinetic analysis of the chlorophyll fluorescence inductions from chloroplasts blocked with 3-(3,4-dichlorophenyl)-1,1-dimethylurea. *Biochim. Biophys. Acta*, **502**, 514–23.

HORTON, P., ALLEN, J. F., BLACK, M. T. & BENNETT, J. (1981). Regulation of phosphorylation of chloroplast membrane polypeptides by the redox state of plasto-quinone. *FEBS Lett.*, **125**, 19–6.

HORTON, P. & BLACK, M. T. (1980). Activation of adenosine 5′-triphosphate-induced quenching of chlorophyll fluorescence by reduced plastoquinone. The basis of state I–state II transitions in chloroplasts. *FEBS Lett.*, **119**, 141–4.

HORTON, P. & BLACK, M. T. (1981). Light-dependent quenching of chlorophyll fluorescence in pea chloroplasts induced by adenosine 5′-triphosphate. *Biochim. Biophys. Acta*, **635**, 53–62.

JOLIOT, A. & JOLIOT, P. (1964). Etude cinetique de la reaction photochemique liberant l'oxygène au cours de la photosynthese. *Compt. R. Acad. Sci., Paris, Serie D*, **258**, 4622–5.

KATZ, J. J. & NORRIS, J. R. (1973). Chlorophyll and light-energy transduction in photosynthesis. *Curr. Top. Bioenerg.*, **5**, 41–75.

KATZ, J. J. NORRIS, J. R. & SHIPMAN, L. L. (1976). Models for reaction-center and antenna chlorophyll. *Brookhaven Symp. Biol.*, **28**, 16–55.

KNOX, R. S. (1975). Excitation energy transfer and migration: theoretical considerations. In: *Bioenergetics of Photosynthesis*, ed. Govindjee, pp. 183–221. New York: Academic Press.

KNOX, R. S. (1977). Photosynthetic efficiency and exciton transfer and trapping. In: *Primary Processes of Photosynthesis*, ed. J. Barber, pp. 55–97. Topics in Photosynthesis, vol. 2. Amsterdam: Elsevier/North-Holland.

LAVOREL, J. & ETIENNE, A. L. (1977). *In-vivo* chlorophyll fluorescence. In: *Primary Processes of Photosynthesis*, ed. J. Barber, pp. 203–268. Topics in Photosynthesis, vol. 2. Amsterdam: Elsevier/North-Holland.

MACCOLL, R. & BERNS, D. S. (1979). Evolution of the biliproteins. *Trends Biochem. Sci.*, **4**, 44–7.

MALKIN, S. & KOK, B. (1966). Fluorescence induction studies in isolated chloroplasts. I. Number of components involved in the reaction and quantum yields. *Biochim. Biophys. Acta*, **126**, 413–32.

MATHIS, P. & PAILLOTIN, G. (1981). Primary processes of photosynthesis. In: *The Biochemistry of Plants*, vol. 8, ed. M. D. Hatch & N. K. Boardman, pp. 97–161. London and New York: Academic Press.

MARKWELL, J. P., THORNBER, J. P. & BOGGS, R. T. (1979). Higher plant chloroplasts. Evidence that all the chlorophyll exists as chlorophyll–protein complexes. *Proc. Natl. Acad. Sci., USA*, **76**, 1233–5.

MARSHO, T. V. & KOK, B. (1974). Photosynthetic regulation by cations in spinach chloroplasts. *Biochim. Biophys. Acta*, **333**, 353–65.

MULLET, J. E. & ARNTZEN, C. J. (1980). Simulation of grana stacking in a model membrane system. Mediation by a purified light-harvesting pigment–protein complex from chloroplasts. *Biochim. Biophys. Acta*, **589**, 100–17.

MULLET, J. E., BURKE, J. J. & ARNTZEN, C. J. (1980). Chlorophyll proteins of photosystem 1. *Plant Physiol.*, **65**, 814–22.

MURATA, N. (1969). Control of excitation transfer in photosystem 1. Light-induced change of chlorophyll *a* fluorescence in *Porphyridium cruentum. Biochim. Biophys. Acta*, **172**, 242–51.

MURATA, T. (1981). Crystalline chlorophyll-proteins from *Lepidium virginicum* L. In: *Proceedings of the 5th International Congress on Photosynthesis*, Halkidiki, Greece 1980, vol. 3, ed. G. Akoyunoglou, pp. 397–402. Philadelphia: Balaban International Science Services.

MYERS, J. (1971). Enhancement studies in photosynthesis. *Ann. Rev. Plant Physiol.*, **22**, 289–312.

OLSEN, J. M. (1980). Chlorophyll organisation in green photosynthetic bacteria. *Biochim. Biophys. Acta*, **594**, 33–51.

PAILLOTIN, G., SWENBERG, C. E., BRETON, J. & GEACINTOV, N. E. (1979). Analysis of picosecond laser-induced fluorescence phenomena in photosynthetic membranes utilizing a master equation approach. *Biophys. J.*, **25**, 513–34.

PORTER, G., TREDWELL, C. J., SEARLE, G. F. W. & BARBER, J. (1978). Picosecond time resolved energy transfer in *Porphyridium cruentum*. I. In: the intact alga. *Biochim. Biophys. Acta*, **501**, 232–45.

RUBIN, B. T., BARBER, J., PAILLOTIN, G., CHOW, W. S. & YAMAMOTO, Y. (1981). A diffusional analysis of the temperature sensitivity of the Mg^{2+}-induced rise in chlorophyll fluorescence from pea thylakoid membranes. *Biochim. Biophys. Acta*, **638**, 69–74.

RYRIE, I. J., ANDERSON, J. M. & GOODCHILD, D. J. (1980). The role of the light-harvesting chlorophyll *a/b*–protein complex in chloroplast membrane stacking. *Eur. J. Biochem.*, **107**, 345–54.

SAUER, K. (1979). Photosynthesis – the light reactions. *Ann. Rev. Phys. Chem.*, **30**, 155–78.

SEARLE, G. F. W., BARBER, J., PORTER, G. & TREDWELL, C. J. (1978). Picosecond time-resolved energy transfer in *Porphyridium cruentum*. II. In: the isolated light-harvesting complex (phycobilisomes). *Biochim. Biophys. Acta*, **501**, 246–56.

SEELY, G. R. (1977). Chlorophyll in model systems: clues to the role of chlorophyll in photosynthesis. In: *Primary Processes of Photosynthesis*, ed. J. Barber, pp. 1–53. Topics in Photosynthesis, vol. 2. Amsterdam: Elsevier/North-Holland.

SIEGELMAN, H. W., KYCIA, J. H. & HAXO, F. T. (1976). Peridinin-chlorophyll *a*–proteins of dinoflagellate algae. *Brookhaven Symp. Biol.*, **28**, 162–9.

SONG, P. S. (1978). Molecular topography of solar energy harvesting pigments in marine algae. *Trends Biochem. Sci.*, **3**, 25–7.

SONG, P. S., KOKA, P., PREZELIN, B. B. & HAXO, F. T. (1976). Molecular topology of the photosynthetic light-harvesting pigment complex, peridinin–chlorophyll *a*–protein, from marine dinoflagellates. *Biochemistry*, **15**, 4422–7.

STAEHELIN, L. A. (1976). Reversible particle movements associated with unstacking and restacking of chloroplast membranes *in vitro*. *J. Cell Biol.*, **71**, 136–58.

STEINBECK, K. E., BURKE, J. J. & ARNTZEN, C. J. (1979). Evidence for the role of surface-exposed segments of the light-harvesting complex in cation-mediated control of chloroplast structure and function. *Arch. Biochem. Biophys.*, **195**, 546–57.

TELFER, A., BARBER, J., HEATHCOTE, P. & EVANS, M. C. W. (1978). Variable chlorophyll *a* fluorescence from P700-enriched photosystem 1 particles dependent on the redox state of the reaction centre. *Biochim. Biophys. Acta*, **504**, 153–64.

THORNBER, J. P. (1975). Chlorophyll–proteins: light-harvesting and reaction centre components of plants. *Annu. Rev. Plant Physiol.*, **26**, 127–58.

THORNBER, J. P., ALBERTE, R. S., HUNTER, F. A., SHIOZAWA, J. A. & KAN, K. S. (1976). The organisation of chlorophyll in the plant photosynthetic unit. *Brookhaven Symp. Biol.*, **28**, 132–48.

THORNBER, J. P. & BARBER, J. (1979). Photosynthetic pigments and models for their organisation *in vivo*. In: *Photosynthesis in Relation to Model Systems*, ed. J. Barber, pp. 27–70, Topics in Photosynthesis, vol. 3. Amsterdam: Elsevier/North-Holland.

THORNBER, J. P. & MARKWELL, J. P. (1981). Photosynthetic pigment–protein complexes in plants and bacterial membranes. *Trends Biochem. Sci.*, **6**, 122–5.

THORNBER, J. P., MARKWELL, J. P. & REINMAN, S. (1979). Plant chlorophyll–protein complexes: recent advances. *Photochem. Photobiol.*, **29**, 1205–16.

VAN METTER, R. L. (1977). Excitation energy transfer in light-harvesting chlorophyll a/b protein. *Biochim. Biophys. Acta*, **462**, 642–58.

VERNON, L. P. & SEELY, G. R. (eds.) (1966). *The chlorophylls*. New York: Academic Press.

WASIELEWSKI, M. R., NORRIS, J. R., SHIPMAN, L. L., LIN, C. P. & SVEC, W. A. (1981). Monomeric chlorophyll a enol: evidence for its possible role as the primary electron donor in photosystem 1 of plant photosynthesis. *Proc. Natl. Acad. Sci., USA*, **78**, 2957–61.

WESSELS, J. S. C., VAN ALPHEN-VAN WAVEREN, D. & VOORN, G. (1973). Isolation and properties of particles containing the reaction centre complex of photosystem II from spinach chloroplasts. *Biochim. Biophys. Acta*, **292**, 741–52.

WILLIAMS, W. P. (1977). The two photosystems and their interactions. In: *Primary Processes of Photosynthesis*, ed. J. Barber, pp. 99–147. Topics in Photosynthesis, vol. 2. Amsterdam: Elsevier/North-Holland.

IN-VIVO MICROSPECTROFLUORIMETRY
OF VISUAL PIGMENTS

NICOLAS FRANCESCHINI

Institut de Neurophysiologie et Psychophysiologie, Centre National De La
Recherche Scientifique, 31 Chemin Joseph-Aiguier, 13009 Marseille, France

Introduction

Since chlorophyll fluorescence has been used as an elegant probe of
photosynthesis and has brought major advances in this field over the
past 50 years (review by Papageorgiou, 1975), it is somewhat surprising
that the visual sciences have not yet managed to make extensive use of
fluorescence techniques in spite of their three outstanding properties:
sensitivity, selectivity and simplicity.

In the middle of the last century, when Helmholtz first observed the
fluorescence of the vertebrate retina, one problem encountered was the
rapid facing of the fluorescence emission, a problem that has still not
been solved (Liebman & Leigh, 1969). This difficulty comes from the
inherent property of vertebrate visual pigments, which are irreversibly
bleached by any illumination compatible with fluorescence microscopy.
This difficulty is even more serious under the usually non-physiological
conditions of the retina compatible with observation of its individual
receptors (excised retina lacking the pigment epithelium).

Far from being the privilege of the vertebrate visual system, the
retinal chromophore is omnipresent in the animal world and is incorporated in the visual pigments of invertebrates as well (reviews by
Goldsmith, 1972; Hamdorf, 1979). Various studies of the insect compound eye in the 1970s have shown that this kind of visual system
provides us with a suitable model for analysing the properties of visual
pigments, not only *in situ* (i.e. in their natural membrane environment)
but even *in vivo*, under excellent physiological conditions. In some
cases, and in particular in the compound eye of flies, both microspectrophotometry and intracellular recordings could be made from the
same (identified) cells in the living animal. These *in-vivo* techniques are
complementary to those classically used in studies of vertebrate visual
pigments: *in-vitro* analysis of pigment extracts and *in-situ* photometry or
electrophysiology in the excised retina.

A decisive advantage of insect photosensitive visual pigments lies in
their bistability, a property that they share with other invertebrate visual

pigments. Thus the meta-rhodopsin (M) formed upon illumination of rhodopsin (R) is quasi dark-stable and does not hydrolyse into retinal + opsin as is the case in vertebrates. Subsequent illumination within the M-absorption band simply photoreconverts M into its native R-state. This bistability allows clamping the R-concentration at the desired level by simply using the appropriate chromatic adaptation. As a consequence both electrophysiology and microspectrophotometry can be done at a fixed rhodopsin concentration.

Another advantage of insect compound eyes lies in the convexity of their retinas which makes photoreceptor cells optically more accessible than those of the eye-fundus in vertebrates. Taking advantage of this eye construction, several optical methods were devised over the last decade, which allow one to analyse single visual cells in the living, intact animal (Franceschini & Kirschfeld, 1971*a*, *b*; Franceschini, 1972, 1975).

Recent association of these techniques with epi-fluorescence microscopy led to the discovery that most rhabdomeres in the compound eye of insects and especially flies exhibit fluorescence colours which run the gamut of the visible spectrum (Franceschini, 1977, 1978; Stark, Ivanyshin & Greenberg, 1977; Stark, Stavenga & Kruizinga, 1979; Franceschini, Kirschfeld & Minke, 1981*a*). These phenomena, observed in the living animal are interesting in two respects. Firstly, they appear as natural colour labels which are indicative of each cell's spectral sensitivity (Hardie, Franceschini & McIntyre, 1979). This aspect has been used to map out the spectral organisation of the highly developed fly retina (Franceschini, Hardie, Ribi & Kirschfeld, 1981*b*; Franceschini, 1983). Secondly, these fluorescence techniques provide a new spectroscopic tool for analysing the visual pigments of individual cells under physiological conditions.

Only this second aspect is dealt with in this chapter which is based mainly on our recent microspectrofluorimetrical measurements on the fly photoreceptor cells (Franceschini *et al.*, 1981*b*; Franceschini & Stavenga, 1981; Stavenga & Franceschini, 1981; Stavenga, Franceschini & Kirschfeld, unpublished).

Seven microcuvettes per ommatidium

Each compound eye of the fly *Musca domestica* contains 3000 ommatidia. A distal cross-section of such an ommatidium reveals seven photoreceptor cells arranged in a very characteristic pattern (Fig. 1*a*). The light-gathering organelle of each cell is called the *rhabdomere*, a microvillar structure (R; hatched in Fig. 1*a–c*) in the membrane of

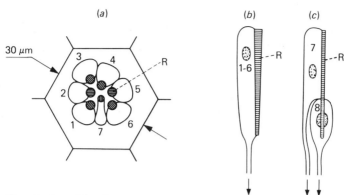

Fig. 1. Organisation of the eight receptor cells within an ommatidium of the compound eye of the fly *Musca domestica*. Schematic drawings based on micrographs by Trujillo-Cenoz & Melamed (1966) and Boschek (1971). Each cell has a rhabdomere (R, hatched in *a*, *b*, *c*) containing the visual pigment(s). Six peripheral cells (nos. 1–6) encircle the two central cells (nos. 7–8) whose rhabdomeres R7 and R8 are arranged one distal to the other so as to form an axial pair. (*b*) and (*c*) are not to scale.

which the visual pigment molecules are buried. The rhabdomere of a cell is long and slender (diameter $\simeq 2\,\mu$m, length $\simeq 200\,\mu$m) and it acts as a kind of absorbing waveguide.

From a spectroscopist's point of view the seven rhabdomeres of an ommatidium are like seven microcuvettes that can be visualised end-on like the rods and cones of an excised vertebrate retina. It is a fortunate property of the fly compound eye to exhibit such neatly separated rhabdomeres. It makes single cell analysis much easier than in the ommatidia of most other insects and crustacea, which possess a *fused rhabdome*. In the latter case our knowledge of the visual pigment of single cells has largely been inferred from the results of intracellular recordings of the receptor potential (reviews in Goldsmith, 1972; Menzel, 1979). In the case of the fly's *open rhabdome*, the more direct *in-situ* microspectrophotometrical approach could naturally be added to the electrophysiological approach (Langer & Thorell, 1966; Stavenga, Zantema & Kuiper, 1973; Stavenga, 1976, 1979; Kirschfeld & Franceschini, 1975; Kirschfeld, Franceschini & Minke, 1977; Stark & Johnson, 1980). Due to the tiny diameter of the beam to be used, end-on microspectrophotometry on single rhabdomeres of the fly is fraught with experimental difficulties as in the case of mammalian rods. An acceptable signal-to-noise ratio can be achieved only by using relatively high intensities but the bistable nature of the invertebrate photopigment makes the bleaching problem less critical than in the case of vertebrates.

The fly rhabdom is not really as 'open' as one would like, and a difficulty exists for the cell no. 8 which, as shown in Fig. 1*c*, has stacked its rhabdomere R8 in the proximal prolongation of the central rhabdomere R7. As a consequence, light transmitted through this tandem microcuvette reports the absorption properties of the pigments contained in both cells without distinction.

The microspectrofluorimetrical approach that was recently developed (Stark *et al.*, 1979; Franceschini *et al.*, 1981*a*; Stavenga *et al.*, unpublished) appears as a third method for analysing the properties of visual pigments in single cells.

In-vivo observation of the fluorescence phenomena within single ommatidia

Combining a non-invasive technique of *optical neutralisation of the cornea* (Franceschini & Kirschfeld, 1971*a*) and the classical technique of *epi-fluorescence microscopy* (Ploem, 1967) we have shown that most rhabdomeres of the living fly are fluorescent when excited within the spectral range relevant to fly vision (Franceschini, 1977, 1978; Franceschini *et al.*, 1981*a*). Figure 2*a* briefly outlines the principle of the method, which consists of applying a clear medium of relatively high refractive index on to the cornea, so as to compensate optically the dioptrical effect of each lenslet. Focussing an epi-fluorescence microscope just below the optically neutralised cornea then reveals the polychrome mosaic of the receptor cells, an example of which is given in Fig. 4. Notice that the blue excitation used here simultaneously reveals four classes of cells: red R1–6; green R7s; black R7s and red R7s. The red fluorescing R7s appear to be the privilege of the male eye, as revealed by fluorescence mappings of the retinal mosaics (Franceschini *et al.*, 1981*b*; Hardie, Franceschini, Ribi & Kirschfeld, 1981).

All these fluorescence emissions apparently stem from different carotenoid-based pigments contained in the rhabdomeres: they all are absent or much reduced in flies (*Musca* and *Drosophila* white, *Calliphora* chalky) whose larvae have been reared on a β-carotene and vitamin-A deficient medium (Franceschini. 1977; Franceschini *et al.*, 1981*a*; Stark *et al.*, 1977, 1979). The pigment molecules radiate isotropically in the depth of the rhabdomere and only a small part of the emitted flux is channelled up to the microscope through the rhabdomeric light-pipe (Franceschini *et al.*, 1981*a*). The patterns seen in Fig. 4 are in fact slightly magnified virtual images of the distal rhabdomere endings. The reason is that the corneal lenslets (refractive index $n \simeq 1.5$) are incompletely neutralised by the immersion medium used (water, $n = 1.33$). The harmlessness of this dew-drop neutralisation enables

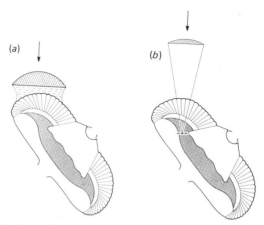

Fig. 2. Two methods used in combination with epi-fluorescence microscopy for viewing the receptor cells in the living insect. (*a*) The cornea is optically neutralised with a drop of water and the microscope (obj. 25 ×/0.65, water immersion) is focused on to the distal rhabdomere endings (after Franceschini & Kirschfeld, 1971*a*; Franceschini, 1975). (*b*) Magnified and superimposed virtual images of rhabdomere patterns as seen when focusing a (low power) microscope at the level of the centre of curvature of the eye ('deep pseudopupil' after Franceschini & Kirschfeld, 1971*b*; Franceschini, 1975).

long-run experiments to be done. In this way for example, we can repeatedly check the emission of the same receptor cells during several days, feeding our little fly every evening with ovomaltine under the epi-fluorescence microscope without changing its posture on the (goniometric) cradle.

One difficulty encountered in the observation of such phenomena is due to the pigmentation of the eye. The ommatidia are shielded from one another by intensely absorbing reddish pigment granules (ommochromes and pteridines) contained in specialised pigment cells. Furthermore the receptor cells themselves contain tiny ommochrome granules (diameter $0.1\,\mu$m) which move towards the rhabdomere upon illumination (Kirschfeld & Franceschini, 1969; Franceschini & Kirschfeld, 1976). All these accessory pigments impair the fluorescence observations since they attenuate the *excitation* light available to the rhabdomeres. In particular the fluorescence intensity decreases considerably over the first few seconds as a consequence of the pigment migration. All this explains why observations and photography of these phenomena are best made using white-eyed mutants which lack the ommochrome pigments.

A second difficulty encountered comes from the pernicious auto-fluorescence of the cornea and of the underlying tissue. The overwhelming intensity of this background fluorescence (which is excited under

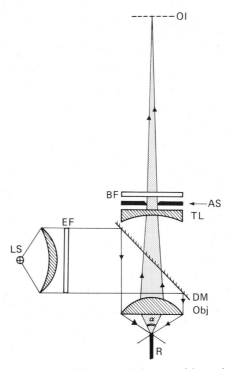

Fig. 3. Principle of the 'aperture filtering' technique used in conjunction with optical neutralisation of the cornea (Fig. 2*a*) for improving the contrast of the *in-vivo* fluorescence observations (Fig. 4). The drawing illustrates the role of aperture stop (AS) which is inserted near the tube lens (TL) of the microscope to reduce selectively the *viewing* aperture down to an angle α without affecting the excitation aperture. The angle $\alpha \simeq 30°$ is the aperture of the directive beam of light radiated by a rhabdomere (R) as a consequence of its light-guide property (the facet lens has not been represented as it is optically neutralised). Using a viewing aperture larger than α dramatically reduces the contrast in collecting much of the corneal and eye-tissue autofluorescence without increasing the fluorescence intensity of the rhabdomeres by so much (OI = objective image, the eyepiece of the microscope is not represented; BF = barrier filter; EF = excitation filter; DM = dichroic mirror of the epifluorescence illuminator; LS = light source: Xe 75 W or Hg 100 W arc lamp; Obj = objective 25 ×/0.65, water immersion'.

blue and UV light as well) would have spoiled the photographic contrast in Fig. 4, had we not used an artifice to overcome it. The idea is based upon the different directional properties of the emitted lights. Whereas the corneal emission is isotropic, the rhabdomeres radiate a highly directive beam of light towards the overlying facet, as a consequence of their light-guide property. Care was thus taken to reduce selectively the *viewing* aperture of the microscope down to an angle $\alpha \simeq 30°$ (Fig. 3) by inserting an aperture stop (AS) near the tube lens (TL) of the epi-fluorescence illuminator. In this way the aperture of the illumination

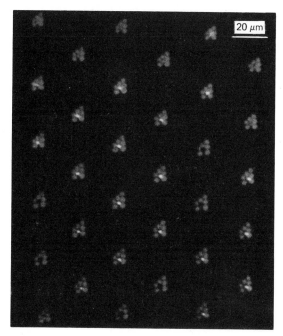

Fig. 4. *In-vivo* observation of the fluorescence colours exhibited by the seven distal rhabdomere endings (cf. Fig. 1*a*) under blue excitation (*Musca domestica* white-eyed mutant). This retinal patch comprises about 30 ommatidia and hence only about 1% of the whole fly retina. Photographic processing of the original colour slide with colour filters allowed us to discriminate the three rhabdomere colours on this black and white picture according to the following code: bright central spots = green-fluorescing R7s, called R7y; grey spots = red-fluorescing R1–6 (plus some red-fluorescing R7s called R7r); black central spots = non-fluorescing R7s, called R7p. All these fluorescence colours probably stem from carotenoid pigments as they are practically absent in β-carotene and vitamin A-deprived flies.

path is kept high, as required for an intense excitation of the rhabdomeres.

In-vivo microspectrofluorimetry (MSF) on the deep pseudopupil (DPP)

Even with the aforementioned techniques of contrast enhancement, quantitative analysis of the fluorescence emission from as tiny a structure as a single rhabdomere still would be a difficult task. Fortunately there exists an optical phenomenon of the compound eye which we have called the *deep pseudopupil* (Franceschini & Kirschfeld, 1971*b*) and which lends itself remarkably well to *in-vivo* microspectrofluorimetry by bringing about a substantial increase in signal-to-noise ratio.

A single facet acts as a magnifying glass and gives magnified virtual images of its seven distal rhabdomere endings along its optical axis. The same holds for a neighbouring facet and for the next neighbour, and so on. As the optical axes of all these facets cross one another at the centre of curvature of the eye, there appears at this point a superposition of virtual images of (identical) rhabdomere patterns. This is the *deep pseudopupil* (DPP), which, in the eye of higher diptera, consists of seven large spots observable when focusing a (low power) microscope deep into the eye (Franceschini & Kirschfeld, 1971*b*; Franceschini, 1975). Each of these spots is due to the superposition of virtual images of the same rhabdomere type from all the ommatidia looking into the microscope objective. The DPP of the compound eye thus appears as a built-in spatial averager whose advantages have been exploited several times for *in-vivo* microphotometry of the receptor cells (reviews may be found in Franceschini, 1972, 1975; Goldsmith & Bernard, 1974; Stavenga, 1979).

This technique (Fig. 2*b*) is even simpler and less invasive than the technique of optical neutralisation of the cornea (Fig. 2*a*) as it does not even require application of a neutralising medium. However, as is the case with any kind of averager, care must be taken to average signals having the same significance. Thus the individual colours of the various R7s (Fig. 4) are averaged out on the central spot of the DPP, and the yellow colour of this spot observed under blue excitation in some eye regions does not betray an intrinsic yellow emission from some R7s but rather results from the additive colours of green and red fluorescing R7s. In this context too, it can be understood how insuperable the task would be to try to infer the fluorescence colours of the individual rhabdomeres in compound eyes with fused rhabdomes, whose DPP fluorescence (orange-reddish colour under blue excitation) hopelessly reports the properties of all classes of cells from all classes of ommatidia.

Quantitative fluorescence measurements in the fly so far (Stark *et al.*, 1979; Franceschini *et al.*, 1981*a*; Stavenga *et al.*, unpublished) concern mainly the six peripheral receptor cells, which have the same spectral sensitivity and visual pigments. As they constitute a homogeneous population, the light emission from the whole DPP was collected. Contamination of the spectra by the central cells R7s was avoided by making the measurements in the male fovea, this dedicated part of the compound eye where the R7s have the same visual pigments as the R1–6 rhabdomeres (Franceschini *et al.*, 1981*b*; Hardie *et al.*, 1981).

If N is the number of ommatidia looking into the microscope objective (N is hence the number of superimposed images on the DPP), the overall gain in signal-to-noise ratio achieved when using the whole DPP of the male fovea with respect to using a single rhabdomere is approximately equal to $\sqrt{7N}$. In our experiments the number of ommatidia contributing to the deep pseudopupil was estimated by counting the number of red-lit facets at the surface of the cornea: $N \simeq 63$. This leads to a substantial increase in signal-to-noise ratio by a factor $\sqrt{7 \times 63} = 21$. The DPP appears as a genuine gift to the spectroscopist.

Red emission observed under blue excitation

In 1962, Burkhardt presented pioneer intracellular recordings from dipteran photoreceptor cells. The spectral sensitivity he most frequently encountered had two maxima of similar height, one in the ultra-violet, the other in the blue-green part of the spectrum. Recovery of intracellularly stained cells (McCann & Arnett, 1972; Hardie, 1979) later confirmed Burkhardt's hypothesis that such a spectral type was characteristic of receptor cells nos. 1–6.

The odd appearance of a dual-peak spectral sensitivity in a photoreceptor cell has led to many conjectures since Burkhardt's original finding and it recently became likely that this in fact was due to the presence of two pigments housed in the same rhabdomere. However, as discussed in the next sections the ultra-violet pigment probably does not play the same role as the blue-green absorbing rhodopsin.

In this section we discuss the properties and origin of the red fluorescence emission which emanates from rhabdomeres R1–6 (Fig. 4).

The photosensitive pigment of fly receptor cells nos. 1–6 shows the typical bistability of invertebrate visual pigments. Its basic properties, as known from the work of Hamdorf and coworkers (1973) and Stavenga and coworkers (1973) are briefly outlined here (Fig. 5). Rhodopsin R_{490} shows maximal absorption in the blue-green part of the spectrum ($\lambda_{max} = 490\,nm$). It is photo-interconvertible with a meta-rhodopsin M_{580} which is highly absorbing in the orange part of the spectrum ($\lambda_{max} = 580\,nm$). Both R and M are, under normal conditions (white light), in a steady photodynamic equilibrium. Whereas the conversion $R \to M$ is coupled to the transduction mechanism generating a receptor potential, the back conversion $M \to R$ is not. This explains why the $490\,nm$ sensitivity band of receptor cells nos. 1–6 is well accounted for by the absorption property of their rhodopsin.

A classical $R \to M$ photoconversion can be shown using adaptation with monochromatic lights. In Fig. 6 the light flux transmitted by the

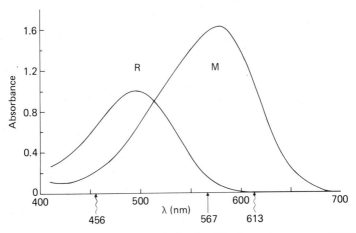

Fig. 5. Fly rhodopsin (R) and meta-rhodopsin (M) absorption spectra (*Calliphora erythrocephala*) calculated from difference spectra after correction for waveguide effects (from Stavenga, 1974). The straight and wavy arrows indicate the wavelengths used in Fig. 6 for monitoring meta-rhodopsin concentration and for inducing a change in photo-equilibrium, respectively.

DPP was monitored in the M-absorption range (λ_{test} = 567 nm). After a red adaptation (λ = 613 nm) which brings practically all the molecules into the R-state, a step of blue light (λ = 456 nm) induces a partial photoconversion R → M as attested by the decrease in transmission (due to an increase in absorption) at λ = 567 nm (Fig. 6a). When starting from a meta-rhodopsin-rich equilibrium on the other hand, a step of red light (λ = 613 nm) shifts the pigment back to the R-state as attested by the increase in transmission (decrease in absorption) at the test wavelength λ = 567 nm (Fig. 6b).

If in a similar set of experiments the light emission from the DPP in the range λ > 665 nm, is measured, a distinct fluorescence signal is seen to parallel the photoconversion R → M with the same time course (Fig. 6c, d). Blue light creates M and increases the fluorescence intensity (Fig. 6c), red light regenerates R and leads to a vanishing of the fluorescence signal (Fig. 6d). One explanation for this behaviour is that fly R1–6 meta-rhodopsin M fluoresces in the red part of the spectrum. Fluorescence of meta-rhodopsin has been recently detected using rhabdoms isolated from crayfish (Cronin & Goldsmith, 1981).

Thus far the fluorescence spectrum of fly meta-rhodopsin M has not been measured and the data obtained (Stavenga *et al.*, in press) concern a closely related species which has been called M' (Franceschini *et al.*, 1981*a*), and whose fluorescence is in fact responsible for the red colour of the rhabdomeres R1–6 (Fig. 4). M' is created when the intensity used

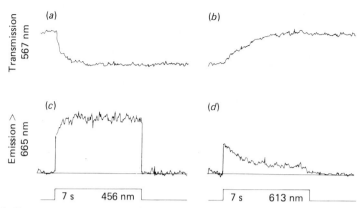

Fig. 6. Changes in transmission at 567 nm, that is in the meta-rhodopsin absorption range (*a, b*) and parallel changes in red emission subsequently observed on the rhabdomeres (*c, d*) under the same blue (*a, c*) and red (*b, d*) adaptations. Measurements made on the deep pseudopupil (DPP) of *Calliphora erythrocephala* (chalky). At the moderate intensities used here, several seconds are needed to reach a new photo-equilibrium. (From Stavenga, Franceschini & Kirschfeld, unpublished.)

for the photoconversion R → M is about 1000 times higher than that used in the experiment of Fig. 6. Under such conditions the new photo-equilibrium R → M is reached very rapidly and M' begins to appear. Therefore, the new substance M' is thought to be created by blue light affecting some short-lived intermediate of the conversion R → M.

Figure 7*a* is a continuous recording of M' appearance (monitored by its fluorescence emission at $\lambda > 610$ nm) under strong blue irradiation ($\lambda = 436$ nm) of the DPP. Figure 7*b* shows that M' can be made to disappear under strong irradiation at a wavelength belonging to the M absorption-band (the 577 nm Hg-line gives a qualitatively similar result). Both lights used in Fig. 7 had been carefully adjusted for the same quantal flux and this clearly shows that the fluorescence yield observed under green light (initial part of Fig. 7*b*) is considerably higher (by a factor of 20–30) than that observed under blue light (end-part of Fig. 7*a*).

Appearance and disappearance of the red fluorescing substance M' is a reproducible and in fact quite spectacular phenomenon. Figure 8 illustrates two successive disappearances of the DPP red emission induced by and photographed under the same green light as in Fig. 7*b*. Before both photographs 8*a* and *c*, M' had been created by prolonged blue irradiation as in Fig. 7*a*.

'Photocreation' and 'photodestruction' of M' according to the wavelength of irradiation provide a convenient means of determining a

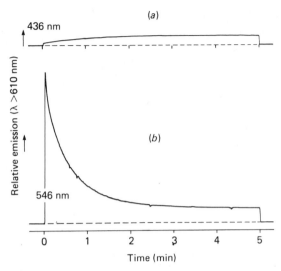

Fig. 7. Slow increase and decrease of the red emission from the rhabdomeres (as measured on the DPP of *Musca domestica*, white-eyed mutant) observed under strong blue (*a*) and green (*b*) irradiations. The quantal flux of both blue and green lights (Hg lines) had been made equal. Note the high level of initial red emission observed when switching on the green light (*b*) after the blue adaptation period.

fluorescence emission spectrum which is not contaminated by other substances in the eye (Stavenga *et al.*, in press). Figure 9 shows raw emission spectra of the DPP obtained under blue excitation. Each curve displays a stable green peak corresponding to the autofluorescence of the cornea and eye tissue. But whereas the initial curve (1) exhibits a minor red peak (probably due to the meta-rhodopsin M, created by the blue excitation light), an intense red peak is seen in curve 2, once M' has been created by prolonged blue adaptation (as in Fig. 7*a*). Photodecomposition of M' by prolonged red light yielded the end spectrum (curve 3) which is nearly identical to the original curve 1.

The corrected difference emission spectrum of M', presented in Fig. 10 (curve 1) exhibits a characteristic peak in the red at 660 nm. It is strikingly similar to the emission spectrum of crayfish meta-rhodopsin M (Fig. 10, curve 2) as determined *in situ* by Cronin & Goldsmith (1981).

The excitation spectrum of M' also has been measured by determining the quantum flux required at each wavelength to elicit a criterion emission intensity at $\lambda > 660$ nm. This spectrum has a main band peaking at $\lambda = 570$ nm and a smaller satellite band in the near UV (Fig. 10, curve 3). The 570 nm excitation band is very reminiscent of the meta-rhodopsin M absorption spectrum (curve M of Fig. 5).

The relatively large bandwidth of the analysing monochromator

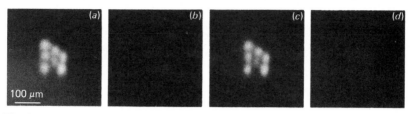

Fig. 8. To illustrate the striking decay of red emission from the rhabdomeres observed under constant green light, as recorded in Fig. 7(*b*). All four photographs taken under the same green light (546 nm): (*a*) after 5 min strong blue irradiation (436 nm); (*b*) after 5 min green irradiation; (*c*, *d*) same sequence as (*a*, *b*) repeated in order to show the reproducibility of the phenomenon (frontal dorsal DPP of a male *Musca*, white).

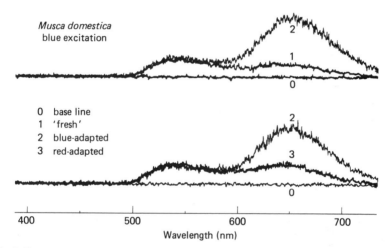

Fig. 9. Raw emission spectra of the rhabdomeres obtained under strong blue light (Xenon arc, band 420–490 nm). The initial curve (1) corresponds to the 'fresh' animal, and shows only a minor red emission beside the greenish autofluorescence of the cornea and eye-tissue (left maximum). Curve 2 recorded after prolonged blue irradiation as in the end-state of Fig. 7(*a*). Curve 3: after prolonged red irradiation (frontal-dorsal DPP of a male *Musca*, white). (From Stavenga, Franceschini & Kirschfeld, unpublished.)

(interference wedge, halfwidth 15 nm) should be kept in mind when judging the diffuseness of the emission spectrum as well as its degree of overlap with the excitation spectrum. However the degree of overlap of absorption and emission bands of retinyl polyenes is known to increase with the viscosity of their environment (Thomson, 1969).

From our study (Stavenga *et al.*, unpublished) as well as from Cronin & Goldsmith's work it can be concluded that meta-rhodopsin M has a distinct red fluorescence, whereas rhodopsin R does not fluoresce detectably, at least up to 700 nm (the limit of our measurement). This

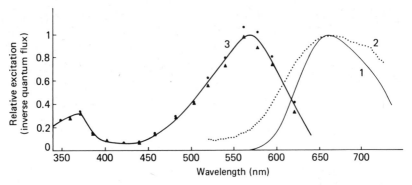

Fig. 10. Curve 1: corrected difference emission spectrum of the rhabdomeres obtained from the data of Fig. 9. Curve 2: emission of crayfish meta-rhodopsin as determined *in situ* in single rhabdoms. (Redrawn from Cronin & Goldsmith, 1981.) Curve 3: excitation spectrum of the red-fluorescing rhabdomeres (*Musca*) determined as the inverse of the quantum flux which is required at each wavelength to elicit a criterion emission above 660 nm. (From Stavenga, Franceschini & Kirschfeld, unpublished.)

result drastically contrasts with the reports of cattle rhodopsin fluorescence (Guzzo & Pool, 1968, 1969). These authors found that both rhodopsin in solution and in dried rod outer segments emit in the range 580–600 nm. However an attempt to reproduce the emission spectrum reported by Guzzo & Pool was unsuccessful (Busch, Applebury, Lamola & Rentzepis, 1972).

The fluorescence properties of crayfish and fly meta-rhodopsin are probably linked to those of the chromophore: whereas fluorescence is seen in all-*trans* retinal, it is practically absent in the 11-*cis* configuration (Balke & Becker, 1967). Bacterio-rhodopsin too, whose chromophore has the all-*trans* configuration, has been shown to fluoresce, but it displays a broadband emission whose peak lies between 700 nm and 800 nm (Lewis, Spoonhover & Perreault, 1976; Govindjee, Becher & Ebrey, 1978).

The new fluorescing compound M' which is formed under intense meta-rhodopsin-creating light has many of the meta-rhodopsin M properties. Its emission spectrum peaks in the red as does that of crayfish M (Cronin & Goldsmith, 1981; compare Fig. 9 curves 1 and 2). Moreover the M' excitation spectrum is very similar to the M absorption spectrum (compare Fig. 10 curve 3 and Fig. 5 curve M). However its UV-side band has a much lower amplitude than would be expected for M on the basis of the sensitising effect brought about by the UV accessory pigment (Minke & Kirschfeld, 1979). When excited by UV light, the M' red emission accordingly would be the result of a direct rather than sensitised fluorescence. Further work, however, will have to show whether M and M' fluorescence excitation spectra, measured on the same animal and the same cells, differ so fundamentally in the near UV.

The intense red emission from M' and the slow appearance and disappearance of this compound, which follow complex kinetics, make it still rather enigmatic. A similar reddish fluorescence has been observed on the fused rhabdom of several insect species under blue excitation (Franceschini *et al.*, 1981*a*).

Irrespective of its detailed physical explanation the red fluorescence can be used to map out the distribution of the spectral receptor types throughout the retinal mosaics. Hence from the homogeneity of the fluorescence colours of rhabdomeres R1–6 observed when screening the retina *in vivo* under the fluorescence microscope, it can be conjectured that all receptor cells nos. 1–6 have the same spectral sensitivity throughout the eye. Furthermore, our conjecture that such a 'red label' would be indicative of a dual-peak spectral sensitivity proved to be correct even for the red-fluorescing central receptor cells, no. R7s of the male fovea (Franceschini *et al.*, 1981*b*; Hardie *et al.*, 1981).

Broadband emission observed under ultra-violet excitation

Whereas only few reports exist dealing with the ultra-violet (UV) sensitivity of the vertebrate eye (see, for example, Eldred & Nolte, 1978; Goldsmith, 1980; Jenison & Nolte, 1980; review in Stark & Tan, 1982), it has been known for 100 years (Lubbock, 1882) that some insects can see UV light (review in Menzel, 1979). However it is only recently that some mechanisms of insect UV-sensitivity have been elucidated. There seem to exist at least two such mechanisms. In the first case, as originally demonstrated for the frontal eye of the Neuropteran *Ascalaphus*, the rhabdomeres contain an ultra-violet-absorbing rhodopsin which, following the classical scheme of invertebrate bistable pigments, is photointerconvertible with a meta-rhodopsin absorbing in another part of the spectrum (Hamdorf, Schwemer & Gogala, 1971; review in Hamdorf, 1979). In the second case, which is illustrated by receptor cells nos. 1–6 of the fly's compound eye, UV-sensitivity is apparently mediated indirectly, via an ultra-violet sensitising pigment that transfers its excitation energy to the rhodopsin. This could explain why these receptor cells are endowed with two spectral maxima, a proper rhodopsin maximum near 490 nm and a 'borrowed' UV-maximum near 350 nm. From a functional standpoint the UV-accessory

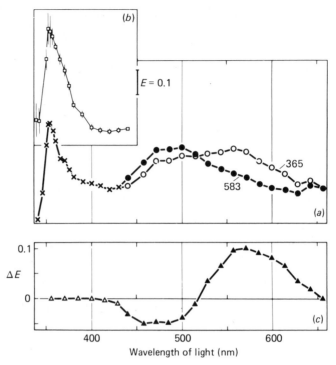

Fig. 11. (*a*) Extinction spectrum of rhabdomeres type R1–6 of *Musca* (white) determined by transmission MSP in an eye-cup preparation. During the measurements the preparation was repeatedly adapted to strong orange (583 nm) or UV (365 nm) lights in order to shift most of the pigment into the rhodopsin or meta-rhodopsin state respectively: (*b*) Photostable UV part of the spectrum; average from six ommatidia. (*c*) Difference spectrum of (*a*). (From Kirschfeld, Franceschini & Minke, 1977.)

pigment would extend the light-harvesting capability of the cell and, in making it panchromatic, increase its detectivity.

The basic MSP experiment upon which the UV-sensitising pigment hypothesis was based is illustrated in Fig. 11. The latter shows a relative extinction spectrum obtained from a group of three rhabdomeres of type R1–6 in a fly ommatidium. A characteristic dual-peak spectrum is observed both after orange (583 nm) and UV (365 nm) adaptations. The difference between these two curves (Fig. 11*c*) betrays the photo-interconversion R $\rightleftharpoons$ M (Fig. 5), as is classically observed when using orange and blue adaptations (Stavenga *et al.*, 1973; Hamdorf *et al.*, 1973; Stavenga, 1976; Stark & Johnson, 1980). The remarkable feature of Fig. 11, however, is that UV-light can readily shift the visual pigment from R to M without depressing the UV-extinction, whose conspicuous stability has been ascertained from other rhabdomeres (Fig. 11*b*).

This and other experiments led us to assume that the rhabdomeres housed an ultra-violet photostable pigment X, harvesting light quanta and transferring its excitation energy to the rhodopsin R (Kirschfeld *et al.*, 1977). Energy transfer has long been known to occur between some accessory pigments and chlorophyll *a* in photosynthesis (Goedheer, 1969). In the context of photography, sensitising dyes are being added to emulsions to selectively increase their sensitivity to a region of the spectrum where the silver halide itself would not absorb (West & Caroll, 1966). In the context of vision, energy transfer from an accessory pigment had already been discussed as a possible mechanism to explain the UV-sensitivity of some visual receptors of arthropods (Chance, 1964; De Voe, Small & Zvargulis, 1969).

The best known mechanism of energy transfer between two molecules is the inductive resonance transfer proposed by Foerster (1951, 1959). In this mechanism a donor molecule in an excited state can pass its excitation energy to an acceptor molecule. This dipole–dipole resonance interaction depends on certain geometric and spectroscopic properties of the donor–acceptor pair and the most stringent conditions to be met are:

(1) the distance D between both molecules must be small enough and a D^{-6}-law makes energy transfer beyond 3–5 nm quasi-impossible (Stryer & Haugland, 1967);

(2) the fluorescence emission spectrum of the donor should largely overlap the absorption spectrum of the acceptor.

In view of this second requirement it has been very interesting to note that fly rhabdomeres R1–6 emit a distinct, broadband fluorescence under UV-excitation (Franceschini, 1977; Stark *et al.*, 1977, 1979; Franceschini *et al.*, 1981*a*).

Several minutes of irradiation with intense UV-light seem to have a deleterious effect upon the UV fluorophores since the DPP soon becomes nearly black, standing out from the bluish autofluorescence of the cornea and eye tissue. Comparing the emissions before and after the UV-bleach allows one to build a difference emission spectrum which, after correction for instrumental responses, yields the spectrum presented in Fig. 12 (Stavenga *et al.*, unpublished).

This corrected spectrum clearly exhibits two peaks, near 470 nm and 660 nm, whose relative heights may vary, resulting in a characteristic pinkish-to-salmon colour of the DPP.

A possible interpretation of this spectrum is that the left emission band (which is extrapolated with a dotted line in Fig. 12) represents the

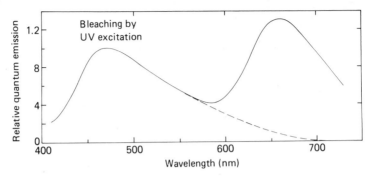

Fig. 12. Corrected emission spectrum of rhabdomeres type R1–6 (*Musca*, white) excited by UV light (365 nm). Difference of the emissions before and after a 6 min UV-bleach which destroys the UV-absorbing fluorophores. The difference between the right peak and the extrapolated left peak (dashed line) fits the red emission spectrum observed under blue excitation (Fig. 10, curve 1). The left peak (h_{max} at 470 nm) corresponds to a bluish-white emission which possibly originates from the ultra-violet sensitising pigment. (From Stavenga, Franceschini & Kirschfeld, unpublished.)

broadband, bluish-white autofluorescence of the ultra-violet sensitising pigment, whereas the right part of the curve corresponds to the red autofluorescence of M'. The latter was indeed shown to be inducible by UV-light too (Fig. 10 curve 3). In fact, the difference between the right hump of the curve and the dotted line (Fig. 12) nicely fits the pure emission spectrum of M' shown in Fig. 10 curve 1.

Stark *et al.* (1979) using another method for determining a difference emission spectrum in *Drosophila* came to the conclusion that the UV-induced emission was substantial only above 560 nm. Although the excitation spectrum they determined did show a major UV-peak, the conclusion they drew that such a fluorescing substance would qualify well as a UV-sensitising pigment is surprising in view of the low rhodopsin absorption above 560 nm (cf. Fig. 5).

The spectrum shown in Fig. 12, on the other hand, does reveal a substantial emission within the rhodopsin absorption band. Thus, if the source of the bluish-white emission is the UV-accessory pigment, this certainly qualifies well as a donor to the rhodopsin acceptor. A precise excitation spectrum of this fluorescence remains to be determined and compared to the UV part of the R1–6 spectral sensitivity. The UV sensitivity spectrum was recently shown to display a characteristic fine structure not commonly found in visual pigments (Gemperlein, Paul, Lindauer & Steiner, 1980).

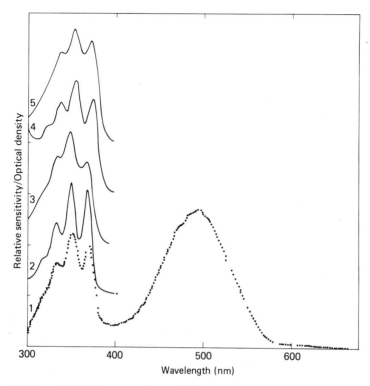

Fig. 13. Curve 1. Relative spectral sensitivity of the fly compound eye (*Calliphora*) measured with high spectroscopic resolution using Fourier spectroscopy combined with electroretinogram recordings (from Gemperlein *et al.*, 1980). 2. Extinction spectrum of phytofluene in hexane (from Koe & Zechmeister, 1952, Fig. 1). 3. Extinction spectrum of retro vitamin A methyl ether (from Oroshnik *et al.*, 1952a, Fig. 4). 4. Extinction spectrum of retro retinol-β-lactoglobulin complex (from Fugate & Song, 1980, Fig. 1B). 5. Corrected fluorescence excitation spectrum of retro-retinol-β-lactoglobulin complex (from Fugate & Song, 1980, Fig. 4B).

Chasing the ultra-violet pigment

Using Fourier spectroscopy associated with electroretinogram recordings, Gemperlein *et al.* (1980) were able to resolve three peaks in the UV part of the spectral sensitivity of the fly's eye, at 332, 350 and 369 nm, and they assigned this fine structure to a polyene, possibly a pentaene. Their spectrum is reproduced in Fig. 13 (curve 1).

A similar fine structure has meanwhile been observed in the spectral sensitivity of individual receptor cells nos. 1–6 measured with intracellular recordings as well as in the extinction spectrum of rhabdomeres R1–6 measured with microspectrophotometry (Kirschfeld *et al.*, 1982).

In the following we examine known substances which possibly could play the role of an ultra-violet-sensitising pigment and which would be characterised by:

(1) a three-peaked absorption spectrum with maxima near 330, 350 and 370 nm;

(2) a UV-induced fluorescence whose emission spectrum would overlap the rhodopsin absorption spectrum.

It first should be noted that although both retinol and retinal fluoresce under UV-excitation (retinol: Hagins & Jenning, 1959; Thomson, 1969; Dalle & Rosenberg, 1970; Radda & Smith, 1970; Chihara *et al.*, 1979; retinal: Balke & Becker, 1967; Moore & Song, 1973; Christensen & Kohler, 1974), none of these chromophores alone is a good candidate in view of their characteristically diffuse absorption spectra which peak at 325 nm and 375 nm respectively.

As a matter of fact, there is an abundance of compounds displaying the required UV-fine structure. They are listed in Table 1 with the position of their three peaks, and some of their spectra have been reproduced in Fig. 13, on top of the fly UV-sensitivity spectrum determined by Gemperlein *et al.* (1980). We now discuss these compounds which belong to three classes of pigments: a C_{40} polyene, various isomers and derivatives of vitamin A, and protein–vitamin A complexes.

Phytofluene

Phytofluene is a naturally-occurring colourless and strongly fluorescent C_{40}-polyene which was discovered by Zechmeister and his associates (Zechmeister & Polgar, 1944; Zechmeister & Sandoval, 1945, 1946). Its main absorption band displays a characteristic fine structure in the UV as shown in Fig. 13 (curve 2). The three maxima occur at 332, 349, and 368 nm in alcohol and are very reminiscent of the three maxima found by Gemperlein and his coworkers (Table 1). Phytofluene is so widespread in the plant kingdom (Zechmeister & Sandoval, 1945; Zechmeister & Karmakar, 1953) that it has been assigned a role in plant metabolism. The idea that it is an intermediate in the biosynthesis of the carotenes was substantiated by the works of Porter & Lincoln (1950) and Koe & Zechmeister (1952). We have not been able to find an emission spectrum of phytofluene in the literature but its intense fluorescence colour has been reported as greenish-grey (Zechmeister & Sandoval, 1945) or bluish-green (Porter & Zscheile, 1946).

Retro-derivatives of vitamin A

It has been known for 50 years that vitamin A can be chemically modified into substances displaying three absorption bands in the UV

Table 1. *Carotenoids and caroteno-proteins which are possible candidates for the ultra-violet-sensitising pigment of fly photoreceptor cells. The positions of their three peaks are compared to those observed on the fine structure of the fly's spectral sensitivity*

	Near UV maxima (nm)	Remarks	References
Spectral sensitivity of the fly's R1–6 cells	332, 350, 369	ERG	Gemperlein *et al.* (1980)
(1) *Phytofluene*	332, 349, 368	In alcohol	Zechmeister & Polgar (1944)
(2) *Retro derivatives of vitamin A*			
Isoanhydro vitamin A	about 330, 350, 367		Shantz *et al.* (1943)
	330, 349, 370		Zechmeister & Sandoval (1946)
	about 330, 347, 367		Isler *et al.* (1947)
	332, 348, 366		Oroshnik (1954)
Anhydro subvitamin A	332, 348, 367		Embree & Shantz (1943)
Rehydro vitamin A	330, 351, 369		Shantz (1950)
('Retro' vitamin A alcohol)			
'Retro' vitamin A methyl ether (*trans*)	333, 349, 369		Oroshnik *et al.* (1952a)
	332.5, 348, 367		Hemley (1979)
			Oroshnik *et al.* (1952a, b)
'Retro' vitamin A acetate (*trans*)	333, 348, 367		Beutel *et al.* (1955)
	330, 346, 367	Excit. spectr.	Fugate & Song (1980)
	352		Schreckenbach *et al.* (1977)
(3) *Vitamin A–protein complexes*			
Cellular retinol binding protein (CRBP)	335, 350, 367	Rat	Ong & Chytil (1978)
	330, 350, 368	Human	Ong (1982)
Retinol–β-lactoglobulin	330, 345, 366		Hemley *et al.* (1979)
	330, 342, 360		Fugate & Song (1980)
'Retro' retinol–β-lactoglobulin	337, 354, 375		Hemley *et al.* (1979)
	335, 352, 371		Fugate & Song (1980)

(Edisbury, Gillam, Heilbron & Morton, 1932). Several distinct vitamin A derivatives have been reported which exhibit a fine structure in their absorption spectra, strikingly similar to that of the fly's UV-spectral sensitivity (Table 1). The first compound is iso-anhydro vitamin A, discovered by Shantz and co-workers (1943). Another compound was called anhydro subvitamin A by Embree & Shantz (1943). In 1950, Shantz described the rehydro vitamin A, which is formed *in vivo* from ingested anhydro vitamin A.

A deeper insight into the molecular organisation responsible for the UV-fine structure of such compounds was provided by Oroshnik, Karmas & Mebane (1952*a,b*). These authors described a new series of vitamin A derivatives: the 'retro' β-ionylidene series which differs from the 'normal' β-ionylidene series in that the conjugated system is displaced by one carbon atom back into the ring:

'normal' β-ionylidene   'retro' β-ionylidene

The 'retro' compounds are formed with great ease; they are more stable and display a much more structured absorption spectrum than the corresponding vitamin A compounds. Curve 3 of Fig. 13 shows the UV-fine structure of 'retro' vitamin A methyl ether, a compound prepared by Oroshnik and co-workers (1952*a, b*). As reported in Table 1, the three peaks appear again at similar wavelengths in 'retro' vitamin A acetate (Beutel, Hinkley & Pollak, 1955; Fugate & Song, 1980) and 'retro' vitamin A alcohol, the latter substance corresponding to the rehydro vitamin A of Shantz (1950).

It had been noted early on (Sobotka, Kann, Winternitz & Brand, 1944) that UV irradiation of vitamin A acetate yields an even more highly fluorescing compound characterised by a three-peaked UV-absorption spectrum similar to that of the aforementioned iso-anhydro vitamin A, whose 'retro' structure was later demonstrated by Oroshnik (1954). More recently, Christensen & Kohler (1973) and Fugate & Song (1980) have presented data about the fluorescence properties of typical 'retro' compounds like anhydro vitamin A and retro vitamin A acetate.

The fine structure observed in the absorption spectrum of the 'retro' compounds is thought to be due to the coplanarity of the ring side-chain systems and to the absence of steric hindrance (Dale, 1954). By contrast the classical absorption spectrum of vitamin A is considered to be atypical. It is hypsochromically displaced by 20–25 nm wih respect to the main peak of retro vitamin A, and the lack of fine structure is attributed to the many possible conformations of the ring side-chain torsional angle due to the single bond between ring and side chain (Christensen & Kohler, 1973; Hemley & Kohler, 1977).

Protein–vitamin A complexes

It has been established that in mammals retinol is carried from the liver to the tissue cells by a specific plasma protein to which it binds with high affinity: the *retinol binding protein* (RBP: Kanai, Raz & Goodman, 1968). Interestingly the retinol fluorescence emission is blue-shifted down to 470 nm upon binding to the RBP (Peterson & Rask, 1971; Muto & Goodman, 1972), and the fluorescence yield is increased by about one order of magnitude with respect to that of retinol in petroleum ether (Futterman & Heller, 1972). The excitation spectrum, however, remains unstructured, still peaking near 330 nm.

Once retinol has reached and entered the target cells it seems to be taken up by a cellular protein discovered more recently (Bashor, Toft & Chytil, 1973), the *cellular retinol binding protein* (CRBP) which is a putative mediator of vitamin A action at the cellular level (Chytil & Ong, 1979). When retinol binds to the CRBP, its spectral properties are considerably altered from those displayed in organic solvents. As seen in Fig. 14, which is reproduced from Ong & Chytil's work (1978), the absorption spectrum of retinol is red-shifted by ~25 nm upon binding to the protein. A characteristic fine structure appears with a main maximum at 350 nm and two secondary maxima or shoulders at 335 nm and 367 nm (Table 1). This unusual three-peaked spectrum is obviously the signature of vitamin A, since long exposure to UV light annihilates the UV absorption (Fig. 14, dotted line) which can then be recovered by reconstitution with all-*trans*-retinol (Ong & Chytil, 1978). The excitation spectrum of the complex displays a similar fine structure and the emission spectrum remains unstructured with a peak at 470–485 nm (uncorrected spectra, Ong & Chyril, 1978; Ong, 1982). Here again a considerable enhancement of fluorescence intensity is observed in the retinol emission upon complexing with the protein (seven- to eight-fold increase; Ong, 1982). Such a fluorescence sensitivity has repeatedly been observed with many retinol–protein complexes (Futterman & Heller, 1972), and it has made retinol an interesting probe for proteins and lipids (Radda & Smith, 1970; Georghiou & Churchich, 1975).

The binding of retinol to model proteins like β-lactoglobulin (BLG) has been thoroughly analysed by Hemley, Kohler & Siviski (1979) and

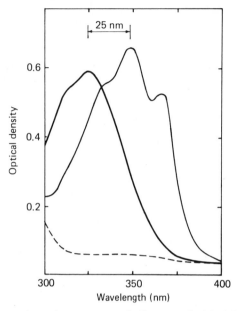

Fig. 14. Bold curve: absorption spectrum of all-*trans* retinol in ethanol. Thin curve: absorption spectrum of cellular retinol binding protein (CRBP) (1.35×10^{-5} M in 0.05 M Tris buffer, pH 7.5). Dashed line: CRBP spectrum recorded after prolonged irradiation with UV light to destroy the retinol (redrawn from Ong & Chytil, 1978, Fig. 2). Note the 25 nm red shift, increased absorbance and appearance of a fine structure in the spectrum of bound retinol.

by Fugate & Song (1980). Both studies combine to show that two distinct complexes can be formed, depending on the relative concentration of vitamin A and protein. Whereas at a retinol protein molar ratio of 1:1 a complex is formed in which the chromophore is all-*trans* retinol, an excess of retinol leads to the formation of another complex in which the chromophore is a 'retro' derivative of vitamin A.

As in the case of CRBP, red shift by ~20 nm and appearance of a fine structure in the absorption spectrum characterise these two complexes. But the most interesting feature is that the absorption spectrum of the 'retro' complex is much more structured than the all-*trans* retinol–protein complex, and we have included it in Fig. 13 (curve 4) together with the corresponding excitation spectrum (curve 5) from Fugate & Song's work (1980). Again a characteristic feature of these retinol–protein complexes is the dramatic increase in fluorescence lifetime (fourfold increase in the all-*trans* complex and even sixfold in the retro-complex) with respect to that of retinol in organic solvents (Fugate & Song, 1980).

The appearance of a fine structure in the retinol spectrum seems to be coupled with the fixing of the torsional angle between ring and side chain, either through a specific external perturbation like that brought about by a protein or by the formation of a definite retro-compound (Hemley *et al.*, 1979).

The typical UV 'finger-print spectrum' again has been observed during reconstitution experiments of the purple complex of bacterio-rhodopsin (Schreckenbach, Walckhoff & Oesterhelt, 1977). These authors have shown that the protein bacterio-opsin induces a planarised conformation of the retinyl moiety, which on one hand is responsible for the observed UV fine structure and on the other hand would facilitate the ultra-violet-mediated photoisomerisation into a retro compound. The UV-fine structure which comes into being when retinol binds to a protein is a remarkable phenomenon and creation of an increased rigidity in the environment of retinol molecules using glycerol or low temperature cannot fully mimic the effect of this protein binding (Georghiou & Churchich, 1975; Fugate & Song, 1980).

In the absence of chemical data concerning the UV accessory pigment, presently we can only rely on the spectroscopic data. As shown in Table 1 and Fig. 13, most of the substances reported above not only have a structured absorption spectrum resembling the fly's UV-sensitivity spectrum but also display a broadband, unstructured fluorescence emission not unlike that measured on the DPP (Fig. 12).

Although data presently available do not allow us to exclude any of these substances, the probable role played by the UV-pigment as an energy donor certainly makes the vitamin A-protein complex the most acceptable candidate since it possesses all of the following properties.

(1) The fine structure of its absorption spectrum is well-pronounced, especially in the 'retro' retinol–protein complex, and the emission spectrum is diffuse and peaks approximately in the R1–6 rhodopsin absorption range.

(2) The marked increase in fluorescence yield observed when the chromophore is bound to the protein would attest to a better capability of energy transfer to a neighbouring chromophore.

(3) The greatly increased stability of the retinol when it is bound to the protein (Futterman & Heller, 1972; Hemley *et al.*, 1979; Fugate & Song, 1980) would make it less vulnerable to damage by the environmental light.

In view of the geometric constraints required by the Foerster model for an efficient energy transfer, an appealing hypothesis would be that the retinol chromophore is bound to the rhodopsin molecule itself, at a site close to the retinal chromophore. Wu & Stryer (1972) have shown that various fluorescent chromophores can be attached at different sites of a rhodopsin molecule and can transfer energy to the retinal chromophore.

Conclusion

Fluorescence spectroscopy can be achieved on identified photoreceptor cells under good physiological conditions (*in vivo*) and at little expense. As a complement to microspectrophotometry and intracellular electrical recordings, microspectrofluorimetry is now adding its specific advantages of simplicity, sensitivity and selectivity to the analysis of visual pigments in single identified photoreceptor cells.

Two simple, non-invasive optical techniques have been associated with epi-fluorescence microscopy. Whereas the technique of optical neutralisation of the cornea allows us to view the phenomena on each single cell behind each single facet (Fig. 4), the technique of the deep pseudopupil (Fig. 8) reveals a local average of the fluorescence emission from each class of cell. The deep pseudopupil (DPP) is a spatial averager which considerably improves the signal-to-noise ratio of the photometric signal and hence the sensitivity of the method. The benefit of this technique is obvious in view of the dimness of the fluorescence emissions. Cronin & Goldsmith (1981) determined a quantum efficiency of only about 10^{-3} for the red fluorescence of crayfish meta-rhodopsin M.

Fly receptor cells nos. 1–6 are so far the only animal photoreceptors to which all three methods now available for an *in vivo* study of visual pigments in single cells have been applied. Compared with microspectrophotometry (MSP) which reports the properties of all kinds of pigments (including screening pigments) contained in a rhabdomere without distinction, microspectrofluorimetry (MSF) allows us to differentiate individual pigments thanks to their characteristic autofluorescence colour. Beside the red emission from meta-rhodopsin M (Fig. 6c. d) whose spectra remain to be analysed (see, however, Cronin & Goldsmith, 1981 for Crayfish M), MSF has revealed a new, red-fluorescing substance M' which has not so far been detected by any other means. Further work is needed to determine precisely the status of this new compound, and in particular to show whether it is a photoproduct of a short-lived intermediate of the photoconversion R → M.

From a functional standpoint, fluorescence of M and M' is probably of no advantage to the receptor cell. It merely witnesses that these molecules suffer some radiative de-excitation, as does the all-*trans* retinal chromophore (Balke & Becker, 1967). On the other hand, absence of fluorescence emission from rhodopsin R makes more sense as any radiative de-excitation of this molecule would be at the expense of the relevant photochemical process.

The rhabdomere fluorescence observed under ultra-violet excitation is most interesting as it may help us to understand the process by which receptor cells nos. 1–6 gain a satellite, ultra-violet peak of sensitivity. The latter is currently thought to originate from a light-harvesting pigment which would transfer its excitation energy to the rhodopsin (Kirschfeld *et al.*, 1977; Kirschfeld, 1981).

Two molecular species seem to be responsible for the pinkish fluorescence observed under UV light. First the red emission of M', second a broadband bluish-white emission which possibly betrays some radiative de-excitation of the ultra-violet sensitising pigment. Here again further work is required to confirm this view. In particular, an excitation spectrum should be measured with high spectroscopic resolution in order to search for a UV-fine structure similar to that revealed by the electrical recordings (Fig. 13, curve 1). Even though energy transfer via inductive resonance according to Foerster's model does not involve donor fluorescence followed by absorption of radiation by the acceptor, some fluorescence of the donor would be expected as a 'spillover' of excitation energy.

The very process of energy transfer from the light-harvesting carotenoid pigments to chlorophyll *a* in the photosynthetic membrane is still a matter of debate (for example: Razi Naqvi, 1980 *contra* Thrash, Fang & Leroi, 1979), especially because of the extremely low quantum yield of β-carotene fluorescence (Tric & LeJeune, 1970; Moore & Song, 1973). Razi Naqvi (1980) proposed that energy transfer from β-carotene to chlorophyll *a*, rather than occurring via a dipole–dipole interaction as in Foerster's model, would take place via an electron exchange according to Dexter's model (1953). Though still depending upon the spectral overlap between donor fluorescence and acceptor absorption, the latter process required an even closer proximity of the donor–acceptor pair.

As discussed in the preceeding section, there is some suggestive evidence that the fine structure observed in the UV-spectral sensitivity of receptor cells nos. 1–6 is due to a vitamin A–protein complex. However, at present no direct chemical data are available to substantiate this hypothesis which is exclusively based on spectroscopic data. Conventional chemical analysis obviously is not very appropriate and one cannot help dreaming of the eighteen kilograms of canned tomato paste from which Zechmeister and Sandoval extracted their phytofluene (see previous section).

One way out of this difficulty possibly lies in pigment substitution experiments which could be attempted *in vivo*, either following vitamin A deprivation or after photodecomposition of the rhabdomere content. Various substances could then be injected into the eye and even into single receptor cells with the hope to recover their original properties. Intracellular injection of an extrinsic fluorescent marker readily allows recovery of the impaled cell *in vivo* without calling upon histology (Franceschini & Hardie, 1980).

The rapidly developing field of *in-vivo* cell microbiochemistry (Taylor & Wang, 1980) makes extensive use of fluorescently labelled molecules as probes of various cell features. The existence of intrinsic fluorophores in photoreceptor cells is welcome and offers a great deal of potential for future studies.

I am indebted to my colleagues D. Stavenga and K. Kirschfeld for permission to include some of our as yet unpublished data (Figs. 6, 9, 10, 12) and to P. S. Song for stimulating discussions. Thanks are due to M. Andre, G. Jacquet, A. Totin-Yvard for technical assistance, and to D. Stavenga and M. Wilcox for critical reading of the manuscript. Permission given by other authors to reproduce their published spectra is gratefully acknowledged. This work was supported by the MPG (West Germany) and by the CNRS, DGRST and Fondation de la Recherche Medicale (France).

References

BALKE, D. E. & BECKER, R. S. (1967). Spectroscopy and photochemistry of all-*trans* retinal and 11-*cis* retinal. *J. Am. Chem. Soc.*, **89**, 5061–2.

BASHOR, M. M., TOFT, D. O. & CHYTIL, F. (1973). *In-vitro* binding of retinol to rat-tissue components. *Proc. Natl. Acad. Sci., USA*, **70**, 3483–7.

BEUTEL, R. M., HINKLEY, D. F. & POLLAK, P. I. (1955). Conversion of vitamin A acetate to retinovitamin A acetate. *J. Am. Chem. Soc.*, **77**, 5166–7.

BOSCHEK, B. (1971). On the fine structure of the peripheral retina and lamina ganglionaris of the fly *Musca domestica. Z. Zell Forsch. Abt. Histochem.*, **118**, 369–409.

BURKHARDT, D. (1962). Spectral sensitivity and other response characteristics of single visual cells in the arthropod eye. *Symp. Soc. Exp. Biol.*, **16**, 86–109.

BUSCH, G. E., APPLEBURY, M. L., LAMOLA, A. A. & RENTZEPIS, P. M. (1972). Formation and decay of prelumirhodopsin at room temperature. *Proc. Natl. Acad. Sci., USA*, **69**, 2802–6.

CHANCE, B. (1964). Fluorescence emission of mitochondrial DPNH as a factor in the ultraviolet sensitivity of visual receptors. *Proc. Natl. Acad. Sci., USA*, **51**, 359.

CHIHARA, K., TAKEMURA, T., YAMAOKA, T., YAMAMOTO, N., SCHAFFER, A. & BECKER, R. S. (1979). Visual pigments. X. Spectroscopy and photophysical dynamics of retinol and retinyl ether. *Photochem. Photobiol.*, **29**, 1001–8.

CHRISTENSEN, R. L. & KOHLER, B. E. (1973). Low resolution optical spectroscopy of retinyl polyenes: low-lying electronic levels and spectral broadness. *Photochem. Photobiol.*, **18**, 293.

CHRISTENSEN, R. L. & KOHLER, B. E. (1974). Excitation spectroscopy of retinal and retinyl polyenes. *Photochem. Photobiol.*, **19**, 401–10.

CHYTIL, F. & ONG, D. E. (1979). Cellular retinol and retinoic acid-binding proteins in vitamin A action. *Fed. Proc.*, **38**, 2510–14.

CRONIN, T. W. & GOLDSMITH, T. H. (1981). Fluorescence of crayfish metarhodopsin studied in single rhabdoms. *Biophys. J.*, **35**, 653–64.

DALE, J. (1954). Empirical relationships of the minor bands in the absorption spectra of polyenes. *Acta Chem. Scandin.*, **8**, 1235–56.

DALLE, J. P. & ROSENBERG, B. (1970). Radiative and non-radiative losses in the retinal polyenes. *Photochem. Photobiol.*, **12**, 151–67.

DE VOE, R. D., SMALL, R. J. W. & ZVARGULIS, J. E. (1969). Spectral sensitivities of wolf spider eyes. *J. Gen. Physiol.*, **54**, 1–32.

DEXTER, D. L. (1953). A theory of sensitized luminescence in solids. *J. Chem. Phys.*, **21**, 836–50.

EDISBURY, J. R., GILLAM, A. E., HEILBRON, I. M. & MORTON, R. A. (1932). Absorption spectra of substances derived from vitamin A. *Biochem. J.*, **26**, 1164–73.

ELDRED, W. D. & NOLTE, J. (1978). Pineal photoreceptors: evidence for a vertebrate visual pigment with two physiologically active states. *Vision Res.*, **18**, 29–32.

EMBREE, N. D. & SHANTZ, E. M. (1943). A possible new member of the vitamins A_1 and A_2 group. *J. Am. Chem. Soc.*, **65**, 906–9.

FOERSTER, T. (1951). *Fluoreszenz organischen Verbindungen*. Göttingen: Vandenhoeck and Ruprecht.

FOERSTER, T. (1959). Transfer mechanisms of electronic excitation. *Disc. Faraday Soc.*, **27**, 7–17.

FRANCESCHINI, N. (1972). Pupil and pseudopupil in the compound eye of *Drosophila*. In: *Processing of Information in the Visual System of Arthropods*, ed. R. Wehner, pp. 75–82. Berlin, Heidelberg & New York: Springer-Verlag.

FRANCESCHINI, N. (1975). Sampling of the visual environment by the compound eye of the fly: fundamentals and applications. In: *Photoreceptor Optics*, ed. A. W. Snyder & R. Menzel, pp. 98–125. Berlin: Springer-Verlag.

FRANCESCHINI, N. (1977). *In-vivo* fluorescence of the rhabdomeres in an insect eye. *Proc. Int. Union Physiol. Sc. XIII.* XXVIIth Int. Congr. Paris, 237.

FRANCESCHINI, N. (1978). Bi-stable and photostable pigments in fly photoreceptors cells: evidence from 'ommatidial fundus fluoroscopy'. *Neuroscience Lett.*, Suppl. 1S, 405.

FRANCESCHINI, N. (1983). Chromatic organisation and sexual dimorphism of the fly retinal mosaic. In: *Photoreceptors*, ed. A. Borsellino & L. Cervetto. New York: Plenum Press.

FRANCESCHINI, N. & HARDIE, R. (1980). *In-vivo* recovery of intracellularly stained cells. *J. Physiol., London*, **301**, 59P.

FRANCESCHINI, N. & KIRSCHFELD, K. (1971a). Etude optique *in vivo* des éléments photorécepteurs dans l'oeil composé de *Drosophila*. *Kybernetik*, **8**, 1–13.

FRANCESCHINI, N. & KIRSCHFELD, K. (1971b). Les phénomènes de pseudopupille dans l'oeil composé de *Drosophila*. *Kybernetik*, **9**, 159–82.

FRANCESCHINI, N. & KIRSCHFELD, K. (1976). Le contrôle automatique de flux lumineux dans l'oeil composé des Diptères: propriétés spectrales, statiques et dynamiques du mécanisme. *Biol. Cybernetics*, **21**, 181–203.

FRANCESCHINI, N. & STAVENGA, D. G. (1981). The ultraviolet sensitizing pigment of flies studied by *in vivo* microspectrofluorometry. *Invest. Ophth. Vis. Sci.*, Suppl. 20, 111.

FRANCESCHINI, N., KIRSCHFELD, K. & MINKE, B. (1981a). Fluorescence of photoreceptor cells observed *in vivo*. *Science*, **213**, 1264–7.

FRANCESCHINI, N., HARDIE, R., RIBI, W. & KIRSCHFELD, K. (1981b). Sexual dimorphism in a photoreceptor. *Nature, London*, **291**, 241–4.

FUGATE, R. D. & SONG, P. S. (1980). Spectroscopic characterization of β-lactoglobulin–retinol complex. *Biochim. Biophys. Acta*, **625**, 28–42.

FUTTERMAN, S. & HELLER, J. (1972). The enhancement of fluorescence and the decreased susceptibility to enzymatic oxidation of retinol complexed with bovine serum albumin, β-lactoglobulin and the retinol binding protein of human plasma. *J. Biol. Chem.*, **247**, 5168.

GEMPERLEIN, R., PAUL, R., LINDAUER, E. & STEINER, A. (1980). UV-fine structure of the spectral sensitivity of flie's visual cells revealed by FIS (Fourier Interferometric Stimulation). *Naturwissenschaften*, **67**, 565–6.

GEORGHIOU, S. & CHURCHICH, J. E. (1975). Nanosecond spectroscopy of retinol. *J. Biol. Chem.*, **250**, 1149–51.

GOEDHEER, J. C. (1969). Energy transfer from carotenoids to chlorophyll in blue-green, red and green algae and greening bean leaves. *Biochim. Biophys. Acta*, **172**, 252–65.

GOLDSMITH, T. (1972). The natural history of invertebrate visual pigments. In: *Handbook of Sensory Physiology*, vol. VII/1 *Photochemistry of Vision*, ed. H. J. Dartnall, pp. 685–719. Berlin: Springer-Verlag.

GOLDSMITH, T. H. (1980). Hummingbirds see near ultraviolet light. *Science*, **207**, 786–8.

GOLDSMITH, T. H. & BERNARD, G. D. (1974). The visual system of insects. In: *The physiology of Insecta*, vol. 2, ed. M. Rockstein, pp. 166–263. New York: Academic Press.

GOVINDJEE, R., BECHER, B. & EBREY, T. G. (1978). The fluorescence from the chromophore of the purple membrane protein. *Biophys. J.*, **22**, 67–77.

GUZZO, A. V. & POOL, G. L. (1968). Visual pigment fluorescence. *Science*, **159**, 312–14.

GUZZO, A. V. & POOL, G. L. (1969). Fluorescence spectra of the intermediates of rhodopsin bleaching. *Photochem. Photobiol.*, **9**, 565–70.

HAGINS, W. A. & JENNINGS, V. (1959). Radiationless migration of electronic excitation of retinal rods. *Discus. Faraday Soc.*, **27**, 180–90.

HAMDORF, K. (1979). The physiology of invertebrate visual pigments. In: *Handbook of Sensory Physiology*, vol. VII/6A, ed. H. Autrum, pp. 145–224. Berlin & Heidelberg: Springer-Verlag.

HAMDORF, K., PAULSEN, R. & SCHWEMER, J. (1973). Photoregeneration and sensitivity control of photoreceptors of invertebrates. In: *Biochemistry and Physiology of Visual Pigments*, ed. H. Langer, pp. 155–66. Berlin & Heidelberg: Springer-Verlag.

HAMDORF, K., SCHWEMER, J. & GOGALA, M. (1971). Insect visual pigments sensitive to ultraviolet light. *Nature, London*, **231**, 458–9.

HARDIE, R. (1979). Electrophysiological analysis of fly retina. I. Comparative properties of R1–6 and R7 and R8. *J. Comp. Physiol.*, **129**, 19–33.

HARDIE, R., FRANCESCHINI, N. & MAC INTYRE, P. (1979). Electrophysiological analysis of fly retina. II. Spectral and polarization sensitivity in R7 and R8. *J. Comp. Physiol.*, **133**, 23–39.

HARDIE, R., FRANCESCHINI, N., RIBI, W. & KIRSCHFELD, K. (1981). Distribution and properties of sex-specific photoreceptors in the fly *Musca domestica. J. Comp. Physiol.*, **145**, 139–52.

HEMLEY, R. & KOHLER, B. E. (1977). Electronic structure of polyenes related to the visual chromophore: a simple model for the observed band shapes. *Biophys. J.*, **20**, 377–82.

HEMLEY, R., KOHLER, B. E. & SIVISKI, P. (1979). Absorption spectra for the complexes formed from vitamin-A and β-lactoglobulin. *Biophys. J.*, **28**, 447–55.

ISLER, O., HUBER, W., RONCO, A. & KOFFLER, M. (1947). Synthèse des vitamin A. *Helv. Chim. Acta*, **30**, 1911–27.

JENISON, G. & NOLTE, J. (1980). An ultra-violet sensitive mechanism in the reptilian eye. *Brain Res.*, **194**, 506–10.

KANAI, M., RAZ, A. & GOODMAN, D. S. (1968). Retinol-binding protein: the transport protein for vitamin-A in human plasma. *J. Clin. Invest.*, **47**, 2025–44.

KIRSCHFELD, K. (1981). Bistable and photostable pigments in microvillar photoreceptors. In: *Sense Organs*, ed. M. S. Laverack & D. J. Cosens, pp. 142–62. London: Blackie.

KIRSCHFELD, K. & FRANCESCHINI, N. (1969). Ein Mechanismus zur Steuerung des Lichtflusses in der Rhabdomeren des Komplexauges von Musca. *Kybernetik*, **6**, 13–22.

KIRSCHFELD, K. & FRANCESCHINI, N. (1975). Microspectrophotometry of fly rhabdomeres. Conference on visual physiology, Günzburg (Germany).

KIRSCHFELD, K., FRANCESCHINI, N. & MINKE, B. (1977). Evidence for a sensitizing pigment in fly photoreceptors. *Nature, London*, **269**, 386–90.

KIRSCHFELD, K., FEILER, R., HARDIE, R., VOGT, K. & FRANCESCHINI, N. (1982). The sensitizing pigment in fly photoreceptors: properties and candidates. *Biophys. Struct. Mechanism*, in press.

KOE, B. K. & ZECHMEISTER, L. (1952). *In-vitro* conversion of phytofluene and phytoene into carotenoid pigments. *Arch. Biochem. Biophys.*, **41**, 236–8.

LANGER, H. & THORELL, B. (1966). Microspectrophotometry of single rhabdomeres in the insect eye. *Exp. cell Res.*, **41**, 673–7.

LEWIS, A., SPOONHOVER, J. P. & PERREAULT, G. J. (1976). Observation of light emission from a rhodopsin. *Nature, London*, **260**, 675–9.

LIEBMAN, P. A. & LEIGH, R. A. (1969). Autofluorescence of visual receptors. *Nature, London*, **221**, 1249–51.

LUBBOCK, J. (1882). On the sense of color among some of the lower animals. *J. Linnean Soc. London, Zool.*, **16**, 121–7.

MCCANN, G. D. & ARNETT, D. W. (1972). Spectral and polarization sensitivity of the dipteran visual system. *J. Gen. Physiol.*, **59**, 534–58.

MENZEL, R. (1979). Spectral sensitivity and colour vision in invertebrates. In: *Handbook of Sensory Physiology*, vol. VII/6A, ed. H. Autrum, pp. 503–80. Berlin & Heidelberg: Springer-Verlag.

MINKE, B. & KIRSCHFELD, K. (1979). The contribution of a sensitizing pigment to the photosensitivity spectra of fly rhodopsin and metarhodopsin. *J. Gen. Physiol.*, **73**, 517–40.

MOORE, T. & SONG, P. S. (1973). Molecular interactions in the ground and excited states of retinal. *Nat. New Biol.*, **243**, 30–2.

MUTO, Y. & GOODMAN, D. S. (1972). Vitamin A transport in rat plasma. Isolation and characterization of retinol-binding protein. *J. Biol. Chem.*, **247**, 2533–41.

ONG, D. E. (1982). Purification and partial characterization of cellular retinal binding protein from human liver. *Cancer Res.*, **42**, 1033–7.

ONG, D. E. & CHYTIL, F. (1978). Cellular retinol-binding protein from rat liver. *J. Biol. Chem.*, **253**, 828–32.

OROSHNIK, W. (1954). The structure of 'isoanhydrovitamin A'. *Science*, **119**, 660.

OROSHNIK, W., KARMAS, G. & MEBANE, A. (1952a). Synthesis of polyenes. I. Retrovitamin A methyl ether. Spectral relationships between the β-ionylidene and retroionylidene series. *J. Am. Chem. Soc.*, **74**, 295–304.

OROSHNIK, W., KARMAS, G. & MEBANE, A. D. (1952b). Synthesis of polyenes. II. Allylic rearrangements and dehydrations in substituted β-ionols. *J. Am. Chem. Soc.*, **74**, 3807–13.

PAPAGEORGIOU, G. (1975). Chlorophyll fluorescence: an intrinsic probe of photosynthesis. In: *Bioenergetics of Photosynthesis*, ed. Govindjee, pp. 319–71. New York: Academic Press.

PETERSON, P. A. & RASK, L. (1971). Studies on the fluorescence of the human vitamin A transporting plasma protein complex and its individual components. *J. Biol. Chem.*, **246**, 7544–50.

PLOEM, O. O. (1967). The use of a vertical illuminator with interchangeable dichroic mirrors for fluorescence microscopy with incident light. *Z. Wiss. Mikrosk.*, **68**, 129–42.

PORTER, J. W. & LINCOLN, R E. (1950). I. Lycopersicon selections containing a high content of carotenes and colorless polyenes. II. The mechanism of carotene biosynthesis. *Arch. Biochem. Biophys.*, **27**, 390–403.

PORTER, J. W. & ZSCHEILE, F. P. (1946). Naturally occurring colorless polyenes. *Arch. Biochem. Biophys.*, **10**, 547–50.

RADDA, G. & SMITH, D. (1970). Retinol: a fluorescent probe for membrane lipids. *FEBS Lett.*, **9**, 287–9.

RAZI NAQVI, K. (1980). The mechanism of singlet–singlet excitation energy transfer from carotenoids to chlorophyll. *Photochem. Photobiol.*, **31**, 523–4.

SCHRECKENBACH, T., WALCKHOFF, B. & OESTERHELT, D. (1977). Studies on the retinal–protein interaction in bacterio-rhodopsin. *Eur. J. Biochem.*, **76**, 499–511.

SHANTZ, E. M. (1950). Rehydro Vitamin A, the compound formed from anhydro Vitamin A *in vivo*. *J. Biol. Chem.*, **182**, 515–24.

SHANTZ, E. M., CAWLEY, J. D. & EMBREE, N. D. (1943). Anhydro ('cyclised') Vitamin A. *J. Am. Chem. Soc.*, **65**, 901–6.

SOBOTKA, H., KANN, S., WINTERNITZ, W. & BRAND, E. (1944). The fluorescence of vitamin A. II. Ultraviolet absorption of irradiated vitamin A. *J. Am. Chem. Soc.*, **66**, 1162–4.

STARK, W. S. & JOHNSON, M. A. (1980). Microspectrophotometry of *Drosophila* visual pigments: determination of conversion efficiency in R1–6 receptors. *J. Comp. Physiol.*, **140**, 275–86.

STARK, W. S. & TAN, K. E. (1982). Ultraviolet light: photosensitivity and other effects on the visual system. *Photochem. Photobiol.*, **36**, 371–80.

STARK, W. S., IVANYSHYN, A. M. & GREENBERG, R. M. (1977). Sensitivity and photopigments of R1–6, a two-peaked photoreceptor, in *Drosophila, Calliphora* and *Musca*. *J. Comp. Physiol.*, **121**, 289–305.

STARK, W. S., STAVENGA, D. G. & KRUIZINGA, B. (1979). Fly photoreceptor fluorescence is related to UV sensitivity. *Nature, London*, **280**, 581–3.

STAVENGA, D. G. (1974). Visual receptor optics, rhodopsin and pupil in fly retinula cells. Thesis, Groningen.

STAVENGA, D. G. (1976). Fly visual pigments. Difference in visual pigments of blowfly and dronefly peripheral retinula cells. *J. Comp. Physiol.*, **111**, 137–52.

STAVENGA, D. G. (1979). Pseudopupils of compound eyes. In: *Handbook of Sensory Physiology*, vol. 7/6A, ed. H. Autrum. Berlin, Heidelberg & New York: Springer-Verlag.

STAVENGA, D. G. & FRANCESCHINI, N. (1981). Fly visual pigment states, rhodopsin R 490, metarhodopsin M and M′ studied by transmission and fluorescence microspectrophotometry *in vivo*. *Invest. Ophthalm. Vis. Sci. Suppl. 20*, **3**, 111.

STAVENGA, D. G., ZANTEMA, A. & KUIPER, J. (1973). Rhodopsin processes and the function of the pupil mechanism in flies. In: *Biochemistry and Physiology of Visual Pigments*, ed. H. Langer, pp. 175–80. Berlin & Heidelberg: Springer-Verlag.

STRYER, L. & HAUGLAND, R. P. (1967). Energy transfer: a spectroscopic ruler. *Proc. Natl. Acad. Sci., USA*, **58**, 719–26.

TAYLOR, D. L. & WANG, Y. L. (1980). Fluorescently labelled molecules as probes of the structure and function of living cells. *Nature, London*, **284**, 405–10.

THOMSON, A. J. (1969). Fluorescence spectra of some retinyl polyenes. *J. Chem. Phys.*, **51**, 4106.

THRASH, R. J., FANG, H. L-B. & LEROI, G. E. (1979). On the role of forbidden low-lying excited states of light-harvesting carotenoids in energy transfer in photosynthesis. *Photochem. Photobiol.*, **29**, 1049–50.

TRIC, C. & LEJEUNE, V. (1970). Les carotènes fluorescent-ils? *Photochem. Photobiol.*, **12**, 339–43.

TRUJILLO-CENOZ, O. & MELAMED, J. (1966). Electron microscope observations on the peripheral and intermediate retinas of Dipterans. In: *The Functional Organization of the Compound Eye*, Part 4, *Integration of Visual Input*, ed. G. G. Bernhard, pp. 339–361. Oxford: Pergamon Press.

WEST, W. & CARROLL, B. M. (1966). Spectral sensitivity and the mechanism of spectral sensitization. In: *The Theory of the Photographic Process*, 3rd edn, ed. K. Mees & T. M. James, pp. 233–76. London: Macmillan.

WU, C. W. & STRYER, L. (1972). Proximity relationships in rhodopsin. *Proc. Natl. Acad. Sci., USA*, **69**, 1104–8.

ZECHMEISTER, L. & KARMAKAR, G. (1953). The occurrence of phytofluene in green plant organs. *Arch. Biochem. Biophys.*, **47**, 160–4.

ZECHMEISTER, L. & POLGAR, A. (1944). On the occurrence of a fluorescing polyene with a characteristic spectrum. *Science*, **100**, 317–18.

ZECHMEISTER, L. & SANDOVAL, A. (1945). The occurrence and estimation of phytofluene in plants. *Arch. Biochem. Biophys.*, **8**, 426–30.

ZECHMEISTER, L. & SANDOVAL, A. (1946). Phytofluene. *J. Am. Chem. Soc.*, **68**, 197–201.

CONFORMATIONAL AND FUNCTIONAL CHANGES INDUCED IN VERTEBRATE RHODOPSIN BY PHOTON CAPTURE

MARC CHABRE

Laboratoire de Biologie Moleculaire et Cellulaire, Departement de Recherche Fondamentale, CENG. 38041 Grenoble, France

Introduction

To the question 'what are the effects induced in rhodopsin by the capture of a photon?', the tentative answers have been often limited to the description of the effects of illumination on the spectral properties of the chromophore. In all visual pigments the chromophore, a retinal molecule or a retinal derivative, constitutes only of the order of 1% of the total mass of the protein molecule in which it is embedded: molecular weights of most of the known visual pigments are of the order of 40 000. Although the retinal molecule is an essential component of the pigment which defines the coupling with light and triggers all the subsequent processes, it is certainly not the only part of the macro-molecule that is modified upon photon capture. Indeed, ample evidence exists that the capture of a photon and the subsequent dark reactions lead to the formation, at the surface of the protein, of specific sites which are recognised by other soluble or peripherally membrane-bound proteins. Some of these photo-induced sites are apparently distant from the chromophore itself by more than 5 mm. The information of the capture of a photon, first stored as a conformational transition of the chromophore, has therefore to be transmitted to these peripheral sites by long-distance conformational changes through the protein. Whether these light-induced changes and the enzymatic activities they initiate are directly on the path of the phototransduction process, or are only part of regulation and adaptation mechanisms is still strongly debated. It has, however, been recently established that one of these photo-induced processes, which activates a GTP-dependent enzyme responsible for the activation of a cyclic GMP (cGMP) phosphodiesterase and therefore the control of cGMP level in the retinal rod cells, is fast enough to be a good candidate for the phototransduction mechanism (Liebman & Pugh, 1979; Kuhn, Bennett, Michel-Villaz & Chabre, 1981). If the functional evidence for changes induced by light is solid, the demonstration and determination by physical measurements of the related conformational changes remain very fragmentary, as we shall discuss later.

The spectroscopic approaches, which concentrate on the studies of the light-induced effects on the chromophore itself and its interaction with its immediate binding site, have been very fruitful in some visual systems. (For an example of the very elegant spectroscopic measurements which take full advantage of the optics of the insect compound eye, see Franceschini, this volume.) However, for the biochemical and structural study of a complete pigment molecule with its predominant protein component, the rod cells of the vertebrate retina with their highly segregated outer segments offer decisive advantages both for *in-situ* biophysical measurements and for isolation and purification of the pigment molecules in significant quantities. Rhodopsin from these rod outer segments is the visual pigment molecule about which the best information has been obtained. Whether the observations made on this 'model' visual pigment molecule are to be generalised to invertebrate visual systems may be questionable, but there are close analogies. The primary photochemical event, characterised by the isomerisation of retinal and the formation of a first batho-chromic photoproduct seems common to all visual systems. Later, in the dark decay of this bathoproduct, the chromophore is released from the protein molecule in the vertebrate pigment, in contrast to its permanent binding in the invertebrate pigment. But this difference may not be very significant in relation to the mechanism of the phototransduction process in the two systems: it is clear that in the vertebrate pigment the detachment of the chromophore is a slow reaction which occurs after the initiation of visual excitation. It is relevant only to the regeneration cycle, which is certainly not the same in the vertebrate and invertebrate systems. In view of their different morphologies, the difference in the persistence of pigment-binding does not imply a basic difference between the two initial phototransduction mechanisms. That similar processes are induced in the protein parts of the two types of pigment upon illumination had been recently demonstrated in an elegant experiment (Eyring *et al.*, 1980): upon illumination, octopus rhodopsin activates the phosphodiesterase enzymatic complex extracted from rod outer segments of cattle, just as vertebrate rhodopsin does. The same sites of binding and activation must, therefore, be created upon illumination at the surface of both types of pigment. One must not, however, overlook the fact that the protein parts of the vertebrate and the invertebrate pigments differ, at least in their molecular weights and in their location and interaction with their respective membrane environments. The ordered packing of

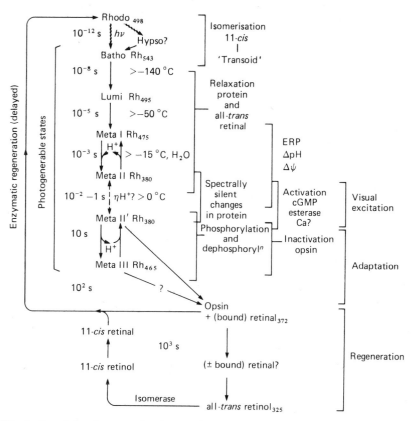

Fig. 1. Correlation between the decay scheme of rhodopsin, defined by the spectral characteristics of the retinal chromophore, and the associated events in the protein, on the disk membrane and in the rod cell. ERP is the Early Receptor Potential, which is generated at the level of the disk membrane. ΔpH denotes a change of pH which can be observed on disk membrane preparations and $\Delta\psi$ a change of interfacial electrical potential also observable on disk membrane preparations. The signals are observed only upon strong illumination which bleaches a large proportion of the rhodopsin molecules, and their amplitudes are strictly proportional to the number of rhodopsin molecules bleached. They reveal changes in the protein structure, but are probably not physiologically significant in the visual transduction process.

the invertebrate pigment in the apparently rigid microvillar membrane contrasts sharply with the high mobility of vertebrate rhodopsin in the very fluid rod disk membrane.

In this chapter I will deal exclusively with vertebrate rhodopsin. I shall first review briefly the knowledge, still very limited, acquired on the structure of the pigment in its native dark-adapted state. The effects of illumination will then be discussed in successive steps: the primary photochemical event on the chromophore; the interaction of the chromophore with its binding site in the protein after its transconformation; the structural changes detected in the protein itself; and the functional changes induced in the pigment molecule and its photo-induced interactions with other proteins in the rod outer segment. (Fig. 1).

The structure of vertebrate rhodopsin

Except for a high molecular weight protein present in small amounts at the rim of the disks (Papermaster, Schneider, Zorn & Kraehenbul, 1978), rhodopsin is virtually the only intrinsic protein present *in* the rod disk membrane. There are, however, several fairly abundant proteins *on* the membrane or in the outer segment cytoplasm, which have been overlooked until recently. Rhodopsin is not packed in an ordered way in the membrane unlike, for example, bacteriorhodopsin in the purple membrane (see Hildebrand, this volume). Rather, individual rhodopsin molecules float freely in the lipid bilayer, with a high degree of lateral and rotational mobility. The only constraint is that the rotation axis remains perpendicular to the membrane plane. This high mobility may be of great importance in allowing fast sequential interactions of a photo-excited rhodopsin molecule with other proteins on the membrane, but it is a very undesirable feature when one is interested in structure. There is no tendency to two-dimensional crystallisation *in situ*, and crystallography is the only method available for high resolution structural determinations. Progress may come in this direction as preliminary evidence has been reported recently (Scott, McCaslin & Corless, 1981) for a two-dimensional ordering obtained upon detergent treatment of the membrane. This limited ordering is, however, still far from a real crystal. In its absence, small pieces of structural information have been gathered from as many biophysical and biochemical techniques as possible, and a consensus has been reached on a model of the type shown in Fig. 2 (see Hubbell & Fung, 1979; Hargrave *et al.*, 1980; Albert & Litman, 1978; Michel-Villaz, Saibil & Chabre, 1979). The main features are: (i) the protein is transmembrane; (ii) a large part of the peptide chain is embedded in the hydrophobic layer of the membrane, and this part probably includes the retinal binding site; (iii) only a short segment of the polypeptide chain protrudes into the intra-diskal space; this is the N-terminal end with two attached carbohydrate chains; (iv) a larger proportion of the protein protrudes into the cytoplasm and bears the sites of interaction with soluble enzymes and with peripheral membrane proteins.

The transmembrane part of the protein seems to be internally very

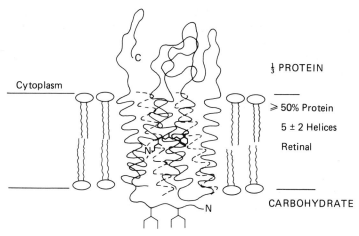

Fig. 2. The structure of rhodopsin in the disk membrane. C and N are the C- and N-terminals of the protein.

hydrophobic: this was demonstrated by the observation that a large proportion of the peptide protons are not exchangeable with deuterium when D_2O is substituted for H_2O in the surrounding medium (Osborne & Nabedrik-Viala, 1977). Bleaching *in situ* does not accelerate the exchange, which hardly seems compatible with the idea of an ionic channel being formed through the protein. More than 50% of the polypeptide chain of the protein is in the α-helical conformation and α-helical segments, oriented preferentially in the direction perpendicular to the membrane plane, constitute the major component of the transmembrane hydrophobic core (Fig. 3). This was proposed as the explanation for the large diamagnetic anisotropy of the rod outer segments (Chabre, 1978; Worcester, 1978), and has been confirmed by polarised infra-red studies of the absorption band of the peptide C=O and N—H bonds of rhodopsin *in situ*, in magnetically-oriented rod suspensions (Michel-Villaz *et al.*, 1979). The transmembrane part of rhodopsin seems therefore to have some structural analogy with the better known structure of bacteriorhodopsin. But this apparent analogy may be of restricted significance since α-helical segments ordered perpendicular to the membrane seems to be a preferred conformation for a transmembrane polypeptide in all membrane proteins.

It is worthwhile stressing that upon illumination, bacteriorhodopsin will not activate any phosphodiesterase system (Eyring *et al.*, 1980).

The retinal chromophore, which is itself hydrophobic, is expected to be inserted in the hydrophobic core of the protein. From fluorescence transfer studies with probes attached at various places on the surface of

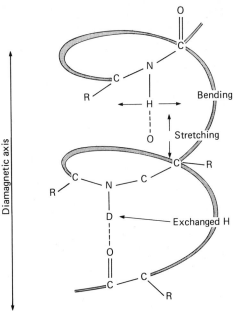

Fig. 3. Detection of the orientation of α-helix segments in the rhodopsin molecule. The diamagnetic anisotropy which causes the orientation of rods in magnetic fields results directly from the diamagnetic anisotropy of these α-helical segments in the rhodopsin molecule. This indicates that a large proportion of the rhodopsin structure must be constituted by transmembrane α-helical segments. This is confirmed by infra-red linear dichroism measurements which detect the orientation of the C=O and N—H bonds in the polypeptide chain. In an α-helix the C=O bonds preferentially absorb infra-red light which is polarised parallel to the helix axis, and the N—H bonds preferentially absorb infra-red light polarised perpendicularly to the helix axis. As the measurements are performed in D_2O, there is also an N—D absorption peak, which results from the H—D exchange, favoured in the hydrophilic part of the protein. The fact that this peak is not dichroic indicates that these hydrophilic parts of the rhodopsin molecule on either side of the disk membrane have no preferential orientation of their polypeptide chain.

the protein, a location close to the intra-diskal side of the membrane has been proposed (Wu & Stryer, 1972), but there were uncertainties about the exact locations of the probes. Recent measurements of diffusion-enhanced fluorescence transfer from long-lived probes in solution favour a location near the centre of the membrane, distant from both aqueous surfaces (Thomas & Stryer, 1982). The binding point of the chromophore on the polypeptide chain has been now identified through recent progress in its sequencing (Wang, McDowell & Hargrave, 1980): it is on lysine 53′ located in a hydrophobic segment of the polypeptide but not far from the very hydrophilic C-terminal end which extends into the cytoplasm (Fig. 2). Allowing for the unknown orientation of the lysine itself, this is compatible with a retinal molecule located near the centre of the membrane and attached to the first transmembrane α-helix.

The sequence is known for a substantial part of the polypeptide molecule, starting from both ends and including a long section around the retinal-binding point, but the tertiary structure is still unknown. This is required in order to build significant models for the binding site which may include sections of polypeptide chains that are far apart in the sequence. There are expectations that the complete sequence will soon be obtained, either by the classical proteolytic method which is very difficult for hydrophobic segments, or through the DNA-sequencing approach. The mRNA which codes for bovine rhodopsin has been isolated and purified (Schechter *et al.*, 1979) and various research groups are attempting the transcription and cloning of this DNA. The example of bacteriorhodopsin has shown that, for a hydrophobic protein, much structural information may be obtained from the knowledge of its complete sequence and secondary structure.

Conformational changes upon illumination

The hypothesis that the isomerisation of retinal is the primary event was proposed more than 20 years ago and, although generally accepted, had not been strictly demonstrated until very recently. There were good arguments, but no absolute proof that the isomerisation caused the first detectable spectral transition to batho-rhodopsin, the chromophore being not extractable from this early state, but only from the much later meta-II state. The dogma was strongly challenged a few years ago when picosecond techniques demonstrated that the transition to batho-rhodopsin was very fast even at very low temperature: 36 ps at 4 K (Peters, Applebury & Rentzepis, 1977). Is a real isomerisation with nuclear movements possible in such a short time at this low temperature? As an alternative hypothesis it was proposed that the primary event was the translocation of a proton along a chromophore of fixed conformation, the process being favoured by a tunnelling mechanism (Fig. 4). The isomerisation would then be a later consequence of this first phenomenon. There is of course always an ambiguity about how 'primary' an event is: batho-rhodopsin, defined through its spectral absorption, is only the first ground-state photoproduct, which is stable indefinitely at low temperature; it has to be reached through short-lived electronic excited states. The challenge proved to be fruitful, as it induced a wealth of new studies on the conformation of retinal in rhodopsin and in batho-rhodopsin. These studies confirmed the original isomerisation

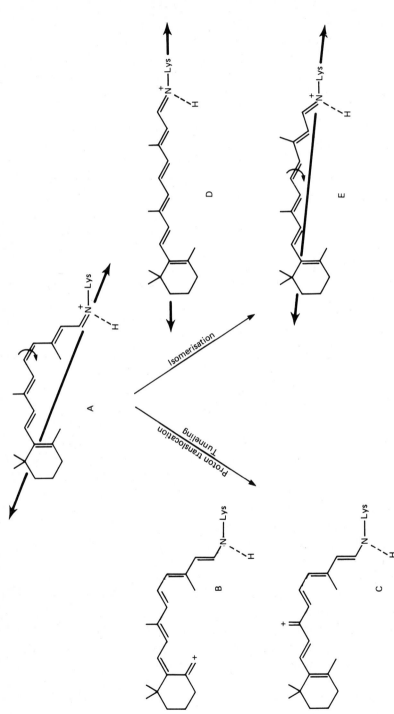

Fig. 4. Models for the initial photochemical event in the retinal chromophore during the transition from unbleached rhodopsin to batho-rhodopsin. An isomerisation from the initial 11-*cis* conformation (A) to a 'distorted trans' conformation (E) is the model best supported by the latest experiments. The change of orientation of the absorbing dipole of the chromophore in such a transition is not as large as for a complete cis-trans isomerisation from conformation A to conformation D. The arrows indicate the approximate dipole directions. The proton translocation hypothesis (scheme B or C),

hypothesis and yielded additional information about the exact conformation of the chromophore: at the early batho stage, the isomerisation is not complete and the molecular movement of the chromophore upon the first transition is of smaller amplitude than one would expect for an isomerisation between the free 11-*cis* and the all-*trans* conformation of the molecule. The main evidence comes from resonance Raman studies, which demonstrate that the conformation of the chromophore in batho-rhodopsin is of the all-*trans*-type but is strongly distorted (Eyring *et al.*, 1980). Moreover, in a picosecond resonance Raman measurement it was demonstrated that this conformation appears within less than 20 ps (Hayward, Carlsen, Siegman & Stryer, 1981).

A simple phenomenological way to look for a conformational change of the chromophore is to study the eventual change of orientation of its absorbing dipole within the pigment molecule. As retinal is a linear molecule, the dipolar absorption anisotropy should be closely related to the geometry of the conjugated double bonds of the polyene chain. In the all-*trans* conformation, for example, the dipole should be aligned with the straight chain, and a *cis–trans* isomerisation changing the geometry of the chain should also change the dipole orientation (Fig. 4).

To detect such changes one needs to orient the pigment molecule, which can be done by orienting the rods in a magnetic field. One needs also to block the rotational motion of the rhodopsin molecules, which is achieved by cooling down to liquid nitrogen temperature. In this frozen state there remains, however, a rotational disorder of the pigment molecules around the orientation axis. The angle that the absorbing dipole of the chromophore makes with the membrane plane (azimuthal angle) is estimated from measurements of the linear dichroism of the sample. The observation axis is perpendicular to the orientation axis of the rods (Fig. 5). The change of linear dichroism upon the photoconversion of rhodopsin to batho-rhodopsin measures the azimuthal angular shift of the dipole upon this transition. To detect the component in the membrane plane (polar component) of this angular shift, one needs to observe the rods axially. The polar components of the dipoles are then randomly oriented around the observation axis. A flash of linearly polarised light, directed along the rod axis, creates an oriented subpopulation of the batho-rhodopsin molecules. The orientation of this subpopulation not only depends on the direction of plane polarisation of the flash, but also on the polar angular shift which may occur upon the transition of rhodopsin to batho-rhodopsin. One sees easily, for example, that if this polar angular shift were 90°, the photoselected batho-rhodopsin population would be oriented in the direction perpendicular

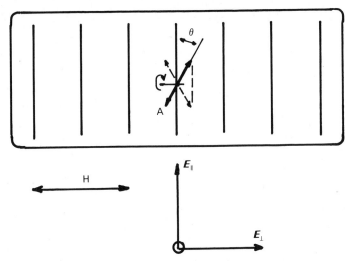

Fig. 5. Orientation of rod outer segments in a magnetic field H and measurements of the angular shift of chromophores in rhodopsin. At room temperature the rhodopsin molecules rotate freely around the direction normal to the membrane plane. For a chromophore fixed into the protein, only the azimuthal angle θ between the absorbing dipole A and the membrane plane is accessible. Its value and its eventual change upon a given transition are estimated on the basis of linear dichroism measurement with a transverse light beam: this is the comparison of the chromophore absorbance for the two orthogonal directions of the electrical field vector: $E_{\parallel}$ and $E_{\perp}$ as shown. At low temperature however the rotation of the rhodopsin molecules in the membrane may be blocked. One may then detect the polar component (in the plane of the membrane) of the angular shift of the chromophore upon its isomerisation in the fixed rhodopsin molecule. This is done by photo-selection experiments with polarised light coaxial with the rods.

to the plane of polarisation of the exciting light. The orientation of the photoselected batho-rhodopsin population is studied by measuring the rates of its photoreversal by polarised light flashes of suitable wavelength and of various orientations of their plane polarisation.

We have in our group developed such measurements on suspensions of intact rod outer segments oriented in a magnetic field of very high intensity which allows cooling to liquid nitrogen temperature without loss of alignment. We measured a rotation of 11 ± 3 degrees for the main absorbing dipole of retinal upon the rhodopsin to batho-rhodopsin transition (Michel-Villaz, Roche & Chabre, 1982). The main component of the rotation is within the membrane plane around the orientation axis, which is what one would expect for the *trans* isomerisation of an 11-*cis* retinal molecule where both segments of the polyene chain would originally be in a plane parallel to the membrane. This is indeed the orientation of the chromophore in rhodopsin since the two absorbing dipoles, which define the plane of symmetry of the molecule to the

11-*cis* conformation, appear to be both nearly parallel to the membrane (Liebman, 1972).

The amplitude of the angular shift upon this initial transition to batho-rhodopsin is much smaller than would be expected from the simple picture of the transition between undistorted 11-*cis* and all-*trans* isomers of a chromophore having its bulky end fixed into the protein. However, the angular shift measured is consistent with results from resonance Raman studies, which indicate that a change of conformation does occur but is small. At this stage the chromophore is highly constrained by the protein site which has probably not yet changed its conformation and so restricts the displacement of the chromophore. This is correlated with a large energy uptake in the transition: the difference in free energy between the 11-*cis* and all-*trans* conformations of retinal is negligible when the molecules are in solution—but an energy uptake of 35 kcal is needed to induce the same conformational change in a retinal molecule bound to its site in the rhodopsin molecule (Cooper, 1979). This is more than 60% of the energy of the absorbed photon and is a very high efficiency for a photo-energetic conversion. The function of a visual pigment however, is to memorise and transmit the information of the capture of a photon, not to store its energy. In the later dark reactions this stored energy induces conformational changes in the protein which relax the strain on the chromophore.

The conformational changes induced by the chromophore in the protein

The dark decay reactions which lead from the primary batho state to the enzymatically active meta-II state extend over a period of the order of a millisecond at physiological temperature. This would be long enough to allow for major rearrangements in the protein structure. For the first two transitions however, to lumi-rhodopsin and meta I, the spectral changes of the chromophore have not yet been correlated to any specific conformational changes in the protein. This may be due to the lack of a proper probe, and reflects the difficulties encountered in looking for small perturbations in transient states which have to be studied at low temperature. Upon the batho–lumi transition, the azimuthal orientation of the chromophore dipole shifts back to the angle observed initially in rhodopsin. No detectable change of orientation of the chromophore occurs upon the lumi–meta-I transition.

It is only with respect to the meta I–meta II transition that changes definitely related to the protein part of the pigment have been charac-terised: perturbations of the near and far ultra-violet (UV) absorption and circular dichroism spectra have been correlated with this transition. However, the interpretation of those perturbations in terms of specific structural changes is hampered by our lack of accurate knowledge of the three-dimensional structure of the protein. One has also to be very careful about the physical state of the pigment in the membrane: large 'structural' changes, which have often been claimed to occur upon bleaching rhodopsin purified in various detergents or other artificially reconstituted systems, are not observed *in situ*. This difference reflects only the fact, known for more than twenty years, that the pigment is much less stable in the meta-II state and so is easily denatured in this state when transferred in an artificial environment. The main conclusions from the latest studies using intact membrane preparations are presented here. Conformational changes of the protein, involving long sections in the secondary structure, have not been proven and are very unlikely. Local conformational changes involving a few aromatic residues do occur: the best-documented one is the perturbation of one tryptophan in a very hydrophobic environment, which has a red-shifted absorption spectrum. The light-induced changes in the near UV spectrum indicate that upon the meta I–meta II transition the environment of this tryptophan and that of several tyrosine residues becomes more hydrophilic (Rafferty, Muellenberg & Shichi, 1980). Linear dichroism measurements in the UV have demonstrated that the perturbation corresponds to a rotation of one tryptophan residue (Fig. 6) which may directly correlate with the rotation of the chromophore also observed at this stage (Chabre & Breton, 1979). This specific tryptophan is probably very close to the retinal chromophore in the protein-binding site. As for the three-dimensional structure, and in particular that of the α-helix bundle in the transmembrane region, major rearrangements seem to be excluded. A large part of the perturbation in the far-UV spectrum may be accounted for by the aromatic residues. No significant change could be observed in the infra-red (IR) linear dichroism (Michel-Villaz *et al.*, 1979), nor in the magnetic anisotropy (Chabre, 1978)—these are not, however, very sensitive methods and, for example, a small rotation of one transmembrane α-helix segment would not be detected.

What about the most classical structural approach, that of X-ray diffraction? Interpretation of data from the earliest investigations led to the proposal that the whole rhodopsin molecule would sink significantly into the membrane upon bleaching. This was later contested. Yet careful measurements made on retina strips (Corless, 1972); on intact, physiologically-active retinas (Chabre & Cavaggioni, 1973); and on isolated rod outer segments oriented in a magnetic field (Chabre, 1975)

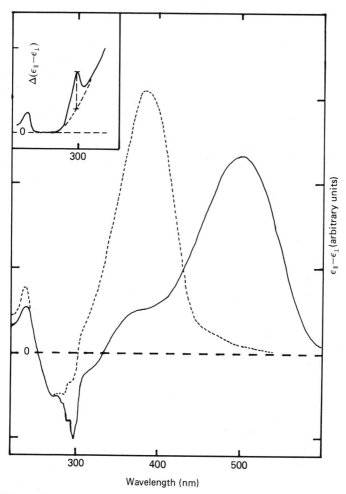

Fig. 6. Change of orientation of the retinal chromophore and of a tryptophan residue in rhodopsin, upon the transition to the meta II state, as detected by linear dichroism measurements on oriented rods. The spectra represent directly the difference of absorbances $\varepsilon_\parallel$ and $\varepsilon_\perp$ for the two orthogonal directions of polarisation defined by their electrical vectors $E_\parallel$ and $E_\perp$ in Fig. 5. The continuous line is the spectrum of dark-adapted rods and the dotted line that of the same sample after the photoinduced transition. The increased amplitude of the main chromophore peak, which shifts from 500 to 380 nm, indicates that upon this transition the chromophore dipole has rotated towards the membrane plane. The change in the sharp structure around 300 nm, seen more clearly in the difference spectrum in the insert, reveals the concomitant perturbation of a tryptophan residue.

have consistently indicated the occurrence of small variations of scattering density on the cytoplasmic side of the membrane upon illumination. This change of state has been confirmed by neutron diffraction (Saibil, Chabre & Worcester, 1976), a technique which also establishes the observed change to be due to a shift in the distribution of protein mass. Whether this reflects a change of conformation of rhodopsin itself is now doubtful, because all of these measurements have been performed on intact organelles to take advantage of the ordered structure of the rod, and the former contain other proteins. Kuhn (1980) has demonstrated recently that the photo-excitation of rhodopsin *in-situ* modifies the binding of fairly abundant peripheral membrane proteins to the disk membrane–in particular binding of a GTP-dependent protein that we shall denote G-protein. This modification of binding must result from a conformational change of the pigment molecule itself to create the photo-induced interaction site. However, the overall shift of protein mass on the cytoplasmic side of the membrane is now thought to reflect the displacements and changes of binding sites of the loosely bound peripheral proteins rather than the actual conformational change of the pigment.

This new interpretation arises from an improved knowledge of the biochemical composition of the rod cells, and from studies of light-scattering transients, which I will discuss later.

The outcome of all the biophysical approaches to the determination of the light-induced structural changes in rhodopsin is not very impressive: very little has been definitely characterised besides the conformation changes of the chromophore itself and some perturbations in its immediate environment. The changes in the protein are probably too subtle and our knowledge of its three-dimensional structure too incomplete to allow their characterisation by non-specific physical techniques. This does not imply that such subtle changes would be insignificant in terms of functional activity; indeed activation of enzymatic systems often result from a very small displacement of a few amino-acid residues at a critical site, which are usually detectable only through high-resolution crystallographic studies. The emphasis on the movement of tryptophan and aromatic residues here, as well as in many other cases where crystallographic determinations are not available, is more likely to reflect the relatively easy spectroscopic determination of these residues than their predominant role in the conformational change. The displacement of charged residues of the protein in the vicinity of the chromophore clearly plays a predominant role in energy storage and in the dark reactions, since changes of 0.2 to 0.3 nanometres in the separation of a pair of charges involve a considerable amount of energy (Honig, Ebrey, Callender, Dinur & Ottolenghi, 1979). But there are no direct spectroscopic signals from these charged residues when they do not interact with a chromophore. One may imagine that such charge displacements, propagated at the cytoplasmic surface of the protein, are responsible for the creation of the site of interaction between the photo-excited pigment molecule and the peripheral proteins. This would not be detectable by spectroscopy if, as seems to be the case, the chromophore is buried deep inside the protein.

Functional changes induced by light and their structural implications

Biochemical approaches, which detect the structural changes through their functional consequences, have been recently more successful than direct structural approaches. The first indication that the capture of a photon induces in rhodopsin a structural change that is specifically recognised by another protein was indeed the demonstration nearly ten years ago that bleached rhodopsin is phosphorylated by a soluble kinase present in the rods; this kinase is inactive on the unbleached pigment (Kuhn & Dreyer, 1972; Bownds, Dawes, Miller & Stahlman, 1972). Seven sites of phosphorylation have been localised on threonine and serine residues packed near the hydrophilic C-terminal end of the protein (Hargrave *et al.*, 1980). It has also recently been observed (H. Kuhn, private communication) that this hydrophilic segment is more susceptible to proteolysis by trypsin after illumination. The retinal-binding lysine is on the first hydrophobic segment linked to this hydrophilic terminal segment and distant by about 40 residues from the phosphorylated region. One may then conceive that the isomerisation of retinal would 'pull' on the lysine and displace slightly the polypeptide chain leading to an uncovering of its end. Phosphorylation of illumin-ated rhodopsin is, however, too slow a process, even *in vivo* (Kuhn, 1974) to account for the transmission of visual excitation and it is probably involved rather in the turning off, or the adaptation, of the response. Indeed it has now been demonstrated that the proteolytic cleavage of this terminal segment (Kuhn & Hargrave, 1981), does not inhibit the creation on photo-excited rhodopsin of a binding site for the peripheral G-protein, which is probably the major effect of light on the pigment. This G-protein is present in the rods with a stoichiometry of 1:10 with respect to rhodopsin; in the dark it is membrane-bound, but this binding is weak and can be released at low ionic strength. Kuhn

(1980) made the key observation that illumination of the system initiates another, tighter binding, which is insensitive to the ionic strength, but which can be reversed by the presence of GTP. Fung, Hurley & Stryer (1981) demonstrated that photo-excited rhodopsin 'activates' the G-protein (for which they propose the name 'transducin') by catalysing the exchange on to it of a molecule of GTP for a molecule of GDP. The activated G-protein will in turn activate a cyclic GMP-phosphodiesterase present also on the membrane, which controls the level of cyclic GMP in the cell. This scheme provides for a large amplification, one photo-excited rhodopsin being able to activate in sequence a large number of G-proteins. At this point the mobility of rhodopsin in the plane of the membrane is probably essential for the successive encounters with many G-proteins.

Although the proposed scheme drawn in Fig. 7 could be established on the basis of binding studies necessitating the separation of membrane fractions from soluble fractions by centrifugation and other biochemical techniques (for example, radioactive labelling, column separation, GTPase activity), such biochemical methods have only a slow resolution time. In order to determine the significance of this process for visual transduction, it is essential to establish its kinetics and stoichiometry. Kuhn *et al.* (1981) have recently demonstrated that light-scattering changes in the far-red induced by flash illumination of rhodopsin in the disk membrane provide sensitive signals, with high time resolution, suitable for an analysis of the kinetics and stoichiometry of the first steps of the proposed scheme. Changes of light-scattering in membrane suspensions, and even in intact cells, have already been reported, but the relation of these signals to proteins other than rhodopsin was not suspected. Using reconstituted systems, with purified rhodopsin and purified G-protein, it was demonstrated that two types of signal are strictly related to the light-triggered interaction between photo-excited rhodopsin (R^*) and G-protein. The clue was that the presence and the saturation characteristics of the signals depend strictly on the presence and the amount of G-protein in the reconstituted system. In the absence of GTP, a first type of signal is related to the binding step of G to R^*-protein; this saturates when the number of photo-excited rhodopsin molecules is equal to the number of G-protein molecules present in the sample, indicating a 1:1 stoichiometric binding. In the presence of GTP, another type of signal is related to the 'activation' of G-protein by GTP; this is catalysed by R^* and is followed by the release of G-protein from R^*. This signal saturates at a very low level of illumination since each molecule of rhodopsin is able to activate in sequence at least 100

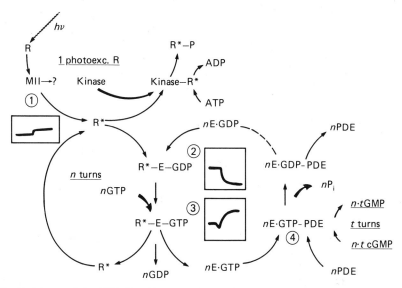

Fig. 7. Scheme of the GTP-dependent cascade reaction through which one photo-excited rhodopsin molecule can induce the hydrolysis of a large number of cyclic nucleotide (cGMP) in a retinal rod outer segment (adapted from Liebman & Pugh, 1979; Fung *et al.*, 1981; Kuhn *et al.*, 1981).

Notations: R = dark-adapted rhodopsin; R* = active state of rhodopsin reached after photo-excitation, probably at the meta-II stage (Fig. 1); E = GTP-dependent enzyme (transducin), 'activated' by R* which catalyses the exchange of a bound GDP molecule for an energy-rich GTP molecule. In the E–GTP state this enzyme can activate one molecule of phosphodiesterase. A slow spontaneous hydrolysis of the bound GTP brings E back to the inactive E–GDP state. PDE = phosphodiesterase which, when activated by E–GTP, hydrolyses specifically cGMP. Kinase: an ATP dependent kinase which is inactive on R but phosphorylates specifically R*. This comparatively slow process is probably an inactivation of R*; *n* is the number of E molecules that one R* can activate in sequence before its own inactivation. *t* is the turn-over of the phosphodiesterase activated by E. One photo-excited rhodopsin molecule will therefore induce the hydrolysis of $n \times t$ cyclic nucleotide molecules.

Light-scattering changes correlated with various steps of the cascade are shown in the insert as transmittance changes measured at 720 nm (Kuhn *et al.*, 1981). (1) Signal observed with rhodopsin alone. (2) Signal observed with the complete system (unbleached or reconstituted) in the absence of GTP. The reaction is then blocked, each R* molecule remaining permanently bound to the first E–GTP molecule it encounters. (3) Signal observed with the complete system in the presence of GTP and related to the activation of many E molecules by each R*. These signals are convenient to measure the kinetics and stoichiometry of the various steps of this amplifying cascade.

G-protein molecules. Both signals take place within a 100 ms at 20 °C and correspond to turbidity changes of the order of 10^{-2} or less.

When rhodopsin is the only protein left in purified disk membranes, one observes a single, smaller signal, corresponding to a decrease in light-scattering. The kinetics of this signal follows the meta I–meta II transition, and it always remains proportional to the amount of rhodopsin bleached by the flash, without any saturation. The signal is, therefore, related to a change occurring in rhodopsin, but not necessarily to a conformational change of the protein: it may reflect only the change of refractive index related to the anomalous dispersion change which results from the shift of the chromophore absorption band from 500 nm for unbleached rhodopsin down to 380 nm for the meta–II state. We do not have satisfactory explanations for the physical origin of the signals related to the interaction of rhodopsin with the G-protein. The fact that these signals are relatively insensitive to the state of the membrane (from nearly intact rods to sonicated membranes) and to the presence of various ionophores, indicates that they reflect molecular rearrangements rather than overall osmotic responses of the vesicles. The light-induced decrease of turbidity in presence of GTP might be due to an actual release to the cytosol of a subunit of the G-protein after its activation (Kuhn, 1981). However, increase of turbidity observed in absence of GTP relating to the formation of the R*–G-protein complex remains puzzling. Since the G-protein is already bound to the membrane before illumination, there is no change in the total mass of the light-scattering object, only the type of interaction of the G-protein with the membrane is modified. As rhodopsin alone (when G-protein is absent) does not give a significant signal upon similar flashes, the change of structure must concern essentially the G-protein upon its binding to R*. It is unrewarding to speculate further on the structural origin of such signals, which are indeed very small, although the very high sensitivity of the simple technique allows reliable and reproducible measurements of transient changes smaller than 10^{-3}. The long wavelength and the small amplitude of the effects will make attempts at physical interpretation very difficult. More might be learned about the nature of the events by studying the dependence of these light-scattering transients upon the orientation of the disk membrane with respect to the incident light beam. The fast kinetics resolved by this competent technique, and the stoichiometry that is reached in this short time, indicate that the cascade reaction initiated by the R*–G-protein interaction is fast enough and provides enough amplification to be involved in the transduction mechanism.

Where does the elusive structural change which creates the interaction site with G-protein take place in the rhodopsin molecule? It has to be on the cytoplasmic surface, since the G-protein (a peripheral membrane protein) does not penetrate the disk membrane. The only information so far available comes again from the proteolytic studies of Kuhn & Hargrave (1981): the formation of the binding site, which is insensitive to the removal of the terminal segment containing the phosphorylation sites, is inhibited by a further cleavage of rhodopsin into two major fragments. This second cleavage is nearer to the C-terminal end, in which the retinal binding site is located, than to the N-terminal end. But when the protease-sensitive region was specifically labelled by transglutaminase insertion of a fluorescent probe, the fluorescence transfer efficiency indicated that this site is about 6 nm distant from the chromophore (Pober, Iwanij, Reich & Stryer, 1978). Even taking into account the uncertainty inherent in such estimates, this suggests that a part of the protein very distant from the chromophore is involved in the formation of the site of interaction with the G-protein. The light-induced conformational change of the chromophore must therefore react over long distances on the structure of the protein.

It may be worthwhile to note that the proteolytic cleavage, which is able to inhibit this important process, does not induce any spectral change of the chromophore, nor any absorption or circular dichroism changes in the UV; neither does it inhibit the spectral regenerability of bleached rhodopsin by 11-*cis* retinal. Thus we are led to seriously question the significance of these various parameters with respect to the functionally significant changes in the protein.

Beyond the primary action on the chromophore conformation, the best documented effects of the capture of a photon by rhodopsin are the creation of specific sites at the cytoplasmic surface of the protein: these sites are either enzymatically active for the GTP-dependent activation of the G-protein, or enzymatically sensitive to the phosphokinase. It is therefore not surprising that biochemical and enzymological approaches have been more successful than purely physical techniques in the detection of such effects. It remains now to elucidate the mechanisms through which the conformational change of the chromophore induces these relatively long distance effects in the protein. This will probably require a much better understanding of the three-dimensional structure of the protein which at present no-one knows how to achieve.

Is the activation of the G-protein by photo-excited rhodopsin really the first step in the sequence of events leading to the cell hyperpolarisation and thence to visual excitation (as implied by the name 'transducin' for the GTP-dependent protein), or do we still need to look for an entirely different process? Rhodopsin is a relatively small molecule and, because it is certainly engaged in the rapid activation of a large number of 'transducin' molecules immediately after photo-excitation, it seems unlikely that it would simultaneously fulfill an entirely different function. But these are open questions for debate and further research.

This chapter is not a complete review of the abundant literature in the field. I apologise for not being able to quote all the relevant work, and I wish to acknowledge the various collaborators who have made possible the work in which I have been involved: Drs Bennett, Breton, Cavagionni, Michel-Villaz, Saibil and Worcester. I thank in particular Dr H. Kuhn for extensive discussions and for his very fruitful collaboration in the recent work on light-scattering. This contribution was written while on leave in the Department of Structural Biology at Stanford and I wish to thank Dr L. Stryer for his hospitality and for very stimulating discussions.

References

ALBERT, A. D. & LITMAN, B. J. (1978). Independent structural domains in the membrane protein bovine rhodopsin. *Biochemistry*, **17**, 3893–3900.

BOWNDS, D., DAWES, J., MILLER, J. & STAHLMAN, M. (1972). Phosophorylation of frog photoreceptor membranes induced by light. *Nat. New Biol.* **237**, 125–7.

CHABRE, M. (1975). X-ray diffraction studies of retinal rods. I. Structure of the disc membrane, effects of illumination. *Biochim. Biophys. Acta*, **382**, 322–35.

CHABRE, M. (1978). Diamagnetic anisotropy and orientation of α helices in frog rhodopsin and Meta II intermediate. *Proc. Natl. Acad. Sci., USA*, **75**, 5471–6.

CHABRE, M. & BRETON, J. (1979). Orientation of aromatic residues in rhodopsin. Rotation of one tryptophan upon the Meta I → Meta II transition after illumination. *Photochem. Photobiol.* **30**, 295–9.

CHABRE, M. & CAVAGGIONI, A. (1973). Light induced changes of ionic fluxes in the retinal rod. *Nat. New Biol.* **244**, 118–20.

COOPER, A. (1979). Energy uptake in the first step of visual excitation. *Nature, London*, **282**, 531–3.

CORLESS, J. M. (1972). Lamellar structure of bleached and unbleached rod photoreceptor membranes. *Nature, London*, **237**, 229–231.

EYRING, G., CURRY, B., MATHIES, R., FRANSEN, R., PALING, I. & LUGTENBURG, G. (1980). Interpretation of the resonance Raman spectrum of Bathorhodopsin based on visual pigment analogs. *Biochemistry*, **19**, 2410–18.

FUNG, B. K-K., HURLEY, J. B. & STRYER, L. (1981). Flow of information in the light-triggered cyclic nucleotide cascade of vision. *Proc. Natl. Acad. Sci., USA*, **78**, 152–6.

HARGRAVE, P. A., FUNG, S. L., McDOWELL, J. H., MAS, M. T., CURTIS, D. R., WANG, J. K., JUSCZAK, E. & SMITH, D. P. (1980). The partial primary structure of bovine rhodopsin and its topography in the retinal rod cell disc membrane. *Neurochem. Intern.* **1**, 231–44.

HAYWARD, G., CARLSEN, W., SIEGMAN, A. & STRYER, L. (1981). Retinal chromophore of rhodopsin photoisomerises within picoseconds. *Science*, **211**, 942–6.

HONIG, B., EBREY, T., CALLENDER, R. H., DINUR, U. & OTTOLENGHI, M. (1979). Photoisomerisation, energy storage, and charge separation. A model for light energy transduction in visual pigment and bacteriorhodopsin. *Proc. Natl. Acad. Sci., USA*, **76**, 2503–7.

HUBBELL, W. L. & B. B. K. FUNG (1979). In "Membrane Transduction Mechanisms," R. A. Cone and J. Dowling (eds.), Raven Press, NY.

KUHN, H. & DREYER, W. J. (1972). Light dependent phosphorylation of rhodopsin by ATP. *FEBS Lett.* **20**, 1–6.

KUHN, H. (1974). Light dependent phosphorylation of rhodopsin in living frog. *Nature, London*, **250**, 588–90.

KUHN, H. (1980). Light and GTP regulated interaction of GTPase and other proteins with bovine photoreceptor membranes. *Nature, London*, **283**, 587–9.

KUHN, H. (1981). Interaction of rod cell proteins with disk membrane: influence of light ionic strength and nucleotides. In *Current Topics in Membrane and Transport*, vol. 15, pp. 171–201. New York: Academic Press.

KUHN, H., BENNETT, N., MICHEL-VILLAZ, M. & CHABRE, M. (1981). Interaction between photoexcited rhodopsin and GTP binding protein: kinetics and stoichiometric analysis from light-scattering changes. *Proc. Natl. Acad. Sci., USA*, **78**, 6873–7.

KUHN, H. & HARGRAVE, P. A. (1981). Light induced binding of GTPase to bovine photoreceptor membranes. Effects of limited proteolysis of the membrane. *Biochemistry*, **20**, 2410–17.

LIEBMAN, P. (1972). Microspectrophotometry of Photoreceptors. In *Handbook of Sensory Physiology*, vol. VII/I Photochemistry of vision (edited by Dartnall, H. S. A.), pp. 481–528. Berlin: Springer-Verlag.

LIEBMAN, P. A. & PUGH, E. N. (1979). The control of phosphodiesterase in rod disc membranes. Kinetics possible mechanisms and significance for vision. *Vision Res.* **19**, 375–80.

MICHEL-VILLAZ, M., ROCHE, C. & CHABRE, M. (1982). Orientational change of the absorbing dipole of retinal in rhodopsin upon the transitions to bathorhodopsin, lumirhodopsin and isorhodopsin. *Biophys. J.* **37**, 603–16.

MICHEL-VILLAZ, M., SAIBIL, H. R. & CHABRE, M. (1979). Orientation of rhodopsin α-helices in retinal rod outer segment membranes studied by infra-red linear dichroism. *Proc. Natl. Acad. Sci., USA*, **76**, 4405–8.

OSBORNE, H. B. & NABEDRIK-VIALA, E. (1977). The hydrophobic heart of rhodopsin revealed by an infra red ^{1}H–^{2}H exchange study. *FEBS Lett.* **84**, 217–20.

PAPERMASTER, D. S., SCHNEIDER, B. G., ZORN, M. A., & KRAEHENBUHL, J. P. (1978). Immunocytochemical localisation of a large intrinsic membrane protein to the incisures and margin of frog rod outer segment discs. *J. Cell Biol.* **78**, 415–25.

PETERS, K., APPLEBURY, M. & RENTZEPIS, P. M. (1977). Primary photochemical event in vision: proton translocation. *Proc. Natl. Acad. Sci., USA*, **74**, 3119–23.

POBER, J. S., IWANIJ, V., REICH, E. & STRYER, L. (1978). Transglutaminase catalysed insertion of a fluorescent probe into the protease sensitive region of rhodopsin. *Biochemistry*, **17**, 2163–9.

RAFFERTY, C. N., MUELLENBERG, C. G. & SHICHI, H. (1980). Tryptophan in bovine rhodopsin. Its content, spectral properties and environment. *Biochemistry*, **19**, 2145–51.

SAIBIL, H., CHABRE, M. & WORCESTER, D. L. (1976). Neutron diffraction studies of retinal rod outer segment membranes. *Nature, London*, **262**, 266–70.

SCHECHTER, J., BURSTEIN, R., ZEMELL, R., ZIV, E., KANTOR, F. & PAPERMASTER, D. (1979). Messenger RNA of opsin from bovine retina. Isolation and partial sequence of the in-vitro translation product. *Proc. Natl. Acad. Sci., USA*, **76**, 2654–8.

SCOTT, B. L., MCCASLIN, D. R. & CORLESS, J. M. (1981). Two dimensional crystals in detergent-extracted disk membranes from frog rod outer segment. *Biophys. J.* **33**, 293a (abstract).

THOMAS, D. D. & STRYER, L. (1982). The transverse location of the retinal chromophore of rhodopsin in rod outer segment disc membranes. *J. Mol. Biol.* **154**, 145–58.

WANG, J. K., MCDOWELL, J. H. & HARGRAVE, P. A. (1980). Site of attachment of 11-*cis* retinal in bovine rhodopsin. *Biochemistry*, **19**, 5111–17.

WORCESTER, D. L. (1978). Structural origin of diamagnetic anisotropy in proteins. *Proc. Natl. Acad. Sci., USA*, **75**, 5475–7.

WU, C. W. & STRYER, L. (1972). Proximity relationship in rhodopsin. *Proc. Natl. Acad. Sci., USA*, **69**, 1104–7.

THE POSSIBLE ROLE OF RHODOPSIN AND THE MICROVILLUS IN LIGHT ADAPTATION OF THE PHOTORECEPTORS OF AN INSECT

SHAHRDAD RAZMJOO AND KURT HAMDORF

Department of Animal Physiology, University of Ruhr at Bochum, 4630 Bochum 1, GFR

Introduction

This chapter is not a review of the subject of light adaptation.* It is, rather, a focus upon the idea that the sensitivity of invertebrate photoreceptors depends largely on the rhodopsin content. This simplistic, but not inconsistent, idea is developed further by re-interpreting the published evidence in the light of a model which was presented by the authors some time ago, and additional evidence which appears here for the first time.

But first, what do we mean by adaptation, and how do we measure it? For our purpose, empirical observations on photoreceptors lead us to the following definition, which differentiates adaptation from its wider, and especially its ecological, nuances. When a system that is subjected to an unvarying input stimulus manifests varying physiological outputs because its 'status' has changed (this must be brought about by additional inputs), the quantitative *relativity* of the different states (that is, the outputs, measured for example electrophysiologically) is a measure of what we mean here by adaptation. It should be noted that this definition implicitly excludes the role played, in animal photoreceptors, by structures such as the pupil and the screening pigments, for these defeat the unvarying-input constraint. Thus the adaptation to which we refer is a modification of sensitivity. Given that in animal photoreceptors a reduction in the membrane depolarising form of the photopigment, rhodopsin, results in reduced sensitivity (for example, in *Calliphora*, Hamdorf, 1970, 1971; in *Limulus*, Lisman & Strong, 1979), it follows that the quantitative variation of this substance must play a role in adaptation, whether or not there may be other contributing factors.

* For a comprehensive review of light and dark adaptation the reader is referred to a recent article by Autrum (1981).

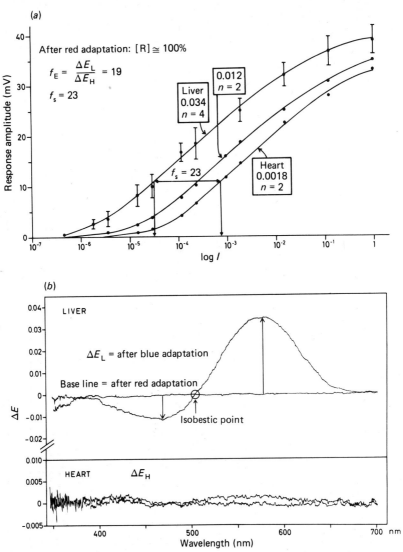

Fig. 1. (*a*) Transfer functions, plotted from amplitude transients of late receptor potentials in *Calliphora*, recorded intracellularly. The stimuli were 30 ms pulses of monochromatic light, 498 nm, close to the isobestic point of the R–M system. The flies were either liver- or heart-fed (H and L). All preparations were pre-adapted at 614 nm, ensuring 100% rhodopsin fraction. Note that with increasing photopigment content in the receptor membrane (corresponding to increasing extinction values of 0.0018, 0.012 and 0.034), the curves shift to lower stimulus intensities. The sensitivity difference factor determined in this way, f_s, is approximately equal to the spectrophotometrically determined extinction factor, f_E. (*b*) Difference spectra of liver-fed and heart-fed *Calliphora*. After the conversion of all M580 to R490 by red-adapting both of the dissected eyes – one in the reference beam, the other in the measuring beam – the baseline was recorded. Then the R:M ratio of the same eye which had been electrophysiologically measured (now in the

Adaptation and the amplitude of the response

The literature abounds in evidence for the modification of sensitivity as a result of the variation in rhodopsin content, and here the example of *Calliphora* suffices to illustrate the point. Electrophysiological recordings, from *Calliphora* that were raised on meat either from bovine heart or liver, reveal different sensitivities as seen from the transfer functions (input light intensity: output response amplitude) in Fig. 1(*a*) (for *Drosophila*, see Stark & Zitzmann, 1976), where the amplitude of the transient of the LRP* has been taken as the measure of sensitivity. In these particular experimental animals the sensitivity difference is seen to be somewhat more than a logarithmic unit. This difference is due to the fact that the receptors of these animals contain different amounts of photopigment, and this is evident from the spectrophotometric results shown in Fig. 1(*b*).

Apart from the variation in the rhodopsin content in *different* animals, the direct proportionality between sensitivity and the rhodopsin content can be demonstrated also in any *individual* animal, since the fraction of the photopigment in the rhodopsin state can be manipulated, within limits, by monochromatic illumination: the equilibrium between rhodopsin and metarhodopsin being a function of the wavelength (*Calliphora*: Stavenga, Zantema & Kuiper, 1973; Hamdorf, Paulsen & Schwemer, 1973; Rosner, 1975; *Drosophila*: Ostroy, Wilson & Pak, 1974). This is illustrated, again through spectrophotometry, in Fig. 2(*a*). Red light induces a rhodopsin fraction of ≃100%, whereas blue light reduces this to ≃20%, and the resulting loss of sensitivity is reflected in the shift of the transfer function to higher intensities (shown in Fig. 2(*b*) for both liver-bred and heart-bred *Calliphora*). Thus with reference to our original definition that adaptation is a modification of sensitivity, we

---

* The late receptor potential (LRP) is the depolarising response of the cell. This unfortunate terminology is the result of attempts to distinguish it from an earlier-occurring, almost stimulus-coincident, potential change named the early receptor potential (ERP, first described by Brown & Murakami, 1964; review: Cone & Pak, 1971). The ERP is thought to be due not to cellular depolarisation, but to charge migrations within the photopigment molecule as the result of photon absorption; thus it is the earliest electrophysiologically measurable event.

---

measuring beam) was altered by saturating exposure to the blue-adapting light (470 nm), after which the spectral extinction difference was recorded; that is, the absorption difference between the red- and the blue-adapted conditions. The ratio $\Delta E_L/\Delta E_H$, determined at 580 nm, is directly proportional to the ratio of R content (C_{RL}/C_{RH}) in the L and H flies. Note the amplification difference of 2 in the two differential extinction scales (from Razmjoo & Hamdorf, 1976).

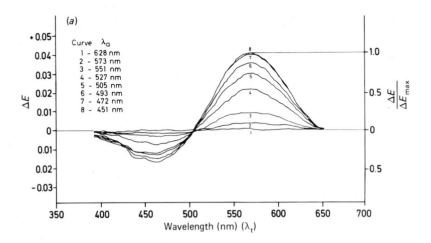

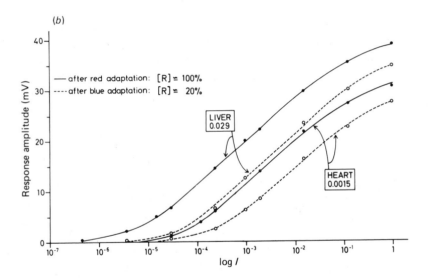

Fig. 2. (*a*) Difference spectrophotometry of the R–M equilibrium in a single eye of *Calliphora erythrocephala* chalky. The baseline was recorded after adapting both eyes (one in the reference, the other in the measuring beam) to 628 nm. After this, the eye in the measuring beam was successively adapted, for 6 min each time, to the wavelengths (λ_a) shown in the inset. The amplitude is proportional, in each case, to the wavelength-induced equilibrium (from Hamdorf, 1979). (*b*) Transfer functions, plotted from amplitude transients of late receptor potentials in *Calliphora*, recorded intracellularly. The functions, after red adaptation (614 nm), and after loss of sensitivity by blue adaptation (473 nm), are shown for both liver-fed and heart-fed flies. The stimulus was as in Fig. 1(*a*) (from Razmjoo & Hamdorf, 1976).

must conclude that the variation in the rhodopsin fraction can serve as an adapting mechanism.

However, these examples are all demonstrations of a 'static' correlation between sensitivity and the rhodopsin content, in that the measurements were carried out before and at the end of some adapting condition, but not *during* the process. It is found that although both before and after the adapting process the proportionality holds, with the onset of either light or dark conditions there is apparently a phase during which sensitivity and the rhodopsin content no longer appear to be directly proportional: one lags behind the other, implying that there is a difference in the *rates* of change of the two. This observation has been taken to demonstrate that the photoregeneration of rhodopsin does not lead to a *proportional* increase in sensitivity (*Limulus* ventral photoreceptors: Fein & DeVoe, 1973; Lisman & Sheline, 1976; Lisman & Strong, 1979; *Balanus*: Hillman, Dodge, Hochstein, Knight & Minke, 1973). For example, in some of the studies cited it was found that whereas all rhodopsin could be photoregenerated in about two seconds (using the ERP to monitor the state of the photopigment) in contrast, the sensitivity of the cell (measured as the amplitude of the LRP) took much longer to return to its initial value, the process having a time constant of some 20 s for the initial exponential part.

If the ERP is an unequivocal and accurate indicator of rhodopsin and metarhodopsin interconversions, these results would seem to demonstrate that the rate of recovery of sensitivity lags behind the rate of photoregeneration of rhodopsin. Apparently convinced of the validity of this index (but see Minke & Kirschfeld, 1980; Stephenson & Pak, 1980), these results have been taken to indicate that there are other adaptive processes operating, in comparison with which the adaptation due to changes in the rhodopsin fraction is an insignificant contribution* (Lisman & Sheline, 1976; using a different approach Naka & Kishida, 1966, came to similar conclusions). To recapitulate the methodology and the reasoning: the ERP was used as a measure of rhodopsin regenerated; the LRP as a measure of sensitivity, and during a phase when the LRP was no longer proportional to the ERP, the sensitivity was interpreted as being no longer proportional to the rhodopsin regenerated.

---

* The literature is also abundant with evidence that calcium may play an important role in the 'membrane processes' of light adaptation. Our lack of attention to this data here is not a dismissal of the role of calcium, but an attempt to explore the extent to which other adaptive mechanisms operate within the system.

Can the apparent conflict between such findings and the well-established 'static' relationship between sensitivity and the rhodopsin fraction mentioned earlier be resolved? That is, in view of the contradictions, can we still maintain that adaptation is largely photopigment related? We can, if we assume the existence of intermediates or coupled processes other than the photopigment interconversions themselves, and that these may become bottlenecks in the regeneration of sensitivity. Such a solution is already incorporated in our photopigment model (Hamdorf & Razmjoo, 1977, 1979; Razmjoo & Hamdorf, 1980) which is shown here in Fig. 3.

Amongst other things, this model proposes two photoregenerative pathways for rhodopsin: $\boxed{R} \leftarrow\!\!\sim \boxed{M}$ and $Ⓡ \leftarrow\!\!\sim Ⓜ$. Were we to accept the validity of the ERP as an index of the interconventions (noting that it is due to charge migrations during the isomerisation process independent of other cellular activity), we should concede that both these suggested reactions would elicit identical ERP signals for identical numbers of rhodopsin molecules photoregenerated. However, only one of the proposed forms of regenerated rhodopsin, $Ⓡ$, has membrane-excitatory properties capable of eliciting the LRP.* Because after the initial excitation metarhodopsin accumulates in the $Ⓜ$ form, any photoregeneration would form only $Ⓡ$, which would have to undergo *metabolic regeneration* before it could assume its active, membrane-excitatory form $\boxed{R}$. It is this latter process, which is metabolic and thus time-dependent, that could become a bottleneck for the regeneration of sensitivity and cause different rates of recovery when monitored by the LRP. If this metabolic step in the photoregeneration process really exists and is a process common to many, if not all, invertebrates, the variation in its rate, dictated by the metabolism and the life-style of the animal, could account for the fact that in the fly this process takes about 10 s or less (at 15 °C) whereas in *Balanus* and *Limulus* (at 25 °C) it can last for about a minute or longer.

Adaptation and the duration of the response

Further insight into the mechanisms of adaptation and the role of the photopigment is gained by considering the *duration* of the LRP as an

---

* If the different forms proposed are truly distinct identities, one may expect to find spectral absorption differences between them. But a search for such differences, for metarhodopsin for example, has not so far yielded conclusive evidence (unpublished results). It could be that these different forms result from weak bonding to, or other association with, energy-rich, or other, neighbouring structures so that the spectral absorption remains largely unchanged.

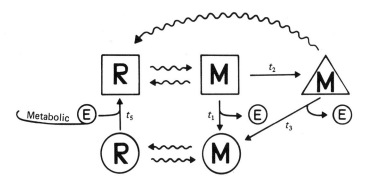

Fig. 3. A 'photopigment' model of insect vision. The late receptor potential is elicited when active rhodopsin $\boxed{R}$ is converted to active metarhodopsin $\boxed{M}$ by photon capture. $\boxed{M}$ will then release 'energy' for membrane excitation through t_1 and end up as inactive metarhodopsin $\textcircled{M}$. When more $\boxed{M}$ has been produced than can be processed through the t_1 pathway, the increased lifetime of $\boxed{M}$, and its resultant increased chemical reactivity, may cause energy sharing with neighbouring structures, bringing about a more stable active state $\triangle\!\!\!\!M$. The latter releases energy at a slower rate, t_3, which is responsible for the phenomenon of the prolonged depolarising afterpotential (PDA). Inactive metarhodopsin $\textcircled{M}$ is converted to inactive rhodopsin $\textcircled{R}$ by photon capture. The latter needs the utilisation of metabolic energy to be converted to the active form $\boxed{R}$. Wavy arrows denote light-induced reactions and straight arrows denote dark ones (from Hamdorf & Razmjoo, 1977).

index of adaptation. Figure 4(*a*) shows the temporal relationship between the input to a dark-adapted photoreceptor (high intensity flash of ≈ 1.5 ms) and the photoreceptor output, an LRP lasting more than a hundred milliseconds. When the same stimulus is given after light adaptation, the duration of the response is greatly shortened (Fig. 4*b*). The employment of a stimulus of high intensity has the effect that, of the three empirical manifestations of light adaptation: the shortening of the latency, the depression of the amplitude and the shortening of the LRP duration (Fuortes & Hodgkin, 1964; Naka & Kishida, 1966), the first two become saturated and show no further change, while the duration still alters considerably (one is tempted to consider this as an 'improvement' in the photoreceptor's temporal resolution).

In the foregoing section on the adaptive mechanism, we have emphasised the role of the rhodopsin fraction. With the evidence of the shortening of the LRP duration following light adaptation, it becomes interesting to ask the question: Does the duration of the LRP reflect the photoactivated fraction of rhodopsin which has produced it? The latter question is pertinent because of the following reasoning: if the variation in the duration of the LRP is a proportional index of adaptation, and if it

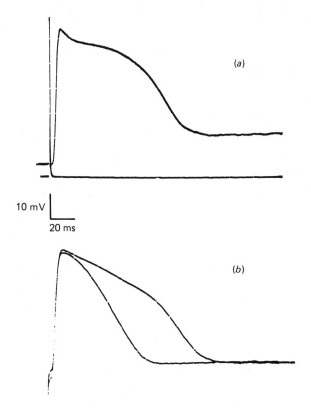

Fig. 4. Intracellular recordings from *Calliphora*, demonstrating the duration of the late receptor potential. (*a*) The stimulus, a 1.5 ms high-intensity flash at 506 nm (~5 × 10⁶–1 × 10⁷ absorbed quanta per receptor), can be seen as a thin line in comparison with the elicited response from a dark-adapted receptor which lasts over 100 ms. (*b*) The late receptor potential with the longer duration is as in (*a*); however, when the flash is repeated ~15 s later, the still-persisting light adaptation due to the first stimulus results in the elicitation of a response with a shorter duration. Before the flashes, the receptor had been adapted to 506 nm. Note that at such stimulus intensities the amplitude varies insignificantly and the latency does not vary at all.

can be demonstrated that the duration is a function of the rhodopsin fraction photoactivated, one may then put the two together and deduce, once again, that light adaptation is a function of the rhodopsin fraction.

The prerequisite that the variation in the duration of the LRP is a proportional index of adaptation has not yet been established because of contradictory evidence such as the following. In the ventral photoreceptors of *Limulus*, an LRP induced during light adaptation is of shorter duration than one induced in the dark, even though the intensities of the stimuli are so arranged that both LRPs have equal amplitudes (Fein & DeVoe, 1973). However, in the lateral photoreceptors of the same

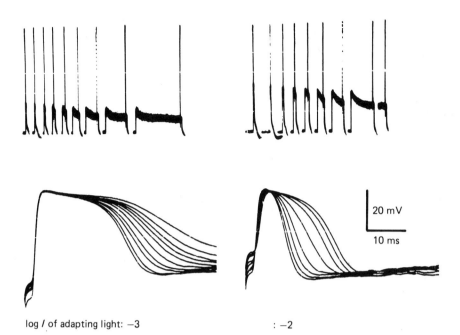

log *I* of adapting light: −3   : −2

Fig. 5. Intracellular recordings from *Calliphora*, showing shortenings of the LRP duration as a function of intensity and duration of light adaptation. Following adaptation at 506 nm, the eye was dark-adapted for 4 min. Then the same stimulus as in Fig. 4 elicited the LRP which, in the group of traces in the lower left, shows the longest duration: the dark-adapted response. Next, this stimulus was repeated at 1 min intervals, each coincident with an adapting background light which was also at 506 nm but at two different intensities for the experiments shown on the right and the left. The adaptation was started 0.125, 0.25, 0.5, 1, 2, 4, 8, 16, and 32 s before the test flash (the LRP after 32 s of adaptation is missing in the group in the lower right). The durations in both cases have thus become successively shorter, although no amplitude and latency modulations can be observed. The upper row, simultaneously recorded on another oscilloscope at slower sweep, shows the steps just described (excluding the dark-adapted response), and although the flashes are 1 min apart, this has not been temporally represented here. Note the different amplitudes of the plateaus due to the different adapting intensities, and that the more intense background light causes a greater shortening of the responses. The downward deflection at the beginning of the faster-sweep traces is the flash artifact.

animal, LRPs which have equal amplitudes are of the same duration, whether induced during light or dark conditions (Benolken, 1962; Fuortes & Hodgkin, 1964). Furthermore, in *Apis*, in the dark period following light adaptation the amplitude of the LRP regains its maximum value in only 10–20 ms, whereas the duration continues to increase for up to 47 min (Baumann, 1975), presenting a confusing situation as to the choice of an index as the two parameters span such different time intervals. In view of such observations it has been more

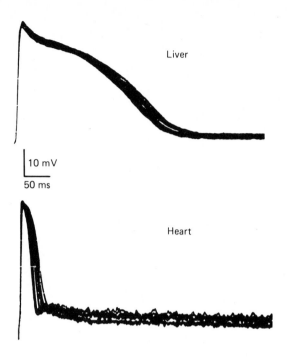

Fig. 6. Intracellular recordings showing LRP duration as a function of dietary conditions in *Calliphora*. Both animals (at similar temperature and having a similar previous history of light and dark conditions) were stimulated repeatedly with flashes (as in Fig. 4) at 30 s intervals. In each case the small scatter in duration is due to light adaptation by the flash itself. Of interest here is the relative difference in duration between the liver-fed and the heart-fed animals, demonstrating a relationship between photopigment content and the LRP duration.

usual to take the LRP amplitude as a measure of sensitivity, although the possibility is acknowledged that the means of information transfer to higher structures may be more than just the amplitude modulation of the LRP (Stieve, 1965; Borsellino, Fuortes & Smith, 1965; Millechia & Mauro, 1969; Stieve, Gaube & Winterhager, 1976).

In *Calliphora*, however, some evidence that the duration of the LRP may represent the extent of light adaptation and, under certain experimental conditions, be a better index of it than either the amplitude or the latency is shown in Fig. 5. Here it can be observed that with the use of stimuli of high intensity (and a weaker adapting light) the duration exhibits a broad range of variation, whereas latency and amplitude do not change at all.

Thus, relying on the duration of the LRP as an index of light adaptation, we may now ask the next question: Is there a *proportionality*

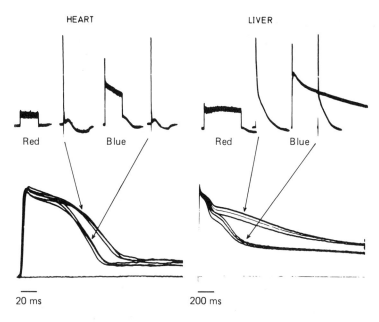

Fig. 7. Duration of the LRP as function of chromatic adaptation in *Calliphora*. The upper row shows the sequence of the experimental steps. Flash stimuli (as in Fig. 4) were given every 2 min, immediately after which an adapting light of alternately red and blue was given (lasting 15 s in the heart-fed and 30 s in the liver-fed animals). The alternate conditions of $\boxed{\text{R}} \simeq 100\%$ (after red) and $\boxed{\text{R}} \simeq 20\%$ (after blue) yield the corresponding responses. Note the rather good reproducibility of the results, although the heart-fed animal was alternately red- and blue-adapted eight times, the procedure lasting some 20 min. The same sequence (repeated six times), does not show similar reproducibility for the liver-fed fly. Such results (often together with depression of the amplitude) are frequently observed in liver-fed flies, and indicate loss of sensitivity (photopigment destruction?) as the result of exposure to strong and repeated test and adapting stimuli. In the case of the liver-fed fly, the upper row shows that although the blue stimulus lasted only 30 s (as with red), the resulting prolonged depolarising afterpotential has prevented immediate repolarisation.

between the duration of the LRP and the photoactivated fraction of rhodopsin eliciting it?

We have already seen the convincing proportionality between sensitivity (measured as the LRP amplitude) and the variation in the photopigment content in flies subjected to different diets. Resorting again to this source, it is observed that flies with the greater photopigment content elicit longer lasting LRPs (Fig. 6). Moreover, when we alter the rhodopsin fraction in a single fly, we obtain the results shown in Fig. 7, which show reproducible differences in LRP durations in the red-adapted and the blue-adapted states (100% and 20% rhodopsin fractions, respectively; see Razmjoo & Hamdorf, 1976).

Light adaptation of insect photoreceptors

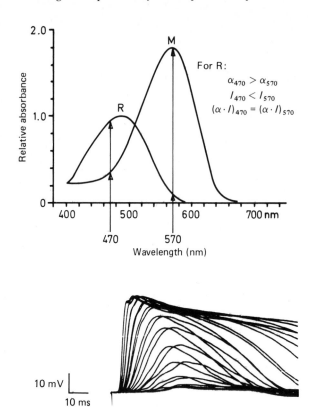

Fig. 8. The establishment of isohelic stimuli at different wavelengths. Rhodopsin absorbs light to a greater extent at 470 nm than at 570 nm. However, by reducing the transmission of a 470 nm band-pass filter it can be ensured that both stimuli activate equal numbers of rhodopsin molecules; that is, despite their wavelength differences the two stimuli are of equal brightness, or 'isohelic'. The methodology is symbolically shown in the inset. The manner in which this is done is by reducing the transmission of one of the filters until the amplitudes of the responses to the two are equal. The accuracy of this matching is greatest at a moderate incident intensity: one that elicits amplitudes in the middle of the straight section of the transfer function where the slope is greatest (Laughlin, 1976), a region in which the validity of the amplitude as an index of the photoactivated rhodopsin finds some documentation (for example, in *Calliphora*, Hamdorf & Razmjoo, 1979). The complete set of responses for both the isohelic stimuli, as a function of increasing stimulus intensity (with the flash of Fig. 4), are also shown in the lower part. Here it may be noticed that although the peaks appear rather well matched, latencies and durations do not. This is because of the lack of adequately long dark intervals between the two measurements, so that one run still shows the after-effects of exposure to the highest intensity flashes of the previous run. (*Calliphora* absorption spectrum courtesy of Dr J. Schewemer.)

An objection to the correctness of these results is the observation that the two adapting stimuli (red and blue) may have differentially adapted the photoreceptor, because of their different intensities alone rather than their wavelengths; for it is evident that the red stimulus is of a lower intensity than the blue one (compare the amplitudes of the plateaus of responses to the adapting lights). We saw in Fig. 5, that the more intense the adapting light, the shorter is the duration of the LRP. This is a valid objection, and the solution is to use adapting lights that have equal intensities. This was done in the following way.

With reference to the absorption probabilities of rhodopsin at 470 nm and 570 nm ($\alpha_{470} > \alpha_{570}$, Fig. 8), the intensities of these stimuli were so arranged ($I_{470} < I_{570}$) that the product of the absorption probability and the intensity for one wavelength was made equal to the product for the other wavelength; that is, both elicited LRPs of equal amplitudes. When the experiments of Fig. 7 were repeated with the two, now 'isohelic', light sources which photoactivated equal fractions of rhodopsin, the results obtained, shown in Fig. 9, convincingly demonstrate a dependence between the fraction of the rhodopsin photoactivated and the duration of the elicited LRP. This enables us to deduce, from the argument presented earlier, that adaptation is proportional to the fraction of the different states of the photopigment.

Some implications of the proposed model

In the preceding section we were again dealing with 'static' conditions, since the responses were elicited well after adaptation had been completed. When we consider the LRP durations *during* the adaptation process, we encounter a contradiction of a well-established finding, namely that the fraction of rhodopsin in *Calliphora* cannot be reduced to less than ~20% of the total photopigment content (Fig. 8a; review; Hamdorf, 1979). For if the longer LRP durations shown in Figs. 7 and 9 are due to the photoactivation of ~100% active rhodopsin, $\boxed{R}$, and the shorter ones due to ~20%, then the much shorter durations *during* adaptation to higher light intensities shown in Fig. 5 would indicate fractions much less than 20%, if we assume that the duration of the response serves as an index of the photoactivated fraction of rhodopsin also *during* the process of light adaptation. Is it therefore possible that during light adaptation the rhodopsin fraction falls to less than ~20%? The photopigment model (Fig. 3) would support such a hypothesis in the following manner. After blue and subsequent dark adaptation, the active rhodopsin fraction is reduced to $\boxed{R} \simeq 20\%$. During light adapta-

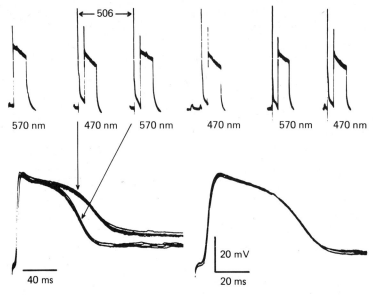

Fig. 9. Duration of the LRP as a function of chromatic adaptation in *Calliphora*. 506 nm flashes (as in Fig. 4) were presented every 2 min. Five seconds after each flash the eye was exposed alternately to either 470 nm or 570 nm adapting light, each for 10 s, after which a dark period of 1 min 45 s elapsed before the next flash induced the corresponding eight responses shown in the lower left. In the upper row notice that due to the conditions of isohelicity (see Fig. 8), the amplitudes of the plateaus to the adapting lights are about the same. The slight, alternate, variations are due to the varying R:M ratio. Also note the excellent reproducibility of the durations of the eight responses. As control, when 506 nm was the adapting light and the isohelic test flashes were alternately given, once every 2 min, the results shown at the lower right were obtained. These durations exactly superimpose (four responses), because the adapting conditions and the isohelic stimuli have together ensured the photoactivation of equal numbers of rhodopsin molecules with each flash.

tion, however, the loss of $\boxed{R}$ can occur at a much faster rate (as a function of intensity and duration of adaptation) than its metabolic regeneration, $\circledR \rightarrow \boxed{R}$. Thus if the latter process is a limiting one, the instantaneous fraction of $\boxed{R}$ could fall well below 20%, however, still maintaining the *collective* fraction of rhodopsin at 20%, thus: $\boxed{R} + \circledR \simeq 20\%$. The model accounts for, on one hand, the spectrophotometrical finding that the total rhodopsin fraction does not fall below ~20%, and on the other hand the deduction from electrophysiology that even during light adaptation the shortening of the LRP duration may be an index of the fraction of rhodopsin photoregenerated.

One test of such a hypothesis is to examine one of its consequences. If the control of the metabolic rate indeed alters the fraction of $\boxed{R}$, then by

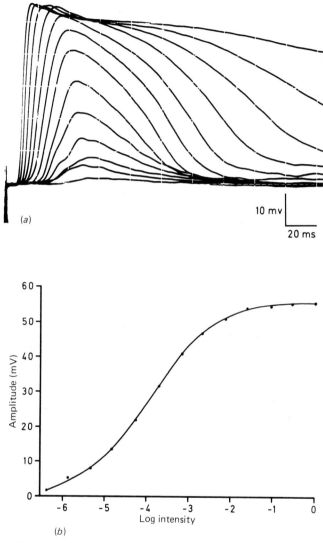

Fig. 10. (*a*) The late receptor potential as function of increasing light intensity in liver-fed *Calliphora*. The stimulus, seen at the left as a dark line, was a 1.5 ms xenon flash at 506 nm. Notice that LRPs greater than 40 mv rejoin the baseline at successively later times: as the amplitude enters the saturation phase, the duration starts to increase. (*b*) The transfer function for the same cell, plotted from the amplitudes shown in (*a*).

severe retardation of this rate (at low temperature, for example) the $\boxed{\text{R}}$ fraction could be drastically reduced, if not actually reduced to zero. Accompanying such reduction should be further shortening of the LRP duration. Furthermore, when the duration shortens no more, with the fraction of ⟨R⟩ continuing to decrease, the *amplitude* of the LRP should next suffer depression. This sequence of events would be in agreement with the common observation (Fig. 10*a*) – but in the reverse order just described – that with increasing light intensity (photoactivating greater numbers of rhodopsin molecules), the LRP amplitude grows to a maximum, after which, with still increasing intensity, the duration of the response starts to lengthen, although, as can be observed in this figure, there is some overlap between the end of the amplitude growth and the start of the expansion of duration, a point which will be dealt with later in more detail.

Integration of the information from the amplitude and the duration of the response: the microvillus hypothesis

When the amplitudes of the LRPs shown in Fig. 10(*a*) are plotted against the light intensities used to generate them, we obtain the photoreceptor transfer function shown in Fig. 10(*b*), a function shown to be common to many physiological transducers (review: Lipetz, 1971). It is generally known that a proportional relationship between intensity and the photoactivated fraction of rhodopsin can be found only in the straight section of such a plot. We shall briefly consider the explanations that have so far been put forward for the deviations from linearity (by this is meant straightness, for the straight section is really logarithmic) at the lower and upper regions of this transfer function.

A quantum bump, the smallest and briefest all-or-none depolarisation (it is about 1 mv in amplitude in *Calliphora*, at 15 °C, and lasts about 50 ms), is the photoreceptor response following the absorption of a single photon (Yeandle, 1958; Fuortes & Yeandle, 1964; Scholes, 1964; see also Hillman, this volume). It is generally accepted that at higher light intensities the amplitude of the LRP results from the amplitude summation of these discrete events. The flat nature of the transfer function at its lower end is a graphic demonstration of the observation that the amplitude of an LRP due to a flash that is, for example, ten times stronger than one which elicits a quantum bump, is *not* ten times larger than the amplitude of a quantum bump (Lillywhite & Laughlin, 1978). This apparent loss of effectivity of photons can be explained in the following way. When single quantum bumps are elicited by weak flashes of light, they do not follow the incident flash after a fixed time interval, but exhibit a scatter in latency. (In *Calliphora* at 15 °C, following a 1 ms pulse of light, the scatter ranges from 50–110 ms with a maximum at around 80 ms; S. Razmjoo & K. Hamdorf, unpublished results.) With increasing light intensity beyond that at which single quantum bumps result, the ensuing bumps do not temporally coincide because of the latency scatter, and thus their amplitudes cannot linearly sum up. At still higher intensities the scatter does not significantly broaden whereas the number of quantum bumps greatly increases, thus leading to appreciable growth in amplitude. Consider one of the smaller LRPs, for example, the 5 mv one in Fig. 10(*a*). It starts at ~30 ms and ends at 120 ms, a span of ~90 ms. The LRP with the amplitude of 40 mv, however, starts at a little over 10 ms earlier but ends also at ~120 ms, a broadening of ~13%, whereas its amplitude has grown some 700%. That is, as the quantum bumps greatly increase in number, but within almost the same range of latency scatter, they temporally sum up more and more, and the transfer function enters the straight (logarithmic) phase, with its particular slope ('voltage gain of transduction', Laughlin, 1976, 1981; for a discussion of the possible advantages, for the fly, in having a logarithmic strategy, see Laughlin & Hardie, 1978).

For the gradual flattening at the upper end of the transfer function several explanations can be found. One is the inevitable result of the self-shunting model of membrane depolarisation (for review, see Lipetz, 1971). It can best be appreciated by imagining two photoreceptors, each having a different EMF across its membrane. Were both to absorb equal numbers of photons, resulting in the opening of equal numbers of channels, the ensuing conductances and thus the extents of depolarisation in the two would not, however, be equal, for the EMFs driving the conductances are different. And so it is within a single photoreceptor: the absorption of a number of photons leads to a conductance which in turn inevitably reduces the initial EMF. In this state, the additional absorption of an equal number of photons and the resultant opening of an equal number of channels elicits a disproportionate increase in conductance, because of a now-reduced EMF. Thus, the ceiling of the transfer function may simply be due to a disappearance of the EMF, that is the final result of the breakdown of the electrochemical gradient across the membrane.

This argument, however, fails to explain why the LRPs start to broaden once the saturation point is reached (Fig. 10*a*), because with the disappearance of the EMF (the ceiling of the transfer function), further conductance should cease. Yet the broadening of the LRPs is an indication that conductance has not stopped. Assuming that the EMF has disappeared, the continued conductance should manifest itself in the form of an overshoot beyond the amplitude of zero EMF. The experi-

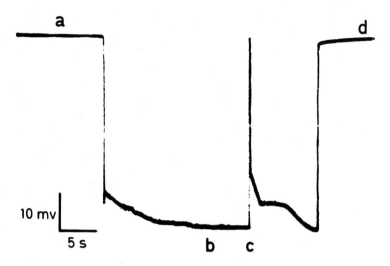

Fig. 11. Sequence *a–d*: the penetration, light stimulation, and withdrawal of a microelectrode from, a *Calliphora* photoreceptor. a: extracellular potential level before penetration, b: after penetration of a photoreceptor, note the potential difference of 50 mv with respect to the previous extracellular position of the electrode, c: LRP elicited by a flash at its highest intensity (as in Fig. 10). Note that the amplitude of the LRP does not overshoot the zero EMF – the level of the extracellular potential, d: withdrawal of the microelectrode from the photoreceptor and the return to the original extracellular potential.

ment illustrated in Fig. 11 demonstrates that no such overshoot occurs. Thus the saturation of the transfer function may not be a consequence of the disappearance of the EMF.

Another approach would be to replace the abscissa in Fig. 10(*b*) with a logarithmic scale of the number of rhodopsin molecules in a rhabdom ($\sim$2 $\times$ 10^8, Dörrscheidt-Käfer, 1972; Hamdorf, 1979). This yields an alternative explanation for the upper flattening of the transfer function, in that it requires increasingly disproportionate amounts of light to engage the last remaining molecules of rhodopsin. But this explanation also fails to account for the broadening of the LRPs with still increasing intensity.

It has also been suggested that the non-linear summation is due to either a declining ratio of channels to photons, or to antagonistic hyperpolarising conductances. This explanation has been given the general title 'declining photon effectiveness' (Laughlin, 1981), an idea which we develop further. We have seen (Fig. 10) that with increasing light intensity, as the growth in amplitude of the LRP reaches the non-linear phase, the response duration begins to increase, and although the two effects share a region of overlap, the amplitude growth does finally come to an end whereas the duration continues to lengthen. It is plausible to consider that the different behaviours originate from two separate mechanisms; and yet we have seen that both the amplitude and the duration of the LRP may have a single source in common, namely the fraction of rhodopsin photoactivated. However, if, with increasing intensity, the prolongation of the LRP duration (beyond the 40 mv response in Fig. 10) is an indication that the rhodopsin fraction is not at an end, the non-linearity and the ceiling in the growth of the amplitude must be due to the exhaustion of something else.

It is interesting to note that systems which are compartmentalised often exhibit redundancy in their input–output relationship. In the *Calliphora* rhabdom the distribution of some 2×10^8 rhodopsin molecules among about 1.4×10^5 microvilli (Hamdorf, 1979) may be considered as such compartmentalisation. If the all-or-none unit that activates the cell is the microvillus and not the rhodopsin molecule itself (the microvillus hypothesis: Hamdorf, 1979), and if it takes only a single photoactivated rhodopsin molecule to fully excite one microvillus, then the remaining rhodopsin molecules on that microvillus would be essentially redundant, serving only to increase the probability of its capturing photons.

At lower light intensities (that is, a sparse photon flux), the probability that the photoactivated rhodopsin molecules are located on different microvilli is large, and thus there is a 1:1 relationship between the input intensity (correctly, the number of *absorbed* quanta, and thus the number of photoactivated rhodopsin molecules) and the number of activated microvilli. If multiple absorptions by the same microvillus did not occur, this relationship would hold until an intensity is applied when all microvilli become activated, each by a single photon. However, multiple absorptions do occur and transition to this point will not be smooth (linear), since before such a point is reached, the number of microvilli absorbing more than a single photon will also increase. Because a microvillus absorbing more than one photon may excite the photoreceptor no more than when absorbing a single photon, there is, in effect, a waste or loss of input. It is this feature that has been termed 'declining photon effectiveness'.

With increasing light intensity the number of multiple absorptions increases (with the Poisson distribution), and thus non-linearity appears before the point at which all microvilli have absorbed a single photon, so that the response shows disproportionately smaller amplitude increases while the duration starts to increase. Figure 10 illustrates that up to the end of the linear section of the transfer function, all LRPs terminate by joining the baseline at the same point, and beyond the point of deviation from linearity the durations of the LRPs begin to lengthen.

Since the function of a microvillus is supposed to be completed after the absorption of the first photon, the depolarising effect due to the absorption of further photons by the same microvillus may have to be deferred (stored) until the microvillus has recovered. (The recovery may be, for example, the readiness for further excitor release, or the replenishment of intermediates, or the readiness of sites for accepting further excitors.) Thus, LRPs due to multiple photon captures would differ from those elicited by single photon absorptions in showing longer durations but not greater amplitudes.

Light adaptation could be similar to such a process, but in reverse. Thus with increasing light adaptation, as the number of active rhodopsin molecules becomes smaller, at first only the LRP duration shortens, until a state is reached at which there is a great probability of finding only one active rhodopsin molecule in each microvillus. Beyond this point, with light adaptation continuing, the number of excitable micro-villi becomes successively reduced and the amplitude diminishes accord-ingly.

However, such a scheme cannot be completely valid for the following reason. We mentioned earlier that we can replace the abscissa of Fig. 10(*b*) with a logarithmic scale of the number of rhodopsin molecules. This may be done by lining up the estimated maximum number of rhodopsin molecules in a rhabdom with the start of the saturation of the function. Accepting that the amplitude of an LRP is a function of the fraction of the photopigment in the rhodopsin state, when one cross-cor-relates the peaks and plateaus of the responses to the background light of Fig. 5 with the transfer function fitted with the scale we have just mentioned,* one obtains the result that the rhodopsin fraction cannot have fallen below 20%. So, either this method suffers from false assumptions, or in view of the microvillus hypothesis just described, it is rather the number of active microvilli that has diminished! If we maintain the microvillus hypothesis with its all-or-none function subject to the capture of a single photon, and bear in mind that in a rhabdom

---

* Such cross-correlation (not shown here) is not straightforward. It should take into account the intensity–time reciprocity relationship, in that, after marking off the amplitude of an LRP plateau on the transfer function and noting its *x*-coordinate, this point should be shifted along the abscissa by a factor, which, when multiplied by the duration of the light pulses used in obtaining the transfer function, yields the duration of the stimulus eliciting the plateau which is being compared.

there are $\sim2 \times 10^8$ rhodopsin molecules but only $\sim 1.4 \times 10^5$ micro-villi, it may be that during light adaptation there is a reduction in the number of the units in *shorter supply* (the microvilli) which causes the reduction in duration. Yet it cannot be refuted that in the static, dark-adapted state, the reduction in the duration of the response is clearly an indication of the reduction in the fraction of active rhodopsin molecules (Figs. 7–9).

In conclusion, taking into account the latter points, we propose that *during* light adaptation the reduction in the duration of the LRP may be due to the reduction in both the active rhodopsin fraction and the number of active (excitable) microvilli. Whereas at the start of light adaptation the loss of sensitivity may be due to the reduction in the number of active microvilli, with prolonged, or more intense, light adaptation the process gradually becomes one of reduction in the number of active rhodopsin molecules. Thus a combination of both effects could account for a great deal, if not all, of sensitivity loss over the entire range of light adaptation.

This work was part of the programme of the SFB 114, supported by the Deutsche Forschungsgemeinschaft.

References

AUTRUM, H. (1981). Light and dark adaptation in invertebrates. In *Handbook of Sensory Physiology*, vol. VII/6C, ed. H. Autrum, pp. 1–91. Berlin, Heidelberg & New York: Springer-Verlag.

BAUMANN, F. (1975). Electrophysiological properties of the honey bee retina. In *The Compound Eye and Vision of Insects*, ed. G. A. Horridge, pp. 53–74. Oxford: Clarendon Press.

BENOLKEN, R. M. (1962). Effects of light- and dark-adaptation processes on the generator potential of the *Limulus* eye. *Vision Res.*, **2**, 103–24.

BORSELLINO, A., FUORTES, M. G. F. & SMITH, T. G. (1965). Visual responses in *Limulus*. *Cold Spring Harbor Symp. Quant. Biol.* **30**, 429–43.

BROWN, K. T. & MURAKAMI, M. (1964). A new receptor potential of the monkey retina with no detectable latency. *Nature, London*, **201**, 622–8.

CONE, R. A. & PAK, W. L. (1971). The early receptor potential. In *Handbook of Sensory Physiology*, vol. I, ed. W. R. Loewenstein, pp. 345–65. Berlin, Heidelberg & New York: Springer-Verlag.

DÖRRSCHEIDT-KÄFER, M. (1972). Die Empfindlichkeit einzelner Photorezeptoren im Komplexauge von *Calliphora erythrocephala*. *J. Comp. Physiol.* **81**, 309–40.

FEIN, A. & DEVOE, R. (1973). Adaptation in the ventral eye of *Limulus* is functionally independent of the photochemical cycle, membrane potential, and membrane resistance. *J. Gen. Physiol.* **61**, 273–89.

FUORTES, M. G. F. & HODGKIN, A. L. (1964). Changes in the time scale and sensitivity in the ommatidia of *Limulus*. *J. Physiol.* **172**, 239–63.

FUORTES, M. G. F. & YEANDLE, S. (1964). Probability of occurrence of discrete potential waves in the eye of *Limulus*. *J. Gen. Physiol.* **47**, 443–63.

HAMDORF, K. (1970). Korrelationen zwischen Sehfarbstoffgehalt und Empfindlichkeit bei Photorezeptoren. *Verh. Dtsch. Zool. Ges.* **64**, 148–57.

HAMDORF, K. (1971). Die Dauer der Dunkeladaptation beim Fliegenauge nach Belichtung mit heterochromatischen Blitzen. *Z. Vergl. Physiol.* **75**, 200–6.

HAMDORF, K. (1979). The physiology of invertebrate visual pigments. In *Handbook of Sensory Physiology*, vol. VII/6A, ed. H. Autrum, pp. 145–224. Berlin, Heidelberg & New York: Springer-Verlag.

HAMDORF, K., PAULSEN, R. & SCHWEMER, J. (1973). Photoregeneration and sensitivity control of photoreceptors of invertebrates. In *Biochemistry and Physiology of Invertebrate Visual Pigments*, ed. H. Langer, pp. 155–66. Berlin, Heidelberg & New York: Springer-Verlag.

HAMDORF, K. & RAZMJOO, S. (1977). The prolonged depolarizing afterpotential and its contribution to the understanding of photoreceptor function. *Biophys. Struct. Mechanism*, **3**, 163–70.

HAMDORF, K. & RAZMJOO, S. (1979). Photoconvertible pigment states and excitation in *Calliphora*; the induction and properties of the prolonged depolarising afterpotential. *Biophys. Struct. Mechanism* **5**, 137–61.

HILLMAN, P., DODGE, F. A., HOCHSTEIN, S., KNIGHT, B. W. & MINKE, B. (1973). Rapid dark recovery of the invertebrate early receptor potential. *J. Gen. Physiol.* **62**, 77–86.

LAUGHLIN, S. (1976). The sensitivity of dragonfly photoreceptors and the voltage gain of transduction. *J. Com. Physiol.* **111**, 221–47.

LAUGHLIN, S. (1981). Neural principles in the peripheral visual systems of invertebrates. In *Handbook of Sensory Physiology*, vol. VII/6B, ed. H. Autrum, pp. 133–280. Berlin, Heidelberg & New York: Springer-Verlag.

LILLYWHITE, P. G. & LAUGHLIN, S. B. (1978). A neglected source of intrinsic noise in photoreceptors. *Proc. Aust. Physiol. Pharmacol. Soc.* **9**, 49 P.

LIPETZ, L. E. (1971). The relation of physiological and psychological aspects of sensory intensity. In *Handbook of Sensory Physiology*, vol. I, ed. W. R. Loewenstein, pp. 191–225. Berlin, Heidelberg & New York: Springer-Verlag.

LISMAN, J. E. & SHELINE, Y. (1976). Analysis of the rhodopsin cycle in *Limulus* ventral photoreceptors using the early receptor potential. *J. Gen. Physiol.* **68**, 487–501.

LISMAN, J. E. & STRONG, J. A. (1979). The initiation of excitation and light adaptation in *Limulus* ventral photoreceptors. *J. Gen. Physiol.* **73**, 219–43.

MILLECHIA, R. & MAURO, A. (1969). The ventral photoreceptor cells of *Limulus*. II. The basic photoresponse. III. A voltage-clamp study. *J. Gen. Physiol.* **54**, 310–51.

MINKE, B. & KIRSCHFELD, K. (1980). Fast electrical potentials arising from activation of metarhodopsin in the fly. *J. Gen. Physiol.* **75**, 381–402.

NAKA, K. I. & KISHIDA, K. (1966). Retinal action potentials during dark and light adaptation. In *The Functional Organization of the Compound Eye*, ed. C. G. Bernhard, pp. 251–66. Oxford: Pergamon Press.

OSTROY, S. E., WILSON, M. & PAK, W. L. (1974). *Drosophila* rhodopsin: photochemistry, extraction and differences in the norp A[P12] phototransduction mutant. *Biochem. Biophys. Res. Commun.* **59**, 960–6.

RAZMJOO, S. & HAMDORF, K. (1976). Visual sensitivity and the variation of total photopigment content in the blowfly photoreceptor membrane. *J. Comp. Physiol.* **105**, 279–86.

RAZMJOO, S. & HAMDORF, K. (1980). In support of the photopigment model of vision in invertebrates. *J. Comp. Physiol.* **135**, 209–15.

ROSNER, G. (1975). Adaptation und Photoregeneration im Fliegenauge. *J. Comp. Physiol.* **102**, 269–95.

SCHOLES, J. H. (1964). Discrete subthreshold potentials from the dimly lit insect eye. *Nature, London*, **202**, 572–3.

STARK, W. S. & ZITZMANN, W. G. (1976). Isolation of adaptation mechanisms of photopigment spectra by vitamin A deprivation in *Drosophila. J. Comp. Physiol.* **105,** 15–27.

STAVENGA, D. G., ZANTEMA, A. & KUIPER, J. W. (1973). Rhodopsin processes and the function of the pupil mechanism in flies. In *Biochemistry and Physiology of Visual Pigments*, ed. H. Langer, pp. 175–80. Berlin, Heidelberg & New York: Springer-Verlag.

STEPHENSON, R. S. & PAK, W. L. (1980). Heterogenic components of a fast electrical potential in *Drosophila* compound eye and their relation to visual pigment photoconversion. *J. Gen. Physiol.* **75,** 353–79.

STIEVE, H. (1965). Interpretation of the generator potential in terms of ionic processes. *Cold Spring Harbor Symp. Quant. Biol.* **30,** 451–6.

STIEVE, H., GAUBE, H. & WINTERHAGER, J. (1976). The effect of long-term light and dark-adaptation and stimulus intensity on the receptor potential of the *Limulus* lateral eye. *Vision Res.* **16,** 1159–68.

YEANDLE, S. (1958). Electrophysiology of the visual system – Discussion. *Am. J. Ophthal.* **46,** 82–7.

THE PHOTOBIOLOGY OF CAROTENES AND FLAVINS

DAVID E. PRESTI

Department of Biology, University of Oregon, Eugene, Oregon 97403, USA

Introduction

Light from the sun is the ultimate source of all biological energy. Living organisms also use light as a source of information about their environment, in ways which range from the simple initiation of physiological processes in bacteria to the complexities of visual perception in animals. The molecules which life has adopted to play the role of photoreceptors perform their functions very efficiently, having large extinction coefficients ($>10^4$–10^5 liter mol^{-1} cm^{-1}) for effective light absorption and high quantum yields (0·1–1) for the photochemical reactions that they undergo.

Carotenes and flavins are not often thought of in the context of photoreception. These molecules occur widely in Nature and subserve a number of functions in addition to the absorption of light for energy or sensory transduction. The photobiology of these two completely unrelated classes of molecule will be dealt with here under one heading because of the possibility that they both function as photoreceptors for a number of physiological responses to blue light in a variety of organisms (Schmidt, 1980; Senger, 1980).

Several examples of such so-called blue-light responses are given in Table 1, and a selection of the detailed action spectra for a number of these light responses are reproduced in Fig. 1. Although the exact wavelengths for the various maxima in the action spectra vary somewhat from organism to organism (not surprisingly, since the various spectra were determined by different experimenters using a variety of procedures), the shape is always basically the same. These spectra are all characterized by a large action band between 400 and 500 nm. This blue-light band has a maximum at approximately 450 nm with subpeaks or shoulders at about 425 and 475 nm. In addition, most of the spectra possess an action band in the ultra-violet with a maximum in the neighborhood of 370 nm. Finally, they all show no effectiveness for light of wavelengths longer than about 500 nm. The similarity among these action spectra has prompted more than one person to speculate that

Table 1. *Examples of physiological responses to blue light*

Response	Organism	References
Sporangiophore phototropism	*Phycomyces blakesleeanus* (fungus)	Curry & Gruen (1959); Delbrück & Shropshire (1960)
Sporangiophore phototropism	*Pilobolus kleinii* (fungus)	Page & Curry (1966)
Coleoptile phototropism	*Avena sativa* (oat)	Shropshire & Withrow (1958); Thimann & Curry (1961)
Phototropism	*Vaucheria geminata* (alga)	Kataoka (1975)
Photomovement	*Euglena gracilis* (protist)	Diehn (1969); Checcucci *et al.* (1976)
Photomovement	*Nitzschia communis* (diatom)	Nultsch (1971)
Photocarotenogenesis	*Phycomyces blakesleeanus* (fungus)	Bergman *et al.* (1973)
Photocarotenogenesis	*Mycobacterium* (bacterium)	Rilling (1964); Howes & Batra (1970)
Photocarotenogenesis	*Neurospora crassa* (fungus)	Zalokar (1955); De Fabo *et al.* (1976)
Photocarotenogenesis	*Fusarium aquaeductuum* (fungus)	Rau (1967)
Enhancement of respiration	*Chlorella* (alga)	Kowallik (1967); Pickett & French (1967)
Avoidance of light	*Physarum nudum* (slime mold)	Bialczyk (1979)
Entrainment of circadian rhythm of conidiation	*Neurospora crassa* (fungus)	Sargent & Briggs (1967)
Entrainment of circadian rhythm of egg hatching	*Pectinophora gossypiella* (moth)	Bruce & Minis (1969)

Entrainment of circadian rhythm of pupae emergence	*Drosophila* (fruit fly)	Frank & Zimmerman (1969); Klemm & Ninnemann (1976)
Chloroplast rearrangement	*Funaria* (moss)	Zurzycki (1972)
Chloroplast aggregation	*Vaucheria sessilis* (alga)	Fischer-Arnold (1963); Blatt & Briggs (1980)
Development of chloroplast structure and function	*Scenedesmus obliquus* (alga)	Senger & Bishop (1972); Brinkmann & Senger (1978)
Inhibition of bioluminescence	*Dissodinium lunula* (dinoflagellate)	Hamman *et al.* (1981)
Retardation of flower opening	*Oenothera lamarckiana* (plant)	Saito & Yamaki (1967)
Changes in leaf protoplasm viscosity	*Helodea densa* (plant)	Virgin (1952)
Stimulation of sexual development	*Nectria haematococca* (fungus)	Curtis (1972)
Stimulation of cell division	*Adiantum* (fern)	Wada & Furuya (1974)
Stimulation of branching and thallus formation	*Scytosiphon lomentaria* (alga)	Dring & Lüning (1975*a,b*)
Stimulation of conidiation	*Trichoderma viride* (fungus)	Gressel & Hartmann (1968); Kumagai & Oda (1969)
Stimulation of stomatal opening	*Vicia faba* (plant)	Hsiao *et al* (1973)
Stimulation of sporangiophore initiation	*Phycomyces blakesleeanus* (fungus)	Bergman (1972); Thornton (1973)

136

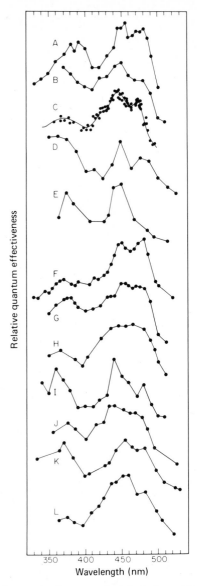

Fig. 1. Examples of action spectra for some physiological responses of several organisms to blue light: (A) phototropism in the fungus *Phycomyces* (Delbrück & Shropshire, 1960); (B) phototropism in the fungus *Pilobolus* (Page & Curry, 1966); (C) phototropism in coleoptiles of the plant *Avena* (Thimann & Curry, 1961); (D) a photomovement response in the motile protist *Euglena* (Diehn, 1969); (E) avoidance of light in the slime mold *Physarum* (Bialczyk, 1979); (F) photocarotenogenesis in the fungus *Neurospora* (De Fabo, Harding & Shropshire, 1976); (G) photocarotenogenesis in the fungus *Fusarium* (Rau, 1967); (H) photocarotenogenesis in the bacterium *Mycobacterium* (Rilling, 1964); (I) stimulation of sexual development in the fungus *Nectria* (Curtis, 1972); (J) enhancement of respiration in the alga *Chlorella* (Pickett & French, 1967); (K) chloroplast rearrangement in the moss *Funaria* (Zurzycki, 1972); and (L) entrainment of the circadian rhythm of pupae emergence in the fruit fly *Drosophila* (Klemm & Ninnemann, 1976).

there might be a common photosensitive mechanism of ancient lineage mediating the multifarious physiological responses to blue light.

The resemblance of these action spectra (Fig. 1) to the absorption spectra (Fig. 2) of riboflavin and some carotenoids, especially β-carotene, has provided the basis for these molecules being the most frequently suggested photoreceptor candidates for blue-light responses. Both riboflavin and β-carotene absorb blue light efficiently; their absorption bands in this region of the spectrum are broad (400–500 nm) and of large extinction at their maxima near 450 nm: approximately $1 \cdot 2 \times 10^4$ liter mol^{-1} cm^{-1} for riboflavin and $1 \cdot 3 \times 10^5$ liter mol^{-1} cm^{-1} for β-carotene.

The riboflavin molecule consists of a tricyclic carbon–nitrogen ring system (isoalloazine) with methyls at the 7 and 8 positions and a ribose sugar attached to the 10-nitrogen (Fig. 3A). This molecule, together with a phosphorylated form, riboflavin 5'-phosphate (flavin mononucleotide, FMN) and a phosphorylated-adenylated form, riboflavin adenine diphosphate (flavin adenine dinucleotide, FAD), are collectively referred to as flavins (Fig. 3B). The carotenoids are a class of molecules built from eight 5-carbon isoprene units in such a manner that the linking of the isoprene units is reversed in the middle of the molecule. This is illustrated by the structural formula for β-carotene (Fig. 3C, D), from which almost all other carotenoids can be formally derived by hydrogenation, dehydrogenation, decyclization, oxidation, or some combination of these processes.

Carotenoids are widely distributed among living organisms, occurring throughout the plant and protist kingdoms, in some bacteria, in many fungi, and in insects, birds, and other animals. These pigments are responsible for many of the brilliant yellow, orange and red colors in fruits, vegetables and the plumage of birds. It is carotenoids that make grapefruits, egg yolks, and buttercups yellow, oranges and carrots orange, tomatoes and lobsters red, and flamingos pink. Except for animals, organisms that contain carotenoids can make their own. Animals have lost the ability to biosynthesize carotenoids and are therefore dependent upon what they eat for a supply. In addition to being used for pigmentation in animals, some of the ingested carotenoids are converted to vitamin A (retinol), a molecule which functions (in the form of an aldehyde of vitamin A called retinal) as the chromophore for the rhodopsin photoreceptors of animal vision, as well as performing several other necessary, but as yet not clearly defined, biological functions. Two molecules of vitamin A are formed by central cleavage and oxygenation of one molecule of β-carotene.

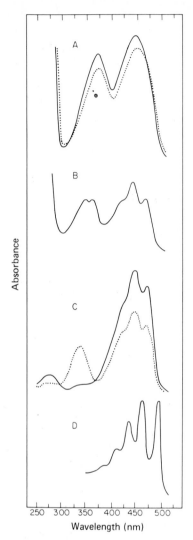

Wavelength (nm)

Fig. 2. Absorption spectra of several flavins and carotenes: (A) riboflavin (——) and FAD (......) in 0.1 M phosphate buffer pH 7 at room temperature (Whitby, 1953); (B) riboflavin in ethanol at 77 K (Sun *et al.*, 1972); (C) all-*trans*-β-carotene (——) and 15,15′-*cis*-β-carotene (......) in hexane at room temperature (Vetter, Englert, Rigassi & Schwieter, 1971); (D) all-*trans*-β-carotene in ethanol at 77 K (Song & Moore, 1974). The absorption spectrum of FMN is virtually identical to that of riboflavin (Whitby, 1953). The vertical absorbance scale is different for each spectrum, except for within the cases when two spectra are shown together (A and C).

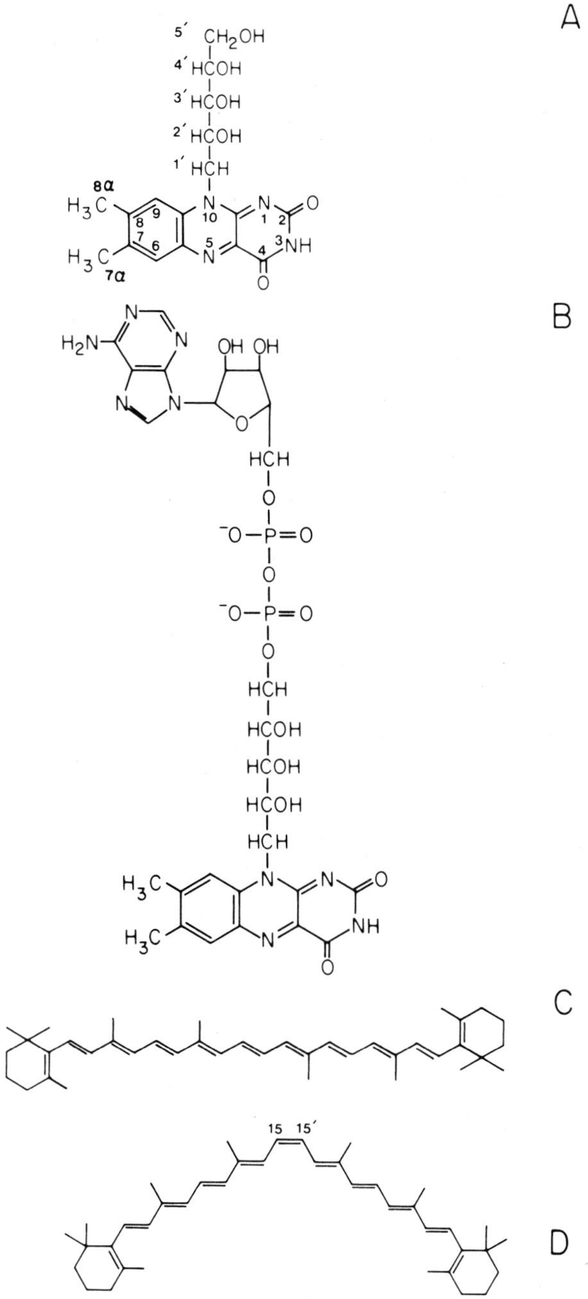

Fig. 3. The molecular structures of (A) riboflavin, showing the conventional numbering of atoms, (B) FAD (flavin adenine dinucleotide), (C) all-*trans*-β-carotene and (D) 15,15′-*cis*-β-carotene. FMN (flavin mononucleotide) is simply riboflavin-5′-phosphate; that is, FAD minus the second phosphate and adenine nucleotide.

Riboflavin (also known as vitamin B_2) is quite possibly present in all living organisms. It is biosynthesized by most bacteria, protists and fungi, and by all plants. Animals cannot make their own riboflavin, and so for them it is an essential element of their diet. Organisms convert riboflavin enzymatically into the phosphorylated and phosphorylated-adenylated forms FMN and FAD; these molecules then function as prosthetic groups for a large variety of enzymes involved in many key metabolic reactions. For example, at least six different flavoproteins are involved in the electron transfer reactions which lead into the mitochondrial respiratory chain. Other flavoproteins are involved in amino acid metabolism and biosynthesis and degradation of nucleotides. The prosthetic groups FMN and FAD are tightly bound to the various apoproteins, in either a noncovalent or covalent fashion. Noncovalent binding probably takes place through ionic interactions between cationic groups of the protein and the anionic phosphates of the coenzyme, and through hydrogen bonding to the nitrogen and oxygen atoms of the flavin. All known cases of covalent attachment occur via the 8α position of the flavin, usually to a cysteine or histidine residue of the protein (Singer & Edmondson, 1974; Edmondson & Singer, 1976).

The most important property of flavins is their ability to undergo a reduction to both a half-reduced ($1e^-$ equivalent) and a fully-reduced ($2e^-$ equivalents) forms. They are thus endowed with great versatility in electron transfer reactions, for a flavoprotein containing a single flavin coenzyme can shuttle either one or two electrons, separately or in tandem, in a catalytic cycle. Many flavoproteins contain two flavins per enzyme molecule, thus providing even more versatility. Such flavoenzymes may function as carriers of $1e^-$, $2e^-$, $3e^-$ or $4e^-$ among several different redox potentials.

Both flavins and carotenes are molecules of ancient origin. Riboflavin is found in the most primitive of living organisms: the nonphotosynthetic, anaerobic, fermenting bacteria. Indeed, flavodoxins, a low molecular weight class of electron-transport flavoproteins present in a variety of microorganisms, are believed to represent some of the oldest extant proteins. Their origin might lie as long as 3.5×10^9 years in the past (Rossman, Moras & Olsen, 1974). Carotenes have probably been around since the early days of oxygen-evolving photosynthesis (2–3×10^9 years ago), for the combination of light, chlorophyll and oxygen requires the presence of carotenes to protect the organism from lethal oxidations due to singlet oxygen (Krinsky, 1971, 1978).

Photophysical and photochemical properties of flavins and carotenes

Photophysical concepts

Most molecules, including all known biological photoreceptors, have electronic ground states in which the electron spins are all paired in an antiparallel fashion, leaving no net-spin magnetic moment. Such a state of no net-spin is called a singlet state. A state having two unpaired electrons (in different orbitals) with the same spin is called a triplet state. Oxygen is one of the few molecules to have a triplet ground state.

When a photoreceptor molecule absorbs a quantum of light of appropriate energy (wavelength), it is raised from its ground state singlet to one of several excited singlet states. From here the excitation energy can do any of several things. An excited species will usually find itself endowed at the moment of its creation with excess vibrational (and rotational) energy, in addition to electronic energy. Under most conditions, this vibrational energy is very rapidly ($<10^{-12}$ s) dissipated in a nonradiative fashion, and the excited species arrives at the lowest vibrational level of one of the singlet states. Higher singlet states (S_n for $n > 1$) rapidly (in $\sim 10^{-12}$ s) relax by a nonradiative process called internal conversion to the lowest excited singlet state S_1. From here the excitation energy may be further dissipated nonradiatively by internal conversion, the species returning to the ground state, S_0. The rate constant for $S_1 \rightsquigarrow S_0$ internal conversion (10^6–10^{12} s^{-1}) is often smaller than that for $S_m \rightsquigarrow S_n (m > n > 1)$ internal conversion ($\sim 10^{12}$ s^{-1}), thus other processes such as fluorescence and intersystem crossing can compete. Fluorescence is the radiative decay of S_1 to S_0 and generally occurs with a rate constant of about 10^6–10^9 s^{-1}. Intersystem crossing is the nonradiative transition between states of different spin multiplicity. The intersystem crossing $S_1 \rightsquigarrow T_1$ (or $S_1 \rightsquigarrow T_n \rightsquigarrow T_1$) is the process by which triplet states are normally populated. Once formed, the lowest triplet state, T_1, can decay to the ground state, S_0, either nonradiatively by intersystem crossing or radiatively by phosphorescence. The emission of a photon via phosphorescence is strictly forbidden on quantum mechanical grounds because it involves the flipping of an electron spin magnetic moment. Accordingly, the rate constant for phosphorescence ($\sim 10^{-2}$–10^4 s^{-1}) is much smaller than that for fluorescence. (Through the quantum mechanical phenomenon of spin–orbit coupling, singlet character is mixed into triplet states and vice versa, so that pure singlet and triplet states, in fact, do not exist. 'Forbidden' transitions between singlet and triplet states can then be viewed as occurring between the respective pure components of the actual hybrid states.)

A useful parameter in the discussion of photophysical and photo-chemical processes is the quantum yield, Φ. The quantum yield for a phenomenon of interest is given by the ratio of the number of times the phenomenon occurs to the number of quanta of radiation absorbed. For example, the quantum yield of fluorescence would be given by:

$$\Phi_f = \frac{\text{Number of molecules which decay by fluorescence}}{\text{Number of quanta of light absorbed}}.$$

The quantum yield is thus a measure of the efficiency of photon use for producing the process of interest.

Flavins

The electronic structure and photophysical properties of riboflavin are known in some detail (Sun, Moore & Song, 1972; Moore, McDaniels & Hen, 1977). The major absorption in the blue-light region of the spectrum is due to the $S_0 \rightarrow S_1$ transition, the structured peak being accounted for by transitions to several vibrational levels of S_1 (Fig. 4). The 370 nm ultraviolet absorption may be assigned to the $S_0 \rightarrow S_2$ transition and the large absorption at 280 nm to the $S_0 \rightarrow S_4$ transition. The energy separation between S_0 and T_1 can be estimated from phosphorescence emission spectra (Sun *et al.*, 1972) to be approximately 600 nm. The quantum yield for intersystem crossing from excited singlet states to T_1 can be as large as 0·7 (varying somewhat with solvent and temperature) (Moore *et al.*, 1977). Fluorescence accounts for the fate of most of the other absorbed energy. The fluorescence lifetime of riboflavin has been measured as 4.2 ns in tris buffer pH 7 (Chen, Vurek & Alexander, 1967) and as 5.6 ns in ethanol (Fugate & Song, 1976). The fluorescence lifetimes of FMN and FAD have been measured as 4·6 and 2·3 ns, respectively, in phosphate buffer pH 7 (Spencer & Weber, 1969). The fluorescence quantum yield of FMN is essentially the same as that of riboflavin (Koziol, 1971), while that of FAD appears to be only about 15–20% that of riboflavin (Weber, 1950; Koziol, 1971). The decreased fluorescence in FAD is interpreted as the result of an equilibrium between an open form and an intramolecularly-complexed form. In the complex, an association between the isoalloxazine ring and the adenine results in quenching of the flavin fluorescence (Weber, 1950; McCor-mick, 1977). The quantum yield for phosphorescence from T_1 in riboflavin is approximately 0·005. The phosphorescence lifetime of the

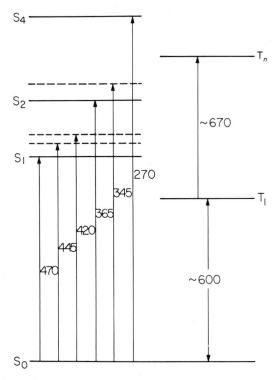

Fig. 4. The electronic energy diagram of riboflavin is shown with the experimental values in nanometers for several major singlet–singlet transitions (assigned according to Sun *et al.*, 1972) and an unassigned triplet–triplet transition (Grodowski, Veyret & Weiss, 1977). The energy separation between S_0 and T_1 is estimated from the flavin phosphorescence emission spectrum (Sun *et al.*, 1972).

triplet state is about 170 ms at 77 K (Sun *et al.*, 1972) and about 90 μs at room temperature (Vaish & Tollin, 1971).

The nature of the solvent in which a free flavin is dissolved can have effects on the locations and magnitudes of absorption maxima. Peaks often shift by several nanometers and occasionally by more than 10 nm. For riboflavin tetrabutyrate (a flavin made more soluble in nonpolar solvents by adding butyl groups to the ribityl side chain), it has been found that the vibrational fine structure of the $S_0 \rightarrow S_1$ transition becomes far better resolved (more like the low temperature spectrum of Fig. 2B) in nonpolar solvents as opposed to polar solvents (Kotaki, Naoi, Okuda & Yagi, 1977). Differences in the locations and magnitudes of absorption maxima are also observed for the chromophore/prosthetic groups of various flavoproteins. Moreover, a chromophore held rigidly within a protein may have its vibrational fine structure somewhat resolved.

The fact that essentially all of the excitation energy of photoexcited riboflavin (and FMN) is accounted for by fluorescence and by intersystem crossing to the relatively long-lived triplet state, indicates that very little energy is dissipated by the extremely rapid nonradiative decay of S_1 to the ground state. This means that this excitation energy is potentially available for initiating photochemical reactions.

Riboflavin is a very reactive photochemical species. For example, photoexcited riboflavin may react by combining with an H_2O or ROH molecule, where R represents an alkyl group or even the ribityl side chain of the flavin itself (Hemmerich, 1976). The most common kind of flavin photochemistry, however, involves the reduction of the flavin molecule by either one electron to form a flavin radical (semiquinone) or two electrons to form a fully-reduced flavin. These photoreduction reactions are generally reversible with respect to the flavin chromophore, whereas the photoaddition of H_2O or ROH is not (Hemmerich, 1976).

A variety of substances, amines and alcohols, for example, may serve as electron donors for the reduction of photoexcited riboflavin. EDTA (ethylenediaminetetraacetic acid) and NADH (nicotinamide adenine dinucleotide) are two commonly used donors for studies of flavin photoreduction chemistry. When no specific electron donors are present, other flavin molecules can fulfill the role if present in sufficient concentration. If not, photoexcited flavin will efficiently dehydrogenate its own ribityl side chain, resulting in irreversible degradation of the molecule to produce a number of photoproducts (Holmström & Oster, 1961; Cairns & Metzler, 1971). Most flavin photoreduction reactions involving an electron donor other than the ribityl side chain of the flavin itself are believed to occur through the relatively long-lived triplet state (Holmström, 1964). Moreover, while most pathways of photochemical dehydrogenation of the ribityl side chain are also believed to involve the flavin triplet state (Holmström, 1964; Song & Metzler, 1967), some degradation products are believed to result from excited single-state photochemistry (Song & Metzler, 1967).

Studies which attempt to distinguish between singlet- and triplet-state photochemical mechanisms often utilize substances which are known to quench certain excited states specifically. For example, silver ions, purines, pyrimidines and potassium iodide (KI) quench the fluorescence of riboflavin (Weber, 1950). Quenching occurs through enhancement of nonradiative decay of excited states. KI also quenches flavin phosphorescence, and thus the lowest triplet state, but at concentrations much less than those required to quench fluorescence (Song & Metzler,

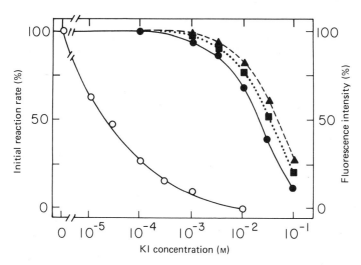

KI concentration (M)

Fig. 5. The effect of potassium iodide (KI) on the singlet and triplet states of flavins (from Vierstra *et al.*, 1981): (O——O) effects of KI on the anaerobic photoreduction of riboflavin by NADH, measured as percentage of the initial reaction rate in the absence of KI. The large decrease in reaction rate at concentrations of KI as small as 10 μM implies that the reduction proceeds via the flavin triplet state (Holmström, 1964; Sun & Song, 1973). The effect of KI on fluorescence emission from the excited singlet state is shown for riboflavin (●——●), FMN (■---■) and FAD (▲---▲). These effects are measured as a percentage of the fluorescence intensity in the absence of KI. Note that millimolar concentrations of KI are necessary before an effect is seen.

1967). Observations that KI quenches certain photochemical reactions at concentrations a thousand times less than those required to quench fluorescence therefore implicate a triplet photochemical mechanism (Holmström, 1964). For example, using KI, the anaerobic photoreduction of riboflavin by NADH has been shown to take place via the riboflavin triplet state (Fig. 5) (Sun & Song, 1973; Vierstra, Poff, Walker & Song, 1981).

Photoexcited riboflavin (RF*), following intersystem crossing to the triplet state, can efficiently transfer energy to the ground (triplet) state of molecular oxygen to form excited singlet oxygen:

$$^1RF \xrightarrow{h\nu} {}^1RF^* \rightsquigarrow {}^3RF^* \xrightarrow{^3O_2} {}^1RF + {}^1O_2^* .$$

Song & Moore (1968) have measured a quantum yield of 0·5 for this reaction. Singlet oxygen is a powerful oxidizing agent and can readily oxidize many substances (Foote, 1976). Thus riboflavin-sensitized photooxidations in the presence of oxygen may procede by one of two routes: (1) interaction of photoexcited riboflavin directly with a substrate to abstract electrons, resulting in reduced riboflavin and oxidized substrate or (2) energy transfer from photoexcited riboflavin to molecular oxygen producing singlet oxygen and regenerating ground state oxidized flavin. The singlet oxygen can then go on to oxidize a substrate.

Photoexcited riboflavin will oxidize indole-3-acetic acid (auxin) (Galston, 1949; Galston & Baker, 1949), which is a plant growth hormone. Galston & Baker (1949) estimated a quantum yield of 0·67 for this reaction, indicating a highly efficient process. This high quantum yield has been subsequently confirmed by other investigators. Song, Fugate & Briggs (1980) found that auxin efficiently quenches riboflavin fluorescence and that the photoreaction between flavin and auxin was unaffected by KI and oxygen, both of which are excellent quenchers of the flavin triplet. From this they concluded that the photoreaction of flavin with auxin proceeds efficiently via the singlet state of the former. Since the absence of oxygen does not affect the photoreaction, it would seem that singlet oxygen is not involved as an intermediate. Oxygen does, however, participate at least in secondary reactions, as the photoproducts of the flavin–auxin reaction are different under aerobic and anaerobic conditions (Song *et al.*, 1980).

Schmidt & Butler (1976) studied the photoinduced electron transfer from flavin to cytochrome *c* in solution. Light causes the photoreduction of the flavin by added electron donors (for example, EDTA). In the absence of an added electron donor other flavin molecules can fulfill the role if substrate amounts are present. Under aerobic conditions the photoreduced flavin reacts with molecular oxygen to form superoxide anion (O_2^-) which in turn can reduce cytochrome *c*. Under anaerobic conditions the photoreduced flavin reduces cytochrome *c* directly.

Thus riboflavin can participate in a variety of photochemical oxidation–reduction reactions. Moreover, riboflavin photochemistry often occurs with quantum yields high enough (that is, in the 0·1–1 range) to make flavins efficient photoreceptor molecules for signal transduction in living organisms.

The interaction of flavins with the amino acids in their flavoprotein-binding sites often has an effect on the quantum yields for various photophysical processes. For example, whereas free flavins are strongly fluorescent, most flavoproteins are nonfluorescent or only very weakly fluorescent. There are, however, a few notable exceptions, among which are lipoamide dehydrogenase and thioredoxin reductase (Ghisla, Massey, Lhoste & Mayhew, 1974). Fluorescence quenching in flavins bound to proteins is believed to arise from complex formation with nearby amino acids, especially tyrosine and tryptophan (McCormick,

1977). Flavin coenzyme–apoprotein complexes can also significantly reduce the quantum yield of intersystem crossing from photoexcited singlet states to triplet states (McCormick, 1977). This appears to be especially the case for flavins that are covalently linked to apoproteins via the 8α position of the flavin (McCormick, Falk, Rizzuto & Tollin, 1975; Edmondson, Rizzuto & Tollin, 1977). Reduced quantum yields for fluorescence and triplet formation indicate that rapid nonradiative pathways for excited singlet-state decay have been enhanced. Thus it seems that many flavoproteins are much less well-suited for participation in photochemical reactions than are free flavins.

Carotenes

Colored carotenes have a large absorption band between 400 and 500 nm. Three components to this band are generally discernible as peaks or shoulders. This absorption band is due to the $S_0 \to S_1$ transition and, as is the case for riboflavin, the structure is accounted for by transitions to several vibrationally-excited levels of S_1 (Fig. 6). The magnitude of this $S_0 \to S_1$ transition is quite large. The all-*trans* isomer of β-carotene absorbs 450 nm light more than ten times more effectively than does riboflavin. Carotene pigments possess an unstructured absorption band in the ultraviolet centered in the region of 345 nm, which arises from the $S_0 \to S_2$ transition. In all-*trans* isomers of carotenes the magnitude of the transition is quite small, whereas, in *cis*-carotenes this ultraviolet absorption band attains considerable size. Because an increase in size of this band is diagnostic of a *trans* → *cis* isomerization, it is often referred to as the '*cis*-peak' (Zechmeister, 1962). *Trans* → *cis* isomerization also results in a decrease in the magnitude of the 400–500 nm peak (Fig. 2C). A relatively small band at about 275 nm in carotene pigments is due to the $S_0 \to S_3$ transition.

A number of reports of carotenoid fluorescence and phosphorescence can probably be attributed to sample impurities. Purified β-carotene shows no phosphorescence in the visible and no fluorescence from the lowest excited singlet state (S_1) (Song, Moore & Sun, 1972). Based on the resolution of the photon-counting spectrofluorometer employed, it can be stated that the quantum yield for S_1 fluorescence in β-carotene is less than 10^{-5} (Song & Moore, 1974). From this it may be estimated that the lowest excited singlet state must decay with a characteristic lifetime of about 10^{-14} s (Song & Moore, 1974). The absence of excited-state Raman spectral features on a picosecond time scale further supports the extremely rapid (10^{-12} s) depopulation of the lowest excited singlet state of β-carotene (Dallinger, Woodruff & Rodgers, 1981). Weak fluor-

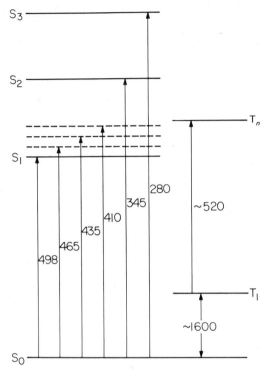

Fig. 6. The electronic energy diagram of β-carotene is shown with the experimental values in nanometers for several major singlet–singlet transitions and an unassigned triplet–triplet transition (Land *et al.*, 1971; Mathis & Kleo, 1973). The energy separation between S_0 and T_1 is estimated according to Fig. 7.

escence ($\Phi < 10^{-3}$) can be observed from β-carotene and other carotenoids and has been assigned to emission from S_2 (Song *et al.*, 1972). These authors point out, however, that although their samples were highly purified, it is not possible to rule out completely the presence of a fluorescent impurity.

The triplet states of the carotene pigments can be populated only by energy transfer from a triplet donor (Mathis & Kleo, 1973) or by pulse radiolysis (Land, Sykes & Truscott, 1971). Triplet states cannot be populated by intersystem crossing following singlet excitation, as the quantum yield for singlet→triplet intersystem crossing is found to be less than 0·001 for β-carotene and related carotenoid molecules (Bensasson, Dawe, Long & Land, 1977).

With no observable phosphorescence emission, it is not possible to measure directly the S_0–T_1 energy separation. Estimates of this energy separation in several carotenes come from measurements of their ability

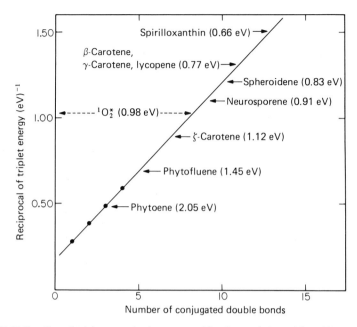

Fig. 7. Estimation of triplet energies for carotenoid polyenes (adapted from Bensasson *et al.*, 1976). The four data points are polyene triplet energies measured by Evans (1960, 1961). An extrapolation of the straight line fit due to Mathis (Mathis & Kleo, 1973) results in estimates of triplet energies for carotenoids within energy ranges that have been experimentally determined. The extrapolation is possible because an electron in a conjugated system behaves approximately like a particle moving in a one-dimensional box. Solution of the Schrödinger equation for such a system yields energy level separations that depend inversely on the length of the box, and thus inversely on the number of conjugated double bonds.

to quench the triplet state of various organic molecules through energy transfer (Bensasson, Land & Maudinas, 1976). These experimentally determined estimates are consistent with carotene triplet energies predicted by extrapolation of experimental data from small polyenes (Fig. 7) (Mathis, quoted in Bensasson *et al.*, 1976).

The estimated triplet energies are also in agreement with the ability of carotenoids to quench the first excited singlet state of molecular oxygen. This quenching is believed to occur via energy transfer from excited singlet oxygen to ground state carotene to form ground state (triplet) oxygen and excited triplet carotene (Foote & Denny, 1968):

$$^1O_2^* + {}^1Car \rightarrow {}^3O_2 + {}^3Car^*$$

The efficiency of this quenching process is found to be a function of the number of conjugated double bonds in the polyene chain. β-carotene, with 11 conjugated double bonds, quenches singlet oxygen with a rate that is diffusion-limited (Foote & Denny, 1968). Carotenoids with nine conjugated double bonds also efficiently quench singlet oxygen (Foote, Chang & Denny, 1970*a*; Mathews-Roth, Wilson, Fujimori & Krinsky, 1974), while a carotenoid with eight conjugated double bonds is two to three times less efficient in the quenching reaction (Mathews-Roth *et al.*, 1974). Finally, carotenoids with seven or fewer conjugated double bonds are several orders of magnitude less efficient in quenching singlet oxygen (Foote *et al.*, 1970*a*; Mathews-Roth *et al.*, 1974). The energy of the first excited singlet state of oxygen is approximately 0.98 eV (Herzberg, 1950), slightly below the estimated triplet energy for a carotenoid containing eight conjugated double bonds and far lower than the triplet energies for carotenoids with seven or fewer such double bonds. Since energy transfer from singlet oxygen to carotene can occur only if the oxygen state lies higher in energy than the lowest triplet state of the carotenoid, a marked drop in the quenching efficiency would be expected for carotenes having fewer than eight conjugated double bonds, exactly as is measured.

With virtually no fluorescence and no intersystem crossing to the long-lived lowest triplet state, it seems that photoexcited carotenes dissipate essentially all of their energy via nonradiative internal conversion processes. These processes occur too rapidly for photochemical reaction to compete effectively, making carotene pigments exceedingly poor photochemical species.

One aspect of photochemistry which many carotenes display is *trans*→*cis* and *cis*→*trans* isomerization (Zechmeister, 1962). Details about the mechanism of such isomerizations are unknown. Foote, Chang & Denny (1970*b*) found efficient isomerization of 15,15'-*cis*-*β*-carotene (Fig. 3D) to all-*trans* *β*-carotene (Fig. 3C) via reactions which involved populating the carotenoid triplet state by energy transfer from a triplet donor or from singlet oxygen. Land *et al.* (1971) observed the formation of *cis*-isomers of *β*-carotene following pulse radiolysis of the *trans*-isomer; their results indicated that the extent of isomerization was considerably greater than the triplet yield from pulse radiolysis, implying that isomerization was not occurring solely from the triplet state. Thus it is possible that polyene isomerization can occur from both the excited singlet and triplet states.

Fong & Schiff (1979) have detected light-induced optical absorbance changes in several strains of the motile alga *Euglena gracilis*, which appear to be due to a *cis*→*trans* isomerization of the major cellular pigment in these strains, *ζ*-carotene. The action spectrum (maxima at

448, 420, 390 and 370 nm) for the production of these absorbance changes is, curiously, not equivalent to the absorption spectrum of *cis* (or *trans*) ζ-carotene. Nor is it equivalent to the typical blue-light action spectrum for phototaxis of *Euglena* (Fig. 1D). Thus, the carotenoid isomerization appears to be sensitized by an unknown photoreceptor. What this means is as yet unknown: it only serves to illustrate an apparent example of light-induced carotenoid isomerization in a living organism.

Flavins and carotenes as possible blue-light receptors

Carotenes were first proposed in the mid-1930s as receptors for phototropism in the fruiting bodies of the fungi *Phycomyces* and *Pilobolus*, and in the coleoptiles of the plant *Avena* (Castle, 1935; Bünning, 1937*a*, *b*, *c*). These proposals were based (quite reasonably) on the fact that the major pigments extracted from the phototropic structures were carotenes whose absorption spectra were very similar to the phototropic action spectra. The presence of polyene pigments in photosensitive oat coleoptiles and in the eyes of animals suggested that they might be functioning as photoreceptors in both cases (Wald & DuBuy, 1936). The carotene photoreceptor hypothesis remained unchallenged for almost 15 years until Galston (1949, 1950; Galston & Baker, 1949) suggested that riboflavin might be playing the role of the photoreceptor for phototropism. This suggestion was motivated by the similarity of the riboflavin absorption spectrum to the action spectra for phototropism in *Avena* and by the ability of riboflavin to sensitize the photooxidation (*in vitro*) of the plant growth hormone (auxin) indole-3-acetic acid, a substance known to stimulate coleoptile growth. Phototropism was believed to result from a differential auxin distribution, arising from migration of auxin from the illuminated to the darkened side of the coleoptile, an idea proposed during the late 1920s by N. G. Chalodny and F. W. Went (see Went & Thimann, 1937). What Galston did in 1949 was to suggest that the auxin distribution in coleoptiles might be modified by selective destruction of indoleacetic acid on the illuminated side of the coleoptile through riboflavin-sensitized photooxidation.

The notion that riboflavin, rather than a carotene, might be playing the role of blue-light receptor has had its ups and downs over the decades. However, much evidence, albeit none of it conclusive, now suggests that riboflavin is probably functioning as the receptor for blue light-mediated physiological responses in many of the organisms that have been studied.

One strong argument providing support for a flavin rather than a carotene photoreceptor is the presence of a band in the ultraviolet (maximum at about 370 nm) in many of the action spectra for blue-light responses (Fig. 1).

Riboflavin has a significant absorption in precisely this region, but carotenoids, in general, do not. Under certain solution conditions, however, some carotenoids can acquire an absorption in the longer wavelength ultraviolet (Hager, 1970), apparently due to stacking interactions between molecules (Song & Moore, 1974). Although such interactions may occur when molecules are in solution, they are not very likely for molecules functioning as physiological photoreceptors. Moreover, β-carotene shows no tendency to form stacked dimers or aggregates, probably because of steric effects of the cyclized ends of the molecule (Song & Moore, 1974).

The 15,15'-*cis*-isomer of β-carotene exhibits an absorption in the ultraviolet at about 340 nm (Fig. 2C). Thus the carotenoid photoreceptor argument can be maintained by postulating that it is a *cis*-isomer which is absorbing the ultraviolet light. However, a careful carotenoid analysis of *Phycomyces* indicated the absence of *cis*-isomers of β-carotene (Presti, Hsu & Delbrück, 1977). Thus, at least in this case, the argument for *cis*-β-carotene photoreceptor cannot be applied.

In several organisms, ultraviolet light of shorter wavelength has also been investigated for physiological effectiveness. The action spectra for phototropism in *Phycomyces* (Curry & Gruen, 1959; Delbrück & Shropshire, 1960), for photocarotenogenesis in *Mycobacterium* (Howes & Batra, 1970) and for chloroplast rearrangement in *Funaria* (Zurzycki, 1972) all show a sharp band of considerable magnitude at about 280 nm. Although riboflavin possesses such an absorption maximum and β-carotene does not (Fig. 2), such action spectrum bands cannot necessarily be taken as additional evidence in support of a flavin photoreceptor. Aromatic amino acids absorb 280 nm light and thus energy transfer from the protein moiety of the photoreceptor to the chromophore might well contribute to action spectrum bands in this wavelength region.

A very strong argument in favor of a non-carotenoid (and therefore, most likely a flavin) photoreceptor being responsible for the blue-light responses of a number of organisms derives from the growth of these organisms under conditions in which their carotenoid content is reduced. Specifically, it has been reported that carotenoid deficiencies in *Phycomyces* (Meissner & Delbrück, 1968), *Neurospora* (Sargent & Briggs, 1967), *Euglena* (Checcucci et al., 1976), *Pilobolus* (Page & Curry, 1966), *Avena* (Bara & Galston, 1968), and *Drosophila* (Zimmer-

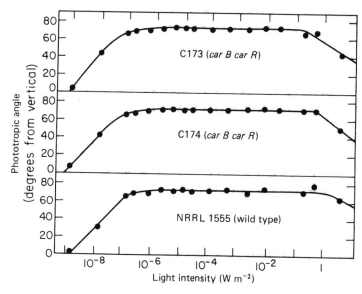

Fig. 8. Phototropic threshold data for two *Phycomyces carBcarR* mutants (which contain no coloured carotenes) and a wild-type strain (Presti *et al.*, 1977). The ordinate gives the angle from the vertical of sporangiophores exposed for 6 h to horizontal blue light (450 ± 50 nm) of the intensity indicated by the abscissa. It is seen that there is no decrease in the sensitivity of carotene-free strains to blue light. These curves also serve to illustrate the wide range of intensity over which phototropism in *Phycomyces* is operative. Like the human eye, *Phycomyces* adapts to light over a range of about 10^{11} in intensity, an extraordinary feat for a single fungal cell.

man & Goldsmith, 1971) had little or no effect on the physiological responses of these organisms to blue light. These studies (using mutants in the cases of *Phycomyces*, *Neurospora*, and *Euglena*, or inhibitors of carotenoid biosynthesis in the cases of *Pilobolus* and *Avena*, or carotenoid-free medium in the case of *Drosophila*), however, have often been criticized as being subject to the limitation that there probably was more than enough residual β-carotene present to serve efficiently as the photoreceptor.

Although the suggestion is that β-carotene (or other related carotenoids) cannot be functioning as the blue-light receptor in these organisms, conclusive proof was lacking until mutants of *Phycomyces* containing no detectable β-carotene were shown to possess phototropic responses to blue light identical to those of the wild type (Presti *et al.*, 1977). *Phycomyces carBcarR* mutants are blocked in all six steps of the biosynthesis of β-carotene from phytoene. Figs. 8 and 9 illustrate that phototropism in these *carBcarR* strains is identical to that in the wild

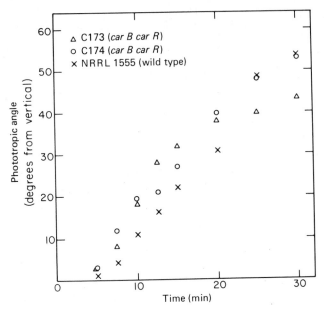

Fig. 9. Time course for the phototropic response of *Phycomyces carBcarR* mutants and a wild-type strain (Presti *et al.*, 1977). Single sporangiophores, initially growing vertically, were preadapted for 30 min to symmetrical blue illumination (450 ± 50 nm) of approximate intensity 10^{-4} W m^{-2} on a rotating turntable. At time zero, rotation was stopped and the phototropic angle measured as a function of time. It is seen that there is no decrease in the sensitivity of carotene-free strains to blue light.

type, both with respect to absolute threshold of the response and to time course of bending. These results definitely rule out the possibility that a carotenoid is functioning as the photoreceptor for phototropism of *Phycomyces* sporangiophores.

The biosynthesis of carotenoids is stimulated by light in a variety of organisms. Presumably this phenomenon has evolved to provide increased amounts of photoprotective pigments to organisms growing in the light. Photocarotenogenesis has been studied in several species of bacteria and fungi (Batra, 1971; Rau, 1976; Harding & Shropshire, 1980). Action spectra have implicated a porphyrin as the photoreceptor in the bacteria *Mycobacterium marinum* (Batra & Rilling, 1964) and *Myxococcus xanthus* (Burchard & Hendricks, 1969). Typical blue-light action spectra, suggesting a flavin or a carotenoid photoreceptor, have been determined for an unidentified species of *Mycobacterium* (Rilling, 1964; Batra & Rilling, 1964; Howes & Batra, 1970), and for the fungi *Fusarium aquaeductuum* (Rau, 1967) and *Neurospora crassa* (Zalokar, 1955; De Fabo *et al.*, 1976). De Fabo *et al.*'s (1976) carefully determined action spectrum for photocarotenogensis in *Neurospora* shows only a very small action band at 280 nm. In addition, the responsiveness to ultraviolet light between 360 and 380 nm is less than would be expected if riboflavin were the photoreceptor. On this basis, these authors argue for a carotenoid receptor mediating photocarotenogenesis in *Neurospora* and propose that the light reaction might be a photoisomerization. Whitaker & Shropshire's (1981) three-point action spectrum for photocarotenogenesis in *Phycomyces* also showed an effectiveness of longer wavelength ultraviolet light that was less than one would expect for a flavin photoreceptor, and prompted them to conclude that in *Phycomyces*, too, β-carotene serves as the receptor for photocarotenogenesis.

In several organisms light has been shown to regulate the synthesis of the carotenoid-pigment precursor phytoene, as well as the biosynthesis of colored carotenoids from phytoene (Harding & Shropshire, 1980). In *Phycomyces* (Jayaram, Presti & Delbrück, 1979) and in *Neurospora* (Schrott, 1980), light-induction of carotene pigments is under (at least) dual control, having components which saturate at low and high light fluence. It is not known if different photoreceptors are involved in the various stages of the regulation of carotene biosynthesis by light.

Inhibitors of flavin photochemistry: singlet or triplet mechanisms

Molecules known to interfere with flavin excited states have been demonstrated to have inhibitory effects on the physiological responses of several organisms to blue light. Potassium iodide (KI), for example, is a molecule known to quench very efficiently the flavin triplet excited state, and at higher concentration to quench also the excited singlet state (Holmström, 1964). The inhibition of the phototactic responses of *Euglena* (Diehn & Kint, 1970; Mikolajczyk & Diehn, 1975) and the phototropic responses of corn seedlings (Schmidt, Hart, Filner & Poff, 1977) by KI have therefore been taken as evidence in support of a flavin photoreceptor in these organisms, and furthermore, that the flavin triplet state is the active species. However, KI begins to quench the flavin singlet state at a concentration of about 1 mM (Fig. 5) (Vierstra *et al.*, 1981), and thus the concentrations of KI needed to obtain significant specific inhibition of phototaxis and phototropism in *Euglena* and corn (>50 mM and >10 mM, respectively) were much too large to distinguish between singlet and triplet reaction mechanisms. Moreover, as KI produces a general inhibition of metabolic processes involving electron transport, a reduction of swimming movement in *Euglena* (Diehn & Kint, 1970; Mikolajczyk & Diehn, 1975) and of tropic movements in corn (Schmidt *et al.*, 1977) results. This makes the determination of a specific effect on photoresponses more difficult.

It has also been found that the phototropism in corn seedlings was specifically inhibited by phenylacetic acid (Schmidt *et al.*, 1977; Vierstra & Poff, 1981), a molecule known to bind covalently to irradiated flavins (Hemmerich, Massey & Weber, 1967), and by azide (Schmidt *et al.*, 1977), an inhibitor of flavin photoreduction (Schmidt & Butler, 1976). These results further support the contention that a flavin is functioning as the receptor for phototropism in corn. Vierstra *et al.* (1981) have shown that millimolar concentrations of xenon, a water-soluble inert gas, specifically quench the triplet excited state of riboflavin without affecting the singlet excited state. Because it is chemically inert, xenon is much preferred to iodide in studies of the specific inhibition of biological photoresponses. Vierstra *et al.* (1981) found that neither the phototropism nor the geotropism of corn seedlings was inhibited in an atmosphere of $Xe:O_2(9:1)$ as compared to an atmosphere of $N_2:O_2(9:1)$. They concluded that the primary photochemical process leading to phototropism in corn probably proceeds from the flavin singlet excited state.

Delbrück, Katzír & Presti (1976) used the relatively intense light provided by a tunable dye laser for the determination of an action spectrum for the sporangiophore growth response in *Phycomyces* between 575 and 630 nm. Their results suggest a new action band, the size and location (595–600 nm) of which are consistent with a $S_0 \rightarrow T_1$ transition in riboflavin. This transition, being spin-forbidden and thus extremely weak, has an oscillator strength of only about 3×10^{-9} that of the $S_0 \rightarrow S_1$ transition in the blue region of the spectrum. Delbrück *et al.* (1976) concluded that the new band in the action spectrum represents the direct optical excitation of the lowest triplet state of riboflavin, suggesting that the photochemistry leading to sporangiophore photo-tropism can occur with a high quantum yield from the flavin triplet state. Two points are worth noting here. One is that singlet and triplet reaction mechanisms need not be mutually exclusive (Birks, 1976). It is possible that a reaction which might normally proceed via the excited singlet state, could also take place via the triplet state, if this state were populated directly. Another point is that the photoreceptor mechanisms leading to tropisms in *Phycomyces* and in plant coleoptiles are likely to be quite different in these phylogenetically distant organisms.

It is possible that the sensitivity to ~600 nm light measured by Delbrück *et al.* (1976) is due to absorption by a second receptor pigment, which could be a different kind of molecule or another form of the flavin receptor. One semiquinone form of riboflavin, for example, has an absorption maximum centered at about 600 nm (Massey & Palmer, 1966). Löser & Schäfer (1980) have interpreted sensitivity of *Phycomyces* phototropism to 605 nm light in terms of a photochromic pigment system with two interconvertible pigment forms absorbing in different spectral regions.

Artificial photoreceptors

Certain organic dye molecules, in the presence of light and oxygen, are known for their ability to sensitize the oxidation of a variety of substances (Spikes, 1977). A number of these photooxidation reactions are believed to take place via the triplet state of the photoexcited dye molecule, and energy transfer producing a singlet oxygen intermediate is often involved (Spikes, 1977). The photosensitizing dyes methylene blue, toluidine blue and neutral red are tricyclic ring systems of carbon, nitrogen and sulfur. Lang-Feulner & Rau (1975) demonstrated that these dyes could act as artificial photoreceptors for photocarotenogene-sis in the fungus *Fusarium aquaeductuum*. Carotenoid production in *Fusarium* is normally sensitive only to blue and longer wavelength ultraviolet light (Fig. 1G). However, following incubation of the fungus with one of these three organic dyes, red light of wavelength 600–700 nm is effective in inducing carotene synthesis (Lang-Feulner & Rau, 1975). All three dyes absorb light in this spectral region, toluidine blue and methylene blue especially so. These latter two dyes were also especially effective in eliciting photocarotenogenesis to red light.

The ability of photosensitizing dyes to function as artificial photo-receptors strongly suggests that an oxidative reaction is the initial step in the light induction of carotene synthesis in *Fusarium*. Flavins are especially efficient mediators of photooxidation reactions, either by direct transfer of electrons from substrate to photoexcited flavin or via the formation of singlet oxygen. Carotenoids are singularly ineffective as participants in redox-type reactions. Although these results strongly support the hypothesis that a flavin is acting as the photoreceptor for photocarotenogenesis in *Fusarium*, there is still the possibility that singlet oxygen formed by energy transfer from a dye molecule could be sensitizing the isomerization of a carotenoid.

Flavin analogue substitution

If a flavin does form the chromophore of the blue-light receptor in a given organism, it might be possible to replace it by a flavin analogue. If this analogue had absorption peaks different from those of riboflavin, and if the analogue-substituted photoreceptor remained functional, one could expect changes in the action spectra for light-regulated physiological responses. The observation of such an action-spectrum change would strengthen the argument in favor of a flavin photoreceptor.

Delbrück and colleagues have experimented with growing *Phycomyces* on a variety of riboflavin analogues. The analogue molecules generally differ from riboflavin by substitutions of one atom for another (for example, carbon for nitrogen or sulfur for oxygen), or replacing one or both methyl groups with other groups or atoms (for example, chlorine, amino, dimethylamino). The flavin analogues tested derive from both synthetic and natural sources and include 1-deaza, 5-deaza, 2-thio, 3-methyl, 7,8-dichloro, 8-amino and 8-dimethylamino riboflavins. The result has been that none of the analogues support growth of a riboflavin auxotroph strain in the absence of riboflavin and that many of the analogues do not even fool the flavin uptake system well enough to enter the organism.

Roseoflavin (so named for its rosey-pink color), the 8-dimethylamino analogue of riboflavin, is produced by a strain of *Streptomyces* bacteria (Otani *et al.*, 1974; Otani, 1976), presumably for antibiotic purposes, and has an absorption spectrum which differs markedly from that of riboflavin (Fig. 10). By culturing a riboflavin auxotroph strain of *Phycomyces* on medium containing roseoflavin and riboflavin in a 14:1 ratio, Otto, Jayaram, Hamilton & Delbrück (1981) showed that roseoflavin is taken up by the mycelium and translocated to the sporangiophores, and that sporangiophores doped in this fashion exhibit changes in their tropic response to light. Their results are in agreement with the differences between the absorption spectra of roseoflavin and riboflavin, and can be interpreted as a substitution of roseoflavin for riboflavin in approximately 80% of the photoreceptor for sporangiophore phototropism and a decrease in the quantum efficiency of the analogue-bearing receptor to about 0·1% of that of the normal receptor.

The spectroscopic characterization of roseoflavin by Song, Walker, Vierstra & Poff (1980) indicates that the fluorescence quantum yield and lifetime of this molecule are smaller than the values for riboflavin, indicating increased nonradiative decay of the excited singlet state. Furthermore, no triplet state population was detected by phosphorescence at low temperature. These observations make roseoflavin kinetically less suitable than riboflavin as a photochemically-active species, thus perhaps accounting for the dramatic decrease in the quantum efficiency of the roseoflavin-substituted photoreceptor in *Phycomyces*.

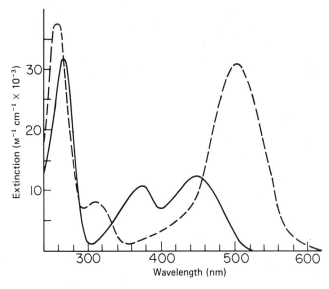

Fig. 10. Absorption spectra of riboflavin (——) and roseoflavin (- - -) in water at room temperature. At 380 nm, roseoflavin has only 0.1 times the absorbance of riboflavin; at 530 nm it has 180 times the absorbance (Otto *et al.*, 1981).

The results of Otto *et al.* (1981) reinforce the contention that riboflavin is the chromophore of the receptor for the photoresponses of the *Phycomyces* sporangiophore. A possibility which cannot, however, be excluded is that roseoflavin is sensitizing sporangiophore phototropism in the same manner that methylene blue and other molecules sensitize photocarotenogenesis in *Fusarium*. If roseoflavin is acting in this manner, actual substitution into the photoreceptor protein would presumably not be required.

Cellular location of blue-light receptors

Even if in some cases there is good reason to believe that the photoreceptor for a particular response to blue light is a flavin, in most instances the specific location of the receptor is not known. However, it is often possible to make some statement about the probable location of blue-light receptors.

In plant coleoptiles, for example, the extreme apex is the most photosensitive part (see Briggs, 1964), suggesting that the photoreceptor is present here in the highest concentration.

In the fungus *Phycomyces,* the sensitivity of both mycelium and sporangiophores to blue light means that photoreceptors must be present in both structures. In sporangiophores, growth is confined to a region called the growing zone, extending from 0·1 mm to about 2 mm below the apical sporangium. To be effective in producing a growth response, light must strike the growing zone, implying that the photo-receptor resides in this zone. Moreover, from studies using narrow beams of light, it has been shown that the photoreceptor molecules are located in a cylindrical shell just inside the cell wall of the sporan-giophore growing zone (Cohen & Delbrück, 1959; Delbrück & Varjú, 1961). Following on earlier work of Castle (1934), Jesaitis (1974) investigated the differential sensitivity of the *Phycomyces* sporan-giophore to light of different polarizations, and concluded that the photoreceptor molecules are rigidly oriented within the cell (in agree-ment with a prediction by Jaffe (1960)). An attractive hypothesis is that the photoreceptor is bound to the plasma membrane of the sporan-giophore. In such a location, the primary photoprocess might involve a modification of membrane conductance.

In the motile protist *Euglena gracilis*, the photoreceptor which mediates movement responses to blue light is believed to be in the paraflagellar body (Checcucci, 1976), a quasi-crystalline structure (Kivic & Vesk, 1972) located within the flagellar membrane near the site of attachment of the flagellum to the cell body. *In-vivo* spectro-fluorometry has indicated that a high concentration of fluorescent flavin is present in the *Euglena* paraflagellar body (Benedetti & Checcucci, 1975; Benedetti & Lenci, 1977). Although this may be suggestive of a flavin photoreceptor in the paraflagellar body, it is far from conclusive, as cellular flavin is not confined to this locale.

Toward a spectroscopic assay

Biological photoreceptor molecules such as the chlorophylls, phyto-chrome, bacteriorhodopsins and animal rhodopsins are distinguished by their characteristic absorption spectra and often are further dis-tinguished by characteristic photochemical reactions which can be readily monitored spectroscopically. In the case of phytochrome, for example, photoreceptor preparations can be photoconverted from a red light-absorbing form (Pr) to a far-red light-absorbing form (Pfr) upon irradiation with red light (~660 nm). Irradiation of Pfr with far-red light (~730 nm) reverses the reaction to regenerate Pr. This spectral property can be used to assay for the presence of phytochrome within a cell, and to monitor the presence of the photoreceptor during isolation and purification.

In the case of blue-light receptors, cellular flavin is not found exclusively in one locale nor does it only play the role of photoreceptor.

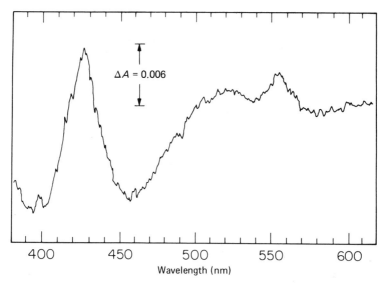

Fig. 11. Blue light-induced absorbance changes measured in mycelium of *Phycomyces* (Lipson & Presti, 1977). The absorbance changes indicate the reduction of a *b*-type cytochrome, according to the peaks at approximately 427, 520 and 555 nm. The broad valley between 440 and 490 nm is suggestive of riboflavin bleaching. Similar blue light-induced absorbance-changes have been measured in whole cells of *Dictyostelium* (Poff & Butler, 1974) and *Neurospora* (Muñoz *et al.*, 1974; Muñoz & Butler, 1975), and in plasma membrane-enriched fractions from *Neurospora* (Brain *et al.*, 1977) and corn (Goldsmith *et al.*, 1980).

Thus, the isolation of the physiological photoreceptor flavin cannot rely on measurement of absorption spectra only, for each of the several dozen varieties of flavins and flavoproteins present in the cell's membranes, cytoplasm and mitochondria will have absorption spectra which are essentially indistinguishable from the action spectra for blue-light physiological responses. In order to select out that specific flavin which is functioning as the physiological photoreceptor, it is necessary to search for additional photochemical properties not characteristic of flavins in general.

One approach to this problem is to look for light-induced optical absorbance changes which might be indicative of a photoreceptor molecule undergoing a photochemical change. Such absorbance changes indicating the reduction of a *b*-type cytochrome (Fig. 11) have been detected in mycelium of the fungi *Phycomyces* (Poff & Butler, 1974; Lipson & Presti, 1977) and *Neurospora* (Muñoz, Brody & Butler, 1974; Muñoz & Butler, 1975) and in cell suspensions of the slime mold *Dictyostelium discoideum* (Poff & Butler, 1974). Action spectra for the

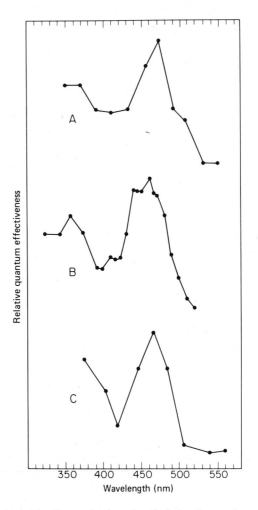

Fig. 12. Action spectra for the production of optical absorbance changes indicative of the reduction of a *b*-type cytochrome in (A) *Phycomyces* (Lipson & Presti, 1977), (B) *Neurospora* (Muñoz & Butler, 1975) and (C) *Dictyostelium* (Poff & Butler, 1974).

production of these absorbance changes suggest that a flavin is functioning as the photoreceptor in all cases (Fig. 12). The photochemical act would be the reduction of the photoexcited flavin by an unknown electron donor. This would produce a decrease in the absorbance at ~450 nm. The reduced flavin would then transfer an electron, either directly or via an intermediate, to cytochrome *b*. This reduction of the cytochrome gives rise to an increased absorbance at about 427 nm (Soret band), 520 nm (β band) and 558 nm (α band) (Fig. 11).

The similarity in the action spectra for the production of the light-induced cytochrome absorbance changes (Fig. 12) and the action spectra for blue-light physiological responses (Fig. 1) makes it attractive to hypothesize that the light-activated flavin–cytochrome electron transport system is, in fact, the phototransduction system for blue-light physiology (Poff & Butler, 1974; Muñoz & Butler, 1975). However, in view of the photoreactivity of flavins and of their many functions within cells, as long as the only connection between the receptor for physiological responses to blue light and the receptor for blue light-induced cytochrome absorbance changes is the similarity of their respective action spectra, it is impossible to say for certain whether the two photoreceptor systems are related.

Lipson & Presti (1977) also discovered that suspensions of HeLa cells, a cell line derived from a human cervical carcinoma, possessed blue light-induced cytochrome absorbance changes. These absorbance changes are very similar to those observed in *Phycomyces, Neurospora* and *Dictyostelium*. As it is highly unlikely that light plays any normal physiological role in HeLa cells, it appears that blue light-induced cytochrome absorbance changes are not specific to organisms with blue-light physiological responses. It should also be noted that *Dictyostelium* is not known to possess any blue light-regulated physiological functions (Poff & Butler, 1975).

Lipson & Presti (1977) quantified the dependence of the light-induced absorbance changes in *Phycomyces* mycelium upon the intensity and duration of actinic light. From these fluence-response data they determined a quantum yield of approximately 0·02 for the flavin-mediated cytochrome reduction, indicating a rather inefficient photoprocess. Since physiologically relevant photoprocesses have generally been found to occur with high quantum yield (0·1–1), this low value does not strongly support a connection between the observed light-induced absorbance changes and the physiological photoreceptor in *Phycomyces*. Interestingly, however, quantum yields calculated by Lipson & Presti (1980) for the flavin-mediated light-induced cytochrome absorbance changes in *Neurospora* and corn are of order unity, and thus indicate a highly efficient photoprocess. This suggests that the absorbance changes in these organisms may possibly be more relevant to photophysiology.

A *b*-type cytochrome has been found to be associated with plasma membrane-enriched fractions from *Phycomyces, Dictyostelium, Neurospora* (Schmidt, Thomson & Butler, 1977) and *Zea mays* (Jesaitis, Heners, Hertel & Briggs, 1977). It was in *Neurospora* that blue light-induced cytochrome absorbance changes were first found to be greatest in a fraction characterized as enriched in plasma membrane (Brain, Freeberg, Weiss & Briggs, 1977). Later studies by Briggs and coworkers have concentrated on purification and characterization of the photoreducible flavin–cytochrome b system from corn coleoptiles (reviewed in Senger & Briggs, 1981). Such a photoreducible cytochrome b, spectroscopically distinct from the total cytochrome content of the organism (as determined by dithionite reduction), has been localized in the plasma membrane fraction (Goldsmith, Caubergs & Briggs, 1980; Leong & Briggs, 1981). The flavin–cytochrome complex is not disrupted during purification and the kinetics of the cytochrome photoreduction are not significantly affected by solubilization of the membrane, suggesting that the flavin and heme are associated with the same protein moiety (Leong & Briggs, 1981). Whether this photoreducible flavin–cytochrome system is actually the photoreceptor for physiological responses to blue light in corn remains to be seen. The efficient photochemical reaction, the specificity of the photoreducible cytochrome b and its location in the plasma membrane argue in favor of its relevance for photophysiology.

Connection between blue light-induced cytochrome absorbance changes and blue-light physiology in *Neurospora* has been reported by Klemm & Ninnemann (1978). They showed that mycelial samples removed from nutrient media and allowed to 'starve' for several hours attained both the capacity to have conidiation stimulated by blue light and possessed a blue-light-induced cytochrome absorbance change, whereas 'unstarved' mycelium would conidiate without blue light and showed no absorbance change. Klemm & Ninneman (1979) then showed that certain aspects of the enzymatic activity of nitrate reductase were correlated with the ability to elicit conidiation with blue light. This enzyme is responsible for catalyzing the reduction of nitrate to nitrite and contains both flavin (FAD) and heme (b-type cytochrome) prosthetic groups (Garrett & Nason, 1967). *Neurospora* nitrate reductase can be photoactivated with blue light; an action spectrum shows that flavin is the photoreceptor activation (Roldán & Butler, 1980). Finally, Ninnemann & Klemm-Wolfgramm (1980) have shown that nitrate reductase from *Neurospora* exhibits blue light-induced cytochrome absorbance changes identical to those observed in the intact organism, and suggest that this enzyme may be the receptor for the physiological responses of *Neurospora* to blue light.

Fluorescence lifetime spectroscopy has been explored by Song & coworkers (Song *et al.*, 1980; Song, 1980) as a possible basis for a flavin photoreceptor assay. While free flavins are quite fluorescent, flavoproteins in general are not, the excited state of flavin being quenched through interactions with aromatic amino acid residues in the protein (McCormick, 1977). The photoreceptor flavin should not have its excited state quenched since it will need to use excitation by light to initiate signal transduction. Thus it is possible that the photoreceptor flavin might be at least slightly fluorescent. Song reasoned further that the fluorescence lifetime might be shortened relative to that of free flavin, indicative of an efficient photochemical mechanism competing with fluorescence. Song *et al.* (1980) examined the plasma membrane-enriched fraction from corn coleoptiles and detected therein a short-lifetime (<1 ns) fluorescent flavoprotein which they suggested might be the receptor flavin for phototropism in corn.

Spectroscopic studies of *Phycomyces* plasma membrane-enriched fractions indicated the presence of fluorescent (probably flavin) species with lifetimes less than 2 ns (Presti, 1978). Whether such a flavin is associated with the physiological photoreceptor cannot be said, as no test is yet available to distinguish specifically the photoreceptor's excited state properties *in vitro*.

Photoreceptor mechanisms

The weight of evidence thus strongly supports the hypothesis that flavins are acting as receptors for the responses of a number of organisms to blue light. One of the most convincing cases is that of phototropism in *Phycomyces*. Here the sensitivity to ultraviolet light, the completely normal phototropism in carotene-free mutants, and the effect of the riboflavin-analogue roseoflavin on the wavelength sensitivity of phototropism provide compelling arguments for a flavin photoreceptor. However, even in those cases where a flavin does appear to be the blue-light receptor, elucidation of the mechanism by which it acts and development of an assay specific for the photoreceptor are still not forthcoming. The fact that riboflavin plays so many roles in living organisms makes it extremely difficult to separate out that particular flavin which is functioning as the physiological blue-light receptor (contrary to the situation for other physiological photoreceptors). Indeed, riboflavin would seem to be both a ubiquitous and an elusive photoreceptor.

Among the many organisms that exhibit physiological responses to blue light there probably exist several, and perhaps even many, different blue-light receptors with differing mechanisms of action. Retinal was discovered and put to use by life as a photoreceptor at several independent points in evolution, the results being retinal-based vision in animals and retinal-based energy and sensory phototransduction in the *Halobacteria*. Given the ubiquity of riboflavin among living organisms, it would not be at all suprising for life to put to use the photochemical properties of this molecule at several independent points in evolution.

The ability of oxidized flavin to undergo reversible light-induced one-and/or two-electron additions makes it a molecule well-suited for potentiating a variety of photochemical oxidation–reduction reactions. Such electron transfer reactions might well lead to the production of light-induced membrane potentials, as are found in photosynthetic electron transport. A variety of metabolic processes may be regulated by membrane potential changes. Moreover, membrane conductance might depend exponentially on membrane potential, thus providing a direct way to amplify the effect of the light stimulus during transduction. Photoreduction of flavin is also believed to result in a conformational change, from a planar to a nonplanar form (Tauscher, Ghisla & Hemmerich, 1973). A conformational change of a flavin chromophore could perhaps induce conformational changes in an associated protein, which might result, for example, in the opening of a transmembrane channel.

Phototropism in plant coleoptiles arises from the light-stimulated migration of auxin from the illuminated to the unilluminated side, rather than directly from the photodestruction of auxin (Briggs *et al.*, 1957; Pickard & Thimann, 1964). Song (1977) has suggested that a conformational change in a flavin photoreceptor might change the conformation of a protein or membrane and result in the release of auxin from a storage site. The auxin would still, however, need to be transported across the coleoptile. The mechanism of phototropism in sporangiophores of *Phycomyces* must be somewhat different. In these giant single cells, unilateral light is focused on the distal side of the sporangiophore, there producing an increase in the growth rate and consequent tropism toward the light. In plant coleoptiles, which are multicellular structures, relatively complex when compared to the single cell of *Phycomyces*, unilateral light exerts its effects on the proximal side. These differences in the mechanism of phototropism may begin even at the level of photoreceptor.

Organisms that have more than one blue-light response may possess several different blue-light receptors. For example, while the photoreceptor for sporangiophore phototropism in *Phycomyces* may be a flavin with a particular mechanism of action, that for induction of carotene synthesis in the mycelium may be another kind of receptor with a different mechanism. That β-carotene plays the role of receptor for photocarotenogenesis is a viable possibility; if it does play such a role, then *cis–trans* isomerizations are the only photochemical mechanisms which one can readily envisage. The long flexible polyene could, if photoisomerized, impose a conformational change on an associated protein as the first step in a transduction process.

An approach to the elucidation of blue-light receptor mechanisms in microorganisms that has yet to realize its potential is the use of specific phototransduction mutants. In the case of *Phycomyces*, a number of photosensory mutants have been obtained and their physiology extensively studied (reviewed by Lipson, 1980). Although none of the mutants thus far obtained appears to be specifically affected in the photoreceptor, characterization of the gene products that are affected will provide a basis for understanding the molecular mechanisms involved in photosensory transduction. (For an example from vision see Stephenson *et al.*, this volume.)

More carotene and flavin photobiology

Carotenes as photosynthetic accessory pigments

Carotenes play a major role in photosynthetic organisms as so-called accessory pigments, absorbing light and efficiently transferring the energy to chlorophyll molecules. Since carotenes have absorption spectra different from that of chlorophylls, the result is an increase in the range of wavelengths available for photosynthesis (for a comparison with vision see Franceschini, this volume). The extremely short lifetime of the carotene excited state ($<10^{-14}$ s) precludes energy transfer from occurring between these molecules and chlorophylls in solution (Dutton, Manning & Duggan, 1943; Song & More, 1974). However, in detergent micelles (Teale, 1958) and in thin films (Sineshchekov, Litvin & Das, 1972), where carotenes can be in close proximity to chlorophyll molecules for relatively long periods of time, efficient energy transfer from the former to the latter can take place.

In photosynthetic systems, carotenes transfer energy to chlorophyll with 100% efficiency (Goedheer, 1969). In such systems, carotenoid and chlorophyll molecules are associated together in specific geometries within proteins. For example, the photosynthetic light-harvesting complex from the marine dinoflagellates *Glenodinium* and *Gonyaulax polydra* contain one molecule of chlorophyll *a* and four molecules of the carotenoid peridinin per pigment–protein complex (M_r ~35 000) (Pré-zlin & Haxo, 1976), while the complex (M_r ~39 000) from the marine dinoflagellate *Amphidinium carterae* contains two chlorophyll *a* molecules and nine peridinin carotenoid molecules (Haxo *et al.*, 1976). These complexes show an efficient energy transfer from photoexcited peridinin to chlorophyll *a* (Prézelin & Haxo, 1976; Haxo *et al.*, 1976). Like β-carotene, free peridinin, or peridinin in denatured dinoflagellate pigment–protein complexes, has a fluorescence quantum yield of less than 10^{-5}, implying a lifetime for the first excited singlet state of less than 10^{-14} s, too short for efficient transfer of excitation energy (Song, Koka, Prézelin & Haxo, 1976). However, the specific association between the peridinins and the chlorophylls in the complexes results in a stabilization of the carotenoid excited state so that light energy absorbed by the carotene can be transferred with 100% efficiency to the chlorophyll (Song *et al.*, 1976; Koka & Song, 1977).

Photomovement behavior in motile photosynthetic microorganisms is often mediated through the photosynthetic system. Indeed, photosynthetic pigments function as the receptors for all photomovement responses of photosynthetic prokaryotes thus far investigated (Nultsch & Häder, 1980). These responses seem to be triggered by the additional energy supply from photosynthetic phosphorylation or by sudden changes in the rate of photosynthetic electron flow (Nultsch & Häder, 1980). Among the photosynthetic eukaryotes, cases where movement responses are mediated by photosynthetic pigments, as well as cases where motile behavior is controlled by a separate photoreceptor system, are known (Nultsch & Häder, 1979). Because of their role as accessory pigments for photosynthesis, absorption of light by carotenoids often contributes to photomovement responses in motile photosynthetic organisms. In such cases, carotenes do not undergo photochemical reactions, but simply pass the light energy on to chlorophylls.

Photoprotection by carotenes

It has long been known that visible light in the presence of molecular oxygen and so-called photosensitizing molecules can damage living organisms. The photosensitizer absorbs light and then transfers energy, via its triplet state, to molecular oxygen. The result is the production of excited singlet oxygen, a powerful oxidizing agent which can create havoc in cells (Foote, 1976). In photosynthetic organisms the role of photosensitizer is filled ably by chlorophyll. Carotenoid molecules play a photoprotective role by efficiently quenching chlorophyll triplet and oxygen singlet excited states (Fujimori & Livingston, 1957; Claes, 1960;

Foote & Denny, 1968; Foote *et al.*, 1970*a*; Krinsky, 1971, 1978). The necessity of carotenoid photoprotection in photosynthetic organisms has been confirmed numerous times since the original observation that mutants of photosynthetic bacteria which contain no carotenoids are killed when exposed to light and oxygen (Griffiths, Sistrom, Cohen-Brazire & Stanier, 1955). In addition to repeated demonstrations of this protective role in photosynthetic organisms, carotenes have been shown to suppress lipid oxidation in illuminated chloroplasts (Takahama, 1978) and to protect chlorophyll from photodecomposition in isolated pigment–protein complexes from marine dinoflagellates (Koka & Song, 1978).

Photoprotection is also useful in nonphotosynthetic organisms, for molecules other than chlorophyll, which have light-inducible triplet states higher in energy than the lowest excited singlet state of molecular oxygen; many sensitize the formation of singlet oxygen. Riboflavin, for example, may act in such a manner (Song & Moore, 1968) and thus mediate oxidative damage in organisms (Pereira, Smith & Packer, 1976; O'Kelley & Hardman, 1979). The problem of singlet oxygen generation is generally less severe in nonphotosynthetic organisms, since the photosensitizers are not present in as high a concentration as chlorophyll is in photosynthetic organisms.

The photoprotective function of carotenoids is an exceedingly important aspect of the photobiology of this class of molecules. It was probably in this role that carotenes first functioned as part of the photosynthetic system, later being also adapted as accessory light-absorbing pigments.

Flavins in photoreactivating enzymes

Damage to cells by ultraviolet light in the wavelength range 220–300 nm is due primarily to the formation of dimers between adjacent pyrimidines in cellular DNA. Such dimers distort the DNA structure, thereby drastically interfering with replication and transcription of the genetic material. Photoreactivation is the splitting of such pyrimidine dimers by enzymes which bind to the damaged DNA and carry out their repair function upon absorbing longer wavelength ultraviolet or visible light. Action spectra for photoreactivation differ among various organisms, suggesting the existence of several kinds of enzyme.

Purified photoreactivating enzyme from the bacterium *Escherichia coli* does not absorb in the spectral region where the photoreactivation action spectrum has its maximum ($\sim$370 nm) (Snapka & Fuselier, 1977). Neither does ultraviolet-irradiated DNA absorb in this region of the spectrum. An appropriate absorption band does, however, appear when

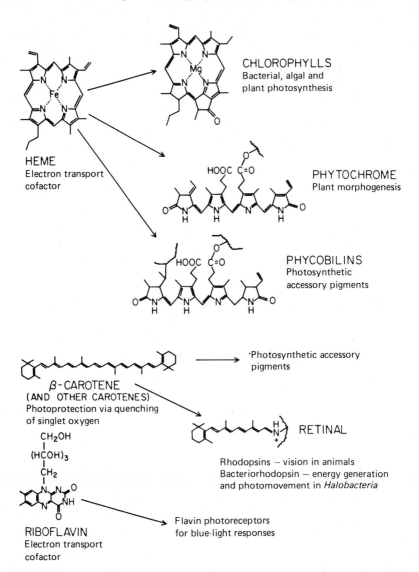

CHLOROPHYLLS
Bacterial, algal and
plant photosynthesis

HEME
Electron transport
cofactor

PHYTOCHROME
Plant morphogenesis

PHYCOBILINS
Photosynthetic
accessory pigments

·Photosynthetic accessory
pigments

β-CAROTENE
(AND OTHER CAROTENES)
Photoprotection via quenching
of singlet oxygen

RETINAL

Rhodopsins – vision in animals
Bacteriorhodopsin – energy generation
and photomovement in *Halobacteria*

Flavin photoreceptors
for blue-light responses

RIBOFLAVIN
Electron transport
cofactor

Fig. 13. Biological photoreceptors have probably evolved from similar molecules that originated to serve other functions.

enzyme is mixed with ultraviolet-irradiated DNA, indicating that it is a complex between the enzyme and a pyrimidine dimer which is the light-absorbing chromophore (Wun, Gih & Sutherland, 1977). In contrast, purified photoreactivating enzyme from the bacterium *Streptomyces griseus* does possess an intrinsic chromophore, the absorption spectrum of which corresponds with the action spectrum for photoreactivation (Eker, 1978). The identity of this chromophore has recently been suggested to be a flavin (7,8-didemethyl-8-hydroxy-5-deazaflavin) derivative (Eker, Dekker & Berends, 1981). Interestingly, 8-hydroxy-5-deazaflavins also occur widely among the methanogenetic bacteria as prosthetic groups of redox enzymes (Eirich, Vogels & Wolfe, 1978, 1979).

Another occurrence of a flavoprotein photoreactivating enzyme is in yeast, where a reduced FAD has been suggested as the light-absorbing chromophore (Iwatsuki, Joe & Werbin, 1980).

Coda

In many cases it seems that life has seized upon molecules that were already functioning in various capacities and adapted them to roles as photoreceptors (Fig. 13). The photosynthetic photoreceptor chlorophylls probably derived from evolutionarily older heme molecules utilized for electron transport in primitive bacteria. The linear tetrapyrroles which form the chromophores of the phycobilins, photosynthetic accessory pigments in cyanobacteria and algae, and of the plant photomorphogenetic receptor phytochrome, may have also had their origin with hemes. The retinal chromophore of the rhodopsin photoreceptors for animal vision and of the bacteriorhodopsin of *Halobacterium* is formed by cleavage of ubiquitous pigment, β-carotene. The carotenoids which play roles as photosynthetic accessory pigments and perhaps function as photoreceptors for several cases of blue light-regulated physiology, probably first evolved as quenchers of singlet oxygen, conferring protection against oxidative damage to organisms growing in the presence of light and oxygen. And riboflavin, probable mediator of a variety of physiological responses to light, first played the role of electron transport cofactor in enzymes like flavodoxins. Life has turned photochemical potentiality into photobiological actuality.

References

BARA, M. & GALSTON, A. W. (1968). Experimental modification of pigment content and phototropic sensitivity in excised *Avena* coleoptiles. *Physiol. Plant.* **21**, 109–18.

BATRA, P. P. (1971). Mechanism of light-induced carotenoid synthesis in nonphotosynthetic plants. In *Photophysiology*, vol. 6, ed. A. C. Giese, pp. 47–76. New York: Academic Press.

BATRA, P. P. & RILLING, H. C. (1964). On the mechanism of photo-induced carotenoid synthesis: aspects of the photoinductive reaction. *Arch. Biochem. Biophys.* **107**, 485–92.

BENEDETTI, P. A. & CHECCUCCI, A. (1975). Paraflagellar body (PFB) pigments studied by fluorescence microscopy in *Euglena gracilis*. *Plant. Sci. Lett.* **4**, 47–51.

BENEDETTI, P. A. & LENCI, F. (1977). *In vivo* microspectrofluorometry of photoreceptor pigments in *Euglena gracilis*. *Photochem. Photobiol.* **26**, 315–18.

BENSASSON, R., DAWE, E. A., LONG, D. A. & LAND, E. J. (1977). Singlet→triplet intersystem crossing quantum yields of photosynthetic and related polyenes. *J. Chem. Soc. Farad. Trans.* I, **73**, 1319–25.

BENSASSON, R., LAND, E. J. & MAUDINAS, B. (1976). Triplet states of carotenoids from photosynthetic bacteria studied by nanosecond ultraviolet and electron pulse irradiation. *Photochem. Photobiol.* **23**, 189–93.

BERGMAN, K. (1972). Blue-light control of sporangiophore initiation in *Phycomyces*. *Planta*, **107**, 53–67.

BERGMAN, K., ELSAVA, A. P. & CERDÁ-OLMEDO, E. (1973). Mutants of *Phycomyces* with abnormal phototropism. *Mol. Gen. Genet.* **123**, 1–16.

BIALCZYK, J. (1979). An action spectrum for light avoidance by *Physarum nudum* plasmodia. *Photochem. Photobiol.* **30**, 301–3.

BIRKS, J. B. (1976). Singlet and triplet mechanisms in photochemistry. *Photochem. Photobiol.* **24**, 287–9.

BLATT, M. R. & BRIGGS, W. R. (1980). Blue-light-induced cortical fiber reticulation concomitant with chloroplast aggregation in the alga *Vaucheria sessilis*. *Planta*, **147**, 355–62.

BRAIN, R. D., FREEBERG, J. A., WEISS, C. V. & BRIGGS, W. R. (1977). Blue light-induced absorbance changes in membrane fractions from corn and *Neuropsora*. *Plant Physiol.* **59**, 948–52.

BRIGGS, W. R. (1964). Phototropism in higher plants. In *Photophysiology*, vol. 1, ed. A. C. Giese, pp. 223–271. New York: Academic Press.

BRIGGS, W. R., TOCHER, R. D. & WILSON, J. F. (1957). Phototropic auxin redistribution in corn coleoptiles. *Science*, **126**, 210–12.

BRINKMANN, G. & SENGER, H. (1978). The development of structure and function in chloroplasts of greening mutants of Scenedesmus. IV. Blue light-dependent carbohydrate and protein metabolism. *Plant Cell Physiol.* **19**, 1427–37.

BRUCE, V. G. & MINIS, D. H. (1969). Circadian clock action spectrum in a photoperiodic moth. *Science*, **163**, 583–85.

BÜNNING, E. (1937*a*). Phototropismus und Carotinoide. I. Phototropische Wirksamkeit von Strahlen verschiedener Wellenlange und Strahlumgasabsorption im Pigment bei *Pilobolus*. *Planta*, **26**, 710–736.

BÜNNING, E. (1937*b*). Phototropismus und Carotinoide. II. Das Carotin der Reizaufnahmezonen von *Pilobolus, Phycomyces* und *Avena*. *Planta*, **27**, 148–58.

BÜNNING, E. (1937*c*). Phototropismus und Carotinoide. III. Weitere Untersuchungen an Pilzen und hoheren Pflanzen. *Planta*, **27**, 583–610.

BURCHARD, R. P. & HENDRICKS, S. B. (1969). Action spectrum for carotenogenesis in *Myxococcus xanthus*. *Journal of Bacteriology*, **97**, 1165–8.

CAIRNS, W. L. & METZLER, D. E. (1971). Photochemical degradation of flavins. VI. A new photoproduct and its use in studying the photolytic mechanism. *J. Amer. Chem. Soc.* **93**, 2272–7.

CASTLE, E. S. (1934). The phototropic effect of polarized light. *J. Gen. Physiol.* **17**, 751–62.

CASTLE, E. S. (1935). Photic excitation and phototropism in single plant cells. *Cold Spring Harbor Symp. Quant. Biol.* **3**, 224–9.

CHECCUCCI, A. (1976). Molecular sensory physiology of *Euglena*. *Naturwissenschaften*, **63**, 412–17.

CHECCUCCI, A., COLOMBETTI, G., FERRARA, R. & LENCI, F. (1976). Action spectra for photoaccumulation of green and colorless *Euglena*: evidence for identification of receptor pigments. *Photochem. Photobiol.* **23**, 51–4.

CHEN, R. F., VUREK, G. G. & ALEXANDER, N. (1967). Fluorescence decay times: proteins, coenzymes, and other compounds in water. *Science*, **156**, 949–51.

CLAES, H. (1960). Interaction between chlorophyll and carotenes with different chromophoric groups. *Biochem. Biophys. Res. Commun.* **3**, 585–90.

COHEN, R. & DELBRÜCK, M. (1959). Photoreactions in *Phycomyces*: growth and tropic responses to the stimulation of narrow test areas. *J. Gen. Physiol.* **42**, 677–95.

CURRY, G. M. & GRUEN, H. E. (1959). Action spectra for the positive and negative phototropism of *Phycomyces* sporangiophores. *Proc. Natl. Acad. Sci., USA*, **45**, 797–804.

CURTIS, C. R. (1972). Action spectrum of the photoinduced sexual stage in the fungus *Nectria haematococca* Berk. and Br. var. *cucurbitae* (Snyder & Hansen) Dingley. *Plant Physiol.* **49**, 235–9.

DALLINGER, R. F., WOODRUFF, W. H. & RODGERS, M. A. J. (1981). The lifetime of the excited singlet state of β-carotene: consequences to photosynthetic light harvesting. *Photochem. Photobiol.* **33**, 275–7.

DE FABO, E. C., HARDING, R. W. & SHROPSHIRE, W., JR. (1976). Action spectrum between 260 and 800 nanometers for the photoinduction of carotenoid biosynthesis in *Neurospora crassa*. *Plant Physiol.* **57**, 440–5.

DELBRÜCK, M. & SHROPSHIRE, W., JR. (1960). Action and transmission spectra of *Phycomyces*. *Plant Physiol.* **35**, 194–203.

DELBRÜCK, M., KATZIR, A. & PRESTI, D. (1976). Responses of *Phycomyces* indicating optical excitation of the lowest triplet state of riboflavin. *Proc. Nat. Acad. Sci., USA*, **73**, 1969–73.

DELBRÜCK, M. & VARJÚ, D. (1961). Photoreactions in *Phycomyces*: responses to the stimulation of narrow test areas with ultraviolet light. *J. Gen. Physiol.* **44**, 1177–88.

DIEHN, B. (1969). Action spectra of the phototactic responses in *Euglena*. *Biochim. Biophys. Acta*, **177**, 136–43.

DIEHN, B. & KINT, B. (1970). The flavin nature of the photoreceptor molecule for photoaxis in *Euglena*. *Physiol. Chem. Phys.* **2**, 483–8.

DRING, M. J. & LÜNING, K. (1975a). Induction of two-dimensional growth and hair formation by blue light in the brown alga *Scytosiphon lomentaria*. *Zeit. Pflanzenphysiol.* **75**, 107–17.

DRING, M. J. & LÜNING, K. (1975b). A photoperiodic response mediated by blue-light in the brown alga *Scytosiphon lomentaria*. *Planta*, **125**, 25–32.

DUTTON, H. J., MANNING, W. M. & DUGGAN, B. M. (1943). Chlorophyll fluorescence and energy transfer in the diatom *Nitzschia closterium*. *J. Phys. Chem.* **47**, 308–13.

EDMONDSON, D. E., RIZZUTO, F. & TOLLIN, G. (1977). The effect of 8α-substitution on flavin triplet state and semiquinone properties as investigated by flash photolysis. *Photochem. Photobiol.* **25**, 445–50.

EDMONDSON, D. E. & SINGER, T. P. (1976). 8α-substituted flavins of biological importance: an updating. *FEBS Lett.* **64**, 255–65.

EIRICH, L. D., VOGELS, G. D. & WOLFE, R. S. (1978). Proposed structure to coenzyme F_{420} from *Methanobacterium*. *Biochemistry*, **17**, 4583–93.

EIRICH, L. D., VOGELS, G. D. & WOLFE, R. S. (1979). Distribution of coenzyme F_{420} and properties of its hydrolytic fragments. *J. Bacteriol.* **140**, 20–7.

EKER, A. P. M. (1978). Some properties of DNA photoreactivating enzyme from *Streptomyces griseus*. In *DNA Repair Mechanisms*, ed. P. C. Hanawalt, E. C. Friedberg & C. F. Fox, pp. 129–132. New York: Academic Press.

EKER, A. P. M., DEKKER, R. H. & BERENDS, W. (1981). Photoreactivating enzyme from *streptomyces griseus*. IV. On the nature of the chromophoric cofactor in *Streptomyces griseus* photoreactivating enzyme. *Photochem. Photobiol.* **33**, 65–72.

EVANS, D. F. (1960). Magnetic perturbation of singlet–triplet transitions. Part IV. Unsaturated compounds. *J. Chem. Soc.* 1735–45.

EVANS, D. F. (1961). Magnetic perturbation of singlet–triplet transitions. Part VI. Octa-1,3,5,7-tetraene. *J. Chem. Soc.* 2566–9.

FISCHER-ARNOLD, G. (1963). Untersuchungen über die Chloroplastenbewegung bei *Vaucheria sessilis*. *Protoplasma* **56**, 495–520.

FONG, F. & SCHIFF, J. (1979). Blue-light-induced absorbance changes associated with carotenoids in *Euglena*. *Planta*, **146**, 119–27.

FOOTE, C. S. (1976). Photosensitized oxidation and singlet oxygen: consequences in biological systems. In *Free Radicals in Biology*, vol. 2, ed. W. A. Pryor, pp. 85–133, New York: Academic Press.

FOOTE, C. S., CHANG, Y. C. & DENNY, R. W. (1970*a*). Chemistry of singlet oxygen. X. Carotenoid quenching parallels biological protection. *J. Amer. Chem. Soc.* **92**, 5216–18.

FOOTE, C. S., CHANG, Y. C. & DENNY, R. W. (1970*b*). Chemistry of singlet oxygen. XI. *Cis-trans* isomerization of carotenoids by singlet oxygen and a probable quenching mechanism. *J. Amer. Chem. Soc.* **92**, 5218–19.

FOOTE, C. S. & DENNY, R. W. (1968). Chemistry of singlet oxygen. VII. Quenching by β-carotene. *J. Amer. Chem. Soc.* **90**, 6233–5.

FRANK, K. D. & ZIMMERMAN, W. F. (1969). Action spectra for phase shifts of a circadian rhythm in *Drosophila*. *Science*, **163**, 688–9.

FUGATE, R. D. & SONG, P.-S. (1976). Lifetime study of photoautomerism of alloxazine and lumichromes. *Photochem. Photobiol.* **24**, 479–81.

FUJIMORI, E. & LIVINGSTON, R. (1957). Interactions of chlorophyll in its triplet state with oxygen, carotene, etc. *Nature, London*, **180**, 1036–8.

GALSTON, A. W. (1949). Riboflavin-sensitized photooxidation of indoleacetic acid and related compounds. *Proc. Nat. Acad. Sci., USA*, **35**, 10–17.

GALSTON, A. W. (1950). Riboflavin, light, and the growth of plants. *Science*, **111**, 619–24.

GALSTON, A. W. & BAKER, R. S. (1949). Studies on the physiology of light action. II. The photodynamic action of riboflavin. *Amer. J. Bot.* **36**, 773–80.

GARRETT, R. H. & NASON, A. (1967). Involvement of a *b*-type cytochrome in the assimilatory nitrate reductase of *Neurospora crassa*. *Proc. Nat. Acad. Sci., USA*, **58**, 1603–10.

GHISLA, S., MASSEY, V., LHOSTE, J.-M. & MAYHEW, S. G. (1974). Fluorescence and optical characteristics of reduced flavins and flavoproteins. *Biochemistry*, **13**, 589–97.

GOEDHEER, J. C. (1959). Energy transfer between carotenoids and bacteriochlorophyll in chromatophores of purple bacteria. *Biochim. Biophys. Acta*, **35**, 1–8.

GOLDSMITH, M. H. M., CAUBERGS, R. J. & BRIGGS, W. R. (1981). Light-inducible cytochrome reduction in membrane preparations from corn coleoptiles. I. Stabilization and spectral characterization of the reaction. *Plant Physiol.* **66**, 1067–73.

GRESSEL, J. B. & HARTMANN, K. M. (1968). Morphogenesis in *Trichoderma*: action spectrum of photoinduced sporulation. *Planta*, **79**, 271–4.

GRIFFITHS, M., SISTROM, W. R., COHEN-BAZIRE, G. & STANIER, R. Y. (1955). Function of carotenoids in photosynthesis. *Nature, London*, **176**, 1211–14.

GRODOWSKI, M. S., VEYRET, B. & WEISS, K. (1977). Photochemistry of flavins. II. Photophysical properties of alloxazines and isoalloxazines. *Photochem. Photobiol.* **26**, 341–52.

HAGER, A. (1970). Ausbildung von maxima im Absorptionsspektrum von Carotinoiden im Bereich um 370 nm; folgen fur die Interpretation bestimmter Wirkungsspektren. *Planta*, **91**, 38–53.

HAMMAN, J. P., BIGGLEY, W. H. & SELIGER, H. H. (1981). Action spectrum for the photoinhibition of bioluminescence in the marine dinoflagellate *Dissodinium lunula*. *Photochem. Photobiol.* **33**, 741–7.

HARDING, R. W. & SHROPSHIRE, W., JR. (1980). Photocontrol of carotenoid biosynthesis. *Ann. Rev. Plant Physiol.* **31**, 217–38.

HAXO, F. T., KYCIA, J. H., SOMERS, G. F., BENNETT, A. & SIEGELMAN, H. W. (1976). Peridinin–chlorophyll *a* proteins of the dinoflagellate *Amphidinium carterae* (Plymouth 450). *Plant Physiol.* **57**, 297–303.

HEMMERICH, P. (1976). The present status of flavin and flavocoenzyme chemistry. *Prog. Chem. Org. Nat. Prod.* **33**, 451–527.

HEMMERICH, P., MASSEY, V. & WEBER, G. (1967). Photo-induced benzyl substitution by flavins by phenylacetate: a possible model for flavoprotein catalysis. *Nature, London*, **213**, 728–30.

HERZBERG, G. (1950). *Molecular Spectra and Molecular Structure. I. Spectra of Diatomic Molecules.* New York: Van Nostrand.

HOLMSTRÖM, B. (1964). The mechanism of the photoreduction of riboflavin. *Ark. Kemi*, **22**, 329–46.

HOLMSTRÖM, B. & OSTER, G. (1961). Riboflavin as an electron donor in photochemical reactions. *J. Amer. Chem. Soc.* **83**, 1867–71.

HOWES, C. D. & BATRA, P. P. (1970). Mechanism of photoinduced carotenoid synthesis: further studies on the action spectrum and other aspects of carotenogenesis. *Arch. Biochem. Biophys.* **137**, 175–80.

HSIAO, T. C., ALLAWAY, W. G. & EVANS, L. T. (1973). Action spectrum for guard cell Rb$^+$ uptake and stomatal opening in *Vicia faba*. *Plant Physiol.* **51**, 82–8.

IWATSUKI, M., JOE, C. O. & WERBIN, H. (1980). Evidence that deoxyribonucleic acid photolyase from baker's yeast is a flavoprotein. *Biochemistry*, **19**, 1172–6.

JAFFE, L. F. (1960). The effect of polarized light on the growth of a transparent cell: a theoretical analysis. *J. Gen. Physiol.* **43**, 897–911.

JAYARAM, M., PRESTI, D. & DELBRÜCK, M. (1979). Light-induced carotene synthesis in *Phycomyces*. *Exp. Mycol.* **3**, 42–52.

JESAITIS, A. J. (1974). Linear dichroism and orientation of the *Phycomyces* photoreceptor. *J. Gen. Physiol.* **63**, 1–21.

JESAITIS, A. J., HENERS, P. R., HERTEL, R. & BRIGGS, W. R. (1977). Characterization of a membrane fraction containing a *b*-type cytochrome. *Plant Physiol.* **59**, 941–7.

KATAOKA, H. (1975). Phototropism in *Vacheria geminata*. I. The action spectrum. *Plant Cell Physiol.* **16**, 427–537.

KIVIC, P. A. & VESK, M. (1972). Structure and function in the Euglenoid eyespot apparatus: the fine structure, and response to environmental changes. *Planta*, **105**, 1–14.

KLEMM, E. & NINNEMANN, H. (1976). Detailed action spectrum for the delay shift in pupae emergence of *Drosophila pseudo-obscura*. *Photochem. Photobiol.* **24**, 369–71.

KLEMM, E. & NINNEMANN, H. (1978). Correlation between absorbance changes and a physiological response induced by blue light in *Neurospora*. *Photochem. Photobiol.* **28**, 227–30.

KLEMM, E. & NINNEMANN, H. (1979). Nitrate reductase – a key enzyme in blue light-promoted conidiation and absorbance change in *Neurospora*. *Photochem. Photobiol.* **29**, 629–32.

KOKA, P. & SONG, P.-S. (1977). The chromophore topography and binding environment of peridinin-chlorophyll *a*–protein complexes from marine dinoflagellate algae. *Biochim. Biophys. Acta*, **495**, 220–31.

KOKA, P. & SONG, P.-S. (1978). Protection of chlorophyll *a* by carotenoid from photodynamic decomposition. *Photochem. Photobiol.* **28**, 509–15.

KOTAKI, A., NAOI, M., OKUDA, J. & YAGI, K. (1977). Absorption and fluorescence spectra of riboflavin tetrabutyrate in various solvents. *J. Biochem.* (Tokyo), **61**, 404–6.

KOZIOL, J. (1971). Fluorometric analyses of riboflavin and its coenzymes. In *Methods in Enzymology*, vol. 18, ed. D. B. McCormick & L. D. Wright, pp. 253–85. New York: Academic Press.

KOWALLIK, W. (1967). Action spectrum for an enhancement of endogenous respiration by light in *Chlorella. Plant Physiol.* **42**, 672–6.

KRINSKY, N. I. (1971). Function (of carotenoids). In *Carotenoids*, ed. O. Isler, pp. 669–716. Basel: Birkhauser Verlag.

KRINSKY, N. I. (1978). Non-photosynthetic functions of carotenoids. *Phil. Trans. Roy. Soc.* **B284**, 581–90.

KUMAGAI, T. & ODA, Y. (1969). An action spectrum for photoinduced sporulation in the fungus *Trichoderma viride. Plant Cell Physiol.* **10**, 287–92.

LAND, E. J., SYKES, A. & TRUSCOTT, T. G. (1971). The *in vitro* photochemistry of biological molecules. II. The triplet states of *β*-carotene and lycopene excited by pulse radiolysis. *Photochem. Photobiol.* **13**, 311–20.

LANG-FEULNER, J. & RAU, W. (1975). Redox dyes as artificial photoreceptors in light-dependent carotenoid synthesis. *Photochem. Photobiol.* **21**, 179–83.

LEONG, T.-Y. & BRIGGS, W. R. (1981). Partial purification and characterization of a blue-light sensitive cytochrome–flavin complex from corn membranes. *Plant Physiol.* **67**, 1042–6.

LIPSON, E. D. (1980). Sensory transduction in *Phycomyces* photoresponses. In *The Blue Light Syndrome*, ed. H. Senger, pp. 110–18. Berlin: Springer-Verlag.

LIPSON, E. D. & PRESTI, D. (1977). Light-induced absorbance changes in *Phycomyces* photomutants. *Photochem. Photobiol.* **25**, 203–8.

LIPSON, E. D. & PRESTI, D. (1980). Graphical estimation of cross sections from fluence-response data. *Photochem. Photobiol.* **32**, 383–91.

LÖSER, G. & SCHÄFER, E. (1980). Phototropism in *Phycomyces:* a photochronic sensor pigment? In *The Blue Light Syndrome*, ed. H. Senger, pp. 244–50. Berlin: Springer-Verlag.

MASSEY, V. & PALMER, G. (1966). On the existence of spectrally distinct classes of flavoprotein semiquinones. A new method for the quantitative production of flavoprotein semiquinones. *Biochemistry*, **5**, 3181–9.

MATHEWS-ROTH, M. M., WILSON, T., FUJIMORI, E. & KRINSKY, N. I. (1974). Carotenoid chromophore length and protection against photosensitization. *Photochem. Photobiol.* **19**, 217–22.

MATHIS, P. & KLEO, J. (1973). The triplet state of *β*-carotene and of analog polyenes of different length. *Photochem. Photobiol.* **18**, 343–6.

MCCORMICK, D. B. (1977). Interactions of flavins with amino acid residues: assessments from spectral and photochemical studies. *Photochem. Photobiol.* **26**, 169–82.

MCCORMICK, D. B., FALK, M. C., RIZZUTO, F. & TOLLIN, G. (1975). Inter- and intramolecular effects of tyrosyl residues on flavin triplets and radicals as investigated by flash photolysis. *Photochem. Photobiol.* **22**, 175–81.

MEISSNER, G. & DELBRÜCK, M. (1968). Carotenes and retinal in *Phycomyces* mutants. *Plant Physiol.* **43**, 1279–83.

MIKOLAJCZYK, E. & DIEHN, B. (1975). The effect of potassium iodide on photophobic responses in *Euglena*: evidence for two photoreceptor pigments. *Photochem. Photobiol.* **22**, 269–71.

MOORE, W. M., MCDANIELS, J. C. & HEN, J. A. (1977). The photochemistry of riboflavin. VI. The photophysical properties of isoalloxazines. *Photochem. Photobiol.* **25**, 505–12.

Muñoz, V., Brody, S. & Butler, W. L. (1974). Photoreceptor pigment for blue light responses in *Neurospora crassa*. *Biochem. Biophys. Res. Commun.* **58**, 332–7.

Muñoz, V. & Butler, W. L. (1975). Photoreceptor pigment for blue light in *Neurospora crassa*. *Plant Physiol.* **55**, 421–6.

Ninnemann, H. & Klemm-Wolfgramm, E. (1980). Blue light-controlled conidiation and absorbance change in *Neurospora* are mediated by nitrate reductase. In *The Blue Light Syndrome*, ed. H. Senger, pp. 238–43. Berlin: Springer-Verlag.

Nultsch, W. (1971). Phototactic and photokinetic action spectra of the diatom *Nitzschia communis*. *Photochem. Photobiol.* **14**, 705–12.

Nultsch, W. & Häder, D.-P. (1979). Photomovement of motile microorganisms. *Photochem. Photobiol.* **29**, 423–37.

Nultsch, W. & Häder, D.-P. (1980). Light perception and sensory transduction in photosynthetic prokaryotes. *Struc. Bond.* **41**, 111–391.

O'Kelley, J. C. & Hardman, J. K. (1979). Flavin compounds as agents for the oxidation of plastocyanin in blue light. *Photochem. Photobiol.* **29**, 829–32.

Otani, S. (1976). Studies on roseoflavin: isolation, physical, chemical and biological properties. In *Flavins and Flavoproteins*, ed. T. P. Singer, pp. 323–7. Amsterdam: Elsevier Scientific.

Otani, S., Takatsu, M., Nakano, M., Kasai, S., Miura, R. & Matsui, K. (1974). Roseoflavin, a new antimicrobial pigment from *Streptomyces*. *J. Antibiot.* **27**, 88–9.

Otto, M. K., Jayaram, M., Hamilton, R. H. & Delbrück, M. (1981). Replacement of riboflavin by an analogue in the blue-light receptor of *Phycomyces*. *Proc. Nat. Acad. Sci., USA*, **78**, 266–9.

Page, R. M. & Curry, G. M. (1966). Studies on phototropism of young sporangiophores of *Pilobolus kleinii*. *Photochem. Photobiol.* **5**, 31–40.

Pereira, O. M., Smith, J. R. & Packer, L. (1976). Photosensitization of human diploid cell cultures by intracellular flavins and protection by antioxidants. *Photochem. Photobiol.* **24**, 237–42.

Pickard, B. G. & Thimann, K. V. (1964). Transport and distribution of auxin during tropistic response. II. The lateral migration of auxin in phototropism of coleoptiles. *Plant Physiol.* **39**, 341–50.

Pickett, J. M. & French, C. S. (1967). The action spectrum for blue-light-stimulated oxygen uptake in *Chlorella*. *Proc. Nat. Acad. Sci., USA*, **57**, 1587–93.

Poff, K. L. & Butler, W. L. (1974) Absorbance changes induced by blue-light in *Phycomyces blakesleeanus* and *Dictyostelium discoideum*. *Nature, London*, **248**, 799–801.

Poff, K. L. & Butler, W. L. (1975). Spectral characterization of the photoreducible *b*-type cytochrome of *Dicytostelium discoideum*. *Plant Physiol.* **55**, 427–9.

Presti, D. (1978). Studies of the blue light receptor in *Phycomyces*. Ph.D. dissertation, California Institute of Technology.

Presti, D., Hsu, W.-J. & Delbrück, M. (1977). Phototropism in *Phycomyces* mutants lacking β-carotene. *Photochem. Photobiol.* **26**, 403–5.

Prézelin, B. B. & Haxo, F. T. (1976). Purification and characterization of peridinin-chlorophyl *a*-proteins from the marine dinoflagellates *Glenodinium* sp. and *Gonyaulax polyhedra*. *Planta*, **128**, 133–41.

Rau, W. (1967). Untersuchungen über die lichtabhangige Carotinoidsyntheses. I. Das Wirkungsspektrum von *Fusarium aquaeductuum*. *Planta*, **72**, 14–28.

Rau, W. (1976). Photoregulation of carotenoid biosynthesis in plants. *Pure App. Chem.* **47**, 237–43.

Rilling, H. C. (1964). On the mechanism of photoinduction of carotenoid synthesis. *Biochim. Biophys. Acta*, **79**, 464–75.

Roldán, J. M. & Butler, W. L. (1980). Photoactivation of nitrate reductase from *Neurospora crassa*. *Photochem. Photobiol.* **32**, 375–81.

ROSSMAN, M. G., MORAS, D. & OLSEN, K. W. (1974). Chemical and biological evolution of a nucleotide-binding protein. *Nature, London*, **250**, 194–9.

SAITO, M. & YAMAKI, T. (1967). Retardation of lower opening in *Oenothera lamarckiana* caused by blue and green light. *Nature, London*, **214**, 1027.

SARGENT, M. L. & BRIGGS, W. R. (1967). The effects of light on a circadian rhythm of conidiation in *Neurospora*. *Plant Physiol.* **42**, 1504–10.

SCHMIDT, W. (1980). Physiological bluelight reception. *Struc. Bond.* **41**, 1–44.

SCHMIDT, W. & BUTLER, W. L. (1976). Flavin-mediated photoreactions in artifical systems: a possible model for the blue-light photoreceptor pigment in living systems. *Photochem. Photobiol.* **24**, 71–5.

SCHMIDT, W., HART, J., FILNER, P. & POFF, K. (1977). Specific inhibition of phototropism in corn seedlings. *Plant Physiol.* **60**, 736–8.

SCHMIDT, W., THOMSON, K. & BUTLER, W. L. (1977). Cytochrome *b* in plasma membrane enriched fractions from several photoresponsive organisms. *Photochem. Photobiol.* **26**, 407–11.

SCHROTT, E. L. (1980). Fluence response relationship of carotenogenesis in *Neurospora crassa*. *Planta*, **150**, 174–9.

SENGER, H. (ed.). (1980). *The Blue Light Syndrome*. Berlin: Springer-Verlag.

SENGER, H. & BISHOP, N. I. (1972). The development of structure and function in chloroplasts of greening mutants of *Scenedesmus*. I. Formation of chlorophyll. *Plant Cell Physiol.* **13**, 633–49.

SENGER, H. & BRIGGS, W. R. (1981). The Blue light receptor(s): primary reactions and subsequent metabolic changes. In *Photochemical and Photobiological Reviews*, vol. 6, ed. K. C. Smith, pp. 1–38. New York: Plenum Press.

SHROPSHIRE, W., JR. & WITHROW, R. B. (1958). Action spectrum of phototropic tip curvature of *Avena*. *Plant Physiol.* **33**, 360–5.

SINESHCHEKOV, V. A., LITVIN, F. F. & DAS, M. (1972). Chlorophyll *a* and carotenoid aggregates and energy migration in monolayers and thin films. *Photochem. Photobiol.* **15**, 187–97.

SINGER, T. P. & EDMONDSON, D. E. (1974). 8α-substituted flavins of biological importance. *FEBS Lett.* **42**, 1–14.

SNAPKA, R. M. & FUSELIER, C. O. (1977). Photoreactivating enzyme from *Escherichia coli*. *Photochem. Photobiol.* **25**, 415–20.

SONG, P.-S. (1977). Molecular aspects of some photobiological receptors. *J. Korean Agri. Chem. Soc.* **20**, 10–25.

SONG, P.-S. (1980). Spectroscopic and photochemical characterization of flavoproteins and carotenoproteins as blue light photoreceptors. In *The Blue-Light Syndrome*, ed. H. Senger, pp. 157–71. Berlin: Springer-Verlag.

SONG, P.-S., FUGATE, R. D. & BRIGGS, W. R. (1980). Flavin as a photoreceptor for phototropic transduction: fluorescence studies of model coleoptile systems. In *Flavins and Flavoproteins*, ed. K. Yagi, pp. 443–53. Tokyo: Japan Scientific Societies Press.

SONG, P.-S., KOKA, P., PRÉZELIN, B. B. & HAXO, F. T. (1976). Molecular topology of the photosynthetic light-harvesting pigment complex, peridinin–chlorophyll *a*-protein, from marine dinoflagellates. *Biochemistry*, **15**, 4422–7.

SONG, P.-S. & METZLER, D. E. (1967). Photochemical degradation of flavins. IV. Studies of the anerobic photolysis of riboflavin. *Photochem. Photobiol.* **6**, 691–709.

SONG, P.-S. & MOORE, T. A. (1968). Mechanism of the photodephosphorylation of menadiol diphosphate: a model for biquantum conversion. *J. Amer. Chem. Soc.* **90**, 6507–14.

SONG, P.-S. & MOORE, T. A. (1974). On the photoreceptor pigment for phototropism and photoaxis: is a carotenoid the most likely candidate? *Photochem. Photobiol.* **19**, 435–41.

SONG, P.-S., MOORE, T. A. & SUN, M. (1972). Excited states of some plant pigments. In

The Chemistry of Plant Pigments, ed. C. O. Chichester, pp. 33–74. New York: Academic Press.

SONG, P.-S., WALKER, E. B., VIERSTRA, R. D. & POFF, K. L. (1980). Roseoflavin as a blue light receptor analog: spectroscopic characterization. *Photochem. Photobiol.* **32**, 393–8.

SPENCER, R. D. & WEBER, G. (1969). Measurements of subnanosecond fluorescence lifetimes with a cross-correlation phase fluorometer. *Ann. New York Acad. Sci.* **158**, 361–76.

SPIKES, J. D. (1977). Photosensitization. In *The Science of Photobiology*, ed. K. C. Smith, pp. 87–112. New York: Plenum Press.

SUN, M., MOORE, T. A. & SONG, P.-S. (1972). Molecular luminescence studies of flavins. I. The excited states of flavins. *J. Amer. Chem. Soc.* **94**, 1730–40.

SUN, M. & SONG, P.-S. (1973). Excited states and reactivity of 5-deazaflavin: comparative studies with flavins. *Biochemistry*, **12**, 4663–9.

TAKAHAMA, U. (1978). Suppression of lipid peroxidation by β-carotene in illuminated chloroplast fragments: evidence for β-carotene as a quencher of singlet molecular oxygen in chloroplasts. *Plant Cell Physiol.* **19**, 1565–9.

TAUSCHER L., GHISLA, S. & HEMMERICH, P. (1973). NMR study of nitrogen inversion and conformation of 1,5-dihydro-isoalloxazines (reduced flavin). *Helvet. Chim. Acta*, **56**, 630–44.

TEALE, F. W. J. (1958). Carotenoid-sensitized fluorescence of chlorophyll *in vitro*. *Nature, London*, **18**, 415–16.

THIMANN, R. V. & CURRY, G. M. (1961). Phototropism. In *Light and Life*, ed. W. D. McElroy & B. Glass, pp. 646–72. Baltimore: Johns Hopkins Press.

THORNTON, R. M. (1973). New photoresponses of *Phycomyces*. *Plant Physiol.* **51**, 570–6.

VAISH, S. P. & TOLLIN, G. (1971). Flash photolysis of flavins. V. Oxidation and disproportionation of flavin radicals. *J. Bioenerg.* **2**, 61–72.

VETTER, W., ENGLERT, G., RIGASSI, N. & SCHWIETER, U. (1971). Spectroscopic methods. In *Carotenoids*, ed. O. Isler, pp. 189–266. Basel: Birkhauser Verlag.

VIERSTRA, R. D. & POFF, K. L. (1981). Mechanism of specific inhibition of phototropism by phenylacetic acid in corn seedlings. *Plant Physiol.*, **67**, 1011–15.

VIERSTRA, R. D., POFF, K. L., WALKER, E. G. & SONG, P.-S. (1981). Effect of xenon on the excited states of phototropic receptor flavin in corn seedlings. *Plant Physiol.*, **67**, 996–8.

VIRGIN, H. I. (1952). An action spectrum for the light induced changes in the viscosity of plant protoplasm. *Physiol. Plant.* **5**, 575–82.

WADA, M. & FURUYA, M. (1974). Action spectrum for the timing of photo-induced cell division in *Adiantum* gametophytes. *Physiol. Plant.* **32**, 377–81.

WALD, G. & DUBUY, H. G. (1936). Pigments of the oat coleoptile. *Science*, **84**, 247.

WEBER, G. (1950). Fluorescence of riboflavin and flavin-adenine dinucleotide. *Biochem. J.* **47**, 114–21.

WENT, F. W. & THIMANN, K. V. (1937). *Phytohormones*. New York: MacMillan.

WHITAKER, B. D. & SHROPSHIRE, W., JR. (1981). Spectral sensitivity in the blue and near ultraviolet for light-induced carotene synthesis in *Phycomyces* mycelia. *Exp. Mycol.*, **5**, 243–52.

WHITBY, L. G. (1953). A new method for preparing flavin-adenine dinucleotide. *Biochem. J.* **54**, 437–42.

WUN, K. L., GIH, A. & SUTHERLAND, J. C. (1977). Photoreactivating enzyme from *Escherichia coli*: appearance of new absorption on binding to ultraviolet irradiated DNA. *Biochemistry*, **16**, 921–4.

ZALOKAR, M. (1955). Biosynthesis of carotenoids in *Neurospora*: action spectrum of photoactivation. *Arch. Biochem. Biophys.* **56**, 318–325.

ZECHMEISTER, L. (1962). *Cis–Trans Isomeric Carotenoids, Vitamins A and Arylopolyenes*. New York: Academic Press.

ZIMMERMAN, W. F. & GOLDSMITH, T. H. (1971). Photosensitivity of the circadian rhythm and of visual receptors in carotenoid-depleted *Drosophila*. *Science*, **171**, 1167–9.

ZURZYCKI, J. (1972). Primary reactions in the chloroplast rearrangements. *Acta Protozool.* **11**, 189–99.

THE MOLECULAR BASIS OF
PHYTOCHROME (Pfr) AND ITS
INTERACTIONS WITH MODEL RECEPTORS

PILL-SOON SONG

Department of Chemistry, Texas Tech University, Lubbock, TX 79409, USA

Introduction

Phytochrome is a blue chromoprotein that plays a vital role as the red-light photoreceptor which controls several diverse morphogenetic and developmental responses in plants, such as leaf movements of *Albizzia*, flowering in shorter and long-day plants, seed germination, stem elongation, and the biosynthesis of chlorophyll, chloroplasts, carotenoids, and anthocyanins (for reviews, see Briggs & Rice, 1972; Pratt, 1978; Rüdiger, 1980).

Higher plants respond to red light with extreme sensitivity (most plants will respond to 3×10^8 photons $m^{-2}s^{-1}$ of red light). This high sensitivity is achieved by virtue of the absorption of red light (660 nm) by one form of phytochrome (Pr), which is converted to the physiologically active form (Pfr) from the light-energized state of the inactive Pr form. This mode of action is illustrated in Fig. 1.

The Pfr form can be converted to the Pr form by far-red light (730 nm). Pfr also reverts to Pr in the dark. The length of the night and the interaction of phytochrome with a time-measuring system (biological clock) cause the seasonal phenomena that can be observed in higher plants.

Unlike the action of other photoreceptors (e.g. rhodopsin in vision), phytochrome-mediated plant morphogenesis and development are slow, usually having a response-time scale of minutes to days. One hypothesis is that red light triggers the transcription of the dormant genes in cellular DNA, thus inducing the synthesis of enzymes necessary for the morphogenetic and developmental responses of higher plants. The fundamental question that awaits an answer is how the red-light signal, perceived by phytochrome, is actually communicated to the chromosomal units within the nucleus, or to the mitochondrial and/or chloroplast DNA (for a review of various red-light responses see Mohr, 1972).

There are many possible models for the manner in which the

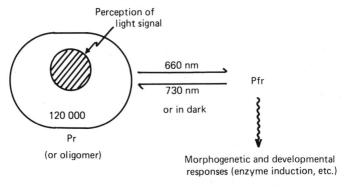

Fig. 1. The scheme for the red light-mediated plant morphogenesis and developmental responses. The shaded circle represents the covalently linked tetrapyrrole chromophore (Pr), which absorbs red light and undergoes transformation to the physiologically active form (Pfr). The molecular weight of the monomeric protein is 120 000.

postulated intracellular communication could be achieved. Five of these models are currently under consideration and are outlined below.

(i)   Pfr passes through the nuclear pore and directly interacts with the chromatin, thus activating the genes. In order for this model to be operative, a specific conformational change of the protein may be required in the phototransformation of Pr to Pfr, where the latter has an elongated shape so that the Pfr–pore-complex can be accommodated. A substantial gross conformational change in going from Pr to Pfr is necessary, since the phytochrome molecule (monomer) with a molecular weight of 120 000 is too large to pass through the pore.

(ii)   Pfr interacts with the nuclear envelope, activating receptor enzyme(s) which then produce signal-carrying chemicals (for example adenylate cyclase and cAMP) to bind at a gene-control site, that is a promotor or operation site.

(iii) The Pfr molecule permeates into the nucleus through a process of fusion into the nuclear envelope.

(iv)   Pfr interacts with its receptor in the cytoplasm and activates it to produce a signal-carrying chemical which is free to pass into the nucleus.

(v)   Pfr may bind specifically to plasma or thylakoid membranes, changing their properties and structure and thus eliciting the relatively fast membrane-mediated responses to red light (Satter & Galston, 1977).

In all these models, the specific interactions between Pfr and its receptor (or membrane) are featured as the initial molecular event, ultimately leading to the expression of red-light signals in higher plants. Herein, we consider this aspect in detail, based on available experimental evidence. I have not attempted to review the literature exhaustively because of the limited space; instead, I have reviewed what I consider to be a reasonable molecular model of the nature of Pfr and its interactions on the basis of experimental data obtained from my laboratory.

What makes Pfr different from Pr?

Although the chemical structure of Pfr has been inferred (Rüdiger, 1980; Lagarias & Rappoport, 1980; Song, 1980a; Sarkar & Song, 1981a), a definitive structure remains to be established. This is attributable to the fact that the Pfr molecule does not have a sufficiently long lifetime for chemical structure determination, since it reverts to the Pr form even in the dark. Nonetheless, it is certain that the chromophore configurations/conformations of the Pr and Pfr forms are similar (there is no gross photo-isomerism). This conclusion can be deduced from spectroscopic analysis (Song, Chae & Gardner, 1979; Song & Chae, 1979) and is based on the fact that the oscillator strength ratios ($f_{Q_{x,y}}/f_{B_{x,y}}$) are approximately unity for both Pr and Pfr. The red shift of λ_{max} for the latter is interpreted in terms of the photo-tautomeric shift of a ring A proton, as illustrated in Fig. 2. However, significant changes in the circular dichroism (CD) spectra of phytochrome accompany the photo-transformation of Pr to Pfr; namely, the induced optical activity of Pfr is substantially weakened in the chromophore absorption region (Song *et al.*, 1979; Song & Chae, 1979).

Upon denaturation, the strong CD signal of Pr is significantly quenched (I. S. Kim & P. S. Song, unpublished results), and the CD sign is inverted (Brandlmeier, Lehner & Rüdiger, 1981). Clearly, the CD inversion results from the chirality of the chromophore at ring A and the helicity of the conjugated ring system. The CD signal of the Pfr form at the chromophore absorption region is negligible compared with that of the Pr form. These results suggest that the induced optical activity due to the Pr chromophore–protein interaction is stronger than that due to the Pfr chromophore–protein interaction. Qualitatively, it can be suggested that hydrophobic and hydrogen bonding to the apoprotein, aside from the covalent linkage, is tighter in the Pr form than in the Pfr form. Other lines of evidence described below are also consistent with this conclusion.

In the proposed reaction mechanism of the Pr → Pfr phototransformation, which is qualitatively supported by solvent D_2O isotope

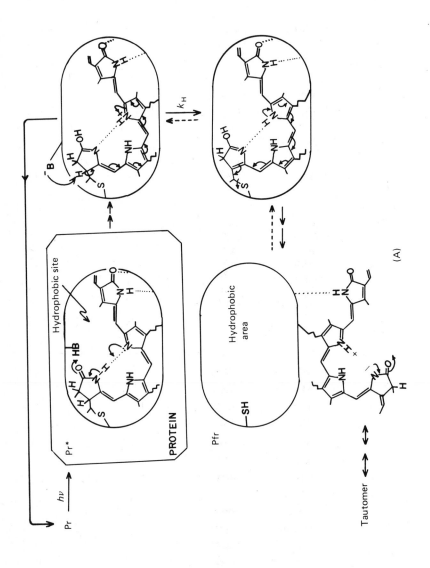

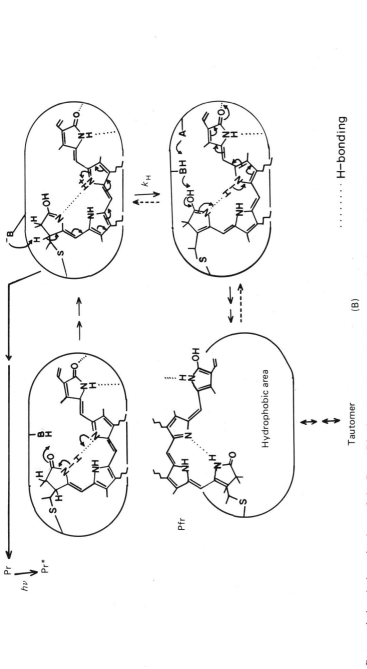

Fig. 2. Proposed chemical mechanisms of the Pr → Pfr phototransformation (modified from Song *et al.*, 1979; Song, 1980*a*; Sarkar & Song, 1981*c*). Mechanism (A) is based on the release of the sulfhydryl group from the thioether linkage and the propionic side chain at ring C is retained, while Mechanism (B) assumes that the thioether linkage is not released in the phototransformation. The main features of the phototransformation are: (*a*) tautomeric shifts of protons (Song *et al.*, 1979; Song & Chae, 1979; Sarkar & Song, 1981*c*); (*b*) C—H bond breaking assisted by a basic group (Sarkar & Song, 1981*a*) and (*c*) exposure of the chromophore binding surface as the result of chromophore relocation/reorientation. (Song *et al.*, 1979; Hahn & Song, 1981.)

effects (Sarkar & Song, 1981*a*), the thioether linkage between the chromophore ring A and apoprotein (with a cysteine residue) is released (Mechanism A, Fig. 2). This is based on the assumption that one of the propionic acid side chains is covalently linked to the polypeptide on both the Pr and Pfr forms (cf. Rüdiger, 1980; Killilea, O'Carra & Murphy, 1980) and that one additional –SH group becomes accessible in the Pfr form (Hunt & Pratt, 1981). In this mechanism the chromophore could possibly loosen the binding of its contact surface on the apoprotein in the Pfr form. On the other hand, Mechanism B (Fig. 2) assumes that the thioether linkage remains intact in the Pfr form and that no additional covalent linkage is present between the propionic acid group and the peptide, as has been argued by Lagarias & Rappoport (1980). In this mechanism, the chromophore becomes loosened from the apoprotein as the result of phototautomeric disruption of the hydrogen bonding at ring D. It is important to note that the chromophore conformation/configuration in Pfr remains essentially identical to that in Pr through appropriate hydrogen bonds (Fig. 2).

To gain insight into the topography of the chromophores in the Pr and Pfr phytochromes, as suggested by analysis of the CD spectroscopy and the mechanisms proposed in Fig. 2, one can compare the relative rates of chemical reactions between the chromophores and externally added reagents, on the assumption that the relative exposure or accessibility of the chromophores largely contribute to the rates of chemical reactions.

The rate constants for oxidation of the Pr and Pfr forms with micromolar concentrations of permanganate have been estimated to be 1.2×10^5 and $1.0 \times 10^6 \, \text{M}^{-1} \text{s}^{-1}$, respectively (Hahn, Kang & Song, 1980). The oxidant at micromolar concentrations apparently did not damage the proteins to any significant extent (Hahn *et al.*, 1980; Jung *et al.*, 1980). Figure 3 shows results of the reduction of Pr and Pfr with sodium borohydride, which is a less drastic agent than permanganate. It can be seen that, as with the permanganate oxidation, the Pfr form reacts with the reducing agent much faster than does the Pr form. These results are consistent with the suggestion that the chromophore in Pfr is exposed, whereas the Pr chromophore is relatively shielded within a chromophoric crevice of the protein. It should be noted that both Pr and Pfr chromophore model compounds undergo oxidation and reduction with permanganate and borohydride, respectively, at about the same rate (Hahn *et al.*, 1980).*

* Hunt & Pratt (1981) used tetranitromethane to modify tyrosyl residues in Pr and Pfr. This reagent is known to oxidize the bilin chromophore rapidly. Thus, the observed bleaching of Pfr with this reagent is consistent with the increased degree of exposure of the chromophore in Pfr, relative to that in Pr.

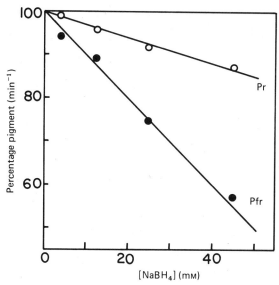

Fig. 3. The borohydride reduction of Pr and Pfr at 298 K, pH 7.8 in 0.1 M phosphate buffer. The ordinate scale represents initial rates of reduction; percentage pigment remaining was calculated after 1 min of reduction. (T.-R. Hahn & P. S. Song, unpublished.)

In addition to the above findings, the Pfr phytochrome exhibits the following characteristics. (i) Cations enhance the pelletability of Pfr (Quail, 1978). (ii) Pfr tends to associate with hydrophobic membrane fragments and membraneous structures (Marmé, 1977). (iii) The charged species, acetylcholine, is released from its bound form, upon phototransformation of Pr to Pfr in mung bean tissue (Jaffe, 1970). (iv) Pfr is more reactive to metal ions, *N*-ethylmaleimide and *p*-mercuribenzoate than is Pr (see review by Pratt, 1978). (v) Pfr is bleached more rapidly than Pr in the presence of 5 M urea (Butler, Siegelman & Miller, 1964). (vi) Pfr preferentially binds to blue dextran agarose via hydrophobic forces (Smith, 1981). (vii) Pfr is considerably less soluble in buffer compared with Pr. These findings are consistent with concepts of increased exposure of the chromophore and hydrophobicity of the Pfr form, compared with the Pr form of phytochrome.

In summary, the picture that emerges from these observations is that the Pr → Pfr phototransformation brings about a certain degree of flexibility to and exposure of the chromophore, resulting in the development of a hydrophobic surface on the Pfr protein. This picture has led us to propose a molecular model for the putative physiologically active form (Pfr) of the photomorphogenetic receptor (Song *et al.*, 1979; Song, 1980*a*). In the next section, this model is described in detail on the basis of additional lines of evidence obtained from recent studies in this laboratory.

The hydrophobic model of Pfr

The hydrophobic model of Pfr is presented in Fig. 4. It accommodates all of the observed characteristics of Pfr, and features the unique role of the Pr → Pfr chromophore phototransformation in the development of a specific hydrophobic surface on the Pfr protein. Two possible alternatives for the Pfr model are shown, since the covalent linkage between the propionic acid group and the peptide remains to be established (Fig. 2).

In the Pfr model (illustrated in Fig. 4), the chromophore of Pfr reorients itself from the original (Pr) chromophore crevice, thus exposing a substantial fraction of the hydrophobic surface. 8-Anilinonaphthalene-1-sulfonate (ANS) is a well known hydrophobic fluorescence probe, which becomes strongly fluorescent upon binding to a hydrophobic site on a protein. If the model illustrated in Fig. 4 realistically represents the Pr → Pfr phototransformation, the Pfr molecule would be expected to bind ANS specifically at the exposed hydrophobic surface. This has been borne out. Thus, ANS complexes with both forms of phytochrome, but exhibits a higher affinity for the Pfr form than for the Pr form, and ANS fluorescence is enhanced in the Pfr form as the result of ANS binding (Hahn & Song, 1981). Table 1 presents the fluorescence enhancement data. Accompanying the fluorescence intensity increases shown in Table 1, moderate increases in fluorescence anisotropy and the lifetime of bound ANS were observed upon phototransformation of Pr to Pfr in the presence of ANS, as expected. These results are explicable in terms of occupancy by ANS of the hydrophobic surface on the Pfr form, in addition to another 'non-specific' ANS site elsewhere on both the Pr and Pfr molecules.

The specific binding of ANS at the hydrophobic site on Pfr is also expected to preferentially affect both the absorption spectrum and the photo- and dark-reversion kinetics of the phytochrome (Pfr). Figure 5 shows the absorption spectra of Pr and its phototransformation to Pfr in the presence of ANS. The binding of ANS affects the Pr absorption spectra only slightly; however, at ANS concentrations greater than 1 mM a moderate degree of bleaching (hypochromism) can be seen (Fig. 5B). At the same time, Pfr produced from Pr is gradually bleached, with increasing concentrations of ANS. At 2 mM ANS, the Pfr absorption is virtually abolished (Fig. 5C). These spectral perturbations by ANS

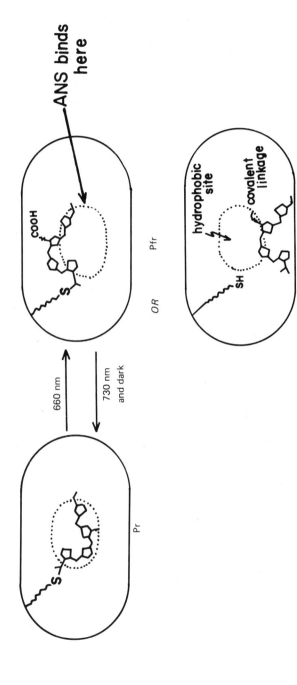

Fig. 4. Two alternatives for the hydrophobic model of the Pfr form of phytochrome. Note that the gross conformations of Pr and Pfr are identical. (Redrawn from Hahn & Song, 1981.)

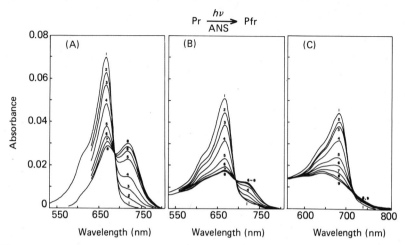

Fig. 5. Changes in the absorption spectra of phytochrome (Pr, 1 μM) in phosphate buffer, pH 7.8, 273 K, as a function of irradiation with 660 nm light (fluence rate at 7.5 W m^{-2}) in the presence (B, C) or absence (A) of ANS. (A), 0 mM ANS; (B), 1.0 mM ANS; (C), 2.0 mM ANS. In each case, spectra 1 to 8 represent progressive irradiation times from 0 s to 92.4 s, respectively, and spectrum 9 in (C) for 210 s. (Redrawn from Hahn & Song, 1981.)

suggest that the binding of ANS at the hydrophobic site of the Pfr molecule causes the chromophore to become fully exposed by forcing the expulsion of the chromophore from its partially occupied crevice, and by disrupting the hydrogen bonding that maintains a semi-circular chromophore conformation (Fig. 4; Song & Chae, 1979). Thus, the binding of ANS to Pfr leads to the spectral perturbation (that is, bleaching), which is similar to the effects of denaturation and binding of ionic detergents (I. S. Kim & P. S. Song, unpublished). It appears that an anionic group, in addition to the hydrophobic moiety, is necessary for the bleaching of the Pfr phytochrome which occurs when compounds such as ANS or anionic detergents bind to the hydrophobic site. The anionic group presumably acts by disrupting the hydrogen bonds between the chromophore and the apoprotein. Once the Pfr chromophore is completely exposed and becomes flexible, it resumes a cyclic conformation which is thermodynamically stable and spectrally hypochromic in the far-red region (Song & Chae, 1979; Song *et al.*, 1979).

To insure that the spectral bleaching of Pfr is not due to the trapping of intermediate(s) produced during the Pr → Pfr phototransformation in the presence of ANS, the reagent can be added to a solution of Pfr which has been produced from Pr in its absence. Figure 6 shows the effect of ANS on the absorption spectra of Pfr. It can be seen that the

Table 1. *Effect of the Pr → Pfr phototransformation on the fluorescence of bound ANS[a]*

Phytochrome (μM)	ANS (μM)	Fluorescence enhancement[b] (%)
1.6	0	0[c]
1.6	390	5.1[c]
1.0	500	6.2[c]
1.0	1000	9.6[c]
0.053	0.5	0[d]
0.352	0.5	3[d]
0.528	0.5	14[d]
0.88	0.5	31[d]

[a] Conditions as described in the caption to Fig. 5.

[b] Defined as $\Delta I_F^{Pr \to Pfr} \times 100/I_F^{Pr}$, where I_F^{Pr} and $\Delta I_F^{Pr \to Pfr}$ represent ANS fluorescence intensities before and after Pr → Pfr phototransformation; $\Delta I_F^{Pr \to Pfr}$ was measured immediately upon Pfr formation from Pr incubated for 2–5 h in the presence of ANS. Average of ten measurements.

[c] Using Pr from 'Affi-gel' procedure with highest purity (virtually free of contaminant proteins).

[d] Using Pr obtained from 'conventional' procedure, and values corrected for ANS fluorescence attributable to contaminant proteins.

far-red absorption band of Pfr is instantly bleached upon addition of ANS, indicating that the bleaching of the Pfr absorption band shown in Fig. 5 is not due to a trapping of intermediates by ANS during the phototransformation of Pr → Pfr. Interestingly, Pr can still be produced by far-red irradiation of a completely bleached solution of Pfr (Fig. 6). This peculiar effect can be explained in terms of a dynamic equilibrium between the fully bleached Pfr–ANS complex and a small amount of free Pfr, which is not detected spectrophotometrically at this low concentration, where only the latter undergoes phototransformation to the Pr form (Hahn & Song, 1981).

The binding of ANS to the hydrophobic crevice of the phytochrome chromophore would be expected to alter the kinetics of the phototransformation and dark-reversion of phytochrome. For example, the forward phototransformation, Pr → Pfr, is expected to be accelerated by ANS, which acts as a competitive activator by aiding the reorientation of the chromophore from the binding crevice. On the other hand, the reverse phototransformation and dark-reversion of Pfr are expected to be inhibited by ANS, which acts as a competitive inhibitor by occupying

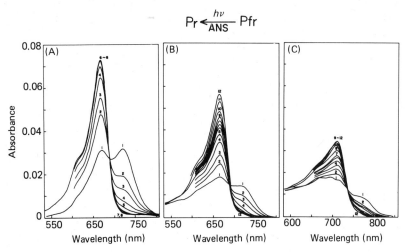

Fig. 6. Changes in the absorption spectra of phytochrome (Pfr, $0.8\,\mu\text{M}$) in phosphate buffer, pH 7.8, 273 K, as a function of irradiation with 730 nm light (fluence at 1.6 kW m^{-2}), in the presence (B, C) or absence (A) of ANS: (A), 0 mM ANS; (B), 1.0 mM ANS; (C), 2.0 mM ANS. In each case, spectra 1 through to 8 represent progressive irradiation times from 0 s to 92.4 s, respectively, and spectra 9 through to 12 from 210 s to 1410 s, respectively. (Redrawn from Hahn & Song, 1981.)

the hydrophobic surface to which the Pfr chromophore must return in reverting to the Pr form of phytochrome. All of these predictions have been experimentally confirmed.

As can be seen from Table 2, ANS up to 1 mM enhances the rate of Pr → Pfr phototransformation, whereas the slow component of the Pfr → Pr photoreversion rates steadily declines with ANS concentrations. Table 3 shows that ANS also inhibits the dark reversion. In both the photo- and dark-reversions of Pfr, only the slow components are inhibited by ANS. This observation suggests that the photochemical reaction responsible for the initial, fast bleaching of the 730 nm absorption band of Pfr is localized in the chromophore, and that the rate-limiting step involves the re-shuffling of the chromophore back to its original crevice. Thus, the latter step is competitively inhibited by ANS.

Further details of the hydrophobic model

In the previous section, the hydrophobic model for the Pfr form of phytochrome was discussed in relation to the effects of ANS on its fluorescence, absorption, and kinetic behaviour of the molecule. In this section, the model is further developed in terms of molecular details which have been revealed by recent studies in my laboratory.

Table 2. *Rate constants obtained from linear regression analyses of the kinetics of the phototransformations of phytochrome (1 μM) in phosphate buffer (pH 7.8) at 273 K with 660 nm (7.5 W m⁻²) and 730 nm (1.6 kW m⁻²) light, as a function of ANS concentration[a]*

	Pr → Pfr		Pfr → Pr		
[ANS] (μM)	$k_0(s^{-1})$	relative k_0	$k_1(s^{-1})$	$k_2(s^{-1} \times 10^2)$	k_2 component[b] (%)
0	0.098	1.0	0.24	0	0
50			0.22	0	0
100			0.22	0	0
200	0.114	1.17	0.27	7.52	22
500	0.123	1.26	0.28	2.23	28
1000	0.131	1.34	0.27	1.97	57
1500	0.095	0.97	0.26	0.97	63
2000	0.072	0.73	0.28	0.98	66

[a] Correlation coefficient for the rate constants listed ranged from 0.997 to 1.00.
[b] k_1 and k_2 for faster and slower components, respectively, and magnitude of the latter as resolved by the peeling procedure.

Table 3. *Rate constants obtained from linear regression analyses of the kinetics of dark-reversion (Pfr → Pr) of phytochrome in phosphate buffer, pH 7.8, at 273 K, in the presence of ANS*

[phytochrome] (μM)	[ANS] (mM)	k_1 ($s^{-1} \times 10^4$)	k_2 ($s^{-1} \times 10^6$)
2.4	0	7.23	7.71
2.1	0.625	7.69	7.60
2.1	1.250	7.70	4.92

First, we examine one crucial feature of the hydrophobic model, namely, the reorientation of the chromophore in going from Pr to Pfr (Figs. 2 and 4). When in the Pfr form, does the chromophore actually reorient or relocate relative to the chromophore-binding crevice of phytochrome? A definitive answer could come from X-ray crystallographic analysis of the structure of phytochrome. Unfortunately, such X-ray analysis is currently unavailable. However, it has been possible to answer the above question affirmatively by means of spectroscopic analysis of the phytochrome molecule (Sarkar & Song, 1982a), as is discussed below.

There is at least one tryptophan residue at or near the chromophore-binding crevice (Song *et al.*, 1979), and it should be possible to determine the relative orientation of the chromophore with respect to the tryptophan (Trp) molecule in both the Pr and Pfr forms of phytochrome. This can be done by measuring the energy transfer from the excited state of tryptophan to the chromophore via the Förster-type mechanism. The critical energy transfer distance (R_0) at which energy transfer is 50% has been calculated from the spectral overlap between the energy emitting donor (tryptophan) and the absorbing acceptor (chromophore). The values of R_0 for the ^{1}Trp–^{1}Pr and ^{1}Trp–^{1}Pfr transfer pairs (where superscript 1 stands for singlet excited state) are approximately equal, namely, 2.8 nm (Song *et al.*, 1979). The values of R_0 for the ^{3}Trp–^{1}Pr ($B_{x,y}$) and ^{3}Trp–^{1}Pr (Q_x) energy transfer processes are 3.128 and 3.634 nm, respectively, whereas those for the ^{3}Trp–^{1}Pfr ($B_{x,y}$) and ^{3}Trp–^{1}Pfr (Q_x) pairs are 3.187 and 3.582 nm, respectively. Two sets of values for R_0 for the energy transfer from the phosphorescent ^{3}Trp to Pr and Pfr are calculated because the phosphorescence spectrum of Trp in phytochrome overlaps with both the near UV ($B_{x,y}$) and visible (Q_x) absorption bands of the phytochrome chromophore. From the calculated values of R_0 based on the spectral overlaps and the assumption that the donor and acceptor transition dipoles are randomly oriented, one can predict that the probability of energy transfer is approximately the same for Pr and Pfr.

If the latter assumption is valid, one expects no major differences in the probability of energy transfer from tryptophan to the chromophore of Pr and Pfr since the critical distances are essentially the same. On the other hand, if the right binding of the Pr chromophore at the protein crevice involving tryptophan residue(s) and the hydrophobic model for Pfr (Fig. 4) are valid (in this case, the donor and acceptor dipoles are not randomly distributed), the probability of energy transfer from tryptophan to the chromophore will be different for Pr and Pfr. This has been tested in terms of energy transfer from the phosphorescent triplet state of tryptophan to the chromophore at 77 K (Sarkar & Song, 1982*a*).

The fluorescence lifetime of Pr at 77 K is less than 1 ns when excited with red light (exclusively absorbed by the chromophore). When the Pr molecule is excited with UV (290 nm is almost exclusively absorbed by tryptophan), the fluorescence emission from the Pr chromophore can be observed with a half-life of 0.17 s. This long fluorescence lifetime therefore indicates that the fluorescence emission at 680 nm recorded with 290 nm excitation is a delayed fluorescence arising from the energy transfer from the phosphorescent tryptophan (half-life 0.2 s) to the Pr

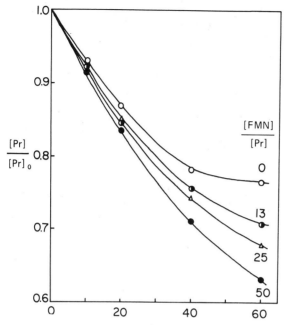

Fig. 7. The phototransformation of phytochrome (Pr, $4\,\mu M$) in 0.1 M phosphate buffer, pH 7.8, with blue light (442 nm, 0.6 W m^{-2}) at 278 K, in the presence of molar excesses of FMN added. [Pr]$_0$ and [Pr] are concentrations before and after irradiation with 442 nm light which is exclusively absorbed by FMN. (Redrawn from Sarkar & Song, 1982*b*.)

chromophore; that is, the fluorescence is sensitized by the triplet tryptophan. A similar delayed fluorescence from Pfr was not observed because of either low fluorescence quantum yield or insensitivity of the detector system used. However, from the fact that the half-life of the tryptophan phosphorescence in Pfr is 1.28 s, it can be concluded that the energy transfer is significantly suppressed in the Pfr molecule.

These results strongly suggest that the relative orientations of the chromophore relative to the tryptophan residue(s) are distinctly different for Pr and Pfr, in support of the model described in the previous section. The phosphorescence of indoles is out-of-plane polarized (Song & Kurtin, 1969). Thus, in the Pfr molecule either the planes of the chromophore and tryptophan have a parallel orientation or the donor/acceptor transition dipoles are nearly perpendicular to each other.

Chromophore reorientation has also been extrinsically probed by FMN (Sarkar & Song, 1981*b*). Figure 7 shows the effect of blue-light irradiation on the phototransformation of Pr in the presence of exogenous FMN, which exclusively absorbs actinic blue light. The rate of

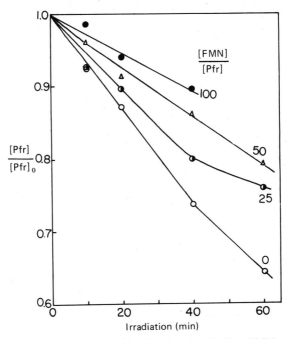

Fig. 8. The phototransformation of Pfr ($4\,\mu$M) in phosphate buffer, pH 7.8, with blue light ($442\,$nm, $0.6\,$W m^{-2}) as a function of molar excess of FMN which exclusively absorbs $442\,$nm light. (Redrawn from Sarkar & Song, 1982b.)

the disappearance of Pr upon irradiation with blue light increases with the increasing molar excess of FMN over Pr. However, at higher concentrations of FMN, the rate slows down and ultimately becomes slower than the control (irradiation of Pr with blue light alone, without added FMN).

Figure 8 shows the effect of blue-light irradiation on the phototransformation of Pfr in the presence of FMN. This rate decreases with increasing concentration of FMN. Phototransformation studies with red and far-red light showed that FMN did not affect the Pr $\rightarrow$ Pfr phototransformation but the Pfr $\rightarrow$ Pr photoreaction was inhibited by about 15% at higher concentrations of FMN.

The above results can be explained in terms of energy transfer within the flavin (sensitizer)/phytochrome (acceptor) complex. The formation of a Pr–FMN complex is readily detected with a sensitive, single beam spectrophotometer (Song, Sarkar, Kim & Poff, 1981), and kinetic studies have established that the flavin moiety is complexed to the Pr apoprotein. Evidence for the energy transfer is also provided by measurements of the fluorescence lifetimes of flavin/phytochrome mix-

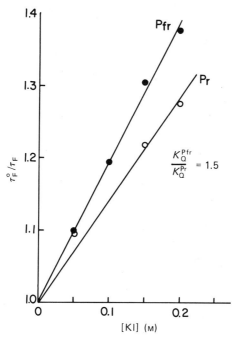

Fig. 9. Stern–Volmer plots for the fluorescence quenching of tryptophan residues in Pr and Pfr at 278 K, as a function of KI concentration. The ratio of the quenching constants for the two plots is 1.5, indicating that the tryptophan residues in Pfr are more exposed than those in Pr. (Redrawn from Sarkar & Song, 1982a.)

tures. For example, a mixture of $3\,\mu M$ Pr and $15\,\mu M$ flavin (as riboflavin or FMN) yields two component lifetimes, one of which shows a 3.34 ns lifetime (4.63 for free flavin). On the other hand, a mixture of $3\,\mu M$ Pfr and $15\,\mu M$ flavin exhibits only one component lifetime (4.89 ns). This result suggests that there is efficient energy transfer from the singlet excited state of flavin to the chromophore of Pr, but not to that of Pfr. This difference can be explained on the basis of the chromophore re-orientation in Pfr, thus making the sensitizer and acceptor transition dipoles orthogonal in the Pfr form. It should be emphasized that the critical energy transfer distances, at which energy transfer from flavin to phytochrome is 50%, are essentially identical for the flavin–Pr and flavin–Pfr pairs: 2.58 and 2.22 nm, respectively. The fact that the energy transfer for the latter pair does not occur suggests that the sensitizer and acceptor transition dipoles involved are orthogonal. This conclusion is, therefore, consistent with the tryptophan data described earlier.

From the foregoing discussion, it can be predicted that the Pr $\rightarrow$ Pfr phototransformation by blue light would accelerate in the presence of

FMN, which absorbs actinic light, while the Pfr → Pr phototransformation under the same conditions would be unaffected by FMN. The former prediction has been confirmed (Fig. 7). However, the Pfr → Pr phototransformation with blue light is inhibited by FMN. This can be readily explained on the basis of a screening effect of bound and free FMN, which may not efficiently transfer excitation energy to the Pfr chromophore. In addition, it is possible that excess FMN competitively inhibits the photoreaction, as the flavin may also occupy the original chromophore binding site vacated upon Pr → Pfr phototransformation (Fig. 4), analogous to the competitive inhibition of the Pfr phototransformation by ANS described earlier (Hahn & Song, 1981).

As a consequence of the chromophore re-orientation away from the binding crevice and tryptophan residue(s), one would expect a selective exposure of tryptophan residues in the Pfr molecule. This prediction is borne out in the data presented in Fig. 9, which shows Stern–Volmer quenching of tryptophan fluorescence as a function of KI concentration. It can be seen that the tryptophan fluorescence of Pfr is quenched by KI 1.5 times more readily than that of Pr, suggesting that the tryptophan residues in the former are markedly more exposed than those in the Pr molecule. A quantitative estimate of the fraction of tryptophan residues exposed can be obtained from a modified Stern–Volmer fluorescence quenching plot, as shown in Fig. 10, which presents plots of the fluorescence quenching data in accordance with the following equation (Lehrer, 1971):

$$I_F^0 / \Delta I_F = (1/f_{acc}) + (1/f_{acc}[KI]K_Q),$$

where I_F^0 is the fluorescence intensity in the absence of KI, ΔI_F represents the decrease in fluorescence intensity as a function of KI concentration, and f_{acc} stands for the fraction of exposed or accessible tryptophan residues. From the data shown, the values of f_{acc} are calculated to be 0.45 and 0.72 in Pr and Pfr, respectively. Thus, the fluorescence quenching results are consistent with the hydrophobic model of phytochrome (Pfr, Fig. 4).

Is there a gross conformational difference between Pr and Pfr?

As pointed out in the Introduction, a gross conformational change of the phytochrome protein upon Pr → Pfr phototransformation may have profound physiological implications. We will now examine the question of whether or not there is a gross conformational change in the phytochrome protein.

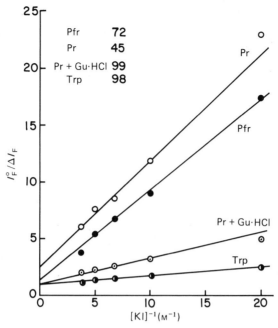

Fig. 10. Double-reciprocal plots for the fluorescence quenching of free tryptophan (Trp) residues in denatured (Pr + guanidine hydrochloride) and native Pr and Pfr in phosphate buffer, pH 7.8, as a function of KI concentration at 278 K. The percentage f_{acc} values are shown. (Redrawn from Sarkar & Song, 1982*a*.) f_{acc} = fraction of exposed Trp residues.

From the previous two sections, it would be expected that the chromophore re-orientation or relocation involved in the phototransformation of phytochrome entails at least local conformational changes of the peptide segments in the vicinity of the chromophore binding site. But, is there a *gross* conformational change of the protein? Is there a marked difference in shape between Pr and Pfr molecules? These questions have been dealt with in a recent study from this laboratory (Song, 1980*b*; Sarkar & Song, 1982*a*).

The Pr molecule has been covalently labelled with one or two molecules of pyrene maleimide (PMI), which specifically reacts with the sulfhydryl group to yield a fluorescent adduct, as shown in Fig. 11. In spite of its long fluorescence lifetime, the PMI label did not yield the rotational relaxation time expected of large molecular weight phytochrome (120 000 as monomer and 240 000 as dimer). The short relaxation times resolved apparently reflect a segmental freedom in the PMI–peptide linkage. Nonetheless, the fluorescence anisotropy, lifetime and rotational relaxation time are virtually identical for Pr and Pfr (Fig. 11). Since these parameters are usually sensitive to local and gross

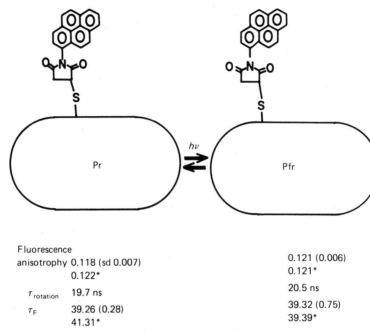

Fluorescence anisotrophy	0.118 (sd 0.007)	0.121 (0.006)
	0.122*	0.121*
$\tau_{rotation}$	19.7 ns	20.5 ns
τ_F	39.26 (0.28)	39.32 (0.75)
	41.31*	39.39*

Fig. 11. Fluorescence anisotropy, rotational relaxation times (τ rotation) and lifetimes (τ_F) of Pr and Pfr labelled with pyrene-N-maleimide (PMI), in 0.1 M phosphate buffer, pH 7.8, at 298 K. 2 PMI per phytochrome. (H. K. Sakar & P.-S. Song, unpublished data.)

conformational environments, the similarity observed between the two forms of phytochrome–PMI suggest that the gross protein conformation remains unchanged in going from Pr to Pfr.

Fluorescamine turned out to be a more useful fluorescence probe for the rotational relaxation measurements, as the amino group-specific fluorescence probe revealed long relaxation times of 100.2 and 101.9 ns for Pr and Pfr respectively, along with shorter component relaxation times (Fig. 12). Again, the results shown in Fig. 12 are virtually identical for Pr and Pfr, suggesting that the shapes of Pr and Pfr proteins are hydrodynamically equivalent.

From these observations, we are led to conclude that the predominant feature in the phototransformation of Pr to Pfr is the chromophore reorientation and the development of a hydrophobic surface in the latter. However, segmental or local conformational changes are likely to result from the chromophore reorientation, as discussed earlier. We have recently recorded the 360 MHz NMR spectra of Pr and Pfr (small molecular weight), in collaboration with Professor Kevin Smith. Preliminary results suggest that several amino acid residues and the chromophore become flexible in the Pfr form. In particular, the tyrosyl

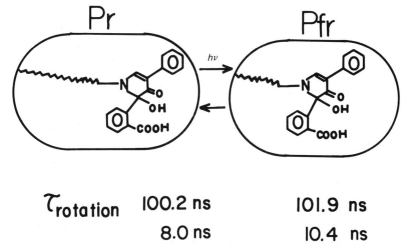

$\mathcal{T}_{\text{rotation}}$ 100.2 ns   101.9 ns

               8.0 ns   10.4 ns

Fig. 12. Rotational relaxation times for Pr and Pfr labeled with fluorescamine (1:2) in 0.1 M sodium phosphate buffer, pH 7.8, at 293 K. Rotational relaxation times were calculated by measuring steady state polarization values of the labeled phytochrome, Pr and Pfr, respectively, in phosphate buffer and in phosphate buffer containing 60% (w/v) sucrose, using Perrin's equation. The second set of values were measured by the differential phase fluorometry method.

and histidyl residues in the Pfr molecule gained a degree of flexibility resulting in the appearance of their proton NMR resonances in the region of 6–9 ppm. The fact that more protons become exchangeable with T^+ in the Pfr solution in T_2O than in the Pr solution (Hahn & Song, 1982) is consistent with the NMR result. These preliminary results support the recent findings that histidyl and tyrosyl residues are more accessible in the Pfr form than in the Pr form (Hunt & Pratt, 1981) and that the tyrosyl hydroxyl proton is exchanged in the Pfr form (Song *et al.*, 1981). An additional line of evidence for the exposure of selective amino acid residues and the chromophore comes from the observation that the photocycling of phytochrome between the Pr and Pfr forms in D_2O enhances the D_2O solvent isotope effects on the fluorescence of Pr, and the rate of photo- and dark-reversion of Pfr (Sarkar & Song, 1981*a*).

Interactions of Pfr with model receptors

As the hydrophobic surface develops on the Pfr protein (Fig. 4), specific interactions between Pfr and its receptor(s) at the hydrophobic surface can be envisaged. While such a Pfr receptor has not yet been found *in vivo*, it is instructive to study the interactions between Pfr and model

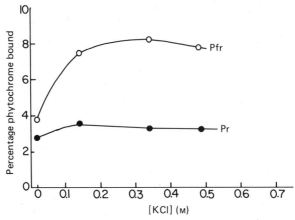

Fig. 13. The binding of phytochrome to unilamellar liposomes as a function of ionic strength (KC1) in 20 mM phosphate buffer, pH 7.2, at 300 K. (Redrawn from Kim & Song, 1981.)

compounds that specifically bind at the hydrophobic area. We have already discussed the specific interactions between the hydrophobic surface and ANS. The proposition that ANS binds at the chromophore site is supported by the fact that dithionite and other reducing agents compete with ANS for the interaction site in accelerating the dark-reversion of Pfr (Hahn & Song, 1981). It should also be noted that ANS does not chemically bleach the chromophore itself.

There is a growing body of evidence that the initial event in phytochrome action involves interactions between Pfr and cellular membrances (Marmé, 1977). However, the nature of the interactions between phytochrome and membranes is not well understood. In this section, a detailed study (Kim & Song, 1981) of the binding properties of phytochrome using liposomes as a model system is reviewed.

Cholesterol in multilamellar liposomes enhances the binding of both the Pr and Pfr forms of phytochrome, with a preferential binding of Pfr over Pr. An examination of the binding of Pfr to unilamellar liposomes suggests that the phytochrome-bound liposomes undergo fusion to produce larger-diameter liposomes (Kim & Song, 1981). The binding of phytochrome to unilamellar liposomes is also drastically affected by ionic strength (Fig. 13). The binding of Pfr generally increases with ionic strength, whereas the binding of Pr slowly declines, especially at lower temperatures (for example, 276 K). These results suggest that the interaction of Pfr with liposomes has a hydrophobic character, but the Pr binding is not exclusively an electrostatic interaction.

To probe the binding topography of the phytochrome–liposome, the

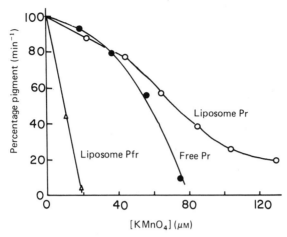

Fig. 14. The initial rates of oxidation of the phytochrome chromophore as a function of potassium permanganate in 20 mM phosphate buffer, pH 7.2, at 276 K. (I. S. Kim & P. S. Song, unpublished.)

photo- and dark-reversion of lipsosome-bound Pfr can be followed. While the phototransformation of bound Pr to Pfr is enhanced, the reversions are significantly inhibited by the binding (Kim & Song, 1981). Thus, these kinetic behaviors are similar to those of the ANS-bound phytochrome, as described earlier. One can conclude that the hydrophobic surface of Pfr is involved in the binding to the lipid phase of liposomes. It is also noteworthy that, while the binding facilitates the forward phototransformation of phytochrome and retards the reversion of Pfr, the accelerating effect of dithionite on the dark-reversion of bound Pfr is minimal, compared with its effect on free Pfr. Thus the binding of Pfr to liposomes makes it difficult for dithionite to reach its reaction site which may be buried in the Pfr–liposome complex (Kim & Song, 1981).

If the hydrophobic site of Pfr is involved in the binding to the lipid bilayer core of liposomes, it is possible that the chromophore may be exposed in the aqueous phase. Figure 14 shows the degradation of the chromophore in free and bound phytochrome molecules. It can be seen that the chromophore in liposome-bound Pfr is more readily oxidized by permanganate than is the chromophore in liposome-bound or free Pr. Similar results have been obtained with borohydride reduction. The rapid oxidation of the chromophore in the Pfr–liposome is indicative of an exposed chromophore which is readily accessible to oxidation.

Interestingly, there is no preferential binding of Pfr to oat protoplasts (right-side-out) (Kim & Song, 1981). Furthermore, the binding of Pfr to the protoplast exterior is inhibited at higher ionic strength, suggesting that it is electrostatic in nature. This is an indication that the hydrophobic model of Pfr may be specific to certain membranes.

Concluding remarks

Although many facets of the phytochrome phototransformation and the interactions between Pfr and its receptor(s) remain to be established, one significant feature has been revealed: the Pfr molecule is characterized by a hydrophobic surface that becomes exposed as the result of the phototransformation of phytochrome from Pr. This hydrophobic surface exhibits specificities in its interactions with model receptor compounds and liposomes. In summary, I present the side view of this model in Fig. 15 (see Fig. 4 for the top view).

In this review, we have focused our attention on the question of chromophore re-orientation in the phototransformation of phytochrome. Such a chromophore re-orientation is analogous to the ligand binding and dissociation processes of proteins. As mentioned earlier, it would be expected that the chromophore re-orientation entails at least local conformational changes of the peptide segments in the chromophore crevice and its vicinity. However, a conformational change away from the chromophore crevice and its vicinity (Hunt & Pratt, 1981; Pratt, 1982) is unlikely without chromophore movement. The fact that the modification of amino acid residues on the Pfr form by chemical reagents (for example, histidyl residues) does not quench the photoreversibility of phytochrome (Hunt & Pratt, 1981) cannot be used as evidence against the chromophore re-orientation model, since no kinetic analysis of the chemical modification effects has been carried out.

The phytochrome-mediated movement of *Mougeotia* chloroplasts in response to polarized light can be best explained in terms of chromophore re-orientation (Haupt & Weisenseel, 1976), thus lending support for the proposed model described in this paper.

A number of biologically significant consequences arise from the proposed model. These include (*a*) strong spectral perturbations (including bleaching) of the Pfr absorption spectrum (relative to that in free solution) as the result of interactions between Pfr and its receptor component, (*b*) a shift of the photostationary equilibrium as the result of the modification of rates for the photo- and dark-transformations of bound phytochrome, and (*c*) non-specific hydrophobic substances in

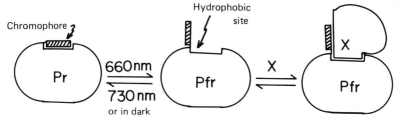

Fig. 15. The side view of the hydrophobic model of Pfr. X represents the Pfr receptor which is yet to be identified. (Redrawn from Song *et al.*, 1979; Song, 1980*a*.)

plant cells are likely to bind to Pfr, resulting in spectral denaturation of the chromophore, and perhaps leading to changes in the sensitivity of plants to light stimuli.

The work described in this article is supported by the Robert A. Welch Foundation (D-182) and the National Science Foundation (PCM79-06806). Technical contributions of Dr In-Soo Kim, Mr Hemanta K. Sarkar, and Mr Tae-Ryong Hahn are greatly appreciated.

References

BRANDLMEIER, T., LEHNER, H. & RÜDIGER, W. (1981). Circular dichroism of the phytochrome chromophores in native and denatured state. *Photochem. Photobiol.*, **34**, 69–73.

BRIGGS, W. & RICE, H. (1972). Phytochrome: chemical and physical properties and mechanism of action. *Ann. Rev. Plant Physiol.*, **23**, 293–334.

BUTLER, W. L., SIEGELMAN, H. W. & MILLER, C. O. (1964). Denaturation of phytochrome. *Biochemistry*, **3**, 851–57.

HAHN, T. R., KANG, S. S. & SONG, P. S. (1980). Difference in the degree of exposure of chromophores in the Pr and Pfr forms of phytochrome. *Biochem. Biophys. Res. Commun.*, **97**, 1317–23.

HAHN, T. R. & SONG, P. S. (1981). Hydrophobic properties of phytochrome as probed by 8-anilinonaphthalene-1-sulfonate fluorescence. *Biochemistry*, **20**, 2602–9.

HAHN, T. R. & SONG, P. S. (1982). Molecular topography of phytochrome as deduced from the tritium-exchange method. *Biochemistry*, **21**, 1394–99.

HAUPT, W, & WEISENSEEL. M. H. (1976). Physiological evidence and some thoughts on localised responses. Intracellular localisation and action of phytochrome. In: *Light and Plant Development*. ed. H. Smith. pp 63–74. London: Butterworths.

HUNT, R. E. & PRATT, L. H. (1981). Physicochemical differences between the red- and far-red-absorbing forms of phytochrome. *Biochemistry*, **20**, 941–45.

JAFFE, M. (1970). Evidence for the regulation of phytochrome-mediated processes in bean roots by the neurohumor, acetylcholine. *Plant Physiol.*, **46**, 768–77.

JUNG, J., SONG, P. S., SWANSON, R., PAXTON, R., EDELSTEIN, M. & HAZEN, E. (1980). Molecular topography of the phycocyanin photoreceptor from *Chroomonas* species. *Biochemistry*, **19**, 24–32.

KILLILEA, S. D., O'CARRA, P. & MURPHY, R. F. (1980). Structures and apoprotein linkages of phycoerythrobilin and phycocyanobilin. *Biochem. J.*, **187**, 311–20.

KIM, I. S. & SONG, P. S. (1981). Binding of phytochrome to liposomes and protoplasts. *Biochemistry*, **20**, 5482–9.

LAGARIAS, J. C. & RAPPOPORT, H. (1980). Chromopeptides from phytochrome. The structure and linkage of the Pr form of the phytochrome chromophore. *J. Am. Chem. Soc.*, **102**, 4821–28.

LEHRER, S. S. (1971). Solute perturbation of protein fluorescence. The quenching of the tryptophyl fluorescence of model compounds and of lysozyme by iodide ion. *Biochemistry*, **10**, 3254–63.

MARMÉ, D. (1977). Phytochrome: membranes as possible sites of primary action. *Ann. Rev. Plant Physiol.*, **28**, 173–98.

MOHR, H. (1972). *Lectures on Photomorphogenesis.* New York: Springer-Verlag.

PRATT, L. H. (1982). Phytochrome: the protein moiety. *Ann. Rev. Plant Physiol.*, **33**, 557–82.

PRATT, L. H. (1978). Molecular properties of phytochrome. *Photochem. Photobiol.*, **27**, 81–105.

QUAIL, P. H. (1978). Irradiation-enhanced phytochrome pelletability in *Avena*: in vivo development of a potential to pellet and the role of Mg^{2+} in its expression in vitro. *Photochem. Photobiol.*, **27**, 147–53.

RÜDIGER, W. (1980). Phytochrome, a light receptor of plant photomorphogenesis. *Struct. Bond.*, **40**, 101–40.

SARKAR, H. K. & SONG, P. S. (1981a). Phototransformation and dark reversion of phytochrome in deuterium oxide. *Biochemistry*, **20**, 4315–20.

SARKAR, H. K. & SONG, P. S. (1981b). Abstract of the Bischofsmais Symposium on Light-Mediated Plant Development, April, 1981.

SARKAR, H. K. & SONG, P. S. (1982a). Phototransformation of phytochrome as probed by the intrinsic tryptophan residues. *Biochemistry*, **21**, 1967–72.

SARKAR, H. K. & SONG, P. S. (1982b). Blue light induced phototransformation of phytochrome in the presence of flavin. *Photochem. Photobiol.*, **35**, 243–6.

SATTER, R. & GALSTON, A. (1977). Light, clocks and ion flux: an analysis of leaf movement. In: *The Chemistry and Biochemistry of Plant Pigments*, ed. T. Goodwin, pp. 681–735.

SMITH, W. O. JR. (1981). Probing the molecular structure of phytochrome with immobilized Cibacron blue 3GA and blue dextran. *Proc. Natl. Acad. Sci., USA*, **78**, 2977–80.

SONG, P. S. (1980a). Molecular aspects of photoreceptor function: phytochrome. In: *Photoreception and Sensory Transduction in Aneural Organisms*, ed. F. Lenci & G. Colombetti, pp. 235–40. New York: Plenum Press.

SONG, P. S. (1980b). What makes Pfr physiologically active? Proc. Intern. Congress on Photobiology, Strasbourg, July 21–26, 1980, Abstract No. 100.

SONG, P. S. & CHAE, Q. (1979). The transformation of phytochrome to its physiologically active form. *Photochem. Photobiol.*, **30**, 117–23.

SONG, P. S. CHAE, Q. & GARDNER, J. (1979). Spectroscopic properties and chromophore conformations of the photomorphogenic receptor: phytochrome. *Biochim. Biophys. Acta*, **576**, 479–95.

SONG, P. S. & KURTIN, W. E. (1969). A spectroscopic study of the polarized luminescence of indoles. *J. Am. Chem. Soc.*, **91**, 4892–906.

SONG, P. S., SARKAR, H. K., KIM, I. S. & POFF, K. L. (1981). Primary photoprocesses of undegraded phytochrome excited with red and blue light at 77 K. *Biochim. Biophys. Acta*, **635**, 369–82.

HALOBACTERIA: THE ROLE OF RETINAL–PROTEIN COMPLEXES

EILO HILDEBRAND

Institut für Neurobiologie, Kernforschungsanlage Jülich, D-5170 Jülich, GFR

Introduction

For a long time the only known retinal–protein complexes were the visual pigments of animals, and it was rather surprising when Oesterhelt & Stoeckenius (1971) discovered a 'rhodopsin-like' retinal–protein compound in the extremely halophilic bacterium *Halobacterium halobium*. Because of its similarity to the rhodopsins they called the substance bacteriorhodopsin. But it soon became apparent that bacteriorhodopsin functions as a light-driven proton pump, that is as a light-energy converter, rather than as a photosensory pigment analogous to rhodopsin (Oesterhelt & Stoeckenius, 1973). Since that time bacteriorhodopsin has been intensively studied in many laboratories all over the world, and it is presently one of the best-known membrane proteins.

More recently another 'rhodopsin' has been found in *Halobacterium*, which functions primarily as a light-driven sodium pump. In addition retinal–protein complexes are known to play a role in photosensory processes in *Halobacterium*, thus by means of retinal pigments the organism can use light as a source of information about its environment.

Bacteriorhodopsin

Bacteriorhodopsin forms irregular patches, the so-called purple membrane, within the cell membrane of *Halobacterium* (Fig. 1). The pigment molecules are arranged in the purple membrane in a rigid two-dimensional hexagonal lattice. Bacteriorhodopsin is the only protein of the purple membrane and represents about 75% of its dry weight; the rest is made up by lipids. Bacteriorhodopsin can be easily isolated and purified and has been studied by various biochemical and biophysical methods. The chromophore, retinal, either in its 13-*cis* form or in the *trans* configuration, is covalently bound as a protonated retinylidene Schiff base to the ε-amino group of a lysine residue of the protein, bacterio-opsin (Fig. 2). Bacterio-opsin consists of 247 amino

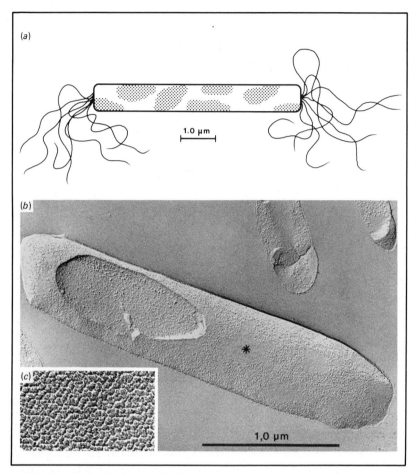

Fig. 1. *Halobacterium halobium.* (*a*) Schematic drawing which shows purple membrane patches and bipolar flagellation. (From Hildebrand, 1977.) (*b*) Electron micrograph of a freeze-fractured preparation. The star indicates one of the purple membrane patches. (*c*) Purple membrane structure at higher magnification. (Courtesy of Dr W. Schröder.)

acid residues, the sequence of which has recently been determined (Ovchinnikov, Abdulaev, Feigina, Kiselev & Lobanov, 1979); retinal is attached to lysine-216 of the amino acid chain (Bayley *et al.*, 1981). The molecular weight of bacteriorhodopsin is 26 500. The chromophore is arranged in the purple membrane in groups of three molecules; each group consists of 7α-helices of 3.5–4 nm in length which traverse the bilipid layer of the cell membrane (Henderson, 1977).

Whilst unprotonated retinylidene Schiff bases exhibit absorption maxima around 370 nm, the main absorption band of bacteriorhodopsin is considerably red-shifted, most probably because of Schiff base

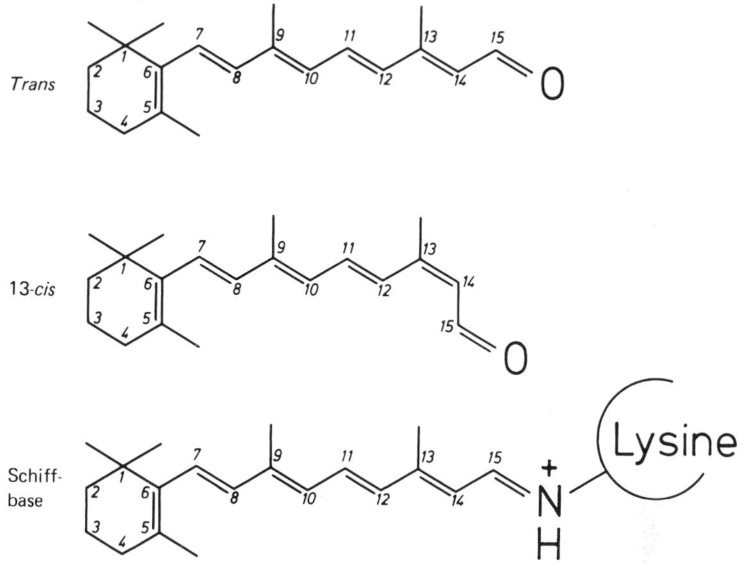

Fig. 2. Retinal isomers of bacteriorhodopsin and protonated retinylidene Schiff-base linkage to a lysine residue of bacterio-opsin.

protonation and non-covalent secondary interactions between the chromophore and aromatic amino acid residues of the protein moiety (for review see Ottolenghi, 1980). Light-adapted bacteriorhodopsin contains all retinal in the *trans* configuration and has its absorption peak at 568 nm. Dark adaptation leads to an equilibrium between the *trans* and the 13-*cis* isomers, but in this way no more than 50% of the retinal can be isomerised and the absorption maximum is shifted to 558 nm. Bacteriorhodopsin, when reconstituted in the dark from bacterio-opsin and 13-*cis* retinal contains exclusively the 13-*cis* isomer and its absorption maximum is at 548 nm (Fig. 3).

When a photon is absorbed by *trans* bacteriorhodopsin, the pigment undergoes a rapid photocycle through several short-lived intermediates (Fig. 4) which have been detected by low temperature spectroscopy and flash spectroscopy. Some similarities of early photoproducts with those of the visual pigment rhodopsin exist with respect to a bathochromic absorption shift. At present it is not clear whether or not a *trans–cis* isomerisation of the retinal takes place in the photocycle. Some recent results point towards a 13-*cis* configuration in the intermediate that absorbs at 411 nm (Braiman & Mathies, 1980; Mowery & Stoeckenius, 1981).

The photocycle of *trans* bacteriorhodopsin is accompanied by a release and uptake of protons. This is considered to be the basis of

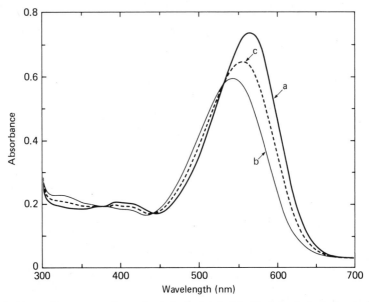

Fig. 3. Absorption spectra of *trans* bacteriorhodopsin (a), 13-*cis* bacteriorhodopsin (b), and of the 1:1 mixture of both obtained after dark adaptation of *trans* bacteriorhodopsin (c). (From Dencher, Rafferty & Sperling, 1976.)

proton translocation (H^+ pump activity) across the membrane, which can be monitored as an acidification of the medium when a suspension of halobacteria is irradiated by light of the appropriate wavelength (for review see Eisenbach & Caplan, 1979). The H^+ gradient thereby established contributes – besides the respiratory activity – to the overall membrane potential of at least 200 mV (inside negative) of the cell (for review see Lanyi, 1978). Some authors consider bacteriorhodopsin as an emergency power generator under conditions of low oxygen supply.

The transmembrane H^+ gradient is used by the cell to drive Na^+ outward transport *via* an antiporter system, and the Na^+ gradient in turn drives amino acid uptake (Lanyi, 1979). The uptake of K^+ (intracellular concentration up to 4 M), however, is driven by the potential gradient, $\Delta\psi$, across the membrane (Garty & Caplan, 1977; Wagner, Hartmann & Oesterhelt, 1978).

The biosynthesis of bacteriorhodopsin from bacterio-opsin and retinal takes place in the so-called brown membrane (Sumper, Reitmeier & Oesterhelt, 1976), where the aggregation of bacteriorhodopsin mono-mers to trimers can also be observed. Monomers are functionally active in H^+ translocation with an efficiency equal to that of crystalline bacteriorhodopsin (Dencher & Heyn, 1979).

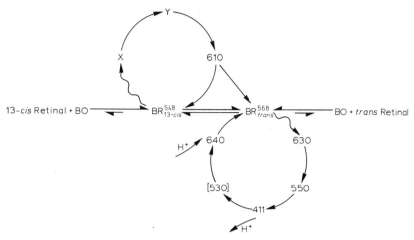

Fig. 4. Reaction cycles of 13-*cis* bacteriorhodopsin and *trans* bacteriorhodopsin and their interrelationship. Intermediate products are characterised by the maxima of their difference absorption spectra (spectrum of the intermediate minus spectrum of the initial pigment: 13-*cis* bacteriorhodopsin or *trans* bacteriorhodopsin, respectively). Wavy lines indicate photoreactions. BR = bacteriorhodopsin, BO = bacterio-opsin. (Modified after Lozier, Bogomolni & Stoeckenius, 1975, and Sperling, Rafferty, Kohl & Dencher, 1979.)

13-*cis* bacteriorhodopsin passes through a separate photocycle when a photon is absorbed (Fig. 4). The only photoproduct which is so far spectroscopically characterised shows a red-shift and returns thermally to the initial form. The 13-*cis* cycle is mutually connected by several pathways with the *trans* cycle (Sperling *et al.*, 1979). The physiological role of the 13-*cis* cycle is not yet known.

The most important aspects of bacteriorhodopsin structure and function have recently been reviewed by Stoeckenius (1980).

Halorhodopsin, P_{588}

The study of mutant strains that are unable to synthesise bacteriorhodopsin has recently led to the discovery of a second light-driven transport system in the cell membrane of *Halobacterium*. Red cells (strain R_1mR) isolated by Matsuno-Yagi & Mukohata (1977), ET 15 cells (Weber & Bogomolni, 1981), and L 33 cells, isolated by Lanyi (Wagner, Oesterhelt, Krippahl & Lanyi, 1981), which contain no detectable amounts of bacteriorhodopsin, lack all light-dependent H^+ extrusion, but show a H^+ uptake upon illumination (Matsuno-Yagi & Mukohata, 1977; Weber & Bogomolni, 1981; Wagner *et al.*, 1981). This H^+ uptake is not blocked but rather is enhanced by uncouplers (proton ionophores) such as FCCP (Fig. 5), and can therefore be regarded as

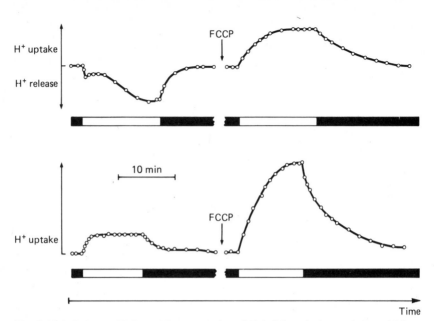

Fig. 5. Light-induced pH changes in a suspension of *H. halobium* before and after addition of the protonophore FCCP (*p*-trifluoromethoxycarbonylcyanide-phenylhydrazone). *Upper trace:* strain ET 1001, carrying bacteriorhodopsin and halorhodopsin. *Lower trace:* strain ET 15, carrying halorhodopsin, but lacking bacteriorhodopsin. Illumination is indicated by the light rectangles below the traces. (After Weber & Bogomolni, 1981.)

passive transport (Weber & Bogomolni, 1981). Lindley & MacDonald (1979) reported that in cell-envelope vesicles prepared from R_1mR cells a membrane potential (inside negative) is created and Na^+ ions are extruded upon illumination. Both membrane potential and H^+ uptake are abolished by valinomycin plus K^+, but Na^+ efflux is not: even in the absence of a potential difference across the membrane. The authors therefore conclude that a light-dependent Na^+ transport, as the primary process, contributes to the membrane potential which in turn drives passive H^+ uptake and ATP synthesis (Wagner *et al.*, 1981). This view has been confirmed by further experiments (Greene & Lanyi, 1979; MacDonald, Greene, Clark & Lindley, 1979).

The action spectrum for light-induced H^+ uptake shows a red shift of about 20 nm with respect to that for bacteriorhodopsin-mediated H^+ release (Matsuno-Yagi & Mukohata, 1977, 1980; Green, MacDonald & Perrault, 1980; Weber & Bogomolni, 1981). The maximum activity is found near 590 nm (Fig. 6). When R_1mR cells are grown in the presence of nicotine, which blocks β-carotene biosynthesis from lycopene (Howes & Batra, 1970) and thereby prevents the synthesis of retinal (Sumper, Reitmeier & Oesterhelt, 1976), the proton uptake is completely abol-

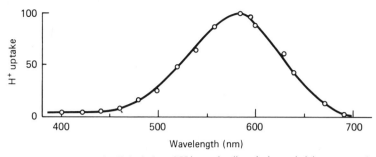

Fig. 6. Action spectrum for light-induced H^+ uptake (in relative units) in a suspension of envelope vesicles prepared from strain ET 15 which contains halorhodopsin, but lacks bacteriorhodopsin. (From Weber & Bogomolni, 1981.)

ished, but can be restored after addition of *trans* retinal (Matsuno-Yagi & Mukohata, 1980). Using the strain ET 15, Lanyi & Weber (1980) were able to detect spectroscopically the pigment associated with the light-driven Na^+ pump. Envelope vesicles containing the pigment were bleached in the presence of hydroxylamine, and the pigment-like bacteriorhodopsin could be reconstituted after adding *trans* retinal to the suspension (Fig. 7). These results clearly demonstrate that retinal is the chromophore. The pigment, which is now called halorhodopsin (Mukohata & Kaji, 1981*a*) or, according to its absorption maximum, P_{588} (Weber & Bogomolni, 1981), obviously represents a second light-energy-converting transport system in *Halobacterium* in addition to bacteriorhodopsin. Its content in R_1 (or wild type) cells is approx. 5% that of bacteriorhodopsin (Lanyi & Weber, 1980). Apparently it does not undergo *trans–cis* isomerisation when kept in the dark (Lanyi & Weber, 1980). The isomeric conformation of retinal in halorhodopsin is not yet known. Recently, Weber & Bogomolni (1981) found that halorhodopsin undergoes a photocycle within about 10 ms, like bacteriorhodopsin, and they could detect four intermediate products with absorption maxima which differ from those of bacteriorhodopsin intermediates.

The physiological role of halorhodopsin is not yet fully understood. Under normal circumstances in wild-type cells its contribution to ATP synthesis seems to be small. Na^+ outward transport is predominantly driven by an H^+/Na^+ antiport which is energised by respiration or by the bacteriorhodopsin-mediated H^+ pump (Luisi, Lanyi & Weber, 1980). Recently Mukohata & Kaji (1981*b*) discussed the possibility that halorhodopsin acting as a light-driven Na^+ pump may have been present first and that bacteriorhodopsin has evolved later as a H^+ pump for intracellular pH regulation. The properties and functions of

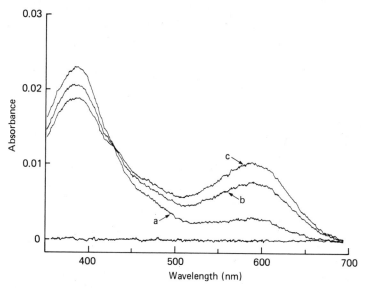

Fig. 7. Reconstitution of halorhodopsin in envelope vesicles from *Halobacterium*, strain ET 15. Vesicles were bleached in the presence of hydroxylamine. The suspension was then divided into two portions and *trans* retinal in methanol was added to the sample cuvette and an equal amount of methanol to the reference cuvette. Difference spectra were obtained 2 min (a), 8 min (b), and 24 min (c) after the addition of retinal. The absorption peak at 380 nm is caused by free retinal. (From Lanyi & Weber, 1980.)

halorhodopsin are reviewed in more detail in a recent article by Lanyi (1981).

Photosensory pigments, PS 565 and PS 370

Halobacteria are motile. They swim equally well in both directions of their long axis by means of bipolarly inserted flagella (Hildebrand & Dencher, 1975). With an interval of 10–50 s the organisms spontaneously reverse their swimming direction. The frequency of these reversal reactions can be modulated by steplike changes of the incident light intensity (Fig. 8). A sudden decrease of yellow-green light or an increase of blue or UV light elicits an extra reversal response within a latent period of a few seconds after the onset of the stimulus (Hildebrand & Dencher, 1975). On the other hand, a sudden increase of yellow-green light or a decrease of blue/UV light leads to a suppression of spontaneous reversals for some ten seconds (Spudich & Stoeckenius, 1979).

An analysis of the spectral sensitivity of the stimulus-evoked reversal response revealed that two different photosystems control the light-

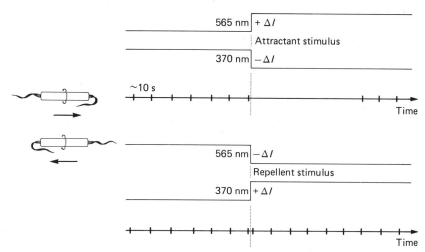

Fig. 8. Modulation of the frequency of reversal responses in *Halobacterium* by intensity changes of yellow-green and UV light. Each vertical bar indicates a directional change of the swimming cell. (For details see text.)

dependent behaviour in *Halobacterium*. According to their prominent sensitivity peaks, we called one of these PS 565 (Fig. 9) and the other PS 370 (Fig. 10). Both photosystems are blocked by growing the cells for several generations in the presence of nicotine, and can be restored within some ten minutes by addition of *trans* retinal or, with some delay, with 13-*cis* and 11-*cis* retinal (Dencher & Hildebrand, 1979). Reconstitution of both photosystems is also possible with the artificial retinal analogue retinal$_2$ (3,4-dehydroretinal) (Sperling & Schimz, 1980); the sensitivity maxima are thereby shifted by about 15 nm to 580 nm and 385 nm, respectively (Fig. 11). From these experiments it seems evident that retinal is involved as the chromophore in both sensory photosystems. However, we are not able so far to isolate or to detect the pigments spectroscopically.

The biological relevance of the photosensory function seems clear in the case of PS 565: this sytem is certainly useful to detect optimal light conditions for photophosphorylation (ATP synthesis), since its spectral sensitivity is very close to the absorption maximum of light-adapted bacteriorhodopsin. In fact, the behavioural responses lead to an accumulation of cells in an area illuminated with white or yellow-green light (Fig. 12A) as can be observed under the microscope (Dencher, 1978) or photometrically detected in a cuvette (Nultsch & Häder, 1978). The opposite effect, namely depletion and avoidance of an irradiated area, can be observed if UV light is used instead of visible light (Fig. 12B),

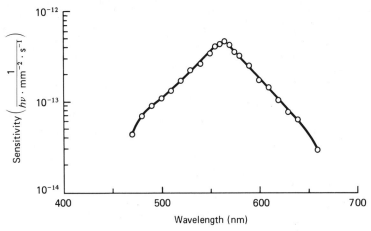

Fig. 9. PS 565 of *Halobacterium* strain R_1. Action spectrum for induced reversal response. Sensitivity is given as the reciprocal of light intensity decrease which was needed to elicit a response after a given latent period (standard response). (After Hildebrand & Dencher, 1975.)

which suggests that PS 370 may serve to protect the organisms from damaging irradiation.

Relationship between retinal pigments in *Halobacterium*

The similarity of the action spectrum of PS 565 in the visible range and the absorption spectrum of *trans* bacteriorhodopsin is striking (Figs. 3 and 9), and for a long time we thought that both pigments were identical, and that bacteriorhodopsin might function not only as a light-energy converter but also as a photosensor (Hildebrand & Dencher, 1975). For PS 370 we proposed another retinylidene protein, probably with weaker non-covalent interactions between retinal and the protein moiety. Energy transfer from aromatic amino acid residues is indicated by a peak around 280 nm (Dencher & Hildebrand, 1979). A UV-absorbing intermediate of the bacteriorhodopsin photocycle was proposed by G. Wagner (personal communication) to be the effective photopigment of PS 370, but for reasons which will be discussed below, we regard this as unlikely.

Recent results of Narurkar & Spudich (1981) seem to support the hypothesis that behavioural responses to visible light are mediated through bacteriorhodopsin and indicate that halorhodopsin may act as a sensory photosystem also. The mutant strain ET 15 S (which lacks spectroscopically detectable amounts of bacteriorhodopsin) shows a single sensitivity peak at 590 nm, whereas the parent strain ET 1001 S,

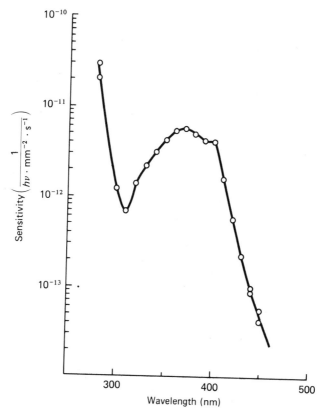

Fig. 10. PS 370 of *Halobacterium* strain R_1L_3. Action spectrum for induced reversal response. Sensitivity is given as the reciprocal of light intensity increase which was needed to elicit a response after a given latent period (standard response). (After Dencher & Hildebrand, 1979.)

which contains both bacteriorhodopsin and halorhodopsin, is also sensitive at 570 nm. However, two facts lead us now to question the hypothesis that PS 565 and bacteriorhodopsin are identical: (1) The mutant L 33 which contains no detectable bacteriorhodopsin, but four times the amount of halorhodopsin present in ET 15 cells (Wagner *et al.*, 1981), shows the main sensitivity maximum at 565 nm and only a small secondary peak near 588 nm (B. Traulich, E. Hildebrand, A. Schimz & G. Wagner, unpublished results). (2) Replacement of retinal by retinal$_2$ in PS 565 in the bacteriorhodopsin-containing strain R_1L_3 shifts the maximal sensitivity to 580 nm (Sperling & Schimz, 1980) whereas the absorption maximum of bacteriorhodopsin$_2$ (retinal$_2$ opsin) was found at 600 nm in both isolated purple membrane preparations and intact bacteria (Schimz, Sperling, Hildebrand & Köhler-Hahn, 1982).

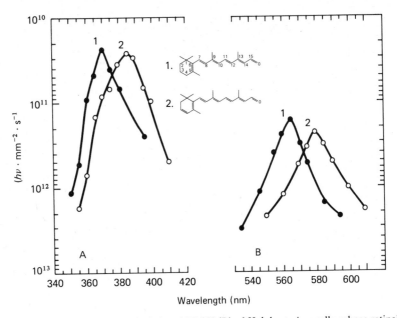

Fig. 11. Action spectra of PS 370 (A) and PS 565 (B) of *Halobacterium* cells, whose retinal synthesis had been blocked by nicotine, after reconstitution with retinal (1) and retinal$_2$ (3,4-dehydroretinal) (2), respectively. Sensitivity is given as light intensity change which is needed to elicit a reversal response after a given latent period (standard response). (After Sperling & Schimz, 1980 and unpublished results.)

We propose therefore that the pigment of PS 565 is a separate retinal–protein complex, different from bacteriorhodopsin and halorhodopsin. At present it cannot be ruled out that, according to the idea of Wagner, an intermediate product of the unknown 565 pigment may function as the sensory photoreceptor molecule in PS 370. It seems possible also that bacteriorhodopsin as well as halorhodopsin – if present in the cells – may contribute to the long wavelength sensory system.

We hope that further separation of the different photosystems will be possible with the help of appropriate mutants and that the sensory photopigments may be identified in the near future.

Concluding remarks

Neither the mechanism of H$^+$ translocation across the cell membrane, mediated by bacteriorhodopsin, nor the mechanism of halorhodopsin-mediated Na$^+$ outward transport, is presently known. Also the mechanism of photosensory transduction in *Halobacterium* is not yet under-

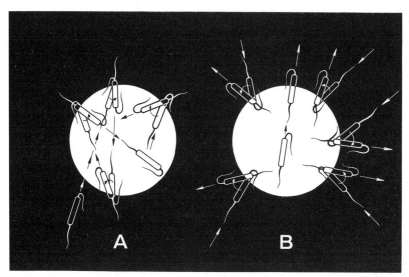

Fig. 12. Accumulation of *Halobacterium* in white or yellow-green light (A) and avoidance of blue or UV light (B) as a consequence of stimulus-induced reversal responses at the visible light/dark or dark/UV boundaries, and the suppression of reversals after crossing dark/visible light or UV/dark boundaries. (Modified after Hildebrand, 1977.)

stood. Some data indicate that the membrane potential might be essential for the photosensory activity (Nultsch & Häder, 1978; Wagner, 1978; E. Hildebrand, unpublished results). Halobacteria respond also to chemical changes in the environment (Schimz & Hildebrand, 1979), and light stimuli and chemical stimuli are integrated by the cell (Spudich & Stoeckenius, 1979). This indicates that at least the mechanism of flagellar rotation control must be common in both the photosensory and chemosensory pathways. Methylation of membrane proteins was found to accompany the action of attractant light and chemo-attractants, while repellent light and chemo-repellents lead to demethylation (Schimz, 1981, 1982). These reactions seem to play a role in *Halobacterium* similar to that which they play in the chemosensory transduction of other bacteria (Springer, Goy & Adler, 1979).

One may ask why photopigments of a similar biochemical nature (retinal–protein complexes) are used for such different functions as light-energy conversion and photosensing. The answer could be that after retinal biosynthesis has been established once in the organism only protein modifications are needed to create pigments with altered properties and functions. Thus it may be most economical to use the same chromophore for functionally different pigment molecules.

I would like to thank B. Traulich and Dr G. Wagner for the consent to cite unpublished results and Dr Angelika Schimz for critically reading the manuscript. Our work on the photosensory properties of *Halobacterium halobium* was supported by the Deutsche Forschungsgemeinschaft (SFB 160).

References

BAYLEY, H., HUANG, K.-S., RADHAKRISHNAN, R., ROSS, A. H., TAKAGAKI, Y. & KHORANA, G. (1981). Site of attachment of retinal in bacteriorhodopsin. *Proc. Natl. Acad. Sci., USA*, **78**, 2225–9.

BRAIMAN, M. & MATHIES, R. (1980). Resonance Raman evidence for an all-*trans* to 13-*cis* isomerization in the proton-pumping cycle of bacteriorhodopsin. *Biochemistry*, **19**, 5421–8.

DENCHER, N. A. (1978). Light-induced behavioural reactions of *Halobacterium halobium*: evidence for two rhodopsins acting as photopigments. In: *Energetics and Structure of Halophilic Microorganisms*, ed. S. R. Caplan & M. Ginzburg, pp. 67–88. Amsterdam: Elsevier/North-Holland Biomedical Press.

DENCHER, N. A. & HEYN, M. P. (1979). Bacteriorhodopsin monomers pump protons. *FEBS Lett.*, **108**, 307–10.

DENCHER, N. A. & HILDEBRAND, E. (1979). Sensory transduction in *Halobacterium halobium*: retinal protein pigment controls UV-induced behavioral response. *Zeit. Naturforsch.*, **34c**, 841–7.

DENCHER, N. A., RAFFERTY, C. N. & SPERLING, W. (1976). 13-*cis* and *trans* bacteriorhodopsin: photochemistry and dark equilibrium. *Berichte der Kernforschungsanlage Jülich*, Nr. 1374, 1–42.

EISENBACH, M. & CAPLAN, S. R. (1979). The light-driven proton pump of *Halobacterium halobium*: mechanism and function. *Curr. Top. Memb. Transp.*, **12**, 165–248.

GARTY, H. & CAPLAN, S. R. (1977). Light-dependent rubidium transport in intact *Halobacterium halobium* cells. *Biochim. Biophys. Acta*, **459**, 532–45.

GREENE, R. V. & LANYI, J. K. (1979). Proton movements in response to a light-driven electrogenic pump for sodium ions in *Halobacterium halobium* membranes. *J. Biol. Chem.*, **254**, 10986–94.

GREENE, R. V., MACDONALD, R. E. & PERRAULT, G. J. (1980). Action spectra determined with tunable dye laser light for light-induced proton efflux and uptake in membrane vesicles of *Halobacterium halobium*. *J. Biol. Chem.*, **255**, 3245–7.

HENDERSON, R. (1977). The purple membrane from *Halobacterium halobium*. *Ann. Rev. Biophys. Bioeng.*, **6**, 87–109.

HILDEBRAND, E. (1977). What does *Halobacterium* tell us about photoreception? *Biophys. Struct. and Mech.*, **3**, 69–77.

HILDEBRAND, E. & DENCHER, N. (1975). Two photosystems controlling behavioural responses of *Halobacterium halobium*. *Nature, London*, **257**, 46–8.

HOWES, C. D. & BATRA, P. P. (1970). Accumulation of lycopene and inhibition of cyclic carotenoids in *Mycobacterium* in the presence of nicotine. *Biochim. Biophys. Acta*, **222**, 174–9.

LANYI, J. K. (1978). Light energy conversion in *Halobacterium halobium*. *Microbiol. Rev.*, **42**, 682–706.

LANYI, J. K. (1979). The role of Na+ in transport processes of bacterial membranes. *Biochim. Biophys. Acta*, **559**, 377–97.

LANYI, J. K. (1981). Halorhodopsin – a second retinal pigment in *Halobacterium halobium*. *Trends Biochem. Sci.*, **6**, 60–2.

LANYI, J. K. & WEBER, H. J. (1980). Spectrophotometric identification of the pigment associated with light-driven primary sodium translocation in *Halobacterium halobium*. *J. Biol. Chem.*, **255**, 243–50.

LINDLEY, E. V. & MACDONALD, R. E. (1979). A second mechanism for sodium extrusion in *Halobacterium halobium*: a light-driven sodium pump. *Biochem. Biophys. Res. Commun.*, **88**, 491–9.

LOZIER, R. H., BOGOMOLNI, R. A. & STOECKENIUS, W. (1975). Bacteriorhodopsin: a light-driven proton pump in *Halobacterium halobium*. *Biophys. J.*, **15**, 955–62.

LUISI, B. F., LANYI, J. K. & WEBER, H. J. (1980). Na$^+$ transport *via* Na$^+$/H$^+$ antiport in *Halobacterium halobium* envelope vesicles. *FEBS Lett.*, **117**, 354–8.

MACDONALD, R. E., GREENE, R. V., CLARK, R. D. & LINDLEY, E. V. (1979). Characterization of the light-driven sodium pump of *Halobacterium halobium*. *J. Biol. Chem.*, **254**, 11831–8.

MATSUNO-YAGI, A. & MUKOHATA, Y. (1977). Two possible roles of bacteriorhodopsin; a comparative study of strains of *Halobacterium halobium* differing in pigmentation. *Biochem. Biophys. Res. Commun.*, **78**, 237–43.

MATSUNO-YAGI, A. & MUKOHATA, Y. (1980). ATP synthesis linked to light-dependent proton uptake in a red mutant strain of *Halobacterium* lacking bacteriorhodopsin. *Arch. Biochem. Biophys.*, **199**, 297–303.

MOWERY, P. C. & STOECKENIUS, W. (1981). Photoisomerization of the chromophore in bacteriorhodopsin during the proton pumping photocycle. *Biochemistry*, **20**, 2302–6.

MUKOHATA, Y. & KAJI, Y. (1981*a*). Light-induced membrane potential increase, ATP synthesis, and proton uptake in *Halobacterium halobium* R$_1$mR catalyzed by halorhodopsin: effects of *N*,*N'*-dicyclohexylcarbodiimide, triphenyltin chloride, and 3,4-di-*tert*-butyl-4-hydroxybenzylidene-malonitrile (SF 6847). *Arch. Biochem. Biophys.*, **206**, 72–6.

MUKOHATA, Y. & KAJI, Y. (1981*b*). Light-induced ATP synthesis dependent on halorhodopsin – pH regulation. *Arch. Biochem. Biophys.*, **208**, 615–17.

NARURKAR, V. & SPUDICH, J. L. (1981). Evidence that the light-driven sodium ion pump is a phototaxis receptor in *Halobacterium halobium*. *Biophys. J.*, **33**, 218a.

NULTSCH, W. & HÄDER, M. (1978). Photoakkumulation bei *Halobacterium halobium*. *Ber. Deutsch. Bot. Gesell.*, **91**, 441–53.

OESTERHELT, D. & STOECKENIUS, W. (1971). Rhodopsin-like protein from the purple membrane of *Halobacterium halobium*. *Nat. New Biol.*, **233**, 149–52.

OESTERHELT, D. & STOECKENIUS, W. (1973). Functions of a new photoreceptor membrane. *Proc. Natl. Acad. Sci., USA*, **70**, 2853–7.

OTTOLENGHI, M. (1980). The photochemistry of rhodopsins. *Adv. Photochem.*, **12**, 97–200.

OVCHINNIKOV, YU. A., ABDULAEV, N. G., FEIGINA, M. YU., KISELEV, A. V. & LOBANOV, N. A. (1979). The structural basis of the functioning of bacteriorhodopsin: an overview. *FEBS Lett.*, **100**, 219–24.

SCHIMZ, A. (1981). Methylation of membrane proteins is involved in chemosensory and photosensory behavior of *Halobacterium halobium*. *FEBS Lett.*, **125**, 205–7.

SCHIMZ, A. (1982). Localization of the methylation system involved in sensory behavior of *Halobacterium halobium* and its dependence on calcium. *FEBS Lett.*, **139**, 283–6.

SCHIMZ, A. & HILDEBRAND, E. (1979). Chemosensory responses of *Halobacterium halobium*. *J. Bacteriol.*, **140**, 749–53.

SCHIMZ, A., SPERLING, W., HILDEBRAND, E. & KÖHLER-HAHN, D. (1982). Bacteriorhodopsin and the sensory pigment of the photosystem 565 in *Halobacterium halobium*. *Photochem. Photobiol.*, **36**, 193–6.

SPERLING, W., RAFFERTY, C. N., KOHL, K.-D. & DENCHER, N. A. (1979). Isomeric composition of bacteriorhodopsin under different environmental light conditions. *FEBS Lett.*, **97**, 129–32.

SPERLING, W. & SCHIMZ, A. (1980). Photosensory retinal pigments in *Halobacterium halobium*. *Biophys. Struct. Mech.*, **6**, 165–9.

SPRINGER, M. S., GOY, M. F. & ADLER, J. (1979). Protein methylation in behavioural control mechanisms and in signal transduction. *Nature, London*, **280**, 179–84.

SPUDICH, J. L. & STOECKENIUS, W. (1979). Photosensory and chemosensory behavior of *Halobacterium halobium*. *Photobiochem. Photobiophys.*, **1**, 43–53.

STOECKENIUS, W. (1980). Purple membrane of halobacteria: a new energy converter. *Acc. Chem. Res.*, **13**, 337–44.

SUMPER, M., REITMEIER, H. & OESTERHELT, D. (1976). Biosynthesis of the purple membrane of halobacteria. *Angewandte Chemie, International Edition in English*, **15**, 187–94.

WAGNER, G. (1978). Halobacterial potassium transport in orange and near UV light. In: *Energetics and Structure of Halophilic Microorganisms*, ed. S. R. Caplan & M. Ginzburg, pp. 335–40. Amsterdam: Elsevier/North-Holland Biomedical Press.

WAGNER, G., HARTMANN, R. & OESTERHELT, D. (1978). Potassium uniport and ATP synthesis in *Halobacterium halobium*. *Eur. J. Biochem.*, **89**, 169–79.

WAGNER, G., OESTERHELT, D., KRIPPAHL, G. & LANYI, J. K. (1981). Bioenergetic role of halorhodopsin in *Halobacterium halobium* cells. *FEBS Lett.*, **131**, 341–5.

WEBER, H. J. & BOGOMOLNI, R. A. (1981). P_{588}, a second retinal-containing pigment in *Halobacterium halobium*. *Photochem. Photobiol.*, **33**, 601–8.

PHOTOSYNTHESIS: TRANSDUCTION OF LIGHT ENERGY INTO CHEMICAL ENERGY

PAUL MATHIS

Service de Biophysique, Département de Biologie, CEN Saclay,
91191 Gif-sur-Yvette Cedex, France

Introduction

Photosynthesis is universally recognised as one of the major biological processes. Its past activity has permitted the accumulation of the precious fossil fuels, and contributed largely to the formation of our atmosphere. Its present activity, by the production of organic matter, is essential for the subsistence of the animal kingdom in general, and of man in particular (food, fuel, wood, fibres, chemicals, etc.); by the consumption of CO_2 and the concomitant evolution of oxygen, photosynthesis helps to stabilise the composition of the atmosphere. Among the various photobiological processes, photosynthesis is unique in its ability to convert into chemical energy a significant part of the solar energy received by the Earth. In ideal cases, the energetic yield of conversion of the free energy of light into chemical free energy by means of photosynthesis can be rather high, and approaches 10%. Under natural conditions, however, the yield is much lower, and is usually between 0.1 and 1%. A world annual production of dry matter corresponding to $3 \cdot 10^{21}$ J of stored energy is nevertheless achieved (Bolton & Hall, 1979). This is ten times the annual world energy use, and it is quite conceivable that mastering the photosynthetic processes may lead to an important increase in the production of plant biomass.

Photosynthesis transforms the energy of light into chemical energy, essentially in the form of sugars. This transformation cannot be effected in one step, but requires a long and complex sequence of reactions in which the energy takes various forms, as shown in Table 1. A primary requirement for efficient photosynthesis is a good light-gathering apparatus; photosynthetic organisms have thus developed molecules with an extended π-electron system (for example: chlorophylls, carotenoids, bile pigments in phycobiliproteins), which permit an efficient capture of light by a so-called 'antenna', described by Barber in this volume. The electronic excitation energy is transferred within the

Table 1. *Energy conversion in the photosynthetic apparatus.* Reactions normally proceed from left to right

			Chemical			
Form of energy	Electromagnetic (photons)	Electronic (Molecular electronic excitation)	Redox (Reduced or oxidised redox components)	Redox and membrane potential	Redox and chemical bonds	Chemical bonds
Structural component		Pigments	Chain of electron carriers	Electron carriers, thylakoid membrane	NADPH, ATP	Sugars
Process of conversion	Absorption of light	Primary photochemistry	Transmembrane H^+ transfer coupled to electron transfer	ATP synthesis by ATP synthetase	Reactions of carbon metabolism	

Reaction centers

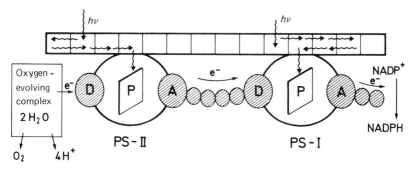

Fig. 1. Functional diagram for photosynthesis in oxygen-evolving organisms. The array of rectangles represents the light-harvesting antenna. The common set of pigments for Photosystems One and Two (PS–I and PS–II) and the chain of electron carriers may have no physical reality. For an explanation of A, D and P see Table 2 and the text.

antenna until it reaches a 'reaction centre', where the primary photochemistry takes place, as described later in this chapter.

In terms of dry weight production, photosynthesis is mainly performed by higher plants and algae, which reduce CO_2 to sugars and oxidise water to molecular oxygen. Some prokaryotic organisms operate similarly (for example, cyanobacteria and *Prochloron*), whereas most of them (for example, purple bacteria and green bacteria) reduce CO_2 at the expense of a mildly reducing substance (Clayton, 1980). These prokaryotes are very useful experimental tools.

Photosynthesis may be divided into two classes of reactions:

(1) *Membrane reactions.* These take place in the photosynthetic membranes (thylakoid or chromatophore), which contain the light-harvesting pigments, the reaction centres, the chain of electron carriers including the oxygen-evolving complex, and the ATP synthetase. Accordingly the reactions proceed from the absorption of light to the synthesis of ATP and the reduction of the terminal electron acceptor ($NADP^+$ or NAD^+). In plants the photosynthetic reactions all occur within organelles called chloroplasts which are bound by another type of membrane constituting a double envelope (Douce & Joyard, 1979).

(2) *Carbon metabolism.* The products of the membrane reactions (ATP and $NADPH_2$), are utilised to reduce CO_2 to sugars in a complex series of reactions which take place in the cytoplasm. These reactions are of the greatest importance for obtaining a good overall yield of photoconversion. They have been adequately reviewed recently (Gibbs & Latzko, 1979; Hatch & Boardman, 1981) and will not be discussed further here.

The membrane reactions of higher plant photosynthesis are schematically depicted as a functional scheme in Fig. 1, and in a slightly more
realistic form in Figs. 8 and 9. Various aspects of these reactions will
now be detailed: the reaction centres, the electron transfer chain, and
the synthesis of ATP. For additional reading see Govindjee (1975),
Barber (1977), Olson & Hind (1977), Clayton & Sistrom (1978), Sauer
(1979), Clayton (1980).

Photosynthetic reaction centres

The reaction centres are the specific sites in the photosynthetic membrane where the primary electron transfer reaction takes place. They
function according to the scheme:

$$DPA \longrightarrow DP^*A \rightarrow DP^+A^- \longrightarrow D^+PA^-.$$

The primary electron donor, P, is excited (*) by energy transfer from
the antenna; it becomes able to react with the primary acceptor, A, and
to donate an electron to A. The primary radical pair (P^+A^-) lasts for
about 100 ps, and the charge separation is stabilised by secondary
electron transfer from A^- to secondary acceptors, and from D to P^+.

Reaction centres have been isolated from purple bacteria (reviewed
by Olson & Thornber, 1979) by treatment with detergents. Each centre
is composed of three polypeptides (molecular weights: 22, 24 and
27×10^3), four molecules of bacteriochlorophyll *a*, two of bacteriopheophytin *a*, two of ubiquinone, one of carotenoid, and one Fe^{2+}
atom. The carotenoid is absent from some mutants without impairing
their photochemistry. Some cytochrome *c* may remain bound to the
isolated centre but this depends on the type of bacterium: in *Chroma-
tium vinosum*, for example, two cytochromes are bound to the purified
centre (for details, see Clayton & Sistrom, 1978).

Oxygen-evolving organisms contain two types of reaction centre,
named Photosystem One (PS-I) and Photosystem Two (PS-II), which
function in series to bring electrons from the terminal electron donor
(water) to the terminal acceptor ($NADP^+$) (Fig. 2). These centres have
not been purified to the same extent as the bacterial ones, but most
preparations contain about 40 chlorophyll molecules per centre. The
reasons for the inability to purify these centres further are not clear: it
may be that a fraction of the light-harvesting antenna is more firmly
bound to c reaction centre than in bacteria, but it is also possible that a
unit with about 40 chlorophylls is the smallest one that has structural
significance in oxygen-evolving organisms.

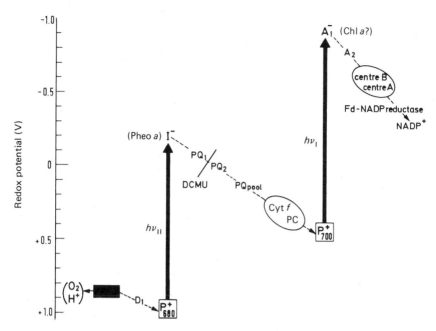

Fig. 2. A scheme with the electron carriers in oxygenic photosynthesis placed according to their redox potentials (Z scheme). Some intermediates may be missing, especially between the plastoquinone (PQ) pool and cytochrome f (Cyt f). PC, plastocyanin; DCMU, inhibitor of electron transfer (diuron); Pheo, pheophytin.

Reaction centres give rise to numerous questions for the biologist: What is their chemical composition; their structural arrangement in terms of molecular organisation; the mode of insertion of reaction centres within the photosynthetic membranes, and the mechanism of the primary reactions (path of the electron, properties of the radical pair, kinetics, mechanism of stabilisation of the charge separation, etc.)? Recent general reviews include: Sauer (1975), Olson & Hind (1977), Blankenship & Parson (1979), Clayton (1980), Mathis & Paillotin (1981). Here I shall focus on the electron carriers and on their sequential arrangement. This sequence is presented in Table 2, in a manner that emphasises the striking similarities between the electron donors of PS-I and of purple bacteria, and between the electron acceptors of PS-II and of purple bacteria (acceptors and donors are defined by reference to the primary charge separation). This presentation also identifies the sequences of electron carriers which are more specific to oxygen-evolving organisms – the high-potential carriers, the oxidation of which permits oxygen evolution – and the low-potential carriers which permit the reduction of NADP$^+$ (an analogous sequence

Table 2. *Sequences of electron carriers in the major three classes of reaction centres*[a]

Reaction centre / e⁻ carrier	D Secondary donor	P Primary donor	Primary acceptor	Secondary acceptor	Tertiary acceptor
PS-I	Plastocyanin Cytochrome f +380 mV	P-700 (Chl a dimer?) +490 mV	A_1 (Chl a ?) −1.0 V	A_2(X) (Iron–sulphur centre?) −730 mV	Iron–sulphur centres A, B −580 mV
Purple bacteria	c Cytochromes ~+350 mV	P-870 (BChl a dimer) +450 mV	I (BPheo a) −600 mV	Q_1 (ubiquinone) ≃0 mV	Q_2 (ubiquinone) ≃+80 mV
PS-II	D_1 (unknown) ~+950 mV	P-680 Chl a (monomer?) +1.1 V(?)	I (Pheo a ?) −600 mV	PQ_1 (plastoquinone) −300 to 0 mV	PQ_2 (plastoquinone) ~+40 mV(?)

[a] The approximate redox potentials (versus NHE) are at pH 7. Electrons move from left to right, in exergonic reactions, from low-redox-potential carriers to high-potential ones. The endergonic reaction from P to the primary acceptor is activated by light.

is found in green bacteria, the reaction centre of which thus resembles closely that of PS-I).

The present photosynthetic organisms have probably evolved from a simple common ancestor which was capable, at least, of performing the primary photochemical charge separation in a single reaction centre. It is still not understood how the system evolved towards the situation found in oxygen-evolving organisms with their two specialised reaction centres (Broda, 1975). Table 2 shows, however, that the PS-I and PS-II centres include features pertaining, respectively, to the donor and to the acceptor side of bacterial centres. The great similarities of the electron acceptors in PS-II and in purple bacteria, and of the electron acceptors in PS-I and in green bacteria, argue against a linear filiation such as: green bacteria → purple bacteria → cyanobacteria or green plants, since the low-potential acceptors would have been lost and found again. Positive conclusions from the observed similarities have not yet been drawn, but it may be that oxygenic photosynthesis is a very ancient form (traces of cyanobacteria may be as old as 3 billion years) and that the modern photosynthetic bacteria have evolved by regression.

Purple bacteria and PS-I reaction centres: electron donors

In these reaction centres the primary donor, P (P-870 in bacteria; P-700 in PS-I) has a rather low oxidation potential: $E_m = +450$ or $+490\,\text{mV}$, respectively. P-870 and P-700 are thus easily maintained in the oxidised state, either chemically or photochemically, and this facilitates spectroscopic studies: absorption and circular dichroism, electron spin resonance, or electron-nuclear double resonance. These studies have led to the conclusion that P is a pair of bacteriochlorophyll *a* (or chlorophyll *a*) molecules, which is oxidised in a one-electron reaction to yield the corresponding radical-cation. The pair has been modelled *in vitro*, particularly by the so-called 'special pair' in which the chlorophyll molecules are held together by one or two water molecules (Katz, Norris, Shipman, Thurnauer & Wasielewski, 1978). It must be admitted, however, that the precise structure of P is still unknown. Indeed some of the spectroscopic data have recently been reinterpreted by Wasielewski, Norris, Shipman, Lin & Svec (1981), who have proposed that P-700, instead of being a pair of chlorophyll *a* molecules, is a single chlorophyll *a* molecule in which ring V assumes an enol configuration.

In purple bacteria, P-870 is reduced by a *c*-type cytochrome. Each reaction centre is associated with one pair of high-potential *c* cytochromes ($E_m \simeq +300\,\text{mV}$) and, in some cases, one pair of low-potential c cytochromes ($E_m \simeq 0\,\text{mV}$). Some of the cytochromes are attached by electrostatic interaction and others by hydrophobic interaction (Dutton & Prince, 1978; van Grondelle, 1978). After purification of the reaction centres the situation is rather variable, since there is a great range of interaction strengths. The low-potential cytochromes donate electrons rapidly to P-870$^+$, with a half-time in the order of a microsecond. Electron donation is slowed down at low temperature, but remains remarkably efficient. *In vivo*, the pool of cytochromes does not come rapidly to a redox equilibrium, probably for structural reasons.

In PS-I, P-700$^+$ is reduced by a pool containing a copper enzyme, plastocyanin, and a c-type cytochrome (cytochrome f). In higher plant chloroplasts, there is good evidence for the sequence: cyt $f \rightarrow$ plastocyanin $\rightarrow$ P-700 (Haehnel, Pröpper & Krause, 1980). Plastocyanin is a rather hydrophilic molecule which has been the object of good structural studies using NMR and X-ray crystallography. It interacts with the reaction centre by electrostatic interactions at a specific site. Under conditions of good interaction, electron transfer takes place from plastocyanin to P-700$^+$ in $10\,\mu s$; this transfer is totally inhibited at low temperature. Cytochrome f is an intrinsic membrane protein. In algae, another cytochrome c is present, cytochrome c-552, which is functionally interchangeable with plastocyanin, as shown by Wood (1978). The concentrations of plastocyanin and cytochrome c-552 are also complementary. The properties of these donors are thus similar, although not identical, to those of the c cytochromes in purple bacteria.

Purple bacteria and PS-II reaction centres: electron acceptors

The sequence of electron acceptors in purple bacteria has been precisely determined, mainly from flash absorption spectroscopy (Clayton & Sistrom, 1978). A special molecule of bacteriopheophytin a, named 'I' for intermediate acceptor, receives an electron from the photo-excited primary donor. After a short flash of light, I is reduced in less than $10 \cdot 10^{-12}\,\text{s}$ in isolated reaction centres. Notwithstanding this short time, there is some evidence that another electron carrier, a specific bacteriochlorophyll, is involved in the sequence of electron transfer from donor P-870* to acceptor I. I$^-$ is then re-oxidised by transferring an electron to a carrier, Q_1, which is a bound ubiquinone molecule (menaquinone in some species), with a $t_{1/2}$ of 150 ps. Q_1^- reduces the next carrier Q_2, another bound ubiquinone, in about $100\,\mu s$, which is 10^6

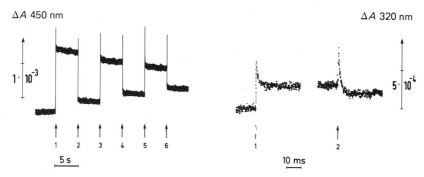

Fig. 3. Absorption changes induced by a sequence of flashes in a suspension of reaction centres of *Rhodopseudomonas sphaeroides* (left trace) and of spinach chloroplasts (right trace). The signals are mostly due to a quinone (ubiquinone or plastoquinone, respectively) radical anion which absorbs at 450 and 320 nm. In the left trace, the fast transients (spikes) are due to P-870$^+$; the stable upward signal is due to the quinone radical anion Q_2^- and the downward stable signal corresponds to the regeneration of Q_2 when even-numbered flashes produce Q_2^{2-}, which is immediately oxidised to Q_2 (see text). Reaction centres concentration $= 1 \mu M$; addition of $100 \mu M$ diaminodurene and $20 \mu M$ ubiquinone; detergent (LDAO) concentration $= 1.5\%$. Unpublished data from Dr A. Vermeglio (see also Vermeglio, 1977). Tris-treated chloroplasts (chlorophyll concentration $= 16 \mu M$); addition of 1 mM sodium ascorbate and $50 \mu M$ *p*-phenylenediamine. (Adapted from Mathis & Haveman, 1977.)

times slower than its own reduction. The initial information about the nature of I was obtained by picosecond flash absorption spectroscopy. It has later been found possible to accumulate I$^-$ by illumination under strongly reducing conditions and to trap it at low temperature. This has permitted refined spectroscopic studies by absorption and by electron spin resonance (ESR). The primary steps, from the state P*IQ$_1$ to the state P*IQ$_1^-$ are undoubtedly of key importance in ensuring that a large fraction ($\simeq 40\%$) of the electronic excitation energy (1.4 eV) is stabilised in the form of redox energy, W. The latter can be expressed as:

$$W = e \times (E_D - E_A),$$

where E_D and E_A are the midpoint redox potentials of the couples P/P$^+$ and Q_1/Q_1^-, respectively, and e is the electronic charge.

The quinone Q_1 is strongly associated (in a non-defined manner) with an iron atom, which is presumed to play a role in the electron transfer to Q_2. Q_1 is a one-electron carrier, but Q_2 is a two-electron carrier. This behaviour is illustrated in Fig. 3 (left trace), in which the stable upward signal is due to the ubiquinone radical anion, Q$^-$. Electron transfer from Q_1 to Q_2 ($\simeq 100 \mu s$) is undetectable in the time scale of the experiment, and the short transient is due to P-870$^+$. The system starts in the state

Table 3. *State of the acceptors in the reaction centre of purple bacteria, and their ESR characteristics (adapted from Wraight, 1979)*

State of electron acceptors	ESR characteristics
PIQ_1^-Fe	Broad singlet
$PI^-Q_1^-Fe$	Doublet
$PIQ_1^{2-}Fe$	No signal
$PI^-Q_1^{2-}Fe$	Narrow singlet
PIQ_1^-	Narrow singlet

(Q_1, Q_2). The state (Q_1, Q_2^-) is attained after each odd-numbered flash; after each even-numbered flash the state (Q_1, Q_2^{2-}) is attained and disappears rapidly by reducing one molecule of ubiquinone in the suspension: $Q_2^{2-} + UQ + 2H^+ \longrightarrow Q_2 + UQH_2$.

At the present state of our knowledge, it is considered that a similar sequence of acceptors operates in PS-II (Fig. 4): pheophytin a, PQ_1 (primary plastoquinone), PQ_2 (secondary plastoquinone). The kinetic and energetic parameters are nearly the same, and it is probable that an iron atom is located between PQ_1 and PQ_2. The two-electron behaviour is illustrated in Fig. 3 (right trace) by data obtained with chloroplasts in which the fully reduced PQ_2 reduces one molecule of a pool of plastoquinone.

The actual experimental data concerning the quinones Q_1, Q_2, and their equivalent in PS-II are far more complicated than described here, and in this respect three points are of special interest:

(1) *ESR properties of the reduced acceptors.* Quinone radical anions usually display an intense and narrow EST line around $g = 2$. In reaction centres the ESR spectrum of Q_1^- is dramatically distorted by a magnetic interaction with the iron atom. Q_2^- can also interact with the iron. When the primary acceptor I is reduced, it can also interact with Q_1^-, in a temperature-dependent manner. A few of the possible states are given in Table 3. The major interest of these ESR studies is that some information about the distances between different paramagnetic species (P^+, I^-, Q_1^-, Q_2^-, Fe^{2+}) can be derived from the magnitude of the magnetic interaction. This is a distinct advantage since there are no other good ways of evaluating intermolecular distances in reaction centres which are not crystallisable. Other fragmentary information about distances can be

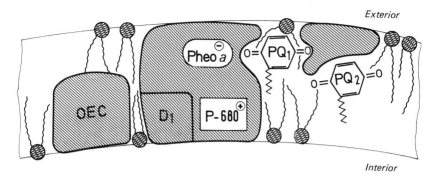

PS-II reaction centre

Fig. 4. Schematic representation of the PS-II reaction centre. OEC = Oxygen-Evolving Complex. The bound plastoquinones PQ_1 and PQ_2 are oversized. Over PQ_1 and PQ_2 is represented the protein which controls electron transfer from PQ_1 to PQ_2 (see text).

obtained by an analysis of the rate of electron transfer (De Vault, 1980), and from diffraction studies with reaction centres (Pachence, Dutton & Blasie, 1981). The ESR data for the electron acceptors of PS-II are somewhat less detailed, but present evidence is in good agreement with the data obtained from purple bacteria (Klimov, Dolan, Shaw & Ke, 1980).

(2) *Thermodynamic properties of the quinones.* Measurements of the redox potentials of Q_1 and Q_2 in purple bacteria revealed a number of unexpected complications, as reviewed by Wraight (1979). In chromatophores, the E_m of Q_1 is pH-dependent, as if the equilibrium were: $Q_1 + H^+ + e^- \rightleftharpoons Q_1H$ ($E_m = -50\,mV$ at pH 7); it becomes pH-independent above pH 9 with $E_m \simeq -180\,mV$ (with some differences between species). In the course of photo-induced Q_1 reduction, the species Q_1^- appears to be non-protonated, even when the pH is well below 9, and this may result from a kinetic barrier to the access of Q_1^- to protons. The functional E_m of Q_1 would thus be as measured at high pH ($\simeq -180\,mV$). However, in isolated reaction centres from *Rhodopseudomonas sphaeroides*, the E_m of Q_1 appears to be pH independent: this suggests that the pH-dependence observed in chromatophores is due to a protonatable group in another membrane component close to Q_1 (perhaps a protein or Q_2 itself). The redox potential of Q_2 is also pH-dependent with a pK of around 9. In PS-II the potential of the primary plastoquinone, PQ_1, has not

been satisfactorily determined. In the literature E_m values vary from $+400\,mV$ to $-300\,mV$ for the PS-II acceptors. Most of the values were obtained from measurements of the chlorophyll fluorescence yield, which is supposed to be related unequivocally to the redox state of PQ_1 (this point may deserve further investigation, as discussed earlier; Mathis, 1981). At pH 7, two waves are generally observed, near $0\,mV$ and $-250\,mV$, the origins of which are not clear. It is often assumed that the PS-II centres are heterogeneous and that this heterogeneity reveals itself in a number of different parameters (for example, redox potential, shape of the fluorescence induction curve, and electrochromic shift of the carotenoids).

(3) *A protein involved in electron transfer from PQ_1 to PQ_2.* A mild digestion of chloroplasts by trypsin completely blocks electron transfer from PQ_1 to PQ_2. This is accompanied by a large increase in the accessibility of PQ_1 to exogenous electron acceptors, and it has thus been suggested that trypsin removes a 'proteinaceous shield' (Renger, 1976). This question has recently been the object of active and successful research (see for example, Mattoo, Pick, Hoffman-Falk & Edelman, 1981; Pfister, Steinback, Gardner & Arntzen, 1981; or Trebst, 1981). It has been shown that a protein of molecular weight 32 000 controls the electron transport from PQ_1 to PQ_2, and is the target of the well-known herbicides triazines, nitrophenols and ureas (such as dichlorophenyl dimethyl urea, DCMU). Several plant species have developed mutants which are herbicide-resistant and it has been shown that these mutants have a modification of this protein, which is incapable of binding some of the herbicides. How this protein can control so finely the electron transfer between the two plastoquinones remains to be determined.

The high-potential carriers: electron donors of PS-II

Electron carriers located on the donor side of PS-II have the task of oxidising water to molecular oxygen (for recent reviews, see Joliot & Kok, 1975; Radmer & Kok, 1975; Amesz & Duysens, 1977; Knaff & Malkin, 1978; Harriman & Barber, 1979; Bouges-Bocquet, 1980; Clayton, 1980; Velthuys, 1980; Mathis & Paillotin, 1981). Under equilibrium conditions, the system $2H_2O \rightleftharpoons 4H^+ + 4e^- + O_2$ has a midpoint potential of $+0.82\,V$ at pH 7.0. In photosynthesis, oxygen is evolved rapidly and irreversibly, and thus the electron carriers must have a midpoint potential well above $+0.82$ V. The driving force in this reaction lies in the primary donor, P-680, which is oxidised in the primary photoreaction: P-680*-I-PQ$_1$ $\rightarrow$ P-680$^+$-I$^-$-PQ$_1$ $\rightarrow$ P-680$^+$-I-PQ$_1^-$. The E_m of P-680 is probably about $+1.0$ V, and it has been impossible to titrate it chemically (antenna pigments are oxidised at lower E_m), and even to keep it oxidised photochemically. Data from flash spectroscopic experiments permit the conclusion that P-680 is a chlorophyll *a*. Furthermore, ESR experiments have led to a proposal that P-680, like P-870 or P-700, consists of a pair of chlorophyll *a* molecules in a special environment or configuration. We can note that, according to this hypothesis, P-700 and P-680 are both pairs of chlorophyll *a*; yet their redox potentials differ substantially, being 0.49 V and around 1.1 V, respectively. This large difference may arise from a different geometry of the pairs of molecules, or from widely different liganding groups. A different proposal has recently been advanced by Davis, Forman & Fajer (1979): namely that P-680 may be a 'ligated' chlorophyll *a* monomer that interacts strongly with its environment, for instance by electrostatic forces.

The mechanism by which photo-oxidation of P-680 leads to oxygen evolution can be considered as one of the major unsolved problems in biology. The chemical species involved are unknown and our information is quite fragmentary. It is usually assumed that the oxygen comes from water: this assumption is very probably true, but unequivocal experimental evidence has not yet been produced. The most striking property of the system is the so-called 'period-4' behaviour: when oxygen-evolving organisms are adapted to darkness for a few minutes, and then excited by short flashes, many of their properties (fluorescence, luminescence, oxidation of cytochrome b_{559}) vary with the flash number according to a period of four. The most important property varying in this manner is the amount of oxygen elicited by one flash (Joliot & Kok, 1975). These results are best interpreted in the S-state model, in which the oxygen-evolving system can adopt four stable states (S_0, S_1, S_2, S_3) and one unstable state (S_4) which returns to S_0 with the liberation of one O_2 molecule (Fig. 5). The S states can thus be considered as oxidation states, differing by one electron equivalent.

What is the nature of the chemical system corresponding to the S states, which we name the Oxygen-Evolving Complex (OEC)? The OEC is often assumed to be a manganese protein (Sauer, 1980), and indeed manganese complexes can have several oxidation states, some of which are highly oxidising. Complexed manganese is also necessary for oxygen evolution and, in many cases, inhibition of oxygen evolution by

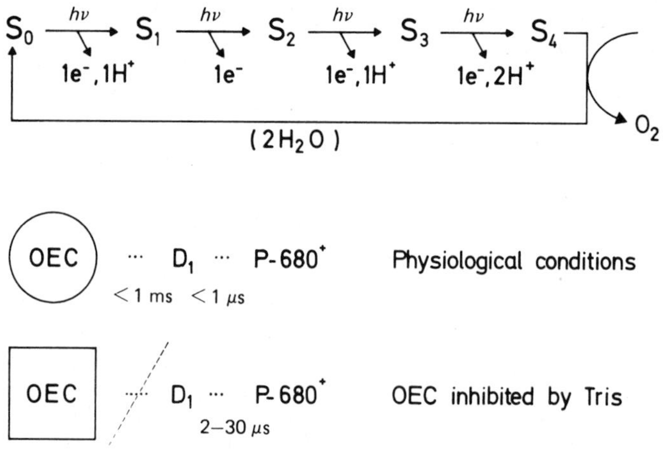

Fig. 5. A scheme of the S states of the Oxygen-Evolving Complex (OEC) which lead to O_2 evolution. In the last step, the ejection of two H^+ may take place as indicated or in parallel with the oxygen evolution. The bottom part indicates, in a simplified manner, the kinetic properties of the high potential carriers.

various treatments parallels the loss of bound manganese. There are some indications that the redox state of manganese is changing with the S state, but this field is highly controversial. The isolation of polypeptides associated with the OEC is of great importance, but several recent reports are still inconclusive. Simplified chemical models of the OEC have been constructed, such as polynuclear manganese complexes, and these may prove helpful in devising experiments for understanding the system.

There is also much uncertainty about the relationships between the primary electron donor P-680 and the OEC. Based on kinetic experiments, some models include several hypothetical electron carriers between P-680 and the OEC; however, a simple model: OEC-D_1 → P-680^+ can account for most of the data (Fig. 5). Under physiological conditions, P-680^+ is reduced rapidly, in less than $1\,\mu s$. A radical is also formed, which may correspond to D_1^+, the decay of which depends on the S state. This radical has a characteristic ESR spectrum, named Signal II_{vf} (vf = very fast). When oxygen evolution is inhibited by treatment with Tris, P-680^+ is reduced more slowly (2–$30\,\mu s$, depending on the pH), probably by the same donor, D_1. An ESR signal, of the same shape as in untreated chloroplasts, is also formed; it is named Signal II_f (f = fast). Flash experiments have revealed that in chloroplasts inactivated by Tris, D_1 has a one-electron capacity. Moreover, it has been shown that the S state influences the rate of reduction of

P-680$^+$ under physiological conditions, as well as affecting other properties of the primary reactions. A short sequence, as shown in Fig. 5, is thus quite reasonable. The chemical identity of D_1 is still unknown, although experiments on subchloroplast particles enriched in the PS-II centre have shown that D_1 is an integral part of the reaction centre and is possibly not a protein (Satoh & Mathis, 1981).

From what is known about the high-potential electron carriers and the OEC, it appears that the photosynthetic processes in higher plants are very similar to those in the primitive cyanobacteria. For example, the rate of electron transfer from D_1 to P-680$^+$ after inhibition of oxygen evolution by Tris is remarkably similar in spinach chloroplasts and in PS-II particles from the cyanobacterium *Phormidium luridum* (Reinman, Mathis, Conjeaud & Stewart, 1981). It seems that the mechanism which evolved more than a billion years ago was near the optimum and has not been modified since. Cyanobacteria may prove to be an important tool for the study of the oxygen-evolving process. Some of them are thermophilic and are relatively insensitive to thermally activated ageing. In addition it has been possible to isolate from them a PS-II complex which is able to carry out induced oxygen-evolution and is devoid of PS-I activity; this complex possesses a small antenna of about 50 chlorophyll *a* molecules.

The low-potential carriers: electron acceptors of PS-I and chlorobacteria

In oxygen-evolving organisms, the electron acceptors of PS-I permit the reduction of NADP$^+$. The system NADP$^+$ + 2e$^-$ + H$^+$ $\rightleftharpoons$ NADPH has an E_m of -0.32 V. Since the electrons move spontaneously from carriers with a more negative to carriers with a less negative potential, the PS-I acceptors must have an E_m below -0.32 V. In the present state of our knowledge, the sequence shown in Fig. 6 is the most probable (for reviews, see Ke, 1978; Clayton, 1980; Mathis & Paillotin, 1981; Avron, 1981). The primary acceptor A_1 is thought to be reduced directly by excited P-700. A_1 is probably a molecule of chlorophyll *a*. This conclusion is based on flash-induced absorption changes in the subnanosecond range, ESR measurements and redox potential determinations (the E_m of chlorophyll *a* is -0.88 V in dimethyl-formamide). The arguments are not definitive; however, no other molecule presently appears to be a better candidate.

'Downhill' from A_1 is an ensemble of three bound iron–sulphur centres named X, Centre A, and Centre B (Malkin & Bearden, 1978). Iron–sulphur proteins are small proteins that contain one or several

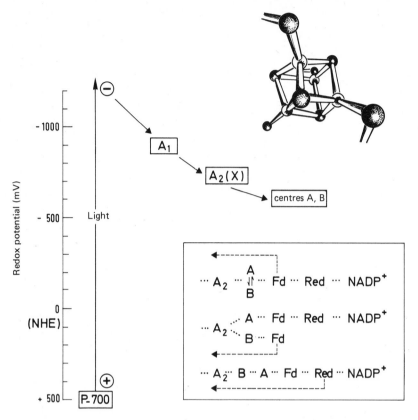

Fig. 6. A scheme of electron transfer on the acceptor side of PS-I. The figure also presents a 4Fe–4S centre and the possible ways of branching of the cyclic electron path on the linear path. A, B, centres A and B; Fd, ferredoxin; Red, ferredoxin–NADP reductase.

clusters generally constituted of 2Fe, 2S or 4Fe, 4S attached by cysteine residues (Fig. 6). They are universally present in bioenergetic structures. Soluble ferredoxins, which are one-electron carriers, are the best known of these molecules. In PS-I, X and the Centres A and B have been well studied by ESR. Centres A and B have spectra typical of iron–sulphur compounds, but X has a quite different spectrum, which makes its assignment as an iron–sulphur centre still tentative. The centre X $(E_m \simeq -730\,\mathrm{mV})$ is undoubtedly a more primary acceptor than Centres A $(E_m \simeq -550\,\mathrm{mV})$ or B $(E_m \simeq -590\,\mathrm{mV})$. These latter two centres have similar E_m values but these may vary from one organism to another; for the moment it is not clear whether the Centres A and B are arranged sequentially or constitute a small pool. All three electron carriers are membrane-bound and have not yet been isolated in

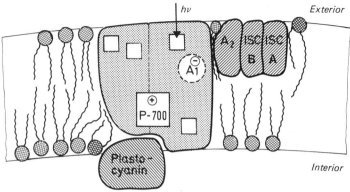

PS-I Reaction centre

Fig. 7. Schematic representation of the PS-I reaction centre and its possible location in the lipid bilayer of the photosynthetic membrane. A_1 and A_2: primary and secondary acceptors, respectively (see Table 2). ISC A and ISC B: iron–sulphur centres A and B, respectively.

a functional state. They probably constitute one single complex since procedures of differential extraction have all failed to extract one centre without the others. X and Centres A or B probably correspond to the chemical species named A_2 and P-430, respectively, in flash absorption experiments at physiological temperature (in contrast, ESR spectra of iron–sulphur proteins can only be obtained at very low temperature). The probable structure of the PS-I reaction centre is presented in Fig. 7.

The next electron carrier is, surprisingly, a soluble ferredoxin, which has long been known to be a part of the electron transfer chain: in most procedures of chloroplast preparation, the soluble ferredoxin is lost and the photoreduction of $NADP^+$, as well as photophorylations catalysed by PS-I, require the re-addition of ferredoxin. This is a small protein having a molecular weight of 10 000, with one 2Fe–2S centre, and a redox potential of $-420\,mV$. Its sequence of amino acids has been determined in many organisms and used for tracing evolutionary patterns. Reduction of $NADP^+$ by reduced ferredoxin is catalysed by a membrane-bound enzyme, ferredoxin–NADP reductase, which is a flavoprotein (molecular weight = 40 000; $E_m = -380\,mV$). An antibody against this protein inhibits electron transfer from ferredoxin to $NADP^+$, showing that the reductase is located on the cytoplasmic side of the thylakoid membrane. Ferredoxin is a one-electron carrier whereas the reduction of $NADP^+$ requires two electrons: the reductase is the link permitting the transitory accumulation of one reducing equivalent.

In addition to the (linear) electron transfer described, PS-I is also known to activate a so-called 'cyclic' electron transfer: electrons are re-injected in the chain between PS-II and PS-I, probably at the level of plastoquinone, permitting a proton translocation. The branching point between linear and cyclic electron flow is still not established (Fig. 6). One proposal places it at the level of the ferredoxin, which is weakly bound to the membrane and may diffuse to another site in the cyclic path. Iron–sulphur Centres A and B may also be involved: the centres may be in some redox equilibrium, with Centre A in the linear chain and Centre B in the cyclic path. Finally, the ferredoxin–NADP reductase may itself be the branching point and direct electron flow to the cycle, when $NADP^+$ is mostly reduced.

In green sulphur bacteria, such as *Chlorobium* or *Prosthecochloris*, a very similar sequence of electron acceptors operates (Olson & Hind, 1977; Clayton, 1980; Swarthoff, 1981). Two bound iron–sulphur centres of low redox potential reduce a ferredoxin, which in turn reduces NAD^+ by means of a ferredoxin–NAD reductase. A porphyrin, possibly bacteriopheophytin *a*, is a more primary acceptor; it may correspond to A_1, possibly a chlorophyll *a*, in PS-I. The resemblance between the reaction centres of PS-I and of green bacteria is remarkable and includes the kinetic behaviour at low temperature. The donors are also quite comparable: P-840 is probably a pair of bacteriochlorophyll *a* molecules, and its E_m ($+300\,mV$) is lowered compared to bacteriochlorophyll *a* in solution. The secondary donor is a *c*-type cytochrome.

The sequences of low-potential electron carriers are of considerable interest for the possible development of hydrogen production by photobiological means. Reduced iron-sulphur proteins can reduce H^+ into molecular hydrogen: $2H^+ + 2X^- \rightarrow H_2 + 2X$. This reaction is catalysed by a hydrogenase. In oxygen-evolving organisms, which are potentially the most interesting since they use water as a source of electrons, there are great practical difficulties because the hydrogenases are generally inactivated by oxygen. Important progress can be expected, however, in the stabilisation of hydrogenases.

General organisation, proton transfer and ATP synthesis

This description of the photosynthetic apparatus has so far been limited to sequences of electron carriers. In oxygen-evolving organisms the sequences are arranged so as to permit an overall electron transfer from H_2O to $NADP^+$ (Figs. 2 and 9). The best model for this organisation is the Z scheme proposed by Hill and Bendall (for reviews in this field, see Trebst, 1974; Crofts & Wood, 1978; Velthuys, 1980; and Avron, 1981).

The classical view is that of a physical chain of electron carriers. In reaction centres, the carriers are certainly arranged in a strict geometry, as shown by all methods of structural analysis. The determinants of this arrangement are still largely unknown, but are probably complex electrostatic and hydrophobic forces. In the electron-transfer sequence there are three points where diffusible molecules are involved, and this means that the 'chain' has a significance that is more statistical or functional than structural. These three points are the pools of plastoquinone, plastocyanin and ferredoxin (Fig. 8). Plastoquinone is a small lipophilic molecule, which can very probably diffuse laterally in the membrane lipid bilayer. Plastocyanin and ferredoxin are, in contrast, rather water-soluble and weakly bound to the membrane: they can probably diffuse inside the thylakoid and in the stroma. The 'chain' can thus be considered as being made of three rigidly organised sub-organelles (not considering the ferredoxin–NADP reductase): PS-I, PS-II (including the OEC) and an oxidoreductase (see below). The three main components may exist in a different stoichiometry and have a somewhat variable relative geometrical arrangement. In addition to this linear electron transfer pathway, there are probably several cyclic ones.

The insertion of the electron carriers has obvious advantages in permitting a good coupling between the light-harvesting antenna and the reaction centres, and in decreasing the probability of wasteful charge recombinations that would lower the quantum efficiency. The membrane location has another important role: to couple a proton translocation, from the outside to the inside, to the electron transfer. Proton transfer results mechanistically from the specific arrangement of carriers involving electrons only (primary partners, cytochromes, plastocyanin, iron–sulphur proteins) and of carriers which involve both electrons and protons (OEC, plastoquinones, $NADP^+$) (Figs. 2 and 9). The overall stoichiometry of H^+ and e^- is still uncertain (it may be variable, too). The scheme of Fig. 9 predicts that one e^- travelling along the sequence will pump two H^+, but the number may be greater than two (see below). In addition to pumping H^+, the electron transfer creates an electric field across the membrane: the arrangement of the reaction centres is such that the primary reactions place negative charges at the outer surface of the membrane and positive charges at the inner surface (Fig. 9). Thus, photo-induced electron transfer sets up a proton electrochemical gradient (see Junge, 1977), which includes an electrical component and a chemical component (H^+ concentration).

In purple bacteria, the electron-transfer chain is also inserted in the

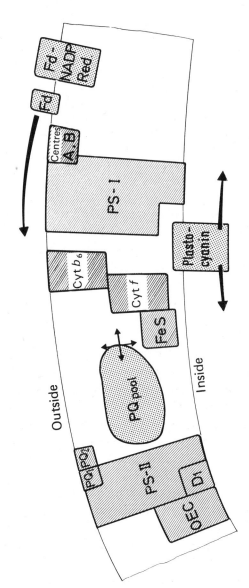

Fig. 8. Schematic representation of electron carriers of oxygenic photosynthetic membranes (for the abbreviations, see text and preceding figures). The carriers are arranged in order to visualise the three major complexes (PS-I, PS-II, and quinone–cytochromes oxidoreductase). The arrows indicate the hypothetical movements of the mobile carriers.

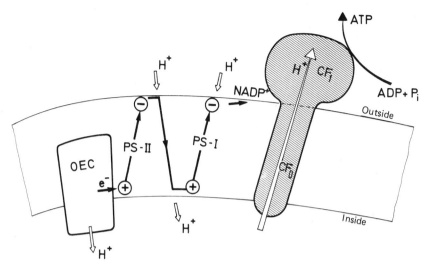

Fig. 9. Schematic view of electron and proton transfer in oxygenic photosynthetic membranes. CF_0 is a proton channel; CF_1 is the ATP synthetase.

membrane so as to pump protons: this is perhaps the main role of the process since the electron transfer is mainly cyclic (Fig. 10). The chain can be considered as made of two main blocks: the reaction centre, the quinone–cytochromes oxidoreductase, and one mobile carrier – cytochrome c_2. A pool of ubiquinone has no obvious role. The discovery of the quinone–cytochromes oxidoreductase has been a major step in our understanding of this system. As shown in Fig. 10, it contains two b cytochromes, a bound quinone, Q_z, and a 'Rieske' iron–sulphur protein (Bowyer, Dutton, Prince & Crofts, 1980). This complex is very similar to a mitochondrial oxidoreductase named Complex III, which can be efficiently coupled to bacterial reaction centres in solution, when some detergent is present (Fig. 10; Packham, Tiede, Mueller & Dutton, 1980). Proton pumping is effected by the reaction-centre bound quinone Q_2 and the oxidoreductase, in a complex manner (Crofts & Wood, 1978; Clayton, 1980). It is supposed that the oxidoreductase elements perform a so-called Q cycle which permits two protons to be pumped per electron transferred. This type of cycle was originally proposed by Mitchell (see discussion by Rich, 1981).

A quinone–cytochromes oxidoreductase also occurs in oxygen-evolving organisms: it is the complex shown in Fig. 8 which includes the cytochromes f and b_6 and a 'Rieske' iron–sulphur protein. Other elements may also function: a bound quinone, another b cytochrome; and a functional complex has been isolated and purified

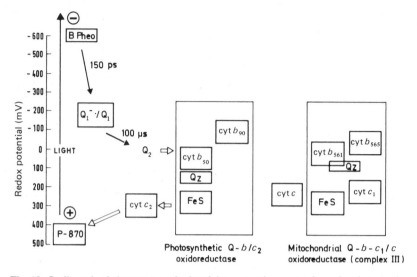

Fig. 10. Cyclic path of electron transfer involving a reaction centre from the photosynthetic bacterium *Rhodopseudomonas sphaeroides* and the quinone–cytochromes oxidoreductase from bovine heart mitochondria (adapted from Packham *et al.*, 1980). The subscripts for the *b* cytochromes refer to their redox potentials (photosynthetic cytochromes) or to the absorption maximum of their α band (mitochondrial cytochromes). Q_z is a bound ubiquinone and FeS is a high-potential ('Rieske') iron-sulphur centre. The vertical position of the components corresponds approximately to their redox potential (vertical scale).

recently (Hurt & Hauska, 1981). All of these components may be required to operate the Q cycle, and this may explain why $H^+ : e^-$ ratios greater than 2 have been measured by some authors.

The membrane potential created by the electron flow is utilised for the synthesis of ATP (see for example, McCarty, 1980; Avron, 1981). Photosynthetic membranes behave, in this respect, like many other membranes involved in a bioenergetic coupling (mitochondria, purple membranes, sarcoplasmic reticulum, bacterial respiratory membranes) (see Lee, Schatz & Ernster, 1979). The chloroplast thylakoid has even provided a good tool for these studies since all the processes can be triggered conveniently by flashes of light; their way of functioning is often considered as one of the best supports for the chemiosmotic theory. It is almost certain that the light-induced membrane potential drives the ATP-synthetase, which is made of a proton channel CF_0 and an active complex CF_1 (Fig. 9). The proton concentration gradient and the electrical field are decreased in the process. ATP is synthesised in the proportion of one molecule per three protons in optimum conditions. It is worth mentioning that the ATP-synthetase can function reversibly and that ATP hydrolysis induces a transmembrane proton translocation and even a reverse (partial) electron flow.

Conclusion

In conclusion it may be useful to compare the membrane reactions of photosynthesis with other photobiological processes.

Recent research has shown that the primary steps of these processes (for example, in photosynthesis, in vision, in purple bacteria and probably in protochlorophyllide reduction) are extremely rapid. The development of pulsed lasers of picosecond duration, and of analytical tools such as flash absorption and fluorescence spectroscopy, resonance Raman spectroscopy, have made it possible to track reactions in the picosecond time-scale (10^{-12} s). In earlier hypotheses, the triplet state of the pigments was often supposed to play an important role, because its longer lifetime (as compared with the singlet excited state) was thought to facilitate photochemical reactions. However, in a few picoseconds the triplet state cannot become populated, and so the reactions must be initiated by the lowest singlet excited state of the pigments. Although this can take place only in a structured environment, it has important advantages for the organism: (i) the state is populated with almost 100% quantum efficiency; (ii) there is more energy available in the singlet than in the triplet excited state; (iii) the short-lived singlet state is not quenched by adventitious quenchers such as oxygen. Triplet states would react with O_2 and produce singlet oxygen, which can induce lethal oxidations in cells. This last advantage of singlets over triplets may be essential and it will be interesting to discover whether less well understood processes, such as phytochrome phototransformation or the 'blue-light' effect, also take place from singlet excited states.

The uniqueness of photosynthesis is essentially determined by the properties of the chlorophylls. These pigments have a large cross-section for the absorption of sunlight (the actual cross-section is enlarged by the antenna and by the accessory pigments); they have a rigid molecular skeleton which minimises the vibrational losses, and they also give rise to redox transitions which involve only one electron ($C \rightarrow C^+ + e^-$; $C + e^- \rightarrow C^-$) and which have mid-point potentials matching other biological molecules. Photosynthesis can thus have its essential energetic features: a quantum efficiency close to unity, and the capacity to convert a significant fraction of the energy of light into chemical free energy. This high energetic yield is obtained in spite of the great complexity of the successive reactions – it is possible that the great number of steps actually minimises the overall loss of free energy. Through these multiple steps, photosynthesis is linked to other basic biological processes: carbon metabolism, respiration, ATP synthesis or hydrolysis in membrane systems; this integration which is also one of the characteristics of photosynthesis, renders its study both complex and so fascinating.

Thanks are due to Dr A. Verméglio for permission to use original data for Fig. 4.

References

AMESZ, J. & DUYSENS, L. N. M. (1977). Primary and associated reactions of System II. In: *Topics in Photosynthesis*, vol. 2: *Primary Processes of Photosynthesis*, ed. J. Barber, pp. 149–85. Amsterdam: Elsevier/North-Holland Biomedical Press.

AVRON, M. (1981). Photosynthetic electron transport and photophosphorylation. In: *The Biochemistry of Plants*. vol. 8: *Photosynthesis*, ed. M. D. Hatch & N. K. Boardman, pp. 163–91. New York: Academic Press.

BARBER, J. (1977). *Topics in Photosynthesis*, vol. 2: *Primary Processes of Photosynthesis*. Amsterdam: Elsevier/North-Holland Biomedical Press.

BLANKENSHIP, R. E. & PARSON, W. W. (1979). Kinetics and thermodynamics of electron transfer in bacterial reaction centers. In: *Topics in Photosynthesis*, vol. 3: *Photosynthesis in relation to model systems*, ed. J. Barber, pp. 71–114. Amsterdam: Elsevier/North-Holland Biomedical Press.

BOLTON, J. R. & HALL, D. O. (1979). Photochemical conversion and storage of solar energy. *Ann. Rev. Energy*, **4**, 353–401.

BOUGES-BOCQUET, B. (1980). Kinetic models for the electron donors of Photosystem II of photosynthesis. *Biochim. Biophys. Acta*, **594**, 85–103.

BOWYER, J. R., DUTTON, P. L., PRINCE, R. C. & CROFTS, A. R. (1980). The role of the Rieske iron–sulfur center as the electron donor to ferricytochrome c_2 in *Rhodopseudomonas sphaeroides*. *Biochim. Biophys. Acta*, **592**, 445–60.

BRODA, E. (1975). *The Evolution of the Bioenergetic Processes*. Oxford: Pergamon Press.

CLAYTON, R. K. (1980). *Photosynthesis: physical mechanisms and chemical patterns*. Cambridge: Cambridge University Press.

CLAYTON, R. K. & SISTROM, W. R. (1978). *The Photosynthetic Bacteria*. New York: Plenum Press.

CROFTS, A. R. & WOOD, P. (1978). Photosynthetic electron-transport chains of plants and bacteria and their role as proton pumps. In: *Current Topics in Bioenergetics*, vol. 7, part A, ed. D. R. Sanadi & L. P. Vernon, pp. 175–244. New York: Academic Press.

DAVIS, M. S., FORMAN, A. & FAJER, J. (1979). Ligated-chlorophyll cation radicals: their function in photosynthetic oxygen evolution. *Proc. Natl. Acad. Sci., USA*, **76**, 4170–9.

DE VAULT, D. (1980). Quantum mechanical tunneling in biological systems. *Quart. Rev. Biophys.*, **13**, 387–564.

DOUCE, R. & JOYARD, J. (1979). Structure and function of the plastid envelope. *Adv. Bot. Res.*, **7**, 1–116.

DUTTON, P. L. & PRINCE, R. C. (1978). Reaction-center-driven cytochrome interactions in electron and proton translocation and energy coupling. In: *The Photosynthetic Bacteria*, ed. R. K. Clayton & W. R. Sistrom, pp. 525–70. New York: Plenum Press.

DUTTON, P. L., PRINCE, R. C. & TIEDE, D. M. (1978). The reaction center of photosynthetic bacteria. *Photochem. Photobiol.*, **28**, 939–49.

GIBBS, M. & LATZKO, E. (1979). *Encyclopedia of Plant Physiology*, vol. 6. *Photosynthetic Carbon Metabolism and Related Processes*. Berlin: Springer-Verlag.

GOVINDJEE (1975). *Bioenergetics of photosynthesis*. New York: Academic Press.

HAEHNEL, W., PRÖPPER, A. & KRAUSE, H. (1980). Evidence for complexed plastocyanin as the immediate electron donor to P-700. *Biochim. Biophys. Acta*, **593**, 384–99.

HARRIMAN, A. & BARBER, J. (1979). Photosynthetic water-splitting process and artificial chemical systems. In: *Topics in Photosynthesis*, vol. 3: *Photosynthesis in Relation to Model Systems*, ed. J. Barber, pp. 243–80. Amsterdam: Elsevier/North-Holland Biomedical Press.

HATCH, M. D. & BOARDMAN, N. K. (1981). *The Biochemistry of Plants: a Comprehensive Treatise*, vol. 8: *Photosynthesis*, ed. P. K. Stumpf & E. E. Conn. New York: Academic Press.

HURT, E. & HAUSKA, G. (1981). A cytochrome f/b_6 complex of five polypeptides with plastoquinol-plastocyanin-oxidoreductase activity from spinach chloroplasts. *Eur. J. Biochem.*, **117**, 591–9.

JOLIOT, P. & KOK, B. (1975). Oxygen Evolution in Photosynthesis. In: *Bioenergetics of Photosynthesis*, ed. Govindjee, pp. 387–412. New York: Academic Press.

JUNGE, W. (1977). Membrane potentials in photosynthesis. *Ann. Rev. Plant Physiol.*, **28**, 503–36.

KATZ, J. J., NORRIS, J. R., SHIPMAN, L. L., THURNAUER, M. C. & WASIELEWSKI, M. R. (1978). Chlorophyll function in the photosynthetic reaction center. *Ann. Rev. Biophys. Bioeng.*, **7**, 393–434.

KE, B. (1978). The primary electron acceptors in green plant photosystem I and photosynthetic bacteria. In: *Current Topics in Bioenergetics*, vol. 7, part A, ed. D. R. Sanadi & L. P. Vernon, pp. 75–138. New York: Academic Press.

KLIMOV, V. V., DOLAN, E., SHAW, E. R. & KE, B. (1980). Interaction between the intermediary electron acceptor (pheophytin) and a possible plastoquinone–iron complex in Photosystem II reaction centers. *Proc. Natl. Acad. Sci., USA*, **77**, 7227–31.

KNAFF, D. B. & MALKIN, R. (1978). The primary reaction of chloroplast Photosystem II. In: *Current Topics in Bioenergetics*, vol. 7, part A, ed. D. R. Sanadi & L. P. Vernon, pp. 139–72. New York: Academic Press.

LEE, C. P., SCHATZ, G. & ERNSTER, L. (1979). *Membrane Bioenergetics*. Reading: Addison-Wesley.

MALKIN, R. & BEARDEN, A. J. (1978). Membrane-bound iron–sulfur centers in photosynthetic systems. *Biochim. Biophys. Acta*, **505**, 147–81.

MATHIS, P. (1981). Primary photochemical reactions in Photosystem II. In: *Proceedings of the 5th International Congress on Photosynthesis*, Halkidiki, Greece, 1980, ed. G. Akoyunoglou, pp. 827–37. Philadelphia: Balaban International Science Services.

MATHIS, P. & HAVEMAN, J. (1977). Analysis of absorption changes in the ultraviolet related to charge-accumulating electron carriers in Photosystem II of chloroplasts. *Biochim. Biophys. Acta*, **461**, 167–81.

MATHIS, P. & PAILLOTIN, G. (1981). Primary processes of photosynthesis. In: *The Biochemistry of Plants*, vol. 8: *Photosynthesis*, ed. M. D. Hatch & N. K. Boardman, pp. 97–161. New York: Academic Press.

MATOO, A. K., PICK, U., HOFFMAN-FALK, H. & EDELMAN, M. (1981). The rapidly metabolized 32 000-dalton polypeptide of the chloroplast is the 'Proteinaceous shield' regulating Photosystem II electron transport and mediating diuron herbicide sensitivity. *Proc. Natl. Acad. Sci., USA*, **78**, 1572–76.

MCCARTY, R. E. (1980). Photosynthetic phosphorylation by chloroplasts of higher plants. In: *Photochemical and Photobiological Reviews*, vol. 5, ed. K. C. Smith, pp. 1–47. New York: Plenum Press.

OLSON, J. M. & HIND, G. (eds.) (1977). *Chlorophyll-proteins, Reaction Centers and*

Photosynthetic Membrances. Brookhaven Symposium no. 28. Brookhaven National Laboratory.

OLSON, J. M. & THORNBER, J. P. (1979). Photosynthetic reaction centers. In: *Membrane Proteins in Energy Transduction*, ed. R. A. Capaldi, pp. 279–340. New York: Marcel Dekker.

PACHENCE, J. M., DUTTON, P. L. & BLASIE, J. K. (1981). The reaction center profile structure derived from neutron diffraction. *Biochim. Biophys. Acta*, **635**, 267–83.

PACKHAM, N. K., TIEDE, D. M., MUELLER, P. & DUTTON, P. L. (1980). Construction of a flash-activated-cyclic electron transport system by using bacterial reaction centers and the ubiquinone–cytochrome $b-c_1/c$ segment of mitochondria. *Proc. Natl. Acad. Sci., USA*, **77**, 6339–43.

PFISTER, K., STEINBACK, K. E., GARDNER, G. & ARNTZEN, C. J. (1981). Photoaffinity labeling of an herbicide receptor protein in chloroplast membranes. *Proc. Natl. Acad. Sci., USA*, **78**, 981–5.

RADMER, R. & KOK, B. (1975). Energy capture in photosynthesis: Photosystem II. *Ann. Rev. Biochem.*, **44**, 409–33.

REINMAN, S., MATHIS, P., CONJEAUD, H. & STEWART, A. (1981). Kinetics of reduction of the primary donor of Photosystem II: influence of pH in various preparations. *Biochim. Biophys. Acta*, **635**, 429–33.

RENGER, G. (1976). Studies on the structural and functional organization of system-II of photosynthesis. *Biochim. Biophys. Acta*, **440**, 287–300.

RICH, P. (1981). A generalized model for the equilibration of quinone pools with their biological donors and acceptors in membrane-bound electron transfer chains. *FEBS Lett.*, **130**, 173–8.

SATOH, K. & MATHIS, P. (1981). Photosystem-II chlorophyll *a*–protein complex: a study by flash absorption spectroscopy. *Photobiochem. Photobiophys.*, **2**, 189–98.

SAUER, K. (1975). Primary events and the Trapping of Energy. In: *Bioenergetics of Photosynthesis*, ed. Govindjee, pp. 116–81. New York: Academic Press.

SAUER, K. (1979). Photosynthesis: The light reactions. *Ann. Rev. Phys. Chem.*, **30**, 155–78.

SAUER, K. (1980). A role for manganese in oxygen evolution in photosynthesis. *Acc. Chem. Res.*, **8**, 249–56.

SWARTHOFF, T. (1981). Thesis, University of Leiden.

TREBST, A. (1974). Energy conservation in photosynthetic electron transport in chloroplasts. *Ann. Rev. Plant Physiol.*, **25**, 423–58.

TREBST, A. (1981). Action mechanism of herbicides in photosynthetic electron transport. In: *Proceedings of the 5th International Congress on Photosynthesis*, Halkidiki, Greece, 1980, ed. G. Akoyunoglou, vol. 6, pp. 507–20. Philadelphia: Balaban International Science Services.

VAN GRONDELLE, R. (1978). A study on primary and cytochrome reactions in bacterial photosynthesis. Thesis, University of Leiden.

VELTHUYS, B. R. (1980). Mechanism of electron flow in Photosystem II and toward Photosystem I. *Ann. Rev. Plant Physiol.*, **31**, 545–67.

VERMEGLIO, A. (1977). Secondary electron transfer in reaction centers of *Rhodopseudomonas sphaeroides*. *Biochim. Biophys. Acta*, **459**, 516–24.

WASIELEWSKI, M. R., NORRIS, J. R., SHIPMAN, L., LIN, C. P. & SVEC, W. A. (1981). Monomeric chlorophyll *a* enol: evidence for its possible role as the primary electron donor in Photosystem-I of plant photosynthesis. *Proc. Natl. Acad. Sci., USA*, **78**, 2957–61.

WOOD, P. M. (1978). Interchangeable copper and iron proteins in algae photosynthesis. *Eur. J. Biochem.*, **87**, 9–19.

WRAIGHT, C. A. (1979). The role of quinones in bacterial photosynthesis. *Photochem. Photobiol.*, **30**, 767–76.

TRANSDUCTION OF LIGHT ENERGY TO ELECTRICAL SIGNALS IN PHOTORECEPTOR CELLS

H. STIEVE

Institut für Neurobiologie der KFA Jülich GmbH, 5170 Jülich, GFR

The causal chain of the transduction process leading from photon absorption by the rhodopsin molecule to the membrane voltage signal is described and the underlying mechanisms are discussed for the individual steps. Mechanisms of gain control (light–dark adaptation) are also discussed in terms of the described causal transduction chain. This is presented separately for photoreceptors of vertebrates and invertebrates.

Receptor potential of the invertebrate photoreceptor resulting from bump summation

The absorption of a photon causes a bump, that is a delayed fluctuation of the membrane potential of the photoreceptor cell (Figs. 1 and 2). A bump is based on a transient fluctuation in ion-specific membrane conductance. These bumps sum up, when evoked by stronger light stimuli, to the smooth receptor potential.

Figure 3 shows that even the linear summation of single voltage-bumps (fluctuations in membrane voltage), recorded following very dim light flashes, results in a signal whose shape resembles that of a receptor potential. The shape of this sum-signal depends strongly – as does the actual receptor potential that is evoked by stronger light flashes – upon the Ca^{2+} concentration. Lowering the external Ca^{2+} concentration from 10 mmol l^{-1} to 250 μmol l^{-1} causes the following changes in the sum-signal: the latency period is prolonged to about 200%, the half-time of the repolarising phase of the signal is lengthened to about 400%, and the (unsaturated) response amplitude is reduced to about 35%.

These changes are mainly due to changes in the statistics of the bump latencies (Fig. 4). Specifically the latency distribution is broadened, and the mean latency is lengthened, by lowering extracellular Ca^{2+} concen-

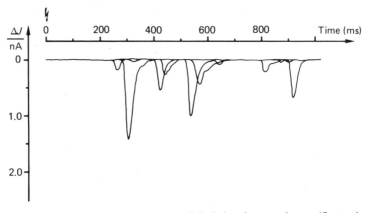

Fig. 1. *Limulus* ventral nerve photoreceptor: light-induced current-bumps (fluctuations of membrane current under voltage clamp conditions). Illumination: repetitive 20 ms, light flashes of 546 nm, about $1.3 \cdot 10^8$ photons cm^{-2} s^{-1}. Superposition of three recordings. Temperature, 15 °C; clamp voltage, -60 mV.

tration. The shape of an individual bump is not changed appreciably (Stieve & Bruns, 1981).

Causal chain of visual excitation

The causal chain leading from photon absorption to bump generation – that is, the transient opening of light-activated channels – is summarised in the flow chart (Fig. 5) for an invertebrate photoreceptor cell (*Limulus* ventral nerve photoreceptor). The situation differs quantitatively among photoreceptors of different invertebrate animals. In the photoreceptors of vertebrates the situation is different, mainly in that photon absorption causes a transient *closing* of light channels.

Apparently rhodopsin is not the molecule that forms the light channel – that is, the ion channel through the cell membrane, whose opening and closing is controlled by light.

Hamdorf & Kirschfeld (1980) showed that in the fly eye, metarhodopsin (M) newly formed following light absorption from rhodopsin (R) must exist for a minimal time of a few milliseconds to induce the generation of a receptor potential. No receptor potential is generated if R is photoregenerated (due to absorption of a second photon) immediately after M has been formed, within a time shorter than this latency period. A reasonable explanation is that a transiently activated state (Rh*) of rhodopsin (an early state of metarhodopsin) triggers the generation of the receptor potential.

Current bumps

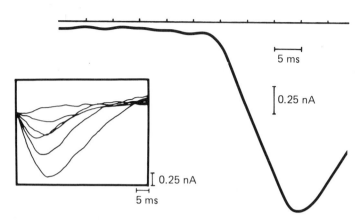

Fig. 2. *Limulus* ventral nerve photoreceptor: current bump following a 20 ms light flash at higher time resolution. Temperature, 15 °C; clamp voltage, −60 mV. Inset: superimposed bump recordings triggered by a current threshold.

The latency period

In the photoreceptors of *Limulus* and other invertebrates the light-induced transition from rhodopsin to metarhodopsin can be considerably shorter than the latency period. Metarhodopsin is formed in the *Limulus* ventral nerve photoreceptor within about 5 ms, whereas the latency period between photon absorption and the beginning of the bump can be much longer and varies greatly (Fig. 4). The variation of bump latency seems to be considerably less in insects (fly: J. G. J. Smakmann & D. G. Stavenga, personal communication; K. Hamdorf & S. Razmjoo, personal communication), and practically absent in verte-brate photoreceptors (Baylor, Lamb & Yau, 1979).

In contrast to the spectroscopically monitored changes of rhodopsin (M. Wilms & H. Stieve, unpublished), the mean latency period and its frequency distribution depend strongly upon Ca^{2+} concentration (Fig. 4).

Invertebrate photoreceptors show after very strong illumination a prolonged depolarising afterpotential (PDA) lasting for several seconds after the formation of metarhodopsin (Hillman, Keen & Winterhager, 1977). This afterpotential effect is also due to bump formation. According to Hamdorf (1979) and Hamdorf & Razmjoo (1979), the PDA in the photoreceptor of the fly is due to the formation of a special

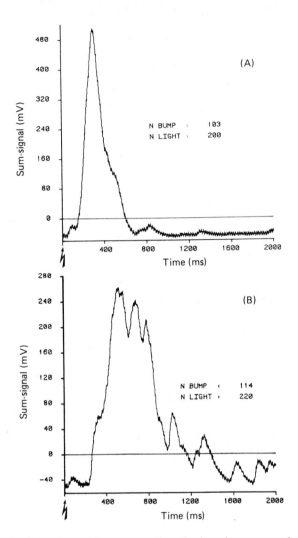

Fig. 3. Sum-signal, resultant of linear summation of voltage-bumps, recorded as fluctuations of membrane voltage of dark-adapted *Limulus* ventral nerve photoreceptor. The bumps evoked by very dim 10 ms white light flashes (E corresponding to about $3 \cdot 10^8$ (550 nm photons) cm^{-2}) were linearly summed by computer. The shape of the sum-signal resembles that of a receptor potential. (A) sum-signal of bumps while the receptor was superfused by physiological saline containing 10 mmol l^{-1} Ca^{2+}; (B) while the receptor was superfused by a saline in which the Ca^{2+} concentration was lowered to 250 μmol l^{-1}. N BUMP, number of individual bumps that were summed; N LIGHT, number of light stimuli. About every second stimulus is followed by a bump.

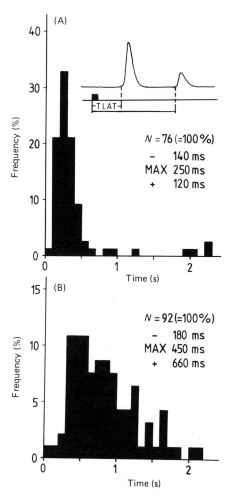

Fig. 4. Frequency distribution of bump latency periods of *Limulus* ventral nerve photoreceptor. The latency periods of voltage bumps (fluctuations of membrane voltage) following very dim 10 ms white light flashes (E corresponding to about $3 \cdot 10^8$ (550 nm photons) cm^{-2}) were measured and their frequencies plotted. The frequency distribution is characterised by the maximum (MAX) and by the lower and the upper half-width ($-$ and $+$). (A) The photoreceptor was superfused by physiological saline containing 10 mmol l^{-1} Ca^{2+}. (B) Same experiment, but during this period the photoreceptor was superfused by saline in which the calcium concentration was lowered to 250 μmol l^{-1}. Temperature, 15 °C; N, number of bumps; the number of stimulating light flashes was 200 (upper and lower curves), that is, about every second stimulus evoked 1 bump.

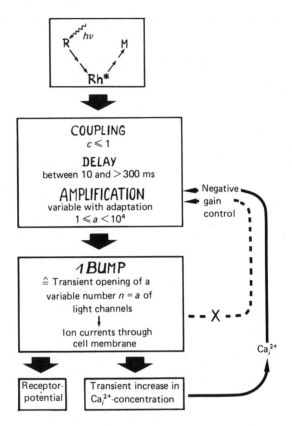

Fig. 5. Flow chart of causal chain of events in visual excitation in the photoreceptor of an invertebrate: *Limulus* ventral nerve photoreceptor. R, rhodopsin in excitable state; M, metarhodopsin; Rh*, light-activated state of rhodopsin molecule. Probability that an absorbed photon causes an R→M transition is about 0.7; probability that an absorbed photon evokes a bump is >0.5.

kind of long-lived activated state of rhodopsin Rh** (different from Rh*).

The duration of the latency period is therefore not determined primarily by the kinetics of the light-induced rhodopsin reactions, but by successive steps coupled to a transient, activated state of rhodopsin. There are several indications that the process determining the latency period (delay in Fig. 5) is different from the successive amplification processes: (*a*) The temperature dependence of the bump latency has a Q_{10} of 4, which is much higher than that of the time course of the bump; the half-width has a Q_{10} of 2.5 (Wong, Knight & Dodge, 1980). (*b*) The external Ca^{2+} concentration affects the bump latencies much more than the bump shape (see above). (*c*) In the *Limulus* ventral nerve photo-receptor the latency period and size of the bumps vary greatly. The duration of the latency period is however not correlated with the bump amplitude or the half-time of bump rise (H. Stieve, R. Barluschke & M. Bruns, unpublished).

Amplification

Bump generation is the result of a causal chain of events with a high amplification. The electrical energy of an average bump of a dark adapted *Limulus* ventral nerve photoreceptor is about 10^{-13} W. This is about 10^6 times greater than the energy of a 550 nm photon by which such a bump can be elicited.

The causal chain involves the recruitment of many molecules. Following the successful absorption of a photon by a single rhodopsin molecule, many molecules (as ion channels in the cell membrane) are affected in the late stages of the causal chain of the transduction process.

In the *Limulus* photoreceptor, this type of amplification can be estimated to be about as high as 10^4 in the maximally dark-adapted state; that is to say, one activated Rh* molecule triggers the transient opening of up to 10^4 light channels for one bump (Wong, 1978; Brown & Coles, 1979; Wong *et al.*, 1980; Stieve & Bruns, 1980*a*). This amplification factor is diminished by light adaptation (gain control) probably maximally to 1 (Wong, 1978). This means that in the maximally light-adapted state, one light activated Rh* molecule induces the transient opening of one light channel.

The frequency distribution of bump amplitudes of the dark-adapted *Limulus* ventral nerve photoreceptor has either a minimum and a maximum, or at least a shoulder (Stieve & Bruns, 1980*a*; parts A1 and B1 of Fig. 12). This shape and its conversion to a monotonic falling curve (Parts A2 and B2 of Fig. 12) due to light adaptation does not seem to be easily reconciled with the assumption of the release of quantised transmitter packages in the transduction process.

The shape of the current-bump (membrane current measured under voltage-clamp conditions) starts with an increasing steepness (Fig. 2). This indicates an acceleration of activation during the initial part of the rising phase of the bump. Such an amplification could be established in a visual cell, for example by a cooperative effect or by enzymatic action, as in an enzyme cascade. The search for cooperative actions in the visual excitation of photoreceptor cells has been negative so far.

An acceleration in activation could also be caused by autocatalysis, but there is no biochemical indication for an autocatalysis in visual cells. Alternatively an enzyme cascade could have the desired properties, such as increasing activation and high amplification, and there are a number of indications for such a mechanism in the vertebrate photoreceptor (see below).

Vertebrate photoreceptors

The transduction mechanism of vertebrate photoreceptors differs markedly from that of invertebrates. The outer photosensory membrane of rods and cones has a high permeability for sodium in the dark (Hagins, Penn & Yoshikami, 1970; Hagins, 1972; Korenbrot & Cone, 1972; Cone, 1973; Yoshikami & Hagins, 1973; Yau, McNaughton & Hodgkin, 1981), and is relatively depolarised under this condition, having a membrane potential of $-10\,mV$ to $-30\,mV$, in, for example, turtle cones (Baylor & Fuortes, 1970; Baylor, Fuortes & O'Bryan, 1971) and *Necturus* (Werblin & Dowling, 1969) – for review see Rodieck (1973). Illumination causes a transient decrease in the conductance of the photosensory membrane, and thus a transient hyperpolarisation (Toyoda, Nasaki & Tomita, 1969; Hagins *et al.*, 1970; Penn & Hagins, 1972; Werblin, 1975). Additionally there are voltage-sensitive conductances in the cell membrane of the rod – and probably cone – inner segment, such as a potassium conductance which is increased upon hyperpolarisation, and tends to increase the hyperpolarisation further (Arden, 1968, 1977; Werblin, 1975; Fain, Quandt & Gerschenfeld, 1977; Bader, Macleish & Schwartz, 1979). Rods and cones differ primarily in that the photosensory membrane of the cones is continuous and infolded all over the outer segment; whereas the rod photosensory membrane is topographically and electrically separated from the outer segment cell membrane, and the disk membrane forms closed flat sacs inside the rod outer segment. This poses the question: By what mechanism is the information of photon absorption by a rhodopsin molecule, located somewhere in the disk membrane, transmitted to the outer cell membrane where the conductance decrease occurs? A reasonable explanation is the assumption of a transmitter diffusing through the cytoplasm (Baylor & Fuortes, 1970; Cone, 1973).

The calcium hypothesis

Yoshikami & Hagins (1971) and Hagins (1972) proposed a 'calcium hypothesis' which is very appealing, especially since it is able to explain

(a) DARK

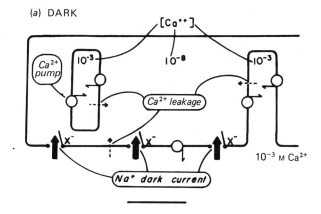

(b) LIGHT

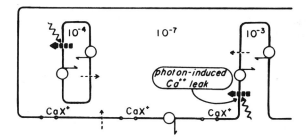

Fig. 6. Mechanism proposed by Hagins and Yoshikami for excitation in vertebrate rods and cones. Values for calcium ion activities in cytoplasm and disks are estimates derived from studies of nerve and muscle, and do not represent actual measurements. (*a*) rod/cone in darkness; (*b*) rod/cone in light (from Hagins, 1972).

the excitation mechanism both in rods and cones (Fig. 6): the calcium concentration, which is low in the cytosol of the outer segment in sustained darkness, is transiently raised by illumination; for calcium is released from the disks in rods, or from the infoldings of the photosensory membrane in cones. The raised intracellular calcium concentration causes calcium binding to the outer cell membrane and this in turn results in a conductance decrease of the cell membrane.

Four types of experiment have been carried out to test this hypothesis (for reviews see also Ostroy, 1977; Stieve, 1979; Brown *et al.*, 1979).

Under physiological conditions

Type 1 experiments investigating the light-induced release of calcium from disks have failed to demonstrate reliably a significant release of calcium, which could be sufficiently large and fast to serve as a transmitter at low intensities of light stimulus (Bownds, Gordon-Walker, Gaide-Huguenin & Robinson, 1971; Neufeld, Miller & Bitensky, 1972; Hendriks, Daemen & Bonting, 1974; Liebman, 1974; Mason, Fager & Abrahamson, 1974; Sorbi & Cavaggioni, 1975; Weller, Virmaux & Mandel, 1975; Hemminki, 1975a, 1975b; Winterhager, 1975; Shevchenko, 1976; Daemen & Bonting, 1977; Daemen, Schnetkamp, Hendriks & Bonting, 1977; Hendriks, van Haard, Daemen & Bonting, 1977; Smith, Fager & Litman, 1977; Szuts & Cone, 1977; Nöll, Stieve & Winterhager, 1979; Smith & Bauer, 1979; Szuts, 1980).

Kaupp & Junge (1977) and Kaupp, Schnetkamp & Junge (1979) found a fast, transient, light-induced release of membrane-bound calcium into the disk volume, but this released calcium did not leave the disk. It may correspond to the transiently released calcium in the invertebrate photoreceptor cell which is involved in the Ca^{2+}/Na^{2+}-binding competition, regulating opening and closing of the light channels, but which has no transmitter function (Stieve, 1974, 1981; Stieve & Bruns, 1978).

Attempts to demonstrate a light-induced intracellular calcium release *in vivo* in the photoreceptor cell have failed so far. Yoshikami, George & Hagins (1980) and Gold & Korenbrot (1980) found in intact retinas a large light-induced Ca^{2+} release from the rod outer segments into the extracellular space.

Type 2. There is a small uptake of calcium from the external medium into the disks (or rod outer segments, respectively) which depends on metabolic energy supplied by ATP (Bownds *et al.*, 1971; Ostwald & Heller, 1972; Winterhager, 1975; Schnetkamp, Daemen & Bonting, 1977; Miki, Kuo, Hayashi & Akiyama, 1980). This uptake is, however, much more difficult to demonstrate experimentally than the much larger calcium uptake into the sarcoplasmic reticulum.

Type 3. Lowering the external calcium concentration of the saline superfusing a cone retina to the nanomolar range is expected, according to the calcium transmitter hypothesis, to render the cones non-excitable. The experimental results, however, are contradictory: whereas Hagins & Yoshikami (1978) found in an iguano retina that cones become non-excitable under these conditions, Arden & Low (1978) observed in the pigeon retina and Bertrand, Fuortes & Pochobratsky (1978) in the turtle retina that cones stayed excitable after prolonged soaking of the retina in calcium-deficient media. These findings do not necessarily contradict the calcium hypothesis, since it may take considerable time and a high concentration of calcium-binding buffer to exhaust the calcium content in an extracellular compartment of the retina, as shown for the crayfish retina (Stieve & Classen-Linke, 1980; Schröder, Frings & Stieve, 1980).

Type 4. Intracellular application of the calcium-buffer EDTA either by ionophoretic injection into toad rod (Brown, Coles & Pinto, 1977) or by vesicle fusion into rat rods (Hagins & Yoshikami, 1977) lowered the sensitivity of the rods as predicted by the calcium hypothesis.

So about nine years after its formulation the calcium hypothesis is still neither proven nor disproven.

Enzyme cascade

There is considerable evidence for light-induced biochemical reactions in the vertebrate rod outer segment, reactions which are possibly involved in the transduction mechanism. There arises a picture, putting these reactions together into a causal chain, which can explain time delay and amplification in the transduction process by means of an enzyme cascade. Figs. 7 and 8 show a tentative scheme of the enzyme cascade in the vertebrate rod outer segment (cattle and frog), modified from H. Kühn (personal communication), summarising results from Liebman & Pugh (1979), Woodruff & Bownds (1979), Bownds (1980), Kühn (1980, 1981) and Fung, Hurley & Stryer (1981). The main steps in this hypothesis are as follows.

Step 1. Photon absorption by rhodopsin causes a change of the rhodopsin molecule into an activated transient state, Rh^*, (probably metarhodopsin II) to which the protein E ('GTPase' or 'GTP binding protein') from the cytoplasm is transiently bound (Kühn, 1980; Kühn, Bennett, Michel-Villaz & Chabre, 1981; Chabre, this volume).

Step 2. Rh^*, while having bound $E \cdot GDP$, catalyses the exchange of GDP for GTP bound to E (Fig. 8). During its life-time Rh^* can bind many $E \cdot GDP$ complexes successively, thus catalysing the GDP/GTP exchange in up to about 500 E molecules (amplification A_1) (Fung & Stryer, 1980).

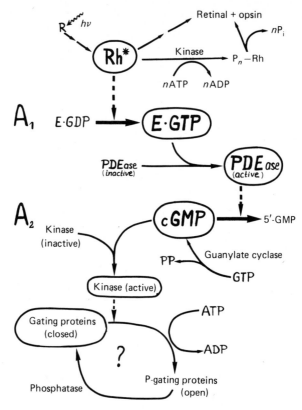

Fig. 7. A tentative scheme describing the proposed light-induced enzyme cascade, leading from light activation of rhodopsin to closing of the sodium channel in the plasma membrane of the vertebrate rod outer segment modified from H. Kühn (personal communication). For explanation see text. A_1 and A_2 are amplification steps. PDE, phosphodiesterase.

Rh* can be deactivated by slow spontaneous decay of active photo-product (for example, the reaction to metarhodopsin III and cleavage to retinal and opsin), or probably faster by its phosphorylation associated with consumption of ATP (Liebmann & Pugh, 1980). This phosphorylation is catalysed by a rhodopsin kinase which also is transiently bound to Rh* to fulfill its catalytic action (Kühn, 1978).

Step 3. The complex $E \cdot GTP$ activates the cGMP-phosphodiesterase (PDEase) (Fung *et al.*, 1981). This activation of PDEase causes a relatively rapid reduction of the cGMP level. One activated PDEase molecule can cause the cleavage of many (up to 100) cGMP molecules (amplification A_2). *In vitro*, one Rh* molecule can cause the hydrolysis

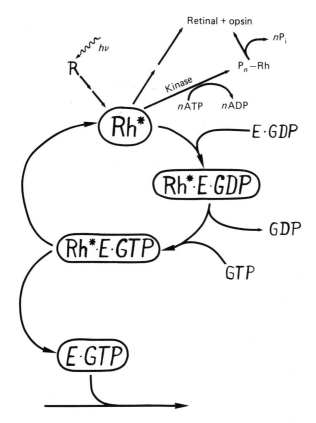

Fig. 8. Reaction scheme describing the formation of E·GTP catalysed by light-activated rhodopsin Rh* (modified from H. Kühn, unpublished; Kühn *et al.*, 1981; Fung *et al.*, 1981). For explanation see text.

of up to between 10^5 and 10^6 cGMP molecules per second (Yee & Liebmann, 1978).

Step 4. It seems conceivable (Bownds, 1980; Polans, Hermolin & Bownds, 1979) that cGMP activates a protein kinase which catalyses the phosphorylation of a gating protein in the plasma membrane. This gating protein could control the gating of the light-dependent sodium-channel. The gate of the sodium-channel should be open when the gating protein is phosphorylated and closed when it is dephosphory-lated. The light-induced drop in cGMP level could cause a decrease in activity (deactivation) of the kinase; dephosphorylation of the gating protein would thereby become more influential, resulting in the closing of sodium-channels.

Such an enzyme cascade could provide an explanation for latency and amplification of the electrical light response, although it is not yet clear which reactions determine the latency period. One critical question is that of the velocity of the described enzymatic reactions. There are indications that the light-induced lowering of the cGMP level may be fast enough for a causal step in the excitatory transduction process (Woodruff & Bownds, 1979 and indirect measurements by Yee & Liebmann, 1978). Raising the cGMP level in the rod, either by applying the PDEase inhibitor IBMX (Lipton, Rasmussen & Dowling, 1977; Leser, 1981), or by intracellular cGMP injection into toad rods (Waloga, Brown & Pinto, 1978; Nicol & Miller, 1978) effects an increase in sensitivity.

It is not clear how calcium ions act in this causal chain of excitation. The enzymatic reactions described (Fig. 7), seem to depend to only a limited extent upon the Ca^{2+} concentration. It could be that the phosphorylation of the gating protein is calcium-dependent. Moreover, Ca^{2+} and cGMP show, in some respects, antagonistic actions on the light response (Lipton, Ostroy & Dowling, 1977; Lipton *et al.*, 1977; Waloga *et al.*, 1978; Leser, 1981). This may suggest the possibility that the transduction process is caused not so much by a light-induced variation of the level in intracellular Ca^{2+} but rather by a light-induced change in cGMP level, which in turn regulates the effectiveness of the internal calcium in closing the sodium channels, for example by changing the affinity of a membrane protein to bind Ca^{2+}. Such a mechanism would explain why the experiments studying the release of calcium from disks showed only small effects as compared to the experiments in which calcium buffers were applied intracellularly. Opening and closing of the sodium-channels may depend upon calcium binding, perhaps in a way similar to that in the invertebrate photoreceptor. There the closing of the light-activated channel is controlled by calcium binding, characterised by a calcium/sodium binding competition in which calcium and sodium have anatagonistic actions (Stieve, 1974, 1981; Stieve & Bruns, 1978).

In invertebrate photoreceptors the biochemical nature of the transduction mechanism is less well understood than in the photoreceptors of vertebrates. Calcium ions here clearly do not act as an excitatory transmitter. There are indications that a GTP-binding protein is involved in the transduction process (Ebrey, Tsuda, Sassenrath, Wert & Waddell, 1980). Injection of relatively unhydrolysable GTP analogues such as GTP-γ-S into the *Limulus* ventral nerve photoreceptor causes an increase in the production of bumps in the dark (Brown, Bolsover & Malbon, 1981; Fein & Corson, 1981).

Bumps can occur even after the *Limulus* ventral nerve photoreceptor has stayed for more than an hour in the dark. It is believed by several authors that these bumps are generated spontaneously and are not evoked by photon absorption with an extremely long delay. If this is true, it may give a clue to the understanding of vertebrate vision.

Phylogenetic reflection

It is most probable that the transduction mechanism of vertebrate photoreceptors has evolved from that of invertebrates. An attractive hypothesis was formulated by Kramer (1977) and Kramer & Widmann (1977) (see also Stieve, 1977). If bumps such as those in *Limulus* photoreceptors were to be generated spontaneously at a high rate in vertebrate photoreceptor in the dark, they would result in a constant depolarisation of the photoreceptor membrane in the dark. Illumination would then cause no substantial further increase in bump rate but, as in invertebrate light adaptation, instead would cause a diminution of bump size and thus a graded light-dependent hyperpolarisation. In this way the vertebrate phototransduction process would be homologous with invertebrate light adaptation (and possibly both could be mediated by calcium ions). It would mean, however, that a different mechanism for light adaptation is needed for vertebrates. Indeed light adaptation of vertebrate photoreceptors shows a number of important differences compared with that of invertebrates (Kramer & Widmann, 1977; Widmann, 1979). Calcium probably does not play a major role in vertebrate light adaptation (Bastian & Fain, 1979); ionised calcium is clearly not the main inhibitory transmitter responsible for desensitisation of vertebrate photoreceptors. A hyperpolarising bump as measured in a vertebrate photoreceptor would then be homologous with a transient light adaptation, and so something quite different from a depolarising bump of an invertebrate photoreceptor.

Adaptation (sensitivity control) in invertebrate photoreceptors

Light adaptation is at least largely a reduction of the amplification factor (gain) in the transduction process (Fig. 5). In invertebrates, as first indicated by experiments of Lisman & Brown (1972, 1975), intracellular calcium ion concentration controls a sensitivity reduction in light adaptation (but see also Razmjoo & Hamdorf, this volume). These reactions have been confirmed by several authors and quantitatively specified (Fein & Lisman, 1975; Fein & Charlton, 1977*a*, *b*, 1978; Stieve & Pflaum, 1978*a*; Stieve & Bruns, 1980*b* and Stieve, 1981). Fig. 9 shows

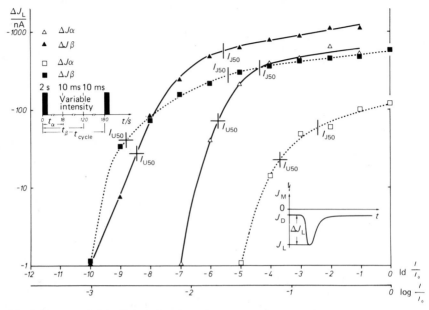

Fig. 9. Response height (light-induced membrane current) versus stimulus intensity curves for two different states of adaptation of a *Limulus* ventral nerve photoreceptor cell superfused by salines containing 10 mmol l^{-1} (solid curve) or 40 mmol l^{-1} Ca^{2+} (dashed curve). The upper left inset describes the stimulus programme which was repeated every 3 min. A strong constant light-adapting 2 s illumination (I_{LA} about 1.8·10^{16} (550 nm photons) cm^{-2} s^{-1}) is followed by two 10 ms test stimuli of variable intensity in order to obtain two stimulus–response curves. The first test stimulus is to record the relative light-adapted α-curve, the second test stimulus the dark-adapted β-curve. The response height ΔJ_{max} of the light induced receptor current (inset lower right) is plotted versus the logarithm to the base 2 (ld) of the intensity of the light stimulus; the cell membrane was clamped to −50 mV; I_0 about 3.7·10^{16} (550 nm photons) cm^{-2} s^{-1}. The stimulus programme (intensity and timing) is identical in both sets of measurements. The response to the 2 s light-adapting illumination was unclamped. Temperature, 15 °C. At each curve the stimulus intensity evoking half-saturation of the amplitude of the membrane current is indicated by I_{J50} and the stimulus intensity evoking half saturation of the amplitude of the voltage response (receptor potential) of the same but unclamped cell under similar stimulus conditions by I_{U50} (see Stieve & Klomfass, 1981).

in a typical experiment how a sensitivity shift due to light adaptation depends on the extracellular Ca^{2+} concentration. Under the experimental conditions described in Fig. 9, the sensitivity of the dark-adapted photoreceptor is not calcium dependent. The sensitivity shift (desensitisation) due to a constant, light-adapting illumination is, however, greater when the external Ca^{2+} concentration is raised, and smaller when the external Ca^{2+} concentration is lowered (Stieve & Bruns, 1980*b*; Stieve & Klomfass, 1981). On the other hand, our recent

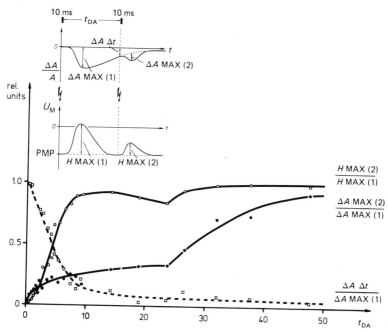

Fig. 10. Time course of dark adaptation of *Limulus* ventral nerve photoreceptor. A 10 ms white light flash of constant intensity (I_0, corresponding to about $7.5 \cdot 10^{17}$ (550 nm photons) $cm^{-2} s^{-1}$) was administered after various delays after a light adapting 10 ms flash of the same intensity. Two kinds of light responses to the test stimulus are plotted: the amplitude, H MAX, of the membrane voltage signal (receptor potential) and ΔA MAX, the amplitude of the light-evoked arsenazo signal (see inset). Additionally, $\Delta A \, \Delta t$, the height of the arsenazo absorption before the beginning of the test stimulus (a measure for the pre-stimulus level of intracellular Ca^{2+} concentration), is plotted. All the values normalised with respect to the signal following the light-adapting flash. The increase in the receptor potential amplitude H MAX shows two phases, only the first is correlated with the pre-stimulus level of the intracellular calcium concentration ($\Delta A \, \Delta t$). The second phase of the recovery of the receptor potential is accompanied by a stronger rise in the intracellular arsenazo signal ΔA MAX. Room temperature.

experiments have shown quite clearly that there is additionally an important *calcium-independent* process of light adaptation in the photo-receptor of the invertebrate *Limulus* (see also O'Day & Lisman, 1981):

(*a*) In one type of experiment K. Nagy measured the time course of dark adaptation using the intracellularly-injected Ca^{2+} indicator arsenazo III (Nagy & Stieve, 1982; for methodology used see Maaz & Stieve, 1980). He compared the amplitude, H MAX, of the receptor potential with two parameters of the arsenazo absorption signal. These are $\Delta A \, \Delta t$, which is a measure of the pre-stimulus level of intracellular free calcium ions just before the test stimulus is applied; and ΔA MAX, which is the amplitude of the transient increase in intracellular Ca^{2+} concentration in response to the test stimulus. In the time course of dark adaptation following a light-adapting flash, two phases of recovery of the amplitude H MAX of the receptor potential, evoked by constant stimuli, can be seen (Fig. 10). The first phase correlates with the decrease of the pre-stimulus level of the intracellular calcium $\Delta A \Delta t$ which had been raised due to the light adaptation. The second phase of the recovery of H MAX, which coincides with the recovery of the light-evoked arsenazo response AA MAX, is not controlled by the intracellular calcium level. Clearly the intracellular calcium level controls only the first phase of dark adaptation.

(b) In another type of experiment by I. Classen-Linke (1981; Classen-Linke & Stieve, 1981), the sensitivity of the *Limulus* ventral nerve photoreceptor was measured electrophysiologically during the course of dark adaptation (Fig. 11). For this, the intensity of a test flash needed to evoke a criterion amplitude of the receptor potential was determined after various delays following a bright light adapting illumination. Here also two phases of dark adaptation can be distinguished, only the first of which is strongly dependent on the extracellular calcium concentration.

In the receptor potential of *Limulus* ventral nerve photoreceptor, two components c_1 and c_2 can be distinguished (Wulff & Müller, 1973; Maaz, Nagy, Stieve & Klomfass, 1981). During the first phase of dark adaptation, the amplitude of the receptor potential is dominated by the component c_1, whereas during the second phase of dark adaptation the receptor potential amplitude is determined mainly by the component c_2. Only the dark adaptation of component c_1 depends strongly upon external calcium concentration.

(c) Finally we measured directly the dependence of bump adaptation upon the calcium concentration by determining the bump amplitude distribution in various states of moderate light adaptation (Stieve & Bruns, 1980a; Stieve & Bruns, 1982). This bump adaptation depends on Ca^{2+}. Raising the extracellular Ca^{2+} concentration augments the effect of light adaptation: the diminution of bump size. Lowering the external Ca^{2+} concentration reduces bump light adaptation; however, this light adaptation cannot be prevented by lowering the extracellular Ca^{2+} concentration. Even at external calcium concentrations as low as <1 nmol l^{-1}, there is still a considerable bump adaptation, although this is less than in physiological saline (Fig. 12). At this low calcium concentration no light-induced transient calcium increase is observable as an arsenazo signal.

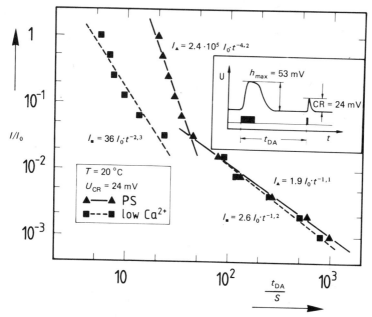

Fig. 11. Time course of sensitivity during dark adaptation of *Limulus* ventral nerve photoreceptor monitored by the intensity of the test stimulus, which evokes a criterion amplitude $U_{CR} = 24$ mV of the receptor potential. The stimulus sequence (inset) was repeated several times while the delay between light-adapting illumination and test stimulus was varied. In the double logarithmic plot the dark adaptation shows two phases, an initial fast one followed by a second slower phase. Only the first phase is markedly influenced by changes in extracellular calcium concentration: on lowering the external calcium concentration from 10 mmol l^{-1} (PS) to 250 μmol l^{-1}, the slope of the first phase is decreased. 1 s light-adapting stimulus; I_{LA} about $4.4 \cdot 10^{16}$ (543 nm photons) cm^{-2} s^{-1}; test stimulus 300 μs; I_0 about $1.4 \cdot 10^{14}$ (543 nm photons) cm^{-2} s^{-1}; temperature, 20 °C.

Our tentative explanation is that dark adaptation is controlled by at least two processes influencing the causal chain of visual transduction: (1) the intracellular calcium concentration controls the amplification factor when the intracellular calcium concentration is higher than a critical level; (2) The amplification factor is also controlled by other mechanisms which become apparent at low calcium concentrations when the influence of intracellular calcium is no longer dominating.

These other sensitivity-controlling factors may influence the reactions in the enzyme cascade of the transduction process; for instance, the limited availability of enzymes or substrates could control the sensitivity.

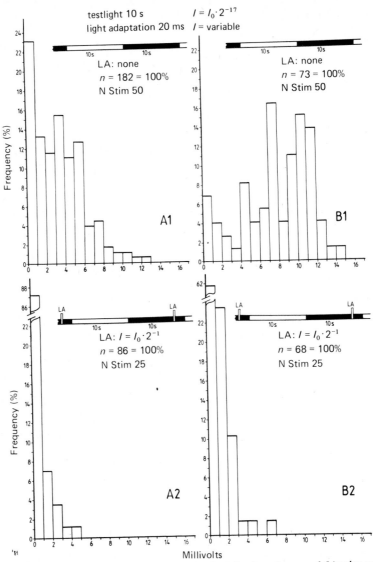

Fig. 12. Frequency distribution of the amplitudes of voltage-bumps of *Limulus* ventral nerve photoreceptor during illumination by weak 10 s bump-evoking stimuli at two different levels of adaptation and two different external calcium concentrations (10 mmol l^{-1}, *left side*, and <1 nmol l^{-1}, *right side*); temperature, 15 °C; I_0, corresponding to about $3.2 \cdot 10^{13}$ (550 nm photons) $cm^{-2} s^{-1}$. In a saline with calcium concentration lowered from 10 mmol l^{-1} to <1 nmol l^{-1} by using 1 mmol l^{-1} of the Ca-buffer EGTA, the average bump size in the dark-adapted state becomes larger, and the diminution in bump size due to light adaptation by the same pre-adapting flash becomes smaller but still significant. With an external calcium concentration as low as <1 nmol l^{-1} no light-induced increase in intracellular Ca^{2+}, monitored by arsenazo, is observed. N Stim = number of stimulus cycles; n = number of light bumps. Only the amplitudes of single bumps or the first of multiple bumps are plotted. About 3% of all bumps observed were riding bumps (riding on top of others).

Vertebrate photoreceptors

This type of gain control seems to be quite important in vertebrate adaptation. There are roughly $3 \cdot 10^6$ GTP-binding protein (E) molecules per rod outer segment in cattle. The ratio between rhodopsin, E, and PDEase molecules is roughly $100 \, R : 10 \, E : 1$ PDEase (Kühn, 1981). This and the fact that one Rh* molecule can affect many E molecules has the consequence that if only 0.1% of the rhodopsin molecules are light-activated, almost all E molecules are already engaged and this causes a tremendous reduction in amplification. This kind of substrate and enzyme limitation may explain the strong dependence of sensitivity upon rhodopsin bleaching (Rushton, 1977), for example bleaching of only 10% of the rhodopsin molecules raises the threshold in the human eye more than 100-fold. This reduction in sensitivity is much larger than would be expected from the lowered probability for photon absorption (Rushton, 1977). The bleaching product of rhodopsin (opsin) also influences the gain of the transduction process: a complete regeneration of rhodopsin is a prerequisite for a complete recovery of sensitivity in the course of dark adaptation (Grabowski, Pinto & Pak, 1972; Grabowski & Pak, 1975; Baylor & Hodgkin, 1974).

The phosphorylation of the light-activated rhodopsin molecule Rh* (Figs. 7 and 8) may also serve to reduce the sensitivity, perhaps also by an inactivation of PDEase (Liebmann & Pugh, 1980).

I wish to thank Hermann Kühn, Peter Hillman, Irmgard Classen-Linke, Inge Ivens and Jochen Winterhager for critical reading and valuable suggestions. Teresa Malinowska is acknowledged for the reference list together with Mechthilde Bruns, Josef Klomfass and Claudia Nieveler for considerable technical help with the manuscript. This study was supported by the DGF (SFB 160).

References

ARDEN, G. B. (1968). The excitation of photoreceptors. *Progr. Biophys. Mol. Biol.*, **19**, part 2, 373–421.

ARDEN, G. B. (1977). Three components of the photocurrent generated in the receptor layer of the rat retina. In *Vertebrate Photoreception*, ed. H. B. Barlow & P. Fatt, pp. 141–58. London: Academic Press.

ARDEN, G. B. & LOW, J. C. (1978). Changes in pigeon cone photocurrent caused by reduction in extracellular calcium activity. *J. Physiol.* **280**, 55–76.

BADER, C. R., MACLEISH, P. R. & SCHWARTZ, E. A. (1979). A voltage clamp study of the light response in solitary rods of the tiger salamander. *J. Physiol.*, **296**, 1–26.

BASTIAN, B. L. & FAIN, G. F. (1979). Light adaptation in toad rods: requirement for an internal messenger which is not calcium. *J. Physiol.* **297**, 493–520.

BAYLOR, D. A. & FUORTES, M. G. F. (1970). Electrical responses of single cones in the retina of the turtle. *J. Physiol.* **207**, 77–92.

BAYLOR, D. A., FUORTES, M. G. F. & O'BRYAN, P. M. (1971). Receptive fields of cones in the retina of the turtle. *J. Physiol.* **214**, 265–94.

BAYLOR, D. A. & HODGKIN, A. L. (1974). Changes in time scale and sensitivity in turtle photoreceptors. *J. Physiol.* **242**, 729–58.

BAYLOR, D. A., LAMB, T. D. & YAU, K. W. (1979). Responses of retinal rods to single photons. *J. Physiol.* **288**, 613–34.

BERTRAND, D., FUORTES, M. G. F. & POCHOBRADSKY, J. (1978). Action of EGTA and high calcium on the cones in the turtle retina. *J. Physiol.* **275**, 419–38.

BOWNDS, M. D. (1980) Biochemical steps in visual transduction: roles for nucleotides and calcium ions. *Photochem. Photobiol.* **32**, 487–90.

BOWNDS, D., GORDON-WALKER, A., GAIDE-HUGUENIN, A. C. & ROBINSON, W. (1971). Characterization and analysis of frog photoreceptor membranes. *J. Gen. Physiol.*, **58**, 225–37.

BROWN, J. E., BOLSOVER, S. R. & MALBON, C. (1981). Is a GTP binding side, analogous to that of vertebrate adenylate cyclase, involved in excitation of *Limulus* ventral photoreceptor? *ARVO, Sarasota 1981*, p. 233.

BROWN, J. E., BROWN, H. M., HESS, B., MÜLLER, P., OESTERHELT, D., PARSON, W., RÜPPEL, H., SPERLING, W. & STIEVE, H. (1979). Rhodopsin mediated processes, group report. In *Light-Induced Charge Separation in Biology and Chemistry*, ed. H. Gerischer & J. J. Katz, pp. 525–51. Life Sciences Research Report 12. Berlin: Dahlem-Konferenzen.

BROWN, J. E. & COLES, J. A. (1979). Saturation of the response of light in *Limulus* ventral photoreceptor. *J. Physiol.* **296**, 373–92.

BROWN, J. E., COLES, J. A. & PINTO, L. G. (1977). Effects of injections of calcium and EGTA into the outer segments of retinal rods of *Bufo marinus*. *J. Physiol.* **269**, 707–22.

CLASSEN-LINKE, I. (1981). Messung der Hell-Dunkel-Adaptation und ihre Beeinflussung durch die extrazelluläre Calciumkonzentration. Elektrophysiologische Messungen am Ventralnerv Photorezeptor von *Limulus polyphemus*. Dissertation. TH Aachen.

CLASSEN-LINKE, I. & STIEVE, H. (1981). Time course of dark adaptation in the *Limulus* ventral nerve photoreceptor – measured as constant response amplitude curve – and its dependence upon extracellular calcium. *Biophys. Struct. Mechanism*, **7**, 336–7.

CONE, R. A. (1973). The internal transmitter model for visual excitation, some quantitative implications. In: *Biochemistry and Physiology of Visual Pigments*, ed. H. Langer, 275–82. Berlin: Springer-Verlag.

DAEMEN, F. J. M. & BONTING, S. L. (1977). Transient light-induced conformational changes in rhodopsin. *Biophys. Struct. Mechanism*, **3**, 117–20.

DAEMEN, F. J. M., SCHNETKAMP, P. P. M., HENDRIKS, T. & BONTING, S. L. (1977). Calcium and rod outer segments. In *Vertebrate photoreception*, ed. H. B. Barlow & P. Fatt, pp. 29–40. London: Academic Press.

EBREY, T., TSUDA, M., SASSENRATH, G., WERT, J. L. & WADDELL, W. H. (1980). Light activation of bovine rod phosphodiesterase by non-physiological visual pigments. *FEBS Lett.* **116**, 217–19.

FAIN, G. L., QUANDT, F. N. & GERSCHENFELD, H. M. (1977) Calcium-dependent regenerative responses in rods. *Nature, London*, **269**, 707–9.

FEIN, A. & CORSON, D. W. (1981). Excitation of *Limulus* photoreceptors by vanadate and by hydrolysis-resistant analogue of guanosine triphosphate. *Science*, **212**, 555–6.

FEIN, A. & CHARLTON, J. S. (1977*a*). A quantitative comparison of the effects of intracellular calcium injection and light adaptation on the photoresponse of *Limulus* ventral photoreceptors. *J. Gen. Physiol.* **70**, 601–20.

FEIN, A. & CHARLTON, J. S. (1977*b*). Increased intracellular sodium mimics some but not all aspects of photoreceptor adaptation in the ventral eye of *Limulus*. *J. Gen. Physiol.* **70**, 601–20.

FEIN, A. & CHARLTON, J. S. (1978). A quantitative comparison of the time-course of sensitivity changes produced by Ca injection and light adaptation in *Limulus* ventral photoreceptors. *Biophys. J.* **22**, 105–13.

FEIN, A. & LISMAN, J. E. (1975). Localized desensitization of *Limulus* photoreceptors produced by light or intracellular calcium ion injection. *Science*, **187**, 1094–6.

FUNG, B. K. K., HURLEY, J. B. & STRYER, L. (1981). Flow of information in the light-triggered cyclic nucleotide cascade of vision. *Proc. Natl. Acad. Sci., USA*, **78**, 152–6.

FUNG, B. K. K. & STRYER, L. (1980). Photolyzed rhodopsin catalyses the exchange of GTP for bound GDP in retinal rod outer segments. *Proc. Natl. Acad. Sci., USA*, **77**, 2500–4.

GOLD, G. H. & KORENBROT, J. I. (1980). Light induced calcium release by intact retinal rods. *Proc. Natl. Acad. Sci., USA*, **77**, 5557–61.

GRABOWSKI, S. R. & PAK, W. L. (1975). Intracellular recordings of rod responses during dark adaptation. *J. Physiol.* **247**, 363–91.

GRABOWSKI, S. R., PINTO, L. M. & PAK, W. L. (1972). Adaptation in retinal rods of axolotl: intracellular recordings. *Science*, **176**, 1240–3.

HAGINS, W. A. (1972). The visual process: excitatory mechanism in the primary receptor cells. *Ann. Rev. Biophys. Bioenerg.* **1**, 131–58.

HAGINS, W. A., PENN, R. D. & YOSHIKAMI, S. (1970). Dark current and photocurrent in retinal rods. *Biophys. J.* **10**, 380–412.

HAGINS, W. A. & YOSHIKAMI, S. (1977). Intracellular transmission of visual excitation in photoreceptors: electrical effects of chelating agents introduced into rods by vesicle fusion. In *Vertebrate Photoreception*, ed. H. B. Barlow & P. Fatt, pp. 97–139. London: Academic Press.

HAGINS, W. A. & YOSHIKAMI, S. (1978). Calcium in excitation of vertebrate rods and cones. *Ann. N.Y. Acad. Sci.* **307**, 545–60.

HAMDORF, K. (1979). The physiology of invertebrate visual pigments. In *Handbook of Sensory Physiology*, VII/6A. *Comparative Physiology and Evolution of Vision in Invertebrates A*, ed. H. Autrum, pp. 145–224. Berlin, Heidelberg & New York: Springer-Verlag.

HAMDORF, K. & KIRSCHFELD, K. (1980). Reversible events in the transduction process of photoreceptors. *Nature, London*, **283**, 859–60.

HAMDORF, K. & RAZMJOO, S. (1979). Photoconvertible pigment states and excitation in *Calliphora*; the induction and properties of the prolonged depolarizing afterpotential. *Biophys. Struct. Mechanism*, **5**, 137–62.

HEMMINKI, K. (1975a). Light-induced decrease in calcium binding to isolated bovine photoreceptors. *Vision Res.* **15**, 69–72.

HEMMINKI, K. (1975b). Localization of ATPase in bovine retinal outer segments. *Exp. Eye Res.* **20**, 79–88.

HENDRIKS, T., DAEMEN, F. J. M. & BONTING, S. L. (1974). XXV. Light-induced calcium movements in isolated frog rod outer segments. *Biochim. Biophys. Acta*, **345**, 468–73.

HENDRIKS, T., VAN HAARD, P. M. M., DAEMEN, F. J. M. & BONTING, S. L. (1977). Calcium binding by cattle rod outer segment membranes studied by means of equilibrium dialysis. *Biochim. Biophys. Acta*, **467**, 175–84.

HILLMAN, P., KEEN, M. E. & WINTERHAGER, J. (1977). Discussion of selected topics about the transduction mechanism in photoreceptors. *Biophys. Struct. Mechanism*, **3**, 183–9.

KAUPP, U. B. & JUNGE, W. (1977). Rapid calcium release by passively loaded retinal discs on photoexcitation. *FEBS Lett.* **81**, 229–32.

KAUPP, U. B., SCHNETKAMP, P. P. M. & JUNGE, W. (1979). Light induced calcium release in isolated intact cattle rod outer segments upon photoexcitation of rhodopsin. *Biochim. Biophys. Acta*, **552**, 390–403.

KORENBROT, J. I. & CONE, R. A. (1972). Dark ionic flux and the effects of light in isolated rod outer segments. *J. Gen. Physiol.* **60**, 20–45.

KRAMER, L. (1977). Discussion to III. Conductivity control of the visual cell membranes, reported by R. M. Meech, K. Hartung & E. Uebags. *Biophys. Struct. Mechanism*, **3**, 153–7.

KRAMER, L. & WIDMANN, T. (1977). Quantitative model for the electrical response of invertebrate and vertebrate photoreceptors. *Biophys. Struct. Mechanism*, **2**, 333–6.

KÜHN, H. (1978). Light-regulated binding of rhodopsin kinase and other proteins to cattle photoreceptor membranes. *Bichemistry*, **17**, 4389–95.

KÜHN, H. (1980). Light- and GTP-regulated interaction of GTPase and other proteins with bovine photoreceptor membranes. *Nature, London*, **283**, 587–9.

KÜHN, H. (1981). Interactions of rod cell proteins with the disk membrane: influence of light, ionic strength, and nucleotides. In *Current Topics in Membranes and Transport*, ed. W. H. Miller, vol. 15. New York: Academic Press, in the press.

KÜHN, H., BENNETT, N., MICHEL-VILLAZ, M. & CHABRE, M. (1981). Interactions between photoexcited rhodopsin and GTP binding protein: kinetic and stoichiometric analysis from light scattering changes. *Proc. Natl. Acad. Sci.*, in the press.

LESER, K. H. (1981). Effect of isobutylmethylxanthine and related drugs on the receptor response (ERG, a-wave) of the frog retina at various extracellular calcium concentrations. *Z. Naturforsch.*, **36c**, 597–603.

LIEBMAN, P. A. (1974). Light-dependent Ca^{2+} content of rod outer segment disc membranes. *Invest. Opthalmol.* **13**, 700–1.

LIEBMAN, P. A. & PUGH, E. N., JR. (1979). The control of phosphodiesterase in rod disk membranes: kinetics, possible mechanisms and significance for vision. *Vision Res.* **19**, 375–80.

LIEBMAN, P. A. & PUGH, E. N., JR. (1980). ATP mediates rapid reversal of cyclic GMP phosphodiesterase activation in visual receptor membranes. *Nature, London*, **287**, 734–6.

LIPTON, S. A., OSTROY, S. E. & DOWLING, J. E. (1977). Electrical and adaptive properties of rod photoreceptors in *Bufo marinus*. I. Effects of altered extracellular Ca^{2+} levels. *J. Gen. Physiol.* **70**, 747–70.

LIPTON, S. A., RASMUSSEN, H. & DOWLING, J. E. (1977). Electrical and adaptive properties of rod photoreceptors in *Bufo marinus*. II. Effects of cyclic nucleotides and prostaglandins. *J. Gen. Physiol.* **70**, 771–91.

LISMAN, J. E. & BROWN, J. E. (1972). The effects of intracellular ionophoretic injection of calcium and sodium ions on the light response of *Limulus* ventral photoreceptors. *J. Gen. Physiol.* **59**, 701–19.

LISMAN, J. E. & BROWN, J. E. (1975). Effects of intracellular injection of calcium buffers on light adaptation in *Limulus* ventral photoreceptors. *J. Gen. Physiol.* **66**, 489–605.

MAAZ, G., NAGY, K., STIEVE, H. & KOMFASS, J. (1981). The electrical light response of the *Limulus* ventral nerve photoreceptor, a superposition of distinct components – observable by variation of the state of light adaptation. *J. Comp. Physiol.* **141**, 303–10.

MAAZ, G. & STIEVE, H. (1980). The correlation of the receptor potential with the light induced transient increase in intracellular calcium concentration measured by absorption change of Arsenazo III injected into *Limulus* ventral nerve photoreceptor cell. *Biophys. Struct. Mechanism*, **6**, 191–208.

MASON, W. T., FAGER, R. S. & ABRAHAMSON, E. W. (1974). Structural response of vertebrate photoreceptor membranes to light. *Nature, London*, **247**, 562–3.

MIKI, N., KUO, C.-H., HAYASHI, Y. & AKIYAMA, M. (1980). Functional role of calcium in photoreceptor cell. *Photochem. Photobiol.* **32**, 503–8.

NAGY, K. & STIEVE, H. (1982). Changes in intracellular calcium ion concentration in the course of dark adaptation measured by arsenazo III in the *Limulus* photoreceptor. *Biophys. Struct. Mechanism*, in press.

NEUFELD, A. H., MILLER, W. H. & BITENSKY, M. W. (1972). Calcium binding to retinal rod disc membranes. *Biochim. Biophys. Acta*, **266**, 67–71.

NICOL, G. D. & MILLER, W. H. (1978). Cyclic GMP injected into retinal rod outer segments increase latency and amplitude of response to illumination. *Proc. Natl. Acad. Sci., USA,* **75,** 5217–21.

NÖLL, G., STIEVE, H. & WINTERHAGER, J. (1979). Interaction of bovine rhodopsin with calcium ions. *Biophys. Struct. Mechanism,* **5,** 43–53.

O'DAY, M. & LISMAN, J. E. (1981). The influence of voltage-dependent conductances on the receptor potential in *Limulus* ventral photoreceptors. *ARVO, Sarasota 1981.* p. 181.

OSTROY, S. E. (1977). Rhodopsin and the visual process. *Biochim. Biophys. Acta,* **463,** 91–125.

OSTWALD, T. J. & HELLER, J. (1972). Properties of magnesium- or calcium-dependent adenosine triphosphate from frog rod photoreceptor outer segment discs and its inhibition by illumination. *Biochemistry,* **11,** 4679–86.

PENN, R. D. & HAGINS, W. A. (1972). Kinetics of the photocurrent in retinal rods. *Biophys. J.* **12,** 1073–94.

POLANS, A. S., HERMOLIN, J. & BOWNDS, M. D. (1979). Light induced dephosphorylation of two proteins in frog rod outer segments. Influence of cyclic nucleotides and calcium. *J. Gen. Physiol,* **74,** 595–613.

RODIECK, R. W. (1973). *The Vertebrate Retina.* San Francisco: W. H. Freeman.

RUSHTON, W. A. H. (1977). Visual adaptation. *Biophys. Struct. Mechanism,* **3,** 159–62.

SCHNETKAMP, P. P. M., DAEMEN, F. J. M. & BONTING, S. L. (1977). Biochemical aspects of the visual process. XXXVI. Calcium accumulation in cattle rod outer segments: evidence for a calcium–sodium exchange carrier in the rod sac membrane. *Biochim. Biophys. Acta,* **468,** 259–70.

SCHRÖDER, W., FRINGS, D. & STIEVE, H. (1980). Measuring Ca-uptake and release by invertebrate photoreceptor cells by a laser microprobe mass spectroscopy. *Scanning Electron Microsc.* 1980/II, 647–50.

SHEVCHENKO, T. F. (1976). Change of calcium ion activity while illuminating the suspension of the fragments of visual cell outer segments. *Biofizika SSR,* **21,** 321–3.

SMITH, H. G., JR. & BAUER, P.-J. (1979). Light-induced permeability changes in sonicated bovine discs: Arsenazo III and flow system measurements. *Biochemistry,* **18,** 5067–73.

SMITH, H. G., JR., FAGER, R. S. & LITMAN, B. J. (1977). Light-activated calcium release from sonicated bovine retinal rod outer segment discs. *Biochemistry,* **16,** 1399–405.

SORBI, R. T. & CAVAGGIONI, A. (1975). Effect of strong illumination on the ion efflux from the isolated discs of frog photoreceptors. *Biochim. Biophys. Acta,* **394,** 577–85.

STIEVE, H. (1974). On the ionic mechanisms responsible for the generation of the electrical response of light sensitive cells. In *Biochemistry of Sensory Functions,* ed. L. Jaenicke, pp. 79–105. Berlin, Heidelberg & New York: Springer-Verlag.

STIEVE, H. (1977). Thoughts on the comparative biology of photosensory function. *Verh. Dtsch. Zool. Ges.* 1–25.

STIEVE, H. (1979). Charge separation by rhodopsin containing photosensory membranes. In *Light induced Charge Separation in Biology and Chemistry,* ed. H. Gerischer & J. Katz, pp. 503–23. Dahlem Konferenz. Weinheim: Verlag Chemie.

STIEVE, H. (1981). Roles of calcium in visual transduction in invertebrates. In *Sense Organs,* ed. M. S. Laverack & D. J. Cosens, pp. 163–83. Glasgow & London: Blackie & Son.

STIEVE, H. & BRUNS, M. (1978). Extracellular calcium, magnesium and sodium ion competition in the conductance control of the photosensory membrane of *Limulus* ventral nerve photoreceptor. *Z. Naturforsch.* **33c,** 574–9.

STIEVE, H. & BRUNS, M. (1980a). Dependence of bump rate and bump size in *Limulus* ventral nerve photoreceptor on light adaptation and calcium concentration. *Biophys. Struct. Mechanism,* **6,** 271–85.

STIEVE, H. & BRUNS, M. (1980b). The sensitivity shift of *Limulus* ventral nerve photoreceptor in light adaptation depending on extracellular Ca^{2+}-concentration. *Verh. Dtsch. Zool. Ges.* 369.

STIEVE, H. & BRUNS, M. (1981). Calcium deficiency in *Limulus* photoreceptors causes a change in the latency distribution of bumps. *Biophys. Struct. Mechanism*, **7**, 344.

STIEVE, H. & BRUNS, M. (1982). Bump latency distribution and bump adaptation of *Limulus* ventral nerve photoreceptor in varied extracellular calcium concentration. *Biophys. Struct. Mechanism*, in press.

STIEVE, H. & CLASSEN-LINKE, I. (1980). The effect of changed extracellular calcium and sodium concentration on the electroretinogram of the crayfish retina. *Z. Naturforsch.* **35c**, 308–18.

STIEVE, H. & KLOMFASS, J. (1981). Calcium dependence of light evoked membrane current signal and membrane voltage signal and their changes due to light adaptation in *Limulus* photoreceptor. *Biophys. Struct. Mechanism*, **7**, 345.

STIEVE, H. & PFLAUM, M. (1978a). The response height versus stimulus intensity curve of the ventral nerve photoreceptor of *Limulus* depending on adaptation and external calcium concentration. *Vision Res.* **18**, 747–9.

SZUTS, E. Z. (1980). Calcium flux disk membranes. Studies with intact rod photoreceptors and purified disks. *J. Gen. Physiol.* **76**, 253–86.

SZUTS, E. Z. & CONE, R. A. (1977). Calcium content of frog rod outer segments and discs. *Biochim. Biophys. Acta*, **468**, 194–208.

TOYODA, J., NASAKI, H. & TOMITA, T. (1969). Light induced resistance changes in single receptors of *Necturus* and *Gekko*. *Vision Res.* **9**, 453–63.

WALOGA, G., BROWN, J. E. & PINTO, L. H. (1978). Effects of cyclic nucleotides and calcium ions on *Bufo* rods. Poster at the Meeting on Visual Sensitivity and Adaptation, Guildford, UK.

WELLER, M., VIRMAUX, N. & MANDEL, P. (1975). Role of light and rhodopsin phosphorylation in control of permeability of retinal rod outer segment discs to Ca^{2+} *Nature, London*, **256**, 68–70

WERBLIN, F. S. (1975). Regenerative hyperpolarization in rods. *J. Physiol.* **244**, 53–81.

WERBLIN, F. S. & DOWLING, J. E. (1969). Organization of the retina of the mudpuppy, *Necturus maculosus*. II. Intracellular recording. *J. Neurophysiol.* **32**, 339–55.

WIDMANN, T. (1979). Ein quantitatives Modell für den Erregungsmechanismus von Photorezeptoren. *Dissertation, Univ. Bayreuth.*

WINTERHAGER, J. (1975). Untersuchung der Calcium-Aufnahme und -Abgabe isolierter Stäbchenaussenglieder und Disks von Rinderretinen. Messungen mit dem Isotop ^{45}Ca. Diplomarbeit, RWTH Aachen.

WONG, F. (1978). Nature of light-induced conductance changes in ventral photoreceptors of *Limulus*. *Nature, London*, **276**, 76–9.

WONG, F., KNIGHT, B. W. & DODGE, F. A. (1980). Dispersion of latencies in photoreceptors of *Limulus* and the adapting-bump model. *J. Gen. Physiol.*, **76**, 517–37.

WOODRUFF, M. L. & BOWNDS, M. G. (1979). Amplitude kinetics, and reversibility of light induced decrease in guanosine 3',5'-cyclic monophosphate in frog photoreceptor membranes. *J. Gen. Physiol.* **73**, 629–54.

WULFF, V. J. & MÜLLER, W. J. (1973). On the origin of the receptor potential in the lateral eye of *Limulus*. *Vision Res.*, **13**, 661–71.

YAU, K.-W., MCNAUGHTON, P. A. & HODGKIN, A. L. (1981). Effect of ions on the light-sensitive current in retinal rods. *Nature, London*, **292**, 502–5.

YEE, R. & LIEBMAN, P. A. (1978). Light-activated phosphodiesterase of the rod outer segment. Kinetics and parameters of activation and deactivation. *J. Biol. Chem.* **253**, 8902–9.

YOSHIKAMI, S., GEORGE, J. S. & HAGINS, W. A. (1980). Light induced calcium fluxes from outer segment layer of vertebrate retinas. *Nature, London*, **286**, 395–8.

YOSHIKAMI, S. & HAGINS, W. A. (1971). Light, calcium and the photocurrent of rods and cones. *Biophys. Soc. Abstr.*, **11**, 47a.

YOSHIKAMI, S. & HAGINS, W. A. (1973). Control of the dark current in vertebrate rods and cones. In *Biochemistry and Physiology of Visual Pigments*, ed. H. Langer, pp. 245–55. Berlin, Heidelberg & New York: Springer-Verlag.

THE PHYSIOLOGY OF PHYTOCHROME ACTION

RICHARD E. KENDRICK

Plant Biology Department, The University, Newcastle upon Tyne
NE1 7RU, UK*

Introduction

Phytochrome, a pigment found in plants, exists in two forms (Pr, red absorbing and Pfr, far-red absorbing) which are convertible by red and far-red light.

$$\text{Pr} \underset{\text{Far-red}}{\overset{\text{Red}}{\rightleftharpoons}} \text{Pfr}$$

Phytochrome controls many aspects of plant growth and development (Borthwick, Hendricks, Parker, Toole & Toole, 1952; Smith, 1975; Smith, 1976a; Kendrick & Frankland, 1976; Satter & Galston, 1976; Pratt, 1978). It can detect the presence or absence of light (for example, in seed germination, Grime & Jarvis, 1975), the spectral quality of the light (Holmes & Smith, 1975; Morgan & Smith, 1978; Smith, 1976b), and it may be capable of detecting irradiance also (Hartmann, 1966; Mancinelli & Rabino, 1978). Phytochrome appears to have evolved in response to a need to detect the spectral quality of light, particularly the light under a vegetation canopy, so that germination can be prevented where light for photosynthesis is limited, or elongation growth can be accelerated so that the plant is able to top the canopy in its fight for survival. Obviously such a pigment acting as a signal transducer could detect the presence of light, but the capacity for the measurement of photon flux may have evolved as a secondary function.

The problem that concerns us here is how does a single pigment carry out these three possible functions, and how is this spectral information transduced into a form to which the plants can respond by modification of growth and development? In other words, what is the molecular mode(s) of phytochrome action?

* Present address: Plant Physiological Research, Agricultural University, Generaal Foulkesweg 72, 6703 BW, Wageningen, The Netherlands.

The old dogma

From analysis of action spectra, phytochrome was proposed as the receptor pigment for the effects of light on many physiological responses in plants, for example seed germination, seedling development and flowering (Borthwick, 1972*a*,*b*). In these cases the responses were initiated by a single pulse of red light lasting a few minutes, and the effects were fully reversible by a subsequent short exposure to far-red light. Are the responses due to the formation of Pfr or to the loss of Pr or both? In other words, which form of phytochrome is physiologically active? Circumstantial data indicated a correlation with Pfr formation, since a small amount of Pfr induces a large response:

$$\text{Pr} \underset{\text{Far-red}}{\overset{\text{Red}}{\rightleftharpoons}} \text{Pfr} \rightarrow \text{Responses}$$

(Borthwick & Hendricks, 1961). However, the late Dr W. S. Hillman (1967, 1972) has shown that there is often a paradox between the phytochrome as Pfr measured spectrophotometrically and the response of the plant.

More recently other workers have proposed that both Pr and Pfr be considered as physiologically active, acting in opposition to each other. In this way Smith (1981) proposes that phytochrome gives plants the capacity to sense the spectral quality of light, irrespective of the total phytochrome present.

$$\begin{array}{c}
\lambda \\
\rightarrow h\nu
\end{array}
\quad
\begin{array}{c}
\text{Pr} \\
\big\Updownarrow \\
\text{Pfr}
\end{array}
\quad
\begin{array}{c}
\xleftarrow{\text{Negative}} \\[1em]
\xrightarrow[\text{Positive}]{}
\end{array}
\quad
\left.\rule{0pt}{3em}\right\}
\begin{array}{l}
\text{Relative} \\
\text{physiological action}
\end{array}$$

Long-term and short-term responses

Initially the phytochrome-mediated responses studied were what might be called 'long term': they involve gross changes in development, such as a change from vegetative growth to floral initiation (Table 1). There is, therefore, a considerable delay between initiation of the phytochrome response (photon capture), and its measured expression. Despite this experimental drawback, positive conclusions can be drawn about rapid effects of phytochrome from carefully designed experiments. An example is the investigation of the time course of 'Pfr' action progressively longer dark intervals by introducing between the exposure of red light and far-red light in order to establish the length of dark

Table 1. *Selected examples of red/far-red reversible responses*

Response	Delay between irradition and response	Reference
Modification of flowering	Days/weeks	Fredericq (1964)
Induction of germination	Hours/days	Kendrick (1976)
Enzyme synthesis	Hours	Dittes *et al.* (1971)
Enzyme activity	Minutes	Oelze-Karow *et al.* (1970)
Chloroplast movement	Minutes	Haumpt (1970)
Ion fluxes	Minutes	Brownlee & Kendrick (1979*a,b*)
Nyctinastic leaf movement	Minutes	Fondeville *et al.* (1966)
Elongation growth	Minutes	Weintraub & Lawson (1972)
Surface potential changes	Seconds	Newman & Briggs (1972)
Transmembrane potential changes	Seconds	Löppert *et al.* (1978)

interval for which the response remains reversible by far-red light (escape time). Despite the long time interval before expression of the response in the case of flowering, the action of phytochrome can be shown to be relatively rapid (from a few seconds to several minutes in some species). In contrast, many examples of seed germination have a requirement for Pfr-action over a period of several hours, while the germination response can be measured just a few hours later (Kendrick & Frankland, 1969).

The wide range of physiological responses under phytochrome control implies that it acts at some major 'crossroad' of metabolism. The suggestion that Pfr was possibly the active form of an enzyme which acted at this 'crossroad' (Hendricks, 1964) has had little supporting evidence – although such a hypothesis could account for signal amplification and the number of responses under control. Subsequently, Mohr (1966*a,b*, 1972) has proposed that development is modified by phytochrome acting to switch genes on and off (Fig. 1). Although new genes are indeed activated as a consequence of phytochrome action, the discovery of more rapid or 'short term' responses such as the nyctinastic leaflet movement in legumes (Fondeville, Borthwick & Hendricks, 1966) led Hendricks & Borthwick (1967) to propose that phytochrome action changes the permeability or other properties of membranes. The leaflet movements, and the attachment of root tips to a negatively charged glass surface (Tanada, 1968), occur in a matter of minutes and were considered too rapid to involve modification of gene expression.

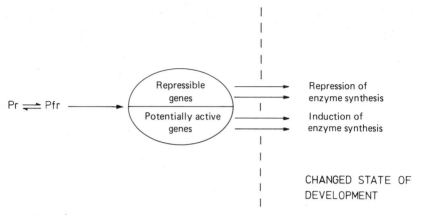

Fig. 1. Mohr's hypothesis to explain the phytochrome control of development.

Where does phytochrome act?

Evidence from physiological experiments

Haupt and his coworkers, investigating chloroplast movement in the alga *Mougeotia* (Haupt, 1970, 1980; Haupt, this volume), have provided evidence for both the site and the mode of action of phytochrome. Using microbeams of polarised light they have established that the single chloroplast moves within the cell (with a half-time of minutes) in such a way that it avoids Pfr. The microbeam need only penetrate the outer layer of the cytoplasm in order to repel a small portion of the chloroplast, demonstrating that the light does not have to be perceived by the chloroplast itself. Red light polarised with the e-vector parallel to the long axis of the cell elicits the response, whereas far-red light is only effective in reversing the response to red light if given with the e-vector perpendicular to the long axis of the cell. Such experiments point to the phytochrome being dichroically oriented in the outer layers of the cytoplasm; the rigid structure to which the phytochrome is attached is suggested to be the plasmalemma. Cells pretreated with far-red light, and then given double flashes of red light of short duration (milliseconds) show only the strong dichroic effect with the second flash (Haupt, Hupfer & Kraml, 1980). Haupt and coworkers suggest that the first red flash effects Pfr formation, which causes an association of other phytochrome molecules (Pr) with rigid membrane receptors. Subsequently the second flash shows the characteristic action dichroism. The fact that phytochrome forms dimers (Briggs & Rice, 1972) may well provide a possible explanation for this effect: the first flash produces Pfr–Pr dimers which associate with the membrane; and being dichroically orientated these can be activated by the second red flash to form active Pfr–Pfr dimers. Action dichroism of polarotrophic growth of fern protonema (Etzold, 1965) and bryophytes (D. J. Cove, personal communication) lends support to the discrete orientation of phytochrome molecules within the cell. Additional evidence for a fixed orientation of phytochrome molecules within the cell comes from etiolated corn coleoptiles, where subsaturating doses of red light, if plane-polarised normal to the longitudinal axis, produce 20% more Pfr than if plane-polarised parallel to the longitudinal axis (Marmé & Schäfer, 1972).

The generation of biopotential changes as a result of phytochrome phototransformation was first implied by experiments with root tips of barley, which in an appropriate medium became attracted to a negatively charged surface (Tanada, 1968). Direct measurement of surface potential led Jaffe (1968, 1970) to propose that a H^+ efflux stimulated by red light accounts for the observed potential change, and that in this specific system acetylcholine acts as a second messenger relaying the signal. Newman & Briggs (1972) and Newman (1974) showed that red and far-red light will modulate surface potential changes in oat coleoptiles, which occur within 15 s of exposure to irradiation. Such changes in surface potential strongly suggest changes in membrane permeability to ions. Racuson & Satter (1975) observed a hyperpolarisation of the membrane potential upon exposure to red light in motor cells of the pulvini of *Samanea samen*, a system which shows phytochrome controlled nyctinastic leaf movements (see Galston, this volume). These changes in potential occur within 90 s of the onset of irradition (Table 2). Racuson (1976) later showed that red light depolarises the membrane potential of oat coleoptile cells, and that the effect is reversed by far-red light, changes being initiated within 5 s of exposure to irradiation. He also established that the red light increases the electrical coupling of coleoptile cells along their longitudinal axis, an effect partially reversed by far-red light.

Changes in membrane potential have also been observed in both short-day (Löppert, Kronberger & Kandeler, 1978), and in long-day species of *Lemna* (Kandeler, Löppert, Rottenberg & Scharfetter, 1980). In both cases red light results in a *transient* depolarisation of membrane potential of mesophyll cells, and far-red light leads to a small but retained hyperpolarisation. However, photosynthesis complicates measurements in these cells, and after its inhibition with DCMU

Table 2. *Reports of phytochrome modulations of membrane potential*

Tissue	Cell	Effect of irradiation[a]		Reference
		Red	Far-red	
Avena sativa	Coleoptile parenchyma	Dep	Hyp	Racuson (1976)
	Coleoptile	Hyp	Dep	Newman & Sullivan (1976)
	Coleoptile	Dep	Hyp	Newman (1981)
Fern gametophyte	Protonemal	Dep	Hyp?	Racuson & Cooke (1981)
Lemna gibba	Mesophyll	Dep	Hyp	Kandeler *et al.* (1980)
Lemna paucicostata 6746	Mesophyll	Dep	Hyp	Löppert *et al.* (1978)
Nitella sp.	Internode	Dep	Hyp	Weisenseel & Ruppert (1977)
Phaseolus aureus	Hypocotyl cortex	Dep	Hyp?	Brownlee (1978)
Samanea samen	Flexor motor	Hyp	Dep	Racuson & Satter (1975)

[a]Dep = depolarization, Hyp = Hyperpolarization.

(3-(3,4-dichlorophenyl)-1, 1-dimethyl urea), a small *retained* depolarisation caused by red light (5–9 mv) is equal in magnitude to the hyperpolarisation caused by far-red light, and the potentials are fully reversible. These changes are initiated about one minute after the start of the actinic irradiation.

Weisenseel & Ruppert (1977) have reported a red-light induced depolarisation of the isolated internode cells of *Nitella*, which is Ca^{2+}-dependent. The lag time was 0.4–3.5 s and reached a steady state within 1–2 min. Although part of the depolarisation observed is connected with photosynthesis, as indicated by use of DCMU as an inhibitor, the smaller effects of far-red light, which are reversible by red light, appear to implicate phytochrome. The authors suggest that these effects could be triggered by a phytochrome modulation of Ca^{2+} influx at the plasmalemma. Subsequently, Dreyer & Weisenseel (1979) have shown that influx of $^{45}Ca^{2+}$ into filaments of the green alga *Mougeotia*, as indicated by autoradiography, is reversible by red/far-red light. These experiments were designed to eliminate complications from photosynthesis: light pulses being followed by a dark interval of 10 min before incubation for 1 min in $^{45}Ca^{2+}$, followed by a 30 min wash in unlabelled medium. The discrete nature of phytochrome action can once again be seen in the autoradiographs where there is a sharp demarcation along a filament between red-irradiated and non-irradiated, or red/far-red irradiated cells. These experiments confirm that the phytochrome-mediated Ca^{2+} influx is initiated after a period of minutes following irradiation, but do not confirm that Ca^{2+} influx is involved in any primary reaction. Yet Ca^{2+} appears to be a common denominator in several phytochrome controlled processes. Ca^{2+} is known to have a marked effect on several metabolic and developmental processes in plants and animals (Baker, 1976; Carafoli, 1977 and Anderson, Charbonneau, Jones, McCann & Cormier, 1980); and Ca^{2+} is therefore a strong contender as a second messenger for signal transmission in phytochrome-controlled processes (see Roux, this volume).

Another rapid response (half-time of minutes) is the nyctinastic movement of leaves in the legume *Mimosa pudica*. Movement in this species induced by seismic shock has been extensively studied by Toriyama (1955), who suggested that the movement involves ion fluxes in the pulvinus and consequent turgor changes. Irradiation of *Mimosa* plants with far-red light at the end of the day retards leaflet closure and the effect is reversible by red/far-red light (Galston & Satter, 1976). Again the involvement of phytochrome with membrane permeability is clear, and confirmed by work with the related species *Albizzia julibrissin* and *Samanea samen* (Satter, Applewhite & Galston, 1974; Galston & Satter, 1976). It is pertinent to note the discrete nature of phytochrome action in this case where the leaflet movement is restricted to the pulvini that are stimulated (Koukkari & Hillman, 1968). Observation of membrane potential changes that precede K^+ fluxes led to the suggestion that phytochrome either controls an ion pump, or opens ion channels, in pulvini motor cells possibly involving a H^+-sucrose cotransport system (Galston, Satter, Lonergan, Morse & Hatch, 1978). Ca^{2+} also appears to be involved in the seismic leaflet movement, again suggesting a possible role of Ca^{2+} in phytochrome action (Vanden Driessche, 1980).

Fluxes of ions have now been measured in several systems and have been shown to be under phytochrome control; the rapidity of altered fluxes suggests that the modulation of membrane properties is an early consequence of phytochrome phototransformation. The observation by Tezuka & Yamamoto (1975), that sections of mung bean hypocotyl hooks show a modulation in K^+ uptake, was refined by Brownlee & Kendrick, who demonstrated that apical and sub-apical sections of the hypocotyl hook show an enhancement and an inhibition of K^+ influx respectively, both being reversible by red/far-red light, and detectable 5 min after irradiation. Pre-loaded sections show similar modulations of K^+ efflux indicating that there is a two-way change in apparent passive permeability of the plasmalemma to K^+ (Brownlee & Kendrick, 1979*a,b*; Brownlee, Roth-Bejerano & Kendrick, 1979). The question arises: are these K^+ fluxes indicative of the molecular events involved at the primary site of phytochrome action, or do they follow a phytochrome-induced process that results in the K^+ moving along an electrochemical potential gradient? The K^+ influx studied is relatively slow and experiments show that the response is fully reversible during the 5–7 min lag before the response occurs. The effects of red light on membrane permeability appear to be discrete, retained and are reversible by far-red light; all consistent with the opening and closing of aqueous pores in the plasma membrane. Therefore the primary ion moved in the low energy mode of phytochrome action cannot be transported as a consequence of energy coupling during production of Pfr, or of cycling of phytochrome during the actinic red irradiation. The opposite responses of different sections of the mung bean hypocotyl point to K^+ influx being a secondary response to phytochrome phototransformation. It was previously proposed that the primary action involves formation of a gated pore which is open as Pfr and closed as Pr (Brownlee *et al.*, 1979). Data consistent with this view come from the water permeability changes in *Mougeotia* (Weinsenseel & Smeibidl, 1973) and *Taraxacum* (Carceller & Sanchez, 1972) which are suggested to be phytochrome-controlled; although Pike (1976) has found no effect on the permeability of pea epicotyl segments to tritiated water.

Although plant tissues act as efficient light pipes (Mandoli & Briggs, 1981), carefully conducted experiments show that phytochrome is present in many parts of the plant. In the case of photoperiodic induction of flowering it is clear that the leaves can be the site of light perception by phytochrome and the signal is relayed, presumably by hormones, to the apex where the following response occurs (see Vince-Prue, 1975 for review). Phytochrome in the primary leaves appear to be able to control hook opening in etiolated beans (De Greef, Caubergs, Verbelen & Moereels, 1976), and phytochrome in both internodes and leaves can control internode extension in light-grown mustard seedlings, although a much longer lag time is found when light is given to the leaves than to the internodes (Morgan, O'Brien & Smith, 1980). Physiological experiments indicate that phytochrome is also present and active in roots (Tanada, 1968; Tepfer & Bonnet, 1972; Racuson & Etherington, 1975). Predictions of the presence of phytochrome in roots as well as other parts of the plant based on physiological evidence have been confirmed by spectrophotometry (Furuya & Hillman, 1964) and immunocytochemistry (Pratt & Coleman, 1974).

Evidence from spectrophotometry and immunocytochemistry

Since direct spectrophotometry is limited to achlorophyllous tissues most studies were initially restricted to etiolated seedlings that are rich in phytochrome. More recently phytochrome measurements have been made by spectrophotometry in light-grown plants depleted of chlorophyll using the herbicide Norflurazon (San 9789), even though the levels of phytochrome present are low (Jabben, 1980). Spectrophotometry reveals that phytochrome is present in highest concentration in regions of most active growth (Furuya & Hillman, 1964; Kondo, Inoue & Shibata, 1973) and this is confirmed by immunocytochemistry (Pratt, Coleman & MacKenzie Jr., 1976). At the subcellular level spectrophotometry is almost impossible with tissue sections (Spruit, 1972; Kendrick & Smith 1976), but immunocyochemistry is capable of visualising the small quantities of phytochrome present. In dark-grown tissue, Coleman & Pratt (1974) have shown that Pr is dispersed throughout the cytosol (Fig. 2). Their pictures suggest that Pr is a soluble protein, although immunocytochemical stain clearly delimits membranes such as

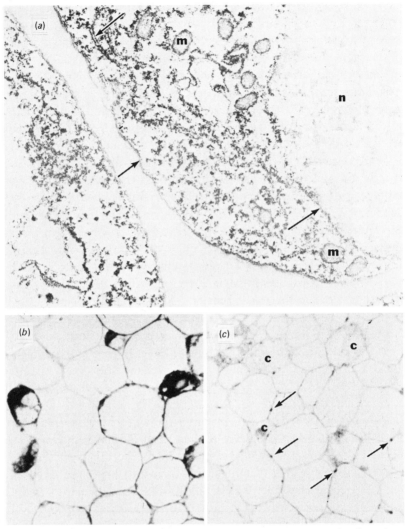

Fig. 2. (*a*) Electron micrograph of oat coleoptile parenchyma cells that have been immunocytochemically stained for phytochrome. Coleoptile fixed prior to light exposure so that phytochrome was present as Pr. Virtually all electron density is associated with phytochrome: n = nucleus, m = mitochondria. Some reaction product appears associated with membranes (arrows). Magnification: × 14000. (From Coleman & Pratt (1974), *J. Histochem. Cytochem.* **22**, 1039–47.) (*b*) Light micrograph of oat coleoptile parenchyma cells immunocytochemically stained for phytochrome. Coleoptiles were fixed at 0°C prior to light exposure so that phytochrome was present as Pr. These cells have large, central vacuoles, but where extensive cytoplasm has been sectioned, intense uniform stain for phytochrome is visible. (*c*) As in (*b*) except that the plants received 8 min of red light immediately prior to fixation so that phytochrome, as Pfr, is now present as discrete areas (arrows). The general cytoplasm, c, is now unstained. Magnification: × 350. (After Mackenzie Jr *et al.*, 1975.)

the plasmalemma, the nuclear membrane and the mitochondrial envelope. Following red irradiation, at least in a number of cereal species, the distribution pattern changes rapidly compared to dark-grown tissue: phytochrome aggregates in discrete centres about the cell, and is not clearly located in any identifiable organelle (Fig. 2*b*,*c*). This phenomenon called, sequestering (MacKenzie Jr., Coleman, Briggs & Pratt, 1975), has recently been confirmed in oats, maize and wheat using a fluorescent antibody technique (Epel, Butler, Pratt & Tokuyasu, 1980). Whether sequestration is a general phenomenon in other species is of interest, since it is a very rapid response to red light, having an escape time of a few seconds at physiological temperatures.

Spectrophotometry shows that phytochrome can be associated with membranes or organelle fractions when they are prepared from various dark-grown tissues (Table 3). Unfortunately the levels are low and are often at the limits of sensitivity of spectrophotometers. Also there are varying degrees of confidence as to the characterisation of the samples which, when coupled with the possibility that the association is non-specific, makes it difficult to conclude with any certainty that phyto-chrome is a membrane protein. Most of the phytochrome in dark-grown tissue is readily solubilised and does not appear to be tightly associated with membranes. At best therefore phytochrome would appear to be a peripheral membrane protein.

The amount of phytochrome as Pr associated with etioplasts (Smith, Evans & Hilton, 1978; Kraak & Spruit, 1980) and mitochondria (Furuya & Manabe, 1976) approximately doubles after red irradiation of crude homogenates. Georgevitch, Cedel & Roux (1977) using purified ^{125}I-labelled phytochrome showed that there were a fixed number of binding sites for Pfr binding on to mitochondria prepared from etiolated oat coleoptiles. A ferritin-labelled antibody revealed that phytochrome was apparently associated with the outer membrane of mitochondria from etiolated oat coleoptiles (Roux, McEntire, Slocum, Cedel & Hale, 1981). However, the same samples show a phytochrome modulation of Ca^{2+} flux, a process that must be regulated at the *inner* membrane. This raises the question of whether the phytochrome measured in these organelles and membrane fractions is native, or is associated with them by non-specific absorption? Also, is this phytochrome responsible for the phytochrome responses measured in such systems?

Evidence from in-vitro *systems*

Isolated subcellular components, such as organelles or membrane fractions, offer a simpler system for the study of phytochrome action;

Table 3. *Selected reports of phytochrome associated with subcellular organelles and membrane fractions*

Tissue	Organelle/membrane	Reference
Avena sativa	Mitochondria – outer membrane	Roux *et al.* (1981)
Cucurbita pepo	Endoplasmic reticulum	Marmé *et al.* (1976)
Glycine max	Endoplasmic reticulum	Williamson *et al.* (1975)
Hordeum vulgare	Etioplast	Evans & Smith (1976*a*)
	Etioplast	Hilton & Smith (1980)
	Mitochondria	Hilton & Smith (1980)
	Etioplast envelope	Evans & Smith (1976*b*)
	Chloroplast envelope	Hilton (1981)
Pisum sativum	Mitochondria	Manabe & Furuya (1975)
	Nuclei	Wagle & Jaffe (1980)
Triticum aestivum	Etioplast	Cooke *et al.* (1975)
Zea mays	Etioplast	Kraak & Spruit (1980)
	Plasma membrane	Marmé *et al.* (1976)

and there are several reports of 'rapid' phytochrome-controlled processes in such systems, and numerous accounts of phytochrome effects *in vitro*. Here we will concern ourselves with just a few of the responses (For a more detailed account of the diversity see Marmé, 1977; Pratt, 1979 and Quail, 1980).

One response is the rapid release of gibberellic acid-like substances from etioplasts taken from wheat (Evans & Smith, 1976*a*) and barley (Cooke, Saunders & Kendrick, 1975). This response has been reproduced in at least three laboratories and in the case of wheat appeared to be a function of the isolated etioplast envelope (Cooke & Kendrick, 1976). Recently Hilton & Smith (1980) have shown with purified organelles that the response is restricted to etioplasts and is not associated with mitochondria. Interestingly Evans & Smith (1976*b*) showed phytochrome associated with etioplast envelopes, but there is as yet no proof as to its physiological significance. Jose (1977) observed rapid phytochrome-controlled ATPase activity in a particulate cell-free extract from *Phaseolus aureus*. This membrane fraction was also shown to contain phytochrome.* Another *in vitro* response is the red light attachment of glycolate oxidase to peroxisomes at pH 9.0 (Roth-Bejerano & Lips, 1978). Red light given to crude homogenates from etiolated barley leaves simultaneously enriches phytochrome associated with the peroxisomal fraction and causes enzyme release. Filipin and cholesterol, both of which inhibit the phytochrome enrichment of the peroxisomal fraction, also inhibit enzyme release, suggesting a causal relationship between the phytochrome associating as Pfr with the peroxisomal fraction and the enzyme release (Roth-Bejerano, 1980). Recently protoplasts from etiolated oat coleoptiles (Hale & Roux, 1980), as well as mitochondria from oats (Roux *et al.*, 1981) have been found to show phytochrome modulations of Ca^{2+} flux; in these systems (see Roux, this volume), which use the dye murexide to measure Ca^{2+} in the surrounding medium, red light enhances efflux of Ca^{2+}. In contrast, in *Mougeotia* red light causes a rapid influx of Ca^{2+} (Dreyer & Weisenseel, 1979). The importance of Ca^{2+} as a second messenger in the control of many developmental processes in both plants and animals has been noted by several workers (Dreyer & Weisenseel, 1979; Roux *et al.*, 1981; Dieter & Marmé, 1980; Haupt, 1980), and they propose that the primary process of phytochrome action involves changes in the level of cytosolic Ca^{2+}, which in turn activates or deactivates the calcium-

---

* The report of modulation of peroxidase activity by Penel, Greppin & Boisard (1976) in a particulate fraction from *Cucurbita pepo*, following pre-irradiation with red light, does not appear to be reproducible (Quail, 1980).

binding protein calmodulin, which is present in plants as well as animals. Specific enzymes, such as the NAD-kinase associated with the mitochondrial membrane (Anderson *et al.*, 1980; Dieter & Marmé, 1980) and an ATPase associated with a microsomal fraction, probably enriched with plasmalemma (Dieter & Marmé, 1981), may be activated through calmodulin in this way. Unfortunately Marmé cannot reproduce the reversible effects reported by Roux *et al.* (1981) with isolated mitochondria from maize: although mitochondria from this tissue pre-treated with 6 h continuous far-red light showed a decrease in both the accumulation of Ca^{2+} in mitochondria and the calmodulin-dependent Ca^{2+} extrusion through the plasma membrane (P. Dieter & D. Marmé, personal communication).

Haupt (1980) has also pointed out the possible significance of Ca^{2+} in this movement of the chloroplast in *Mougeotia* (see Haupt, this volume). In this case a change in cystosolic Ca^{2+} concentration is again inferred as the stimulus (via calmodulin) for an ATPase that results in contraction of actomyosin filaments within the cytoplasm. Following red light, Ca^{2+} could move into the cytoplasm by release from an internal compartment such as the mitochondrion (Haupt, 1980), as well as by enhanced uptake from the surrounding medium (Dreyer & Weinsenseel, 1979).

Experiments performed to date, provide no unequivocal evidence to show that Ca^{2+} is the primary ion moved as a consequence of phytochrome phototransformation. We must therefore consider that other ions may be involved in the primary mode of phytochrome action, with Ca^{2+} moving down an electrochemical potential gradient in a passive manner as a second message. Experiments using calcium-specific ionophores such as A23187 (Pressman, 1976) are now required.

Is phytochrome a membrane constitutent?

Despite its physical dimensions being sufficiently large to allow it to span a biological membrane, phytochrome extracted from etiolated tissue has the characteristic of a soluble protein (Briggs & Rice, 1972; Pratt, 1978). There is little overall change in shape of the protein upon phototransformation as indicated by physical methods, (Pratt, 1978; Smith, Jr., 1981*b*) and by immunology (Pratt *et al.*, 1976), although, Pr and Pfr react differently to various chemical reagents (Pratt, 1978; Hahn & Song, 1981; Song, this volume). Such experiments tell us about the difference between Pr and Pfr and thus may provide an insight into the active centre of the molecule which is presumably in the vicinity of the chromophore. Hahn & Song (1981; Song, this volume) conclude that phototransformation of phytochrome from Pr to Pfr results in the chromophore becoming 'exposed', revealing a hydrophobic site that is available for interaction with a putative receptor.

Phytochrome has been shown to bind to columns of immobilised Cibacron Blue 3GA (affi-gel-blue) and Blue dextran (Smith Jr. & Daniels, 1980; Smith Jr., 1981a; Song, Kim & Hahn, 1982) and is specifically eluted by flavins such as FMN, suggesting a flavin binding site for phytochrome. Protein degradation products of phytochrome lacking the chromophore also bind to the affi-gel-blue columns and are eluted with FMN (Daniels & Smith Jr., 1981). This fact suggests that there are domains of the phytochrome molecule other than that in the vicinity of the chromophore, which show hydrophobicity and are possible sites of interaction with membranes.

Considerable attention has been given in the past ten years to the observation that red light given *in vivo* or *in vitro* increases the amount of phytochrome associated with pellets from crude homogenates of many species, a process referred to as 'red light enhanced phytochrome pelletability' (Boisard, Marmé & Briggs, 1974; Pratt, 1978; Roth-Bejerano & Kendrick, 1979). In some species the process requires Mg^{2+}, but in others appears to occur to a lesser extent in the absence of Mg^{2+}. The *in-vivo* red light enhanced pelletability, which is expressed *in vitro*, can be studied like any other phytochrome response. Initial experiments suggested that pelletability was not reversible by far-red light *in vivo*. However, short flashes of red and far-red light revealed that the process could be reversed (Pratt & Marmé, 1976) as long as the dark interval between red and far-red was no longer than a few seconds, indicating one of the most rapid escape times to date (half-time of 2 s at 25 °C). Several workers have suggested that this process is significant as it indicates that Pfr binds to a receptor site on membranes which sediment within the pellet, but as yet there is no direct proof of this. Nevertheless, the enhanced association of Pfr with membranes as compared to Pr does enable the nature of the phytochrome–membrane interaction to be investigated. One study (Roth-Bejerano & Kendrick, 1979) points to phytochrome associating with a steroid component in the pellet. Another approach to this problem is to investigate by immuno-cytochemistry the association of phytochrome with solid supports encapsulated by natural membrane fractions (Vierstra, Tokuhisa, Newcomb & Quail, 1981).

Most of the studies of purified phytochrome have been from etiolated tissue in which the phytochrome has not at any time been physiologically active. It is interesting to compare the few studies on phytochrome that has been present as Pfr within the tissue. The phytochrome is usually obtained by release from a pellet of red-irradiated tissue, after exposure to far-red light in the absence of Mg^{2+}. This phytochrome appears to elute more rapidly from gel exclusion columns (Grombein & Rüdiger, 1976; Yu & Carter, 1976a,b; Boeshore & Pratt, 1977; Epel, 1980), and to migrate more slowly on an electrophoresis gel. These factors suggest quite large changes in the molecule and are consistent with an increase in molecular weight of 20 000–50 000 (Epel, 1980). The 'larger' molecular weight phytochrome appears to be spectrocopically different, Pfr having an absorption maximum at 9–10 nm longer wavelength, typical of that found *in vivo* (Horwitz & Epel, 1977; Kendrick & Roth-Bejerano, 1978; Epel, 1980, 1981). In order to obtain this wavelength-form of phytochrome, two criteria have to be met: it is present as Pfr, and the presence of pelletable material.

It is therefore possible to speculate that in the absence of Mg^{2+} to stabilise a peripheral membrane protein 'receptor', phytochrome complexes with the 'receptor' protein and is released as a soluble complex. It is clear that future studies of phytochrome isolated from light-grown plants using conventional (Shimazaki, Moriyasu & Furuya, 1981), as well as newer techniques of immunoaffinity purification (Hunt & Pratt, 1979) will be of great interest.

Phytochrome from etiolated tissue can interact with model membranes. Roux & Yguerabide (1973) showed changes in membrane conductance on addition of Pr to a black lipid membrane of oxidised cholesterol. They showed also that red/far-red reversible changes in membrane conductance occurred within 60 s of the irradiation. In preliminary reports Georgevich *et al.* (1976) and Roux, Krauland & Mauk (1978) describe similar experiments with synthetic liposomes which show that Pfr enhances the efflux of K^+ from pre-loaded liposomes two to three times more effectively than Pr. Also there is an apparent specificity for K^+ as compared to Na^+ in this response. Kim & Song (1981) have recently characterised the interaction of phytochrome with multilamellae liposomes, indicating a preferential binding of Pfr. Optimal binding differential between Pr and Pfr was found at neutral pH with liposomes containing 10% cholestrol. However, these workers failed to find differential associations of Pr and Pfr to the outside of protoplasts *in vitro*. Since physiological experiments point to phytochrome being associated with the plasmalemma (Haupt, 1970; Marmé & Schäfer, 1972), the outcome of experiments on binding of phytochrome to inverted plasmalemma vesicles is awaited with interest. Clearly despite the hydrophilic properties of phytochrome as a whole, it can bind both as Pr and Pfr with hydrophobic surfaces. It is therefore quite possible that it could interact with membrane constituents, such as proteins, steroids or lipids, as a peripheral protein, with Pfr forming the more tenacious association with membranes.

How does phytochrome act?

Physiological experiments point overwhelmingly to membranes being the site(s) of phytochrome action. Conflicting with this is the evidence that points to phytochrome being a soluble protein rather than a membrane constituent, and it is therefore unlikely that phytochrome is an intrinsic protein (Pratt, 1978; Quail, 1980). However, other work particularly that of Song and his coworkers, (Hahn & Song, 1981; Song, this volume) shows how a small hydrophobic site in the vicinity of the chromophore could enable phytochrome to interact with one or more membrane constituents. The evidence of modification of phytochrome after red light could indicate its association with a peripheral membrane protein (Epel, 1980). Such a complex could therefore be considered as **functional phytochrome,** and is the phytochrome we should study, in contrast to that from etiolated tissue. The original proposal, that the Pfr is the active form of an enzyme (Hendricks, 1964) has received little direct support over the years, but by analogy could still be essentially correct, phytochrome acting as a catalyst of membrane transport. Phytochrome (Pfr) could be an activator of a membrane ATPase, for example, or directly facilitate the movement of an ionic species across a membrane. Raven (1981) has pointed out that it is unlikely that phytochrome phototransformation is part of an active pumping mechanism since the maximum flux attainable by energy transduction is low (assuming all the phytochrome were uniformly distributed in the plasmalemma). It is therefore more likely that phytochrome acts as a signal transducer by regulating the activity of a membrane porter (Raven & Smith, 1980). However, Kandeler, Löppert, Rottenberg & Scharfetter (1980) argue that red/far-red modulation of membrane potential in *Lemna* involves modulations of an active component of membrane potential and that phytochrome plays a role in the formation of a proton gradient at the plasmalemma. In contrast Brownlee *et al.* (1979) conclude that phytochrome opens and closes aqueous pores in the membrane, in much the same way as Smith proposed earlier (1976*b*).

How fast does phytochrome act? The most rapid responses discussed so far occur within a time scale of seconds. If phytochrome controls membrane channels, the effect of Pfr might be expected to be very rapid. However, phototransformation takes time, and although Pfr starts to appear after 10 ms of photon capture it is not complete until a few seconds later, due to the limiting dark intermediate transitions of the protein and chromophore (Kendrick & Spruit, 1977). For many rapid responses, for example, the *in-vivo* induced red light enhancement of pelletability (Pratt & Marmé, 1976) and modulation of K^+ uptake by mung bean hypocotyl sections (Brownlee & Kendrick, 1979*a,b*), the responses are still reversible by far-red light, despite their rapidity. This strongly suggests that the responses are not a function of continued excitation, but reflect the continued presence of Pfr, which leads to a cumulative response. If phytochrome action is initiated by conformational rearrangements of the protein, these reactions are relatively slow with respect to photon capture. This would be particularly true if Pfr had also to diffuse to target sites on the membrane before action could begin. Estimates suggest, however, that this process would be quite rapid with approximately 100 ms being sufficient, even if phytochrome diffused maximal distances (Quail, 1980). If the rapid potential changes observed in several systems are a reflection of the primary action of Pfr, its immediate removal by an intense flash of far-red light should be reflected in faster changes than for induction by red light, assuming that intermediates between Pfr and Pr are inactive. Detailed kinetic analysis of membrane potential modulations by red and far-red light is therefore worthy of study.

Another question which can be asked is: how much phytochrome does a plant need? Certainly de-etiolated seedlings appear to have a universally high phytochrome content (Frankland, 1972; Hillman, 1972) and exhibit the phenomenon of destruction, which is a loss of photoreversibility, and therefore detectability, by an energy-dependent enzymic process. What is the function of this reaction and is it a possible process that is involved in, or is a consequence of, phytochrome action? Destruction appears to be a function of Pfr (Kendrick & Frankland, 1968) or Pr that has been cycled through Pfr (Chorney & Gordon, 1966; Dooskin & Mancinelli, 1968; MacKenzie Jr., Briggs & Pratt, 1978). Since the rapid phytochrome responses occur in a matter of seconds it is not necessary for phytochrome to undergo destruction in order to act. Nevertheless, the process could be a consequence of phytochrome that has been functional. The reaction does not appear to be restricted to etiolated tissues, as radioimmunological assay has shown that plants grown in a day/night cycle have three times more phytochrome present at the end of the night than at the end of the day (Hunt & Pratt, 1980). Similar observations have now been made for phytochrome in light-grown plants depleted of chlorophyll using Norflurazon (Jabben, 1980). Destruction depletes the tissue of Pfr, as too does the process of dark reversion of Pfr to Pr, resulting in both cases in the tissue being re-primed, capable of detecting the onset of light. Energetically speaking, dark reversion appears to be a much more efficient way of re-setting the phytochrome system, and destruction, which has been shown to involve a degradation of the whole protein (Pratt *et al.*, 1976), would appear wasteful unless it has some other purpose. In terms of a phytochrome-activated membrane porter, the destruction process may be the way of removing the molecule from the membrane. Depletion of active phytochrome by destruction and reversion could therefore lead to the dark adaption of the tissue. The transition from light to dark as well as dark to light could then be measured by the plant as signals for such processes as 'time setting' of rhythmical phenomena.

It is therefore possible that phytochrome has more than one role to play in development. Firstly to indicate the presence or absence of light or perhaps more importantly the transition from light to dark or dark to light – necessary for daily time-setting and entrainment of the biological rhythms as well as inducing de-etiolation of seedlings at the soil surface. Secondly to detect the spectral quality of light, where the red/far red reversibility is important. Smith (1981) considers that it is for this function that phytochrome evolved. Several species have now been shown to modify their rate of elongation-growth inversely with the red/far-red quantum ratio of the light (Morgan & Smith, 1976); in this way plants can continually modify their growth to avoid shade of other plants. Under continuous light of different spectral quality, the total phytochrome depends on the first order rate of destruction and the zero order rate of synthesis. It is predicted that ultimately, irrespective of the wavelength, a similar level of Pfr is reached, although differing amounts of total phytochrome will be present (Schäfer, 1975). However, in the experiments of Morgan *et al.* (1980) the rate of internode extension in plants grown under white light can be modulated within a few minutes by supplementary red and far-red light by optical fibres. Such a rapid response can correlate with any of three parameters: (i) the Pfr:Pr ratio, (ii) the Pfr concentration, (iii) the enhanced rate of cycling between Pr and Pfr by the supplementary light.

In a number of responses of etiolated tissues and seeds, phytochrome has been implicated as the pigment involved in the irradiance-dependent reaction, called the 'high-irradiance reaction' (Mohr, 1972; Hartmann, 1966, Mancinelli & Rabino, 1978). How can phytochrome act as

Smith (1970)

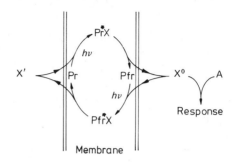

Schäfer (1975)

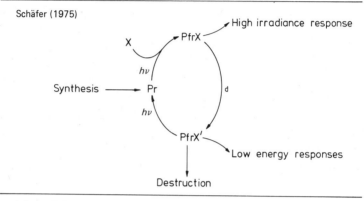

Johnson & Tasker (1979)

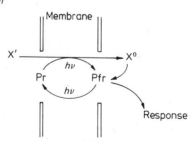

Fig. 3. *Top.* Smith's model of phytochrome action. He proposes that interconversion of Pr and Pfr transports a metabolite X between pools X' and X^0 on either side of a membrane. X^0 initiates phytochrome responses by interaction with substrate A. *Middle.* Schäfer's model of phytochrome action. He proposes that Pfr has two states of interaction with a receptor (X). Under high irradiance conditions PfrX accumulates, whereas under low energy conditions PfrX' accumulates. PfrX and PfrX' are the effector molecules of the high irradiance reaction and low energy reactions respectively. *Bottom.* Johnson and Tasker's model of phytochrome action. They propose phototransformation of Pr to Pfr transports X across a membrane from pool X' to X^0, where X^0 interacts with Pfr to initiate the response.

a signal transducer that activates and deactivates a membrane porter upon phototransformation and also detect irradiance? This third function of phytochrome complicates our simple model of phytochrome action. Could phytochrome also act as an energy transducer?

Johnson & Tasker (1979) proposed a model of phytochrome action in which both Pfr concentration and the cycling rate between Pr and Pfr are important. In this way they propose that photoconversion of Pr to Pfr transports a metabolite X across a membrane (Fig. 3) where it interacts in an additive or multiplicative way with Pfr to produce the response. Such a model could explain the low energy reaction as well as high irradiance responses. In contrast Schäfer (1975) has proposed the low energy reaction and high irradiance responses of phytochrome depend on two forms of association of Pfr with a receptor X. In his model (Fig. 3) under high irradiance conditions a pool of metastable PfrX, which is irradiance dependent, accumulates, whereas in the low energy reaction the relatively stable PfrX′ is the effector. In the complete model both the rate constants of synthesis of Pr and destruction of Pfr are important. Johnson (1980) has attempted to distinguish between these two models of phytochrome action and has provided evidence in favour of the Johnson and Tasker model.

One of the problems of interpreting the high irradiance reaction on the basis of phytochrome alone is the high activity of the blue region of the spectrum (Hartmann, 1966; Hartmann & Unser, 1972). One suggestion is that there is a phytochrome–flavin interaction. Efficient energy transfer between FMN (donor) and Pr (acceptor) when *in vitro* mixtures are irradiated has been shown (Song, Sarkar, Kim & Poff, 1981). Further, Pr→Pfr phototransformation is enhanced whereas Pfr→Pr phototransformation is retarded in such mixtures (Sarker & Song, 1982). This supports the view originally put forward by Siegelman & Hendricks (1957) that flavins are involved in the high irradiance reaction.

In the simplistic model proposed here (Fig. 4) phytochrome (Pfr) is envisaged as forming open functional channels as a result of interaction with a membrane protein(s) with interconversion between Pr and Pfr acting as a gating mechanism. To accommodate the high irradiance reaction it is necessary to propose that such channels can be further activated by the Pr→Pfr transition enhancing flux across the membrane. In this way cycling of phytochrome, which is irradiance-dependent, would facilitate movement across the membrane. This can be viewed as a special case of the Johnson and Tasker model with Pfr and cycling rate acting in an additive manner. It also has features in common

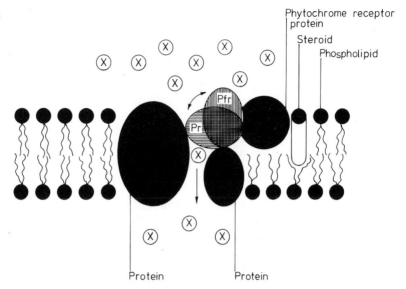

Fig. 4. Simplistic model hypothesis of phytochrome action. Phytochrome is envisaged as a peripheral protein which interacts with a specific phytochrome receptor protein. This phytochrome–membrane complex forms channels which are closed as Pr and open as Pfr. The open channels allow movements of an ion (X) by diffusion across the membrane. Under high irradiance conditions it is proposed that Pr → Pfr phototransformation facilitates the diffusion of X across the membrane.

with the model proposed by Smith (1970). In this way, under high irradiance conditions, phytochrome phototransformation could act by facilitated diffusion, whereas under low energy conditions it operates as a gated channel through which the primary ion can diffuse passively. This model is proposed as a working hypothesis and should not be taken as a physical model. In presenting it I am documenting how responses to phytochrome phototransformation could be realised in physical terms. This simple model accommodates most of our present knowledge of phytochrome.

In a broader context it is proposed that there is only one fundamental mechanism of phytochrome action, which involves ion transport. The diversity of phytochrome action can be accounted for by the nature of the second messengers and the subcellular compartment concerned (Fig. 5). Diverse tertiary modification (such as changed levels and distribution of hormones) ultimately lead to the long term changes in growth and development. The short term responses discussed here that involve membrane phenomena are the manifestation of either the primary action or second messages, and only further research will reveal which.

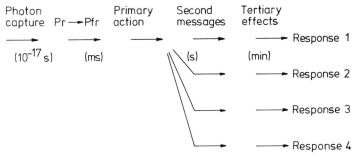

Fig. 5. The wide spectrum of phytochrome responses are explained on one common primary mode of action. Diversity of response is accounted for on the basis of diversity of second messages and the specific subcellular compartment concerned.

References

ANDERSON, J. M., CHARBONNEAU, H., JONES, H. P., McCANN, R. O. & CORMIER, M. J. (1980). Characterization of the plant NAD kinase activator protein and its identification as calmodulin. *Biochemistry*, **19**, 3113–20.

BAKER, P. F. (1976). The regulation of intracellular calcium. In *Calcium in Biological Systems*, ed. C. J. Duncan, pp. 67–88. Cambridge: Cambridge University Press.

BOESHORE, M. L. & PRATT, L. H. (1977). Phytochrome modification and light-enhanced, *in-vivo*-induced phytochrome pelletability *Plant Physiol.* **66**, 550–4.

BOISARD, J., MARMÉ, D. & BRIGGS, W. R. (1974). *In vivo* properties of membrane-bound phytochrome. *Plant Physiol.* **54**, 272–6.

BORTHWICK, H. A. (1972a). History of phytochrome. In *Phytochrome*, ed. K. Mitrakos & W. Shropshire Jr., pp. 3–23. London & New York: Academic Press.

BORTHWICK, H. A. (1972b). The biological significance of phytochrome. In *Phytochrome*, ed. K. Mitrakos & W. Shropshire Jr., pp. 27–44. London & New York: Academic Press.

BORTHWICK, H. A. & HENDRICKS, S. B. (1961). Effects of radiation on growth and development. In *Handbuch der Pflanzenphysiologie*, vol. 16, ed. W. Ruhland, pp. 299–330. Berlin: Springer-Verlag.

BORTHWICK, H. A., HENDRICKS, S. B., PARKER, M. W., TOOLE, E. H. & TOOLE, V. K. (1952). A reversible photoreaction controlling seed germination. *Proc. Natl. Acad. Sci., USA*, **38**, 662–6.

BRIGGS, W. R. & RICE, H. V. (1972). Phytochrome: chemical and physical properties and mechanism of action. *Ann. Rev. Plant Physiol.* **23**, 293–334.

BROWNLEE, C. (1978). Ion fluxes and phytochrome in mung beans. PhD thesis, University of Newcastle upon Tyne.

BROWNLEE, C. & KENDRICK, R. E. (1979a). Ion fluxes and phytochrome in mung bean hypocotyl segments. I. Fluxes of potassium. *Plant Physiol.* **64**, 206–10.

BROWNLEE, C. & KENDRICK, R. E. (1979b). Ion fluxes and phytochrome in mung bean hypocotyle segments. II. Fluxes of Cl^-, H^+ and P_i in apical and sub-hook segments. *Plant Physiol.* **64**, 211–13.

BROWNLEE, C., ROTH-BEJERANO, N. & KENDRICK, R. E. (1979). The molecular mode of phytochrome action. *Sci. Prog., Oxf.* **66**, 217–29.

CARAFOLI, E. (1977). Calcium transport in biological membranes. In *Living Systems as Energy Convertors*, ed. R. Buvet, M. J. Allen & J. P. Massue, pp. 153–74. Amsterdam: North-Holland.

298   *The physiology of phytochrome action*

CARCELLER, M. S. & SANCHEZ, R. A. (1972). The influence of phytochrome in the water exchange of epidermal cells of *Taraxacum officinale*. *Experimentia*, **28**, 364.

CHORNEY, W. & GORDON, S. A. (1966). Action spectrum and characteristics of light activated disappearance of phytochrome in oat seedlings. *Plant Physiol.* **41**, 891–6.

COLEMAN, R. A. & PRATT, L. H. (1974). Electron microscopic localization of phytochrome in plants using an indirect antibody-labelling method. *J. Histochem. Cytochem.* **22**, 1039–47.

COOKE, R. J. & KENDRICK, R. E. (1976). Phytochrome controlled gibberellin metabolism in etioplast envelopes. *Planta*, **131**, 303–7.

COOKE, R. J., SAUNDERS, P. F. & KENDRICK, R. E. (1975). Red light induced production of gibberellin-like substances in homogenates of etiolated wheat leaves and suspensions of intact etioplasts. *Planta*, **124**, 319–28.

DANIELS, S. & SMITH JR., W. O. (1981). The interaction of proteolytically-derived peptides of phytochrome with Cibacron blue 3GA. *Plant Physiol. Suppl.* **67**, 729.

DE GREEF, J. A., CAUBERGS, R., VERBELEN, J. P. & MOEREELS, E. (1976). Phytochrome mediated inner-organ dependence and rapid transmission of the light stimulus. In *Light and Plant Development*, ed. H. Smith, pp. 295–316. London: Butterworths.

DIETER, P. & MARMÉ, D. (1980). Calmodulin activation of plant microsomal Ca^{2+} uptake. *Proc. Natl. Acad. Sci., USA*, **77**, 7311–14.

DIETER, P. & MARMÉ, D. Far red light irradiation of intact corn seedlings affects mitochondrial and calmodulin dependent microsomal Ca^{2+} transport. FEBS Letters (in press).

DITTES, L., RISSLAND, I. & MOHR, H. (1971). On the regulation of enzyme levels (phenylalanine ammonia-lyase) in different organs of a plant (*Sinapis alba*). *Z. Naturforsch.* **26**, 1175–80.

DOOSKIN, R. H. & MANCINELLI, A. L. (1968). Phytochrome decay and coleoptile elongation in *Avena* following various light treatments. *Bull. Torrey Botan. Club*, **95**, 474–87.

DOWNS, R. J. (1956). Photoreversibility of flower initiation. *Plant Physiol.* **31**, 379–84.

DREYER, E. M. & WEISENSEEL, M. H. (1979). Phytochrome-mediated uptake of calcium in *Mougeotia* cells. *Planta*, **146**, 31–9.

EPEL, B. L. (1980). Physiological and biochemical studies of phytochrome activation. In *Photoreceptors and Plant Development*, ed. J. De Greef, pp. 467–79. Antwerp University Press.

EPEL, B. L. (1981). A partial characterization of the long-wavelength activated far red absorbing form of phytochrome. *Planta*, **151**, 1–5.

EPEL, B. L., BUTLER, W. L., PRATT, L. H. & TOKUYASU, K. T. (1980). Immunofluorescence localization studies of the Pr and Pfr forms of phytochrome in the coleoptile tips of oats, corn and wheat. In *Photoreceptors and Plant Development*, ed. J. De Greef, pp. 121–33. Antwerp University Press.

ETZOLD, H. (1965). Der Polarotropismus und Phototropismus de Chloronemen von *Dryopteris filix-mas*. *Planta*, **64**, 254–80.

EVANS, A. E. & SMITH, H. (1976a). Localization of phytochrome in etioplasts and its regulation *in vitro* of gibberellin levels. *Proc. Natl. Acad. Sci., USA*, **73**, 138–42.

EVANS, A. E. & SMITH, H. (1976b). Spectrophotometric evidence for the presence of phytochrome in the envelope membranes of barley etioplasts. *Nature, London*, **259**, 323–5.

FONDEVILLE, J. C., BORTHWICK, H. A. & HENDRICKS, S. B. (1966). Leaflet movement of *Mimosa pudica* indiative of phytochrome action. *Planta*, **69**, 357–64.

FRANKLAND, B. (1972). Biosynthesis and dark transformation of phytochrome. In *Phytochrome*, ed. K. Mitrakos & W. Shropshire Jr., pp. 196–225. London & New York: Academic Press.

FREDERICQ, H. (1964). Conditions determining effects of far red and red irradiations on flowering response of *Pharbitis nil*. *Plant Physiol.* **39**, 812–16.

FURUYA, M. & HILLMAN, W. S. (1964). Observations on spectrophotometrically assayable phytochrome *in vivo* in etiolated *Pisum* seedlings. *Planta*, **63**, 31–42.

FURUYA, M. & MANABE, K. (1976). Phytochrome in mitochondrial and microsomal fractions isolated from etiolated pea shoots. In *Light and Plant Development*, ed. H. Smith, pp. 143–55. London: Butterworths.

GALSTON, A. W. & SATTER, R. L. (1976). Light, clocks and ion flux: an analysis of leaf movement. In *Light and Plant Development*, ed. H. Smith, pp. 159–84. London: Butterworths.

GALSTON, A. W., SATTER, R. L., LONERGAN, T. A., MORSE, M. J. & HATCH, A. M. C. (1978). A mechanistic analysis of light-clock interaction in leguminous pulvini. In *Plant Growth and Light Perception*, ed. B. Deutch, B. I. Deutch & A. O. Gyldenholm. pp. 177–96. University of Aarhus.

GEORGEVICH, G., CEDEL, T. E. & ROUX, S. J. (1977). Use of ^{125}I-labelled phytochrome to quantitate phytochrome binding to membranes of *Avena sativa*. *Proc. Natl. Acad. Sci.*, *USA*, **74**, 4439–43.

GEORGEVICH, G. KRAUNLAND, J. & ROUX, S. J. (1976). Phytochrome interaction with lipid bilayers. *Plant Physiol. Suppl.* **57**, 105.

GRIME, J. P. & JARVIS, B. C. (1975). Shade avoidance and shade tolerance in flowering plants. II. Effects of light on the germination of species of contrasted ecology. In *Light as an Ecological Factor*, vol. 2, ed. G. C. Evans, R. Bainbridge & O. Rackham, pp. 525–32.Oxford: Blackwell.

GROMBIEN, S. & RÜDIGER, W. (1976). On the molecular weight of phytochrome: a new high molecular phytochrome species in oat seedlings. *Hoppe-Seyler's Z. Physiol. Chem.* **357**, 1015–18.

HAHN, T.-R. & SONG, P.-S. (1981). Hydrophobic properties of phytochrome as probed by 8-anilinonaphthalene-1-sulphonate fluorescence. *Biochemistry*, **20**, 2602–9.

HALE, C. C. II & ROUX, S. J. (1980). Photoreversible calcium fluxes induced by phytochrome in oat coleoptile cells. *Plant Physiol.* **65**, 658–68.

HARTMANN, K. M. (1966). A general hypothesis to interpret 'high energy phenomena' of photomorphogenesis on the basis of phytochrome. *Photochem. Photobiol.* **5**, 349–66.

HARTMANN, K. M. & UNSER, I. C. (1972). Analytical action spectroscopy with living systems: photochemical aspects and attenuance. *Ber. Dtsch. Bot. Ges.* **85**, 481–551.

HAUPT, W. (1970). Localization of phytochrome in the cell. *Physiol. Veg.* **8**, 551–63.

HAUPT, W. (1980). Sensory transduction and photobehavior: final considerations and emerging themes. In *Photoreception and Sensory Transduction in Aneural Organisms*, ed. F. Lenci & G. Colombetti, pp. 397–404. New York & London: Plenum Press.

HAUPT, W., HUPFER, B. & KRAML, M. (1980). Blitzlichtinduktion der Chloroplasten-bewegung bei *Mougeotia:* Wirkung unterschiedlicher Spectralbereiche und Polaris-ationsrichtungen. *Z. Pflanzenphysiol.* **96**, 331–42.

HENDRICKS, S. B. (1964). Photochemical aspects of plant photoperiodicity. In *Photophysiology*, vol. 1, ed. A. C. Giese, pp. 305–31. New York: Academic Press.

HENDRICKS, S. B. & BORTHWICK, H. A. (1967). The function of phytochrome in regulation of plant growth. *Proc. Natl. Acad. Sci.*, *USA*, **58**, 2125–30.

HILLMAN, W. S. (1967). The physiology of phytochrome. *Ann. Rev. Plant Physiol.* **18**, 301–24.

HILLMAN, W. S. (1972). On the physiological significance of *in vivo* phytochrome assays. In *Phytochrome*, ed. K. Mitrakos & W. Shropshire, Jr., pp. 573–84. London & New York: Academic Press.

HILTON, J. R. (1981). Phytochrome, control of plastid gibberellin level. PhD thesis, University of Leicester.

HILTON, J. R. & SMITH, H. (1980). The presence of phytochrome in purified barley etioplasts and its *in vitro* regulation of biologically-active gibberellin levels in etioplasts. *Planta*, **148**, 312–18.

HOLMES, M. G. & SMITH, H. (1975). The function of phytochrome in plants growing in the natural environment. *Nature, London*, **254**, 512–14.

HORWITZ, B. A. & EPEL, B. L. (1977). A far-red form of phytochrome exhibiting *in vivo* spectral properties: studies with crude extracts of oats and squash. *Plant Sci. Lett.* **9**, 205–10.

HUNT, R. E. & PRATT, L. H. (1979). Phytochrome radioimmunoassay. *Plant Physiol.* **64**, 327–31.

HUNT, R. E. & PRATT, L. H. (1980). Radioimmunoassay of phytochrome content in green, light-grown oats. *Plant, Cell Environ.* **3**, 91–5.

JABBEN, M. (1980). The phytochrome system in light-grown *Zea mays*. *Planta*, **149**, 91–6.

JAFFE, M. J. (1968). Phytochrome-mediated biopotentials in mung bean seedlings. *Science*, 162, 1016–17.

JAFFE, M. J. (1970). Evidence for the regulation of phytochrome-mediated processes in bean roots by the neurohumor, acetylcholine. *Plant Physiol.* **46**, 768–77.

JOHNSON, C. B. (1980). The effect of red light in the high irradiance reaction of phytochrome. *Plant, Cell Environ.* **3**, 45–51.

JOHNSON, C. B. & TASKER, R. (1979). A scheme to account quantitatively for the action of phytochrome in etiolated and light-grown plants. *Plant, Cell Environ.* **2**, 25–65.

JOSE, A. M. (1977). Phytochrome modulation of ATPase activity in a membrane fraction from *Phaseolus*. *Planta*, **137**, 203–6.

KANDELER, R., LÖPPERT, H., ROTTENBURG, T. H. & SCHARFETTER, E. (1980). Early effects of phytochrome in *Lemna*. In *Photoreceptors and Plant Development*, ed. J. De Greef, pp. 485–92. Antwerp University Press.

KENDRICK, R. E. (1976). Photocontrol of seed germination. *Sci. Prog. Oxf.* **63**, 347–67.

KENDRICK, R. E. & FRANKLAND, B. (1968). Kinetics of phytochrome decay in *Amaranthus* seedlings. *Planta*, **82**, 317–20.

KENDRICK, R. E. & FRANKLAND, B. (1969). Photocontrol of germination in *Amaranthus caudatus*. *Planta*, **85**, 326–9.

KENDRICK, R. E. & FRANKLAND, B. (1976). *Phytochrome and Plant Growth*. London: Edward Arnold.

KENDRICK, R. E. & SMITH, H. (1976). Assay and isolation of phytochrome. In *Chemistry and Biochemistry of Plant Pigments*, 2nd edn, vol. 2, ed. T. W. Goodwin, pp. 334–64. London: Academic Press.

KENDRICK, R. E. & SPRUIT, C. J. P. (1977). Phototransformations of phytochrome. *Photochem. Photobiol.* **26**, 201–14.

KENDRICK, R. E. & ROTH-BEJERANO, N. (1978). Spectral characteristics of phytochrome *in vivo* and *in vitro*. *Planta*, **142**, 225–8.

KIM, I.-S. & SONG, P.-S. (1981). Binding of phytochrome to liposomes and protoplasts. *Biochemistry*, **20**, 5482–9.

KONDO, N., INOUE, Y. & SHIBATA, K. (1973). Phytochrome distribution in *Avena* seedlings measured by scanning a single seedling. *Plant Sci. Lett.* **1**, 165–8.

KOUKKARI, W. L. & HILLMAN, W. S. (1968). Pulvini as the photoreceptors in the phytochrome effect on nyctinasty in *Albizzia julibrissin*. *Plant Physiol.* **43**, 698–704.

KRAAK, H. L. & SPRUIT, C. J. P. (1980). Development and properties of etioplasts as influenced by the phytochrome system. In *Photoreceptors and Plant Development*, ed. J. De Greef, pp. 241–8. Antwerp University Press.

LÖPPERT, H., KRONBERGER, W. & KANDELER, R. (1978). Phytochrome-mediated changes in the membrane potential of subepidermal cells of *Lemna paucicostata* 6746. *Planta*, **138**, 133–6.

MACKENZIE JR., J. M., BRIGGS, W. R. & PRATT, L. H. (1978). Phytochrome photoreversibility: empirical test of the hypothesis that it varies as a consequence of compartmentalization. *Planta*, **141**, 129–34.

MACKENZIE JR., J. M., COLEMAN, R. A., BRIGGS, W. R. & PRATT, L. H. (1975).

Reversible redistribution of phytochrome within the cell upon conversion to its physiologically active form. *Proc. Natl. Acad. Sci., USA*, **72**, 799–803.

MANABE, K. & FURUYA, M. (1975). Distribution and nonphotochemical transformation of phytochrome in subcellular fractions from *Pisum* epicotyls. *Plant Physiol.* **56**, 772–5.

MANCINELLI, A. L. & RABINO, I. (1978). The 'high irradiance responses' of plant photomorphogenesis. *Bot. Rev.* **44**, 129–80.

MANDOLI, D. F. & BRIGGS, W. R. (1981). Photoperceptive region of the low-irradiance responses in etiolated oats. *Plant Physiol. Suppl.* **67**, 735.

MARMÉ, D. (1977). Phytochrome: membranes as possible sites of primary action. *Ann. Rev. Plant Physiol.* **28**, 173–98.

MARMÉ, D., BIANCO, J. & GROSS, J. (1976). Evidence for phytochrome binding to plasma membrane and endoplasmic reticulum. In *Light and Plant Development*, ed. H. Smith, 95–110, London: Butterworths.

MARMÉ, D. & SCHÄFER, E. (1972). On the localization and orientation of phytochrome molecules in corn coleoptiles (*Zea Mays*). *Z. Pflanzenphysiol.* **67**, 192–4.

MOHR, H. (1966*a*). Untersuchungen zur Phytochrominduzierten Photomorphogenese des Senfkeimlings (*Sinapis alba*). *Z.Pflanzenphysiol.* **54**, 63–83.

MOHR, H. (1966*b*). Differential gene activation as a mode of action of phytochrome 730. *Photochem. Photobiol.* **5**, 469–83.

MOHR, H. (1972). *Lectures of Photomorphogenesis*. Berlin: Springer-Verlag.

MORGAN, D. C., O'BRIEN, T. M. & SMITH, H. (1980). Rapid photomodulation of stem extension in light-grown *Sinapis alba*. In *Photoreceptors and Plant Development*, ed. J. De Greef, pp. 517–24. Antwerp University Press.

MORGAN, D. C. & SMITH, H. (1976). Linear relationship between phytochrome photo-equilibrium and growth in plants under stimulated natural radiation. *Nature, London*, **262**, 210–12.

MORGAN, D. C. & SMITH, H. (1978). Simulated sunflecks have large rapid effects on plant stem extension. *Nature, London*, **273**, 534–6.

NEWMAN, I. A. (1974). Electric responses of oats to phytochrome transformation. In *Mechanisms of Regulation of Plant Growth*, Bull. 12, ed. R. L. Bieleski, A. R. Ferguson & M. M. Cresswell. pp. 355–360. Wellington: The Royal Society of New Zealand.

NEWMAN, I. A. (1981). Rapid electric changes in oats, phytochrome and membranes. *Plant Physiol. Suppl.* **67**, 730.

NEWMAN, I. A. & BRIGGS, W. R. (1972). Phytochrome-mediated electric potential changes in oat seedlings. *Plant Physiol.* **50**, 687–93.

NEWMAN, I. A. & SULLIVAN, J. K. (1976). Auxin transport in oats: a model for the electric changes. In *Transport and Transfer Processes in Plants*, ed. I. F. Wardlaw & J. B. Passioura, pp. 153–59. New York: Academic Press.

OELZE-KAROW, H., SCHOPFER, P. & MOHR, H. (1970). Photochrome-mediated repression of enzyme synthesis (lipoxygenase): A threshold phenomenon. *Proc. Natl. Acad. Sci., USA*, **65**,

PENEL, C., GREPPIN, H. & BOISARD, J. (1976). *In vitro* photomodulation of a peroxidase activity through membrane-bound phytochrome. *Plant Sci. Lett.* **6**, 117–21.

PIKE, C. S. (1976). Lack of influence of phytochrome on membrane permeability to tritiated water. *Plant Physiol.* **57**, 185–7.

PRATT, L. H. (1978). Molecular properties of phytochrome. *Photochem. Photobiol.* **27**, 81–105.

PRATT, L. H. (1979). Phytochrome: function and properties. In *Photochemical and Photobiological Reviews*, Vol. 4, ed. K. C. Smith, pp. 59–124. New York: Plenum Press.

PRATT, L. H. & COLEMAN, R. A. (1974). Phytochrome distribution in etiolated grass seedlings assayed by an indirect antibody-labelling method. *Amer. J. Bot.* **61**, 195–202.

PRATT, L. H., COLEMAN, R. A. & MACKENZIE, J. M. JR. (1976). Immunological visualisation of phytochrome. In *Light and Plant Development*, ed. H. Smith, pp. 75–94. London: Butterworths.

PRATT, L. H. & MARMÉ, D. (1976). Red light-enhanced phytochrome pelletability: a re-examination and further characterisation. *Plant Physiol.* **58**, 686–92.

PRESSMAN, B. C. (1976). Biological application of ionophores. *Annu. Rev. Biochem.* **45**, 501–30.

QUAIL, P. H. (1980). Phytochrome: the first five minutes from Pfr formation. In *Photoreceptors and Plant Development*, ed. J. De Greef, pp. 449–66. Antwerp University Press.

RACUSON, R. H. (1976). Phytochrome control of electrical potentials and intercellular coupling in oat-coleoptile tissue. *Planta*, **132**, 25–9.

RACUSON, R. H. & COOKE, T. J. (1981). Electrical changes in fern gametophyte cells during exposure to photomorphogetically-active light. *Plant Physiol. Suppl.* **67**, 731.

RACUSON, R. H. & ETHERINGTON, B. (1975). Role of membrane-bound, fixed charge changes in phytochrome-mediated mung bean root-tip adherance phenomena. *Plant Physiol.* **55**, 491–5.

RACUSON, R. & SATTER, R. L. (1975). Rhythmic and phytochrome-regulated changes in transmembrane potential in *Samanea pulvini*. *Nature, London*, **255**, 408–10.

RAVEN, J. A. (1981). Light quality and solute transport. In *Plants and the Daylight Spectrum*, ed. H. Smith, pp. 375–90. London & New York: Academic Press.

RAVEN, J. A. & SMITH, F. A. (1980). Chemiosmotic viewpoint. In *Plant Membrane Transport: Current Conceptual Issues*, ed. R. M. Spanswick, W. J. Lucas & J. Dainty, pp. 161–78. Amsterdam: Elsevier/North-Holland.

ROTH-BEJERANO, N. (1980). Nature of the phytochrome effect on the binding of glycolate oxidase to peroxisomes *in vitro*. *Planta*, **149**, 252–6.

ROTH-BEJERANO, N. & KENDRICK, R. E. (1979). Phytochrome pelletability in barley. *Plant Physiol.* **47**, 67–72.

ROTH-BEJERANO, N. & LIPS, S. H. (1978). Binding of glycolate oxidase to peroxisomal membrane as affected by light. *Photochem. Photobiol.* **27**, 171–5.

ROUX, S. J., KRAULAND, J. & MAUK, F. (1978). Further characterization of phytochrome conductance effects in artificial bilayer membranes. *Plant Physiol. Suppl.* **61**, 66.

ROUX, S. J., McENTIRE, K., SLOCUM, R. D., CEDEL, T. E. & HALE, C. C. (1981). Phytochrome induces photoreversible calcium fluxes in a purified mitochondrial fraction from oats. *Proc. Natl. Acad. Sci., USA*, **78**, 283–7.

ROUX, S. J. & YGUERABIDE, J. (1973). Photoreversible conductance changes induced by phytochrome in model lipid membranes. *Proc. Natl. Acad. Sci., USA*, **70**, 762–4.

SARKER, H. K. & SONG, P.-S. (1982). Blue light induced phototransformation of phytochrome in the presence of flavin. *Photochem. Photobiol.* **35**, 243–6.

SATTER, R. L., APPLEWHITE, P. B. & GALSTON, A. W. (1974). Rhythmic potassium flux in *Albizzia*: effect of aminophylline, cations, and inhibitions of respiration and protein synthesis. *Plant Physiol.* **54**, 280–5.

SATTER, R. L. & GALSTON, A. W. (1976). The physiological functions of phytochrome. In *Chemistry and Biochemistry of Plant Pigments*, 2nd edn., vol. 1, ed. T. W. Goodwin, pp. 680–735. London & New York: Academic Press.

SCHÄFER, E. (1975). A new approach to explain the high irradiance responses of photomorphogenesis on the basis of phytochrome. *J. Math. Biol.* **2**, 41–56.

SHIMAZAKI, Y., MORIYASU, Y. & FURUYA, M. (1981). Isolation of the red-light-absorbing form of phytochrome from green plant tissues. *Plant Physiol. Suppl.* **67**, 742.

SIEGELMAN, H. W. & HENDRICKS, S. B. (1957). Photocontrol of anthocyanin synthesis in turnip and red cabbage seedlings. *Plant Physiol.* **32**, 393–8.

SMITH, H. (1970). Phytochrome and photomorphogenesis in plants. *Nature, London*, **227**, 665–8.

SMITH, H. (1975). *Phytochrome and Photomorphogenesis*. Maidenhead: McGraw-Hill.

SMITH, H. (ed.) (1976a). *Light and Plant Development.* London: Butterworths.

SMITH, H. (1976b). The mechanism of action and the function of phytochrome. In *Light and Plant Development*, ed. H. Smith, pp. 493–502. London: Butterworths.

SMITH, H. (1981). Function, evolution and action of plant photoresponses. In *Plants and the Daylight Spectrum*, ed. H. Smith, pp. 499–508. London and New York: Academic Press.

SMITH, H., EVANS, A. E. & HILTON, J. R. (1978). An *in vitro* association of soluble phytochrome with a partially purified organelle fraction from barley leaves. *Planta,* **141,** 71–6.

SMITH, JR. W. O. (1981a). Probing the molecular structure of phytochrome with immobilized cibacron blue 3GA and blue dextran. *Proc. Natl. Acad. Sci., USA,* **78,** 2977–80.

SMITH JR, W. O. (1981b). Characterization of the photoreceptor protein, phytochrome. *Photochem. Photobiol.* **33,** 961–4.

SMITH JR., W. O. & DANIELS, S. (1980). Biospecific binding of phytochrome to Cibacron Blue 3GA. *Plant Physiol. Suppl.* **65,** 2.

SONG, P.-S., KIM, I.-S. & HAHN, R.-R. (1982). Purification of phytochrome by affi-gel-blue chromatography: An effect of lumichrome on purified phytochrome. *Anal. Biochem.* (in press).

SONG, P.-S., SARKER, H. K., KIM, I.-S. & POFF, K. L. (1981). Primary photoprocesses of undegraded phytochrome excited with red and blue light at 77 K. *Biochim. Biophys. Acta,* **635,** 369–82.

SPRUIT, C. J. P. (1972). Estimation of phytochrome by spectrophotometry *in vivo*: instrumentation and interpretation. In *Phytochrome*, ed. K. Mitrakos & W. Shropshire, Jr., pp. 78–104. London & New York: Academic Press.

TANADA, T. (1968). A rapid photoreversible response of barley root tips in the presence of 3-indolacetic acid. *Proc. Natl. Acad. Sci., USA,* **59,** 378–80.

TEPFER, D. A. & BONNETT, H. T. (1972). The role of phytochrome in the geotropic behaviour of roots of *Convolvulus arvensis. Planta,* **106,** 311–24.

TEZUKA, T. & YAMAMOTO, Y. (1975). Control of ion absorption by phytochrome. *Planta,* **122,** 239–44.

TORIYAMA, H. (1955). Observations and experimental studies of sensitive plants. VI. The migration of potassium in the primary pulvinus. *Cytologia,* **20,** 367–77.

VANDEN DRIESSCHE, T. (1980). The seismonastic leaf movement of *Mimosa pudica* and the nyctinastic leaf movement. In *Photoreceptors and Plant Development*, ed. J. De Greef, pp. 599–616. Antwerp University Press.

VIERSTRA, R. D., TOKUHISA, J. G., NEWCOMB, E. H. & QUAIL, P. H. (1981). A solid-phase antibody approach to identifying phytochrome bearing structures in particulate fractions. *Plant Physiol. Suppl.* **67,** 727.

VINCE-PRUE, D. (1975). *Photoperiodism in Plants.* Maidenhead: McGraw-Hill.

WAGLE, J. & JAFFE, M. J. (1980). The association and function of phytochrome in pea nuclei. *Plant Physiol. Suppl.* **65,** 5.

WEINTRAUB, R. L. & LAWSON, V. R. (1972). Mechanism of phytochrome-mediated effects of light on cell growth. *Proc. 6th Int. Congr. Photobiol.,* Bochum, Abs. 161.

WEISENSEEL, M. H. & RUPPERT, H. K. (1977). Phytochrome and calcium ions are involved in light-induced membrane depolarization in *Nitella. Planta,* **137,** 225–9.

WEISENSEEL, N. & SMEIBIDL, E. (1973). Phytochrome controls the water permeability in *Mougeotia. Z.Pflanzenphysiol.* **70,** 420–31.

WILLIAMSON, F. A., MORRÉ, D. J. & JAFFE, M. J. (1975). Association of phytochrome with rough-surfaced endoplasmic reticulum fractions from soybean hypocotyl. *Plant Physiol.* **56,** 738–43.

YU, R. & CARTER, J. (1976a). Chromatographic behaviour of phytochrome following its *in vivo* phototransformation. *Plant Cell Physiol.* **17,** 1321–8.

YU, R. & CARTER, J. (1976b). Gel filtration of soluble and solubilized phytochrome. *Ann. Bot.* **40,** 647–9.

THE PHYSIOLOGY OF BLUE-LIGHT SYSTEMS

WERNER SCHMIDT
Universität Konstanz, Fakultät für Biologie, Postfach 5560, 7750 Konstanz, GFR

Introduction

All life on this planet runs on sunlight, in particular the electromagnetic radiation between the wavelengths of 380 and 760 nm, which is visible to the normal human eye. The vision of all other animals, photosynthesis and all kinds of photomorphogenesis are found to occur within a narrow wavelength range, which is slightly wider than that for human vision (Fig. 1). Within this broader wavelength band, there are many physiological responses that are specifically sensitive to blue light. These probably differ appreciably in their sensory transduction pathways, but they have the same classes of photoreceptor pigments: flavins and/or carotenoids.

In the present survey I will confine myself to a few basic principles of these blue-light responses in living organisms. Several recent reviews on this subject are available (Galston, 1974; Presti & Delbrück, 1978; Gressel, 1979; Nultsch & Häder, 1979; Firn & Digby, 1980 and Schmidt, 1980a; Senger & Briggs, 1981), as well as the proceedings of the first conference on Physiological Blue Light Effects held in Marburg, West Germany in 1979 (edited by Senger, 1980). The main questions are: What is the nature of the blue-light receptor pigment? Where is it located in the cell? What are the relevant primary photochemical and photophysical events that initiate the signal chain between receptor and response? What are the secondary metabolic steps which finally lead to an observed response such as phototropic bending, carotenoid synthesis, or changes in circadian rhythms?

Phototropism

The phototropism of both fungi and higher plants, that is their tendency to bend towards light, is the most obvious and earliest studied topic in the field of physiological blue-light responses. Phototropism was first recognised early in the nineteenth century as a typical blue-light response by a worker, who reported that when he placed a flask of port

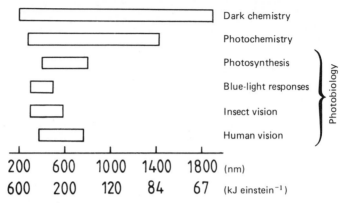

Fig. 1. Energies and wavelength ranges to which various processes are restricted.

wine between a growing plant and light from a window, the plant grew as well as before, but no longer bent towards the light. Diatropism as a kind of phototropic response describes the ability of mature leaves to orient themselves towards the sun, thus gaining an obvious physiological advantage. However, the physiological relevance of phototropic bending of grass coleoptiles or the sporangiphores of *Phycomyces* to light that is slightly unbalanced unilaterally (2% imbalance is sufficient, von Guttenberg, 1959) remains questionable under natural conditions.

It has been well established that phototropic bending is a response to nonhomogeneously distributed light rather than the direction of the incident light. The most photosensitive part of the grass coleoptile is the tip, the first 2 mm being 3×10^5 times more sensitive than the lower part. Some 80 years ago Boysen-Jensen discovered that a signal moved down the unilluminated side of the coleoptile and 25 years later Cholodny and Went proposed a theory for phototropism, which stated that the phototropic curvature of plant organs is the consequence of a light-induced lateral/basipetal redistribution of auxin. Subsequently von Guttenberg demonstrated that a *transverse polarity* of an unknown nature acts as a driving force for the auxin redistribution. However, there remains some inconsistency between this widely accepted theory and recent results on light-induced differential growth in grass coleoptiles reported by Franssen, Cooke, Digby & Firn (1981).

Some 35 years ago Schrank and coworkers performed a series of electrophysiological experiments with *Avena* coleoptiles. Schrank (1947) induced bending of a coleoptile by a transversal current of $10\,\mu A$ applied for 10 min; bending occurred towards the positive pole. The electrically induced transverse polarisation of the coleoptile does not act

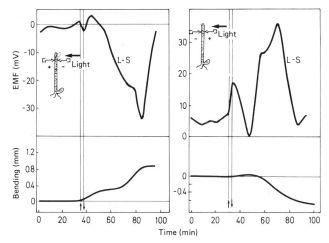

Fig. 2. Electrical (top) and curvature responses (bottom) of *Avena* coleoptiles to two minutes unilateral irradiation, as indicated (vertical lines, ↑ light on, ↓ light off). Bending is measured as deflection of the coleoptile tip from the vertical (positive: towards the light, negative: vice versa); the polarity of the electromotive force (EMF) is sketched (lighted–shadowed side, L–S). *Left:* Light intensity chosen to produce positive curvature; *right:* light intensity chosen to produce negative curvature. Clearly, the convex side of the coleoptile is electrically positive, regardless of the type of curvature (cf. Schmidt, 1980a; after Backus & Schrank, 1951).

on the auxin molecule itself (which is an anion at physiological pHs), as the voltage generated by light is the reverse of that which must be applied externally to obtain the 'correct' curvature. The convex side of the bending coleoptile becomes positive, regardless of the direction of bending with respect to light (first positive or first negative; Fig. 2, after Backus & Schrank, 1951). In addition Webster Jr. & Schrank (1953) verified that the transverse polarisation of freshly decapitated *Avena* coleoptiles, which normally do not bend, could be *electrically* induced and the negative side becomes convex. Related experiments performed by Hartmann (1975) and Hartmann & Schmid (1980), have shown that blue light is capable of inducing strong electrical hyperpolarisation in etiolated hypocotyl hooks of *Phaseolus vulgaris*.

In recent years electrophysiological responses to light have been extensively studied in micro-organisms, revealing the (plasma-) membrane as a locus of the primary (photo-) events (Nultsch & Häder, 1979; Blatt & Briggs, 1980; Blatt, Wessells & Briggs, 1980).

In my discussion of phototropism the revealing experiments of von Guttenberg (1959) need to be mentioned. He showed that when coleoptiles were unilaterally illuminated at 4 °C no bending was observed. However, after transfer to darkness at room temperature the coleoptiles began to bend immediately. This indicates a clearcut separation of the photochemical induction of the physiological response, which requires very little or no activation energy, from the secondary metabolic steps. Additional experiments also showed that the presence of molecular oxygen is an indispensable prerequisite of the photo-induction process, which generally appears true for the blue-light responses.

The nature of the blue-light photoreceptor

Phototropic bending requires the detection of a light gradient within the bending organelle. In oat coleoptiles this gradient is largely produced by shading pigments such as carotenoids, flavonoids, and the little investigated flavin 'FX' (Zenk, 1967*a, b*), rather than by scattering. Action spectroscopy has been a useful tool in determining the chemical nature of a particular photoreceptor, as the action spectrum should resemble (at least in its gross structure) the absorption spectrum of the responsible photoreceptor pigment. This feature is less common for (vectorial) phototropic bending than for scalar responses, because of the modulation of the photoreceptor spectrum by the shading pigment(s). Recently Vierstra & Poff (1979) have devised an interesting approach to overcome this problem. They found that phototropism of corn coleoptiles is virtually unaffected by SAN 9789, a potent inhibitor of carotenoid biosynthesis, when 380 nm light is used (no absorption by carotenoids). Yet when 450 nm light is used a significant reduction of the phototropic effect occurs. This was judged by the authors as evidence that bulk carotenoids are only screening pigments rather than the active photoreceptor.

In recent years many different blue-light responses, in addition to phototropism, have been described and analysed. Their action spectra are not as uniform (Schmidt, 1980*a*) as is usually implied; indeed there is a great diversity, especially with respect to the morphogenetic data (Curry, 1957; Curry & Gruen, 1959; Page & Curry, 1966; Sargent & Briggs, 1967; Gressel & Hartmann, 1968; Klemm & Ninnemann, 1976; de Fabo, Harding & Shropshire, 1976). For example, the action spectra for the phase shift of the pupae emergence of *Drosophila* which is induced by blue light (Klemm & Ninnemann, 1976), and the photo-induction of carotenoid biosynthesis in *Neurospora* (de Fabo *et al.*, 1976) show no, or very small, peaks in the near UV region. It is impossible to decide unequivocally between the putative photoreceptor pigments flavin and carotene solely on the basis of action spectra for several reasons. The near-UV peak of flavins is highly variable in its

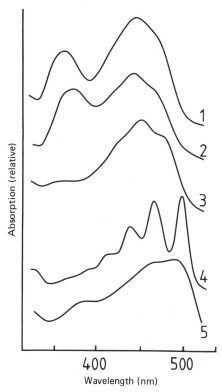

Fig. 3. Comparison of absorption spectra. Curves 1 and 2: 10^{-5} M ethanolic solution of tetra-acetyl-riboflavin, curve 1, room temperature; curve 2, liquid nitrogen temperature. Curves 3 and 4: 10^{-6} M ethanolic solution of β-carotene, curve 3, room temperature; curve 4, liquid nitrogen temperature. Curve 5: absorption of a thin dry film of flavin, evaporated from ethanolic solution.

intensity and wavelength position (Schmidt, 1979, 1980*a*; Fig. 3, curves 1, 2 and 5). Moreover, it has been suggested that the photoreceptor might be the 15-15′-*cis*-isomer of β-carotene, which, in contrast to the *trans* form, shows a UV peak. (This occurs at 330 nm, which is lower than those exhibited by common blue-light action spectra.) Presti, Hsu & Delbrück (1977), claim that there is no *cis*-isomer of β-carotene in *Phycomyces*. However, this conclusion may be unfounded as the mono-*cis*-β-carotene is extremely labile due to uncoupling of the two π-electrons of the central double-bound (H. H. Inhoffen, personal communication), and isomerises to the *trans*-isomer under normal conditions of preparation. Therefore it is extremely difficult to exclude with certainty the presence of *cis*-carotene *in vivo*, where it might be stabilised by an apoprotein (carotenoprotein).

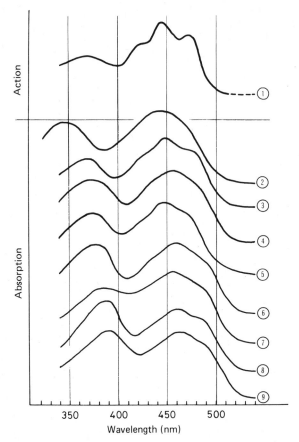

Fig. 4. Absorption spectra of several flavoproteins and the action spectrum of phototropism of the *Avena* coleoptile (curve 1). The spectra are arranged with respect to the position of their UV peaks (after Schmidt, 1980*a*): (2) succinate dehydrogenase, (3) lactate oxidase, (4) D-amino acid oxidase, (5) flavodoxin, (6) old yellow enzyme, (7) ferrodoxin NADP$^+$ reductase, (8) oxynitrilase, (9) L-amino acid oxidase.

In addition to the imprecision of the measurement of physiological properties the spectrum of the photoreceptor pigment depends on several, largely undiscovered, parameters. (1) Relatively small intramolecular changes can alter absorption spectra significantly. (2) Binding of small molecules (coenzymes) to (apo-)proteins has an immobilising effect and may change the absorption spectra. This effect can be mimicked by freezing the chromophore to liquid nitrogen temperature, as demonstrated in Fig. 3. Little effect is observed with flavin (curves 1 and 2) but a large effect is seen with β-carotene (curves 3 and 4), favouring a flavin-photoreceptor. Figure 4 depicts the absorption spec-

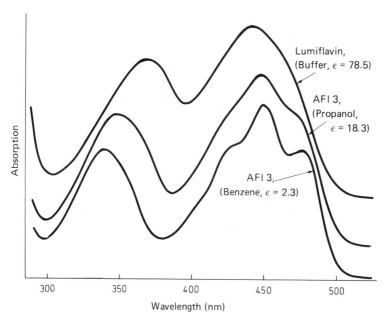

Fig. 5. Absorption spectra of flavin as a function of the polarity of the solvent. Lumiflavin was chosen for the aqueous spectrum, and an amphiphilic flavin (AFl 3, see Fig. 13) for organic solvents (for better solubility).

tra of various flavoproteins, compared with a blue-light action spectrum. All flavoproteins appear to be possible photoreceptor candidates, if we take the variability of action spectra into account. For example, the absorption spectrum of lactate oxidase (curve 3) and the action spectrum of phototropism of the *Avena* coleoptile (curve 1) compare well. (3) Inter-molecular interaction can induce considerable changes in absorption, as exhibited by a thin dry film of flavin, evaporated from ethanolic solution (Fig. 3, curve 5). (4) The solvent may have a large effect upon the absorption of the solute. This is demonstrated by flavins in solvents of different polarities (Fig. 5) – note the strong dependence of the vibrational fine structure and the position of the near-UV peak on the dielectric constant of the solvent. Therefore any attempt to interpret fine details of blue-light action spectra in terms of a positive identification of the photoreceptor pigment is useless. However, the extension of a blue-light action spectrum as that of the 'shadow response' of *Diadema* spines (sea urchin) into the green and yellow regions of the spectrum (Yoshida & Millott, 1959) clearly excludes the (sole) involvement of a flavin or β-carotene photoreceptor in that case (Fig. 6).

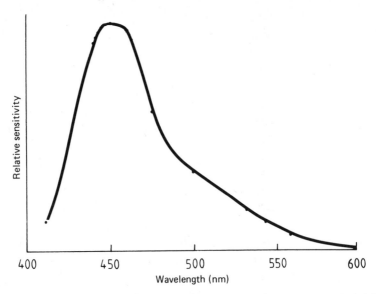

Fig. 6. Action spectrum for the 'shadow response' of *Diadema* spines (sea urchin) (after Yoshida & Millott, 1959).

Replacement of the photoreceptor by external pigments

Generally, the presence of molecular oxygen is a necessary prerequisite for sensory transduction (Schmidt, 1980*a*). Therefore the photo-oxidation of some specific compound could be the primary step of blue-light action. Indeed, the time course of synthesis of various carotenoids in *Fusarium*, as induced by blue-light, can be precisely reproduced by application of hydrogen peroxide (Theimer & Rau, 1970). (This redox hypothesis for the sensory transduction mechanism is strongly supported by the general observation that light induction itself is always independent of temperature, see Schmidt, 1980*a*.) On this basis, a revealing approach to the mechanism of the primary photoreaction was undertaken by Lang-Feulner & Rau (1975) who incubated mycelia of *Fusarium* in the presence of photodynamically active (methylene blue, toluidine blue and neutral red) or inactive (dichlorophenol, malachite green) dyes, and found that only active dyes were capable of acting as artificial photoreceptors. Similarly, Blum & Scott (1933) generated artificial, positive phototropism in wheat roots by the introduction of fluorescein. However, this latter reaction appears to be an artifact caused by unspecific photolysis of essential components on the lighted side.

Otto, Jayarum, Hamilton & Delbrück (1981) succeeded recently, using a flavin-specific permease, in replacing the flavin-photoreceptor for the phototropic response of the sporangiophore of *Phycomyces* with 'roseoflavin', a flavin analogue. The substitution was indicated by an increase of the effectiveness of 529 nm light relative to 380 nm light (the maximum and minimum absorbance of roseoflavin, respectively) in inducing the phototropic response, and by an increase of the light threshold. (Roseoflavin performs its photoreceptor functions with an efficiency of about 0.1% of the natural photoreceptor.)

Absorption changes induced by blue light

Before the nature of the blue-light receptor can be investigated rigorously at a molecular level by the extraction and characterisation of the pigment and analysis of its primary reactions, an unambiguous assay is required, such as the red/far-red photoreversibility of phytochrome.

Dark-reversible light-induced absorption changes (LIACs) have been found *in vivo* in several blue-light-sensitive organisms, exhibiting 'typical' blue-light action spectra (Poff & Butler, 1974; Muñoz & Butler, 1975). Usually these LIACs represent flavin-mediated reduction of a *b*-type cytochrome. Schmidt, Thomson & Butler (1977*a*) have identified *b*-type cytochromes and flavins in plasma membrane-enriched fractions from *Phycomyces, Neurospora* and *Dictyostelium*. (*Neurospora* and *Phycomyces* exhibit physiological blue-light responses. The photoresponses that are controlled by the blue-light receptor in *Dictyostelium* are unknown; see Poff & Butler, 1974.) Recently Goldsmith, Caubergs & Briggs (1980) have optimised the conditions for reproducible photoreactions in membrane fractions prepared from higher plants, and Leong & Briggs (1981) have partially purified and characterised from plasma membranes a cytochrome–flavin complex that is sensitive to blue light. Leong, Caubergs & Briggs (1979) obtained evidence that these flavin–cytochrome moieties may be on the same protein, as is the case for cytochrome b_2 from yeast (Lemberg & Barrett, 1973). From a graphical analysis of published photoreduction kinetics of this unique *b*-type cytochrome, Lipson & Presti (1980) estimated a corresponding quantum efficiency as low as $\varphi = 0.031$ for *Phycomyces*, but a φ value near unity for pellets from corn and *Neurospora*. They conclude that the latter values of φ imply a process that probably has more relevance to photophysiology than the blue-light induced photoreduction of *b* cytochrome from, for example, beef heart mitochondiria (Ninnemann, Strasser & Butler, 1977), *Phycomyces* or even HeLa-cells (Lipson &

Presti, 1977; Presti, this volume), which is probably not physiologically significant (either because of the lack of a physiological blue-light response in the tissue concerned or because of their low quantum efficiency).

Ulaszewski *et al.* (1979) have presented further evidence for the correlation of cytochrome *b* reduction and blue-light responses: various yeast mutants containing *b* cytochromes are inhibited by high irradiances of blue-light, like the wild-type, whereas mutants lacking *b* cytochromes are not affected.

Recent results of Klemm & Ninnemann (1978, 1979) also show a correlation between a LIAC and a light-dependent physiological response in *Neurospora* (light-induced conidiation), probably mediated by nitrate reductase. This NAD(P)H-dependent enzyme of *Neurospora* (as well as of higher plants) is known to possess both a flavin (FAD) and a *b*-type cytochrome with an α-band at 557 nm as prosthetic groups. The activity of the enzyme is regulated by reversible interconversion between the active (oxidised) and inactive (reduced) forms.

The reactivation of the inactive enzyme can be accomplished either by chemical oxidation with ferricyanide, or by irradiation with blue-light (Aparicio, Roldán & Calero, 1976; Roldán & Butler, 1980). The action spectrum provides evidence for the involvement of nitrate reductase as the photoreceptor for the assimilation of nitrate in *Neurospora*.

The natural blue-light-processing redox partner of the *b* cytochrome is undoubtedly flavin rather than carotene. Nearly all known flavin reactions are redox reactions, whereas carotenoids are not redox active at all. This is consistent with the finding that carotenoid-'free' mutants of various blue-light-sensitive organisms are fully responsive to blue-light (for references refer to Presti & Delbrück, 1978).

Probably there are many flavoproteins (rather than free flavins) which can serve as photoreceptor pigments for the various blue-light responses, each initiating its own specific light-response pathway. For example, blue-light induces a sharp, transient, 60% dip in cyclic AMP levels in intact sporangiophores of *Phycomyces* (Cohen, 1974); this can be traced to the action of a blue-light-activated cAMP phosphodiesterase (Cohen & Atkinson, 1978). Another example of a blue-light-activated enzyme is glycine oxidase of *Chlorella* (Schmid & Schwarze, 1969; Fig. 7, *right*). In contrast, glycolate oxidase, one of the key enzymes of photorespiration in higher plants, is inhibited by low intensities of blue-light (Lorimer, Andrews & Tolbert, 1973; Fig. 7, *left*). A detailed discussion of the activation of enzymes by light has been published recently (Hug, 1978).

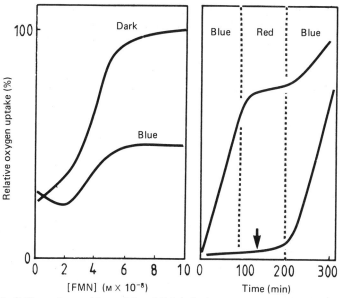

Fig. 7. *Left:* Dependence of the activity of dialysed tobacco glycollate oxidase on blue-light and on the concentration of flavin mononucleotide (after Lorimer *et al.*, 1973). *Right:* Dependence of glycine oxidase activity on blue-light in the colourless *Chlorella* mutant 125. The control was unwrapped after 130 min (arrow) (after Schmid & Schwarze, 1969).

Flavin X ('FX')

In addition to the well-known flavin-mediated photodecarboxylation of auxin *in vitro*, Zenk (1967*a*) obtained clearcut evidence from *Avena* that this process also occurs *in vivo*. He isolated the photoreceptor flavin involved and designated it 'FX'. Nearly all of the flavin in etiolated oat coleoptile is FX which is hydrolysed under mild alkaline and acidic conditions to riboflavin; it is extremely light-sensitive, and does not contain phosphate. Zenk (1967*b*) assumes FX to be a riboflavin ester. Recently S. Ghisla has obtained evidence, by *in-vitro* synthesis, that FX is identical to 5′-malonic acid riboflavin ester (unpublished data). It is important to note that only 10% of FX is protein-bound, the greater part being in a free state: however, Zenk found that only the freely soluble FX was capable of mediating the photodecarboxylation of the hormone, and concluded that this reaction is probably photophysiologically irrelevant.

Recently Hertel, Jesaitis, Dohrmann & Briggs (1980) obtained evidence for a specific binding site for riboflavin and FX on subcellular particles derived from plasmalemma and endoplasmic reticulum from maize coleoptiles and *Curcurbita* hypocotyls: FMN and FAD bound less tightly to these sites, especially in their reduced forms. Since there exist abundant data suggesting a dichroic orientation of the blue-light photoreceptor in the plasma-membrane, these authors proposed that a reversible interaction of the soluble, extractable FX with the specific binding site might be essential for the primary photoprocess. (This appears inconsistent with the conclusion drawn by Zenk that the photocarboxylation of auxin as mediated by the soluble FX, is probably physiologically irrelevant.)

Relatively high concentrations of FX, as characterised by thin-layer chromatography, are also found in mushrooms for which no typical blue-light effect has been described so far. In order to characterise FX chemically from these sources, we have developed a three-step purification procedure (Hemmerich & Schmidt, 1980). An initial spectroscopic characterisation is presented in Fig. 8: it shows corrected fluorescence excitation (= absorption) and emission spectra of FX compared with riboflavin, both in ethanolic solution. Small but characteristic differences are found, as shown by the difference spectra. The available concentration of highly purified FX is only now allowing absorption spectroscopy or photochemical experiments.

Localisation of the blue-light photoreceptor

Experiments with plane-polarised light or with microbeams can elucidate the locus and the specific (often dichroic) embedding of the blue-light photoreceptor pigment in suitable highly transparent objects. In 1934, Castle initiated this type of experiment by the discovery of 'polarotropism' in *Phycomyces*. Later many elegant experiments of this type have been performed, especially on the blue-light-induced movement of chloroplasts (see Haupt, this volume). Medium and low irradiances of light cause chloroplasts to turn their faces, and very high irradiances to turn their edges, towards the light source. It is generally observed that plane-polarised light of moderate intensity impinging on a whole cell, induces the chloroplasts to accumulate near the side walls situated parallel to the e-vector. However, high light intensities have the opposite effect, and the chloroplasts accumulate near the side walls orthogonal to the e-vector. When a microbeam is directed onto the chloroplast itself, no effect is observed, but irradiation of a small area of the cell envelope immediately induces the attachment of the chloroplasts to this region. We have to conclude that the photoreceptor pigment for the orientation of chloroplasts is localised in or near the plasma membrane with its main transition dipole (450 nm) parallel to

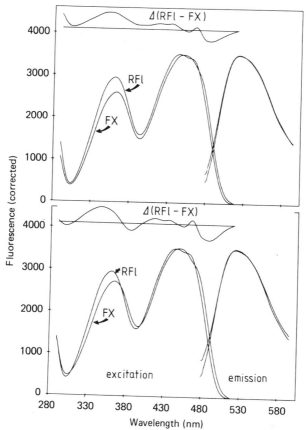

Fig. 8. Fluorescence excitation and emission spectra of FX purified from oat coleoptiles (above) and mushroom (below) compared with those of riboflavin, both in ethanolic solution.

the cell wall. The movement of the chloroplast(s) is probably mediated by actin filaments (Fischer-Arnold, 1963; Mayer, 1964; Zurzycki, 1980).

Two divergent examples exhibiting the oriented localisation of the blue-light receptor are the spores of the imperfect fungus *Botrytis cinera*, and the fern *Osmunda cinnamomea*, respectively (Jaffe & Etzold, 1962). When unilaterally irradiated, *Botrytis* grows from the brighter side and *Osmunda* from the darker side, yet both germinate in the direction of the e-vector of polarised blue-light. The authors conclude that the photoreceptor molecules of *Botrytis* are strongly oriented in a dichroic orthogonal fashion, while those of *Osmunda* are parallel to the cell wall. Taking the angles between the transition dipoles of the flavin nucleus into account (Song, 1969), a crude visualisation of the arrangement of the photoreceptor *in situ* is given in Fig. 9.

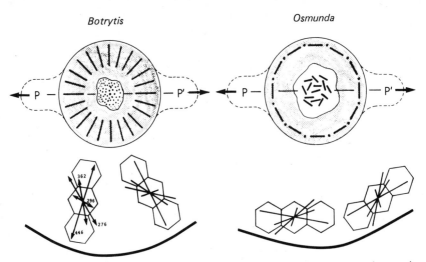

Fig. 9. Visualisation of the localisation of the blue-light photoreceptors for tropic responses, as deduced from experiments with polarised light. *Left*: Spore of the fungus *Botrytis cinera*. *Right*: Spore of the fern *Osmunda cinnamomea*. The transition dipole moments are indicated by short bold lines (dots in front view) (after Jaffe & Etzold, 1962). Assuming that a flavin is the photoreceptor, and taking its transition dipoles as given by Song (1969) into account, a more detailed picture is obtained (lower part of the figure).

Effector response pathway

The molecular details of the sensory transduction chain with the blue-light-receiving pigment on the input side and the final response on the output side remain largely fragmentary. Chemical compounds known to be specific inhibitors of various biochemical processes may help to trace the individual steps of the signal path.

In order to be photochemically active, light must be absorbed by a specific pigment molecule, and for physiological blue-light action the most likely candidate is a flavin, as discussed earlier. Figure 10 depicts the spectral characteristics of the flavin chromophore in the form of a Jablonski diagram (*top*); and as crudely derived from the measured spectra (*bottom*). Sun, More & Song (1972) assign a $\pi \rightarrow \pi^*$ character to all major electronic flavin transitions. Intersystem crossing ($S_1 \rightarrow T_1$) is an efficient process, although it is forbidden on quantum-mechanical grounds ($\varphi = 0.7$, Grodowski, Veyret & Weiss, 1977). Due to its roughly 1000 times longer lifetime it is reasonable to assume that the flavin triplet rather than the singlet is the photochemically active species. Indeed, iodide and azide, known to be effective dynamic triplet quenchers, have been found to be specific inhibitors of phototropism in

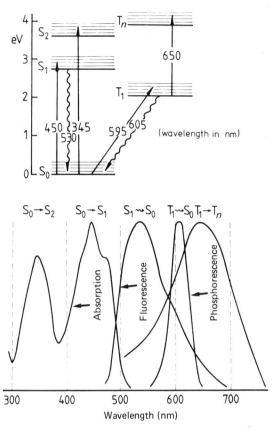

Fig. 10. *Top:* Jablonski diagram for the flavin chromophore. The wavelengths given for absorption and luminescence are approximate average values derived from the spectra shown on the bottom. All spectra are normalised (after Schmidt, 1980*a*).

corn (Schmidt *et al.*, 1977) and the 'inverse phobic response' of *Euglena* (Mikolajczyk & Diehn, 1975).

Another approach has been made by Delbrück, Katzir & Presti (1976), who successfully induced phototropism in *Phycomyces* with yellow/orange laser light; they interpreted this as strong evidence for the involvement of the flavin triplet. A related experiment was described by Schmidt *et al.* (1977): when maize coleoptiles are exposed to phototropically-active light, simultaneous irradiation with 100-times stronger, but phototropically-inert, light will suppress phototropism by up to 35%. These authors assume an effective decrease of the lifetime of the lowest triplet state by rapid triplet–triplet turnovers. However, in contrast to these findings, Vierstra & Poff (1979) found no significant inhibition of

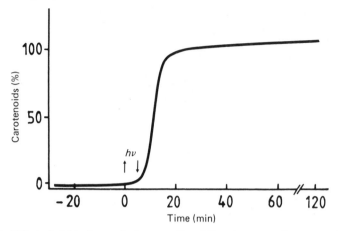

Fig. 11. Effect of dithionite applied to mycelia of *Fusarium aquaeductuum* at different times before and after photo-induction of carotenogenesis The inhibitor was added at the time indicated on the abscissa and removed 30 min later. Carotenoids were measured after several hours (after Theimer & Rau, 1970).

phototropism in corn by xenon, which they interpret to mean that the flavin triplet is not involved in phototropism.

The excited flavin singlet has to dispose of its electronic energy within a few nanoseconds, returning back to its ground state. In solution, part of the excitation energy is unspecifically dissipated to the solvent, the rest is emitted as fluorescence. Therefore, fluorescence can be used as a probe to monitor the photoreceptor micro-environment and dynamics. Song *et al.* (1980) reported significant quantities of plasma-membrane-bound flavin unique to the phototropically active tip of the maize coleoptile, with a much shorter fluorescence lifetime than other flavins.

The physiological photoreceptor flavin bound to a protein and/or a membrane, will alter its micro-environment as it relaxes to the ground state. Changes in a number of parameters such as conformational state, acid–base properties, redox potential, dipole moment or polarisability might initiate sensory transduction. Photodissociation of a flavoprotein, analogous to the action of rhodopsin in rods, has also to be considered.

Clearly (as in vision) the energy of the impinging blue-light serves to trigger an existing signal chain. As already mentioned, blue-light can be effectively replaced by hydrogen peroxide to initiate carotenogenesis in *Fusarium*, including the three phases: (i) induction, (ii) protein synthesis and (iii) accumulation of carotenoids (Theimer & Rau, 1970). Consistently, the blue-light induction is completely abolished by dithionite, which is consistent with the suggestion that the primary reactions are redox in character. However, a few minutes *after* the blue-light pulse the system has escaped sensitivity to dithionite, indicating that the signal chain has lost its redox character and attained a different quality (Fig. 11). A period of protein synthesis has been identified by studies with cycloheximide and chloramphenicol (Spurgeon, Turner & Harding, 1979; Schmidt, 1980*a*). The third phase of carotenoid accumulation has been explored by observing the sequential increase of various carotenoids following induction. Bindl, Land & Rau (1970) conclude that photoregulation takes place between the farnesylpyrophosphate and the coloured carotenoids in *Fusarium*, because the synthesis of sterols is *not* increased after illumination.

More recent data obtained by Jayaram, Presti & Delbrück (1979) and Schrott (1980) indicate a dual light control of β-carotene synthesis in *Phycomyces*; namely (i) regulation at a level not requiring new RNA and/or protein synthesis, and thus probably involving enzyme activation by light, and (ii) regulation at the level of transcription and translation.

Artificial membrane/flavin systems

As already discussed, there is strong evidence that the blue-light photoreceptor is a plasma-membrane-bound flavin, and that the primary events are redox reactions. Unfortunately, in contrast to the well understood 'isotropic flavin chemistry' (the chemistry of free flavins in solution; Hemmerich, 1976; Bruice, 1976), very little is known about 'anisotropic flavin chemistry' (the chemistry of the bound flavin), which solely applies to biological systems.

It is well known that the flavin nucleus has four active positions (C4a, N5, C8 and C10a) which mediate almost exclusively all flavin (redox) reactions. In isotropic flavin chemistry the flavin 'self-contact' is always the fastest reaction, and is largely controlled by the diffusion rate (Barman & Tollin, 1972). This can, however, be ignored in flavoproteins in which the flavin nuclei are bound in a highly anisotropic manner, and a large number of flavoproteins appear to be strongly membrane-oriented under conditions where steric restrictions are able to control the specific flow of flavin substrates. From these considerations it is clear that vesicle-bound flavins may serve as simple model systems for the blue-light photoreceptors.

It is possible to anchor flavin within the membrane by means of long aliphatic chains attached to positions N3, C7 and C10, thereby exposing the four active positions optionally (Schmidt, 1979; Schmidt, 1980*a, b*; Schmidt, 1981*a, b*; Schmidt & Hemmerich, 1981; Michel & Hemmerich, 1981; Michel, Schmidt & Hemmerich, 1982), and mimicking the specific

(Outside)

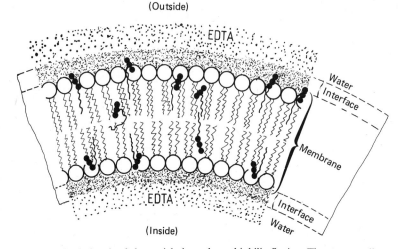

(Inside)

Fig. 12. Idealised sketch of the vesicle-bound amphiphilic flavins. The average distance between the flavin nuclei is about 6.0 nm. In the crystalline state of the membrane they are mainly localised near the lipid/water interface, but sink deeper into the more hydrophobic parts of the membrane upon phase transition. The maximal flavin area density on the membrane is 2–4×10^{12} molecules cm^{-3}, corresponding to a ratio [flavin]:[lipid] = 1–$2:100$ (after Schmidt, 1979).

binding of the coenzyme to the apoprotein (Fig. 12). In this case the flavin 'self-contact' is decreased to about 2% of the isotropic case (Schmidt, 1981*a*).

The molecular structures of the different amphiflavins are depicted in Fig. 13. For example, AFl 3 may serve as a model for flavodoxin with its positions 2 and 3 buried within a hydrophobic pocket, whereas positions 7 and 8 are exposed to the medium (Watenpaugh, Sieker & Jensen, 1973). In AFl 7, the hydrophobic benzene portion of the flavin molecule seems to be buried within the membrane with positions 2 and 3 being exposed, analogous to riboflavin-binding eggwhite flavoprotein (Blanckenhorn, 1978). Flavins are capable of undergoing both $1e^-$ and $2e^-$ input/output redox reactions ('transformase'; Hemmerich, Massey, Mechel & Schug, 1982; Hemmerich, Knappe, Kramer & Traber, 1980), based on structural parameters inherent in the flavin moiety and the regulatory influence of hydrogen bridges extending from the apoprotein to either position 1/2 α ($2e^-$ transfer) or position 5 ($1e^-$ transfer). (Cytochrome b is a typical $1e^-$ redox reagent.)

The marked biophysical and biochemical differences obtained so far with the vesicle-bound flavins compared with common isotropic flavin chemistry, can themselves justify these investigations – even if a simple correlation with primary effects of physiological blue-light action is not

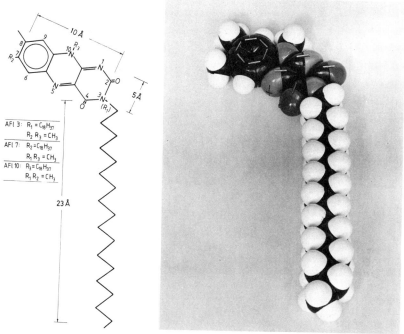

Fig. 13. *Left:* Molecular structure of the three amphiphilic flavins incorporated into an artificial membrane vesicle made from various saturated lecithins. The long aliphatic chains of the amphiflavins ($C_{18}H_{37}$) do not disturb the redox properties of the flavin nucleus. *Right:* A model for the same molecule (AFl 3).

yet feasible. For example, the fluorescence quantum efficiency, motility and localisation of vesicle-bound flavins depend strongly on the phase of the membrane (gel or liquid-crystalline) and on their mode of anchoring (on the locus of substitution of the aliphatic chain).

The pK of the N1-proton in isotropic solution is 6.3, this is dramatically increased by 3.5 units upon membrane-binding (Schmidt, 1980b). Figure 14 demonstrates the corresponding change of the N3-proton from pK = 9 to about 12 (the anion does not fluoresce)) (Schmidt & Hemmerich, 1981).

As expected, the rates of photoreduction and autooxidation of vesicle-bound flavins depend also on the specific amphiphilic flavin used (the locus of substitution of the aliphatic chain affects exclusively the mode of anchoring: all properties of the various amphiflavins remain unchanged under *isotropic* conditions), and on the phase of the membrane. In the liquid-crystalline state of the membrane there is little photoreduction by externally given substrates (electron donors). However, below the

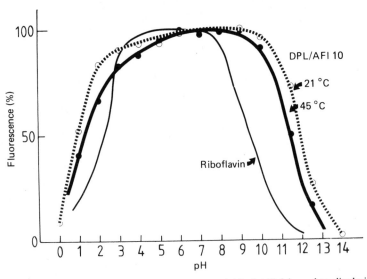

Fig. 14. Dependence of the fluorescence quantum yield of AFl 3 bound to dipalmitoyl-lecithin (DPL) vesicles on the pH and the phase of the membrane (21°C: gel-form; 45°C: liquid crystalline form), compared with aqueous riboflavin. The decrease of fluorescence in the acidic range is due to 'proton quenching' and in the alkaline range is due to the pK of the N3-proton (the anion does fluoresce) (after Schmidt & Hemmerich, 1981).

phase-transition temperature all amphiflavins show appreciable photo-reduction by some *internal* donor (possibly the secondary hydrogen atom of the glycerol backbone of the lecithin molecule; Schmidt & Hemmerich, 1981), in addition to an enhanced photoreduction rate with external electron donors. The quantum efficiency of flavin photoreduc-tion by this intrinsic donor as a function of temperature is depicted in greater detail in Fig. 15, and compared to the isotropic photoreduction of lumiflavin by EDTA. A surprisingly complicated pattern is observed, indicating that the location of the flavin nucleus (as assayed by the availability of the intrinsic donor) depends strongly on temperature. This probably reflects a correlation with the 'prephase transition' of the vesicle rather than with the main phase transition.

In contrast to photoreduction, the reoxidation of vesicle-bound amphiflavin by molecular oxygen proceeds ten times faster in the liquid-crystalline state (halflife 0.05 s) than in the gel state (halflife 0.5 s) of the membrane (Fig. 16). This is interpreted as strongly phase-dependent permeability of the membrane/water interface (its signifi-cance has often been underestimated) (Schmidt & Hemmerich, 1981).

We are presently in the process of studying transmembrane transport of redox equivalents, as mediated by the different amphiflavins. This

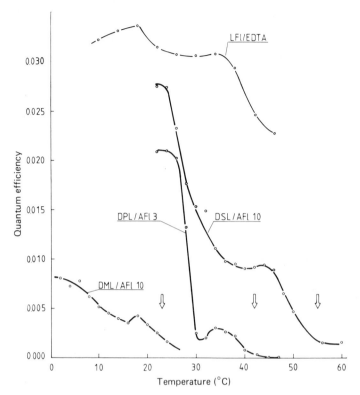

Fig. 15. Quantum efficiencies of photoreduction as a function of temperature in various flavin/vesicle systems. DML/AFl 10: amphiphilic flavin with the aliphatic chain at position 10, bound to dimyristoyl-lecithin vesicles (phase-transition temperature, $T_c = 23\,°C$). DPL/AFl 3: amphiphilic flavin with the aliphatic chain at position 3, bound to dipalmitoyl-lecithin vesicles ($T_c = 41\,°C$). DSL/AFl 10: amphiphilic flavin with the aliphatic chain at position 10, bound to distearoyl-lecithin vesicles ($T_c = 55\,°C$). The corresponding isotropic photoreduction of lumiflavin by exogenous EDTA is also shown. The arrows mark the phase-transition temperatures of DML, DPL and DSL-vesicles (after Schmidt & Hemmerich, 1981).

may serve as a model for the still unknown mode of action of (ubi-)quinones (Trumpower & Landeen, 1977; Futami, Hurt & Hauska, 1979). Preliminary results indicate once again the dependence of these processes on parameters such as membrane phase, ionic strength or mode of anchoring of the flavin. (In this context the review by Berns (1976) on artificial membranes serving as models for photobiological systems is recommended reading.) One remark is pertinent: significant phase transitions are commonly observed with artificial membranes composed of one or only a few types of lipid, usually upon temperature change or change in ionic strength. Phase transitions of biological membranes are rarely observed on a macroscopic scale. However, this

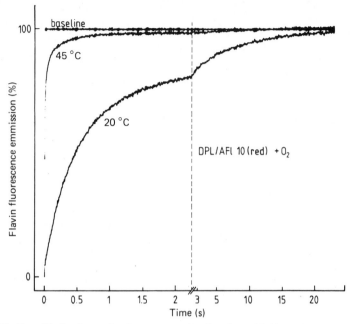

Fig. 16. Re-oxidation by molecular oxygen of reduced amphiphilic flavin 10 (Fig. 13) bound to DPL vesicles, measured by a stopped flow apparatus below (20 °C) and above (45 °C) the phase-transition temperature, T_c. Kinetics are approximately first order with halflives of 0.5 s (20 °C) and about 0.05 s (45 °C). The baseline was taken 0.5 h after these measurements were made (after Schmidt & Hemmerich, 1981).

does not exclude the possibility that phase transitions, undetectable by present assays, can occur in microscopic domains of biological membranes and thereby control enzyme activity or membrane permeability. Indeed preliminary experiments on the reconstitution of monoamine oxidase, a strictly membrane-bound flavin, into a membraneous environment demonstrate a clearcut modulation of the enzyme activity by various parameters; this in turn, could generate a membrane potential, as observed in phototropism.

I thank Professor Dr Helmut Beinert and the editors for critically reading this manuscript. This work was financially supported by the Deutsche Forschungsgemeinschaft (SFB 138, A1).

References

APARICIO, P. J., ROLDÁN, J. M. & CALERO, F. (1976). Blue light photoreactivation of nitrate reductase from green algae and higher plants. *Biochem. Biophys. Res. Commun.*, **70**, 1071–7.

BACKUS, G. E. & SCHRANK, A. R. (1951). Electrical and curvature responses of the avena coleoptile to unilateral illumination. *Plant Physiol.*, **27**, 251–62.

BARMAN, B. G. & TOLLIN, G. (1972). Kinetics and equilibria in partially reduced flavin solutions. *Biochemistry*, **25**, 4760–5.

BERNS, D. S. (1976). Photosensitive bilayer membranes as model systems for photobiological processes. *Photochem. Photobiol.*, **24**, 117–39.

BINDL, E., LAND, W. & RAU, W. (1970). Untersuchungen über die lichtabhängige Carotinoidsynthese. VI. Zeitlicher Verlauf der Synthese der einzelnen Carotenoide bei *Fusarium aquaeductuum* unter verschiedenen Induktionsbedingungen. *Planta*, **94**, 156–74.

BLANCKENHORN, G. (1978). Riboflavin binding eggwhite flavoprotein: the role of tryptophan and tyrosine. *Eur. J. Biochem.*, **82**, 155–60.

BLATT, M. & BRIGGS, W. (1980). Blue-light-induced cortical fiber reticulation concomitant with chloroplast aggregation in the alga *Vaucheria sessilis*. *Planta*, **147**, 355–62.

BLATT, M., WESSELLS, N. & BRIGGS, W. (1980). Actin and cortical fiber reticulation in the siphonaceous alga *Vaucheria sessilis*. *Planta*, **147**, 363–75.

BLUM, H. F. & SCOTT, K. G. (1933). Photodynamically induced tropism in plant roots. *Plant Physiol.*, **8**, 525–36.

BRUICE, T. C. (1976). Models and flavin catalysis. In: *Progress in Bioorganic Chemistry*, vol. 4, ed. E. T. Kaiser & F. J. Kedzy, pp. 1–87. New York, London, Sidney & Toronto: Wiley.

CASTLE, E. S. (1934). The phototropic effect of polarized light. *J. Gen. Physiol.*, **17**, 751–62.

COHEN, R. J. (1974). Cyclic AMP levels in *Phycomyces* during a response to light. *Nature, London*, **251**, 144–6.

COHEN, R. J. & ATKINSON, M. M. (1978). Activation of *Phycomyces* adenosine 3′, 5′-monophosphate phosphodiesterase by blue light. *Biochem. Biophys. Res. Commun.*, **83**, 616–21.

CURRY, G. M. (1957). Studies on the spectral sensitivity of phototropism. P D Thesis, Harvard University, Cambridge, Massachusetts.

CURRY, G. M. & GRUEN, M. E. (1959). Action spectra for the positive and negative phototropism of *Phycoymces* sporangiophores. *Proc. Natl. Acad. Sci., USA*, **45**, 797–804.

DE FABO, E. C., HARDING, R. W. & SHROPSHIRE, JR., W. (1976). Action spectrum between 260 and 800 nanometers for the photoinduction of carotenoid biosynthesis in *Neurospora crassa*. *Plant. Physiol.*, **57**, 440–5.

DELBRÜCK, M., KATZIR, A. & PRESTI, D. (1976). Responses of *Phycomyces* indicating optical excitation of the lowest triplet state of riboflavin. *Proc. Natl. Acad. Sci., USA*, **37**, 1969–73.

FIRN, A. D. & DIGBY, J. (1980). The establishment of tropic curvatures in plants. *Ann. Rev. Plant Physiol.*, **31**, 131–48.

FISCHER-ARNOLD, G. (1963). Untersuchungen über die Chloroplastenbewegung bei *Vaucheria sessilis*. *Protoplasma*, **56**, 495–520.

FRANSSEN, J. M., COOKE, S. A., DIGBY, J. & FIRN, R. D. (1981). Measurement of differential growth causing phototropic curvature of coleoptiles and hypocotyls. *Z. Pflanzenphysiol*, **103**, 207–16.

FUTAMI, A., HURT, E. & HAUSKA, G. (1979). Vectorial redox reactions of physiological quinones. I. Requirement of a minimum length of the isoprenoid side chain. *Biochim. Biophys. Acta*, **547**, 583–96.

GALSTON, A. W. (1974). Plant photobiology in the last half century. *Plant Physiol.*, **54**, 427–36.

GOLDSMITH, M. H. M., CAUBERGS, R. J. & BRIGGS, W. R. (1980). Light-inducible cytochrome reduction in membrane preparations. *Plant Physiol.*, **66**, 1067–73.

GRESSEL, J. (1979). Blue light photoreception. Yearly review. *Photochem. Photobiol.*, **30**, 749–54.

GRESSEL, J. & HARTMANN, E. (1968). Morphogenesis in *Trichoderma*. Action spectrum of photoinduced sporulation. *Planta*, **79**, 271–4.

GRODOWSKI, M. S., VEYRET, B. & WEISS, K. (1977). Photochemistry of flavins. II. Photophysical properties of alloxazines and isoalloxazines. *Photochem. Photobiol.*, **26**, 341–52.

HARTMANN, E. (1975). Influence of light on the bioelectric potential of the bean (*Phaseolus vulgaris*) hypocotyl hook. *Physiol. Plant.*, **33**, 266–75.

HARTMANN, E. & SCHMID, K. (1980). Effects of UV and blue light on the biopotential changes in etiolated hypocotyl hooks of dwarf beans. In: *The Blue Light Syndrome*, ed. H. Senger, pp. 221–37. Heidelberg: Springer-Verlag.

HEMMERICH, P. (1976). The present status of flavin and flavoenzyme chemistry. In: *Progress in the Chemistry of Inorganic Natural Products*, vol. 33, ed. W. Herz, H. Grisebach & G. W. Kirby, pp. 451–527. Vienna & New York: Springer-Verlag.

HEMMERICH, P., KNAPPE, W.-R., KRAMER, H. E. A. & TRABER, R. (1980). Distinction of $2e^-$ and $1e^-$ reduction modes of the flavin chromophore as studied by flash photolysis. *Eur. J. Biochem.*, **104**, 511–20.

HEMMERICH, P., MASSEY, V., MICHEL, H. & SCHUG, CH. (1982). Scope and limitation of single electron transfer in biology. In: *Structure and Bonding*, vol. 48, ed. Dunitz, J. D., Goodenough, J. B., Hemmerich, P., Ibers, J. A., Jørgensen, C. K., Neilands, J. B., Reinen, D., Williams, R. J. P., pp. 94–123. Berlin, Heidelberg & New York: Springer-Verlag.

HEMMERICH, P. & SCHMIDT, W. (1980). Bluelight reception and flavin photochemistry. In: *Photoreception and Sensory Transduction in Aneural Organisms*, ed. F. Lenci & G. Colombetti, pp. 271–83. New York: Plenum Press.

HERTEL, R., JESAITIS, A., DOHRMANN, U. & BRIGGS, W. (1980). *In vitro* binding of riboflavin subcellular particles from maize coleoptiles and Curcurbita hypocotyls. *Planta*, **147**, 312–19.

HUG, D. H. (1978). The activation of enzymes with light. In: *Photochem. Photobiol. Reviews*, ed. K. C. Schmid, pp. 1–35. New York & London: Plenum Press.

JAFFE, L. & ETZOLD, H. (1962). Orientation and locus of tropic photoreceptor molecules in spores of *Botrytis* and *Osmunda*. *J. Cell Biol.*, **13**, 13–31.

JAYARAM, M., PRESTI, D. & DELBRÜCK, M. (1979). Light-induced carotene synthesis in *Phycomyces*. *Exp. Mycol.*, **3**, 42–52.

KLEMM, E. & NINNEMANN, H. (1976). Detailed action spectrum of the delay shift in pupae emergence of *Drosophila pseudo-obscura*. *Photochem. Photobiol.*, **24**, 369–71.

KLEMM, E. & NINNEMANN, H. (1978). Correlation between absorbance changes and a physiological response induced by blue light in *Neurospora*. *Photochem. Photobiol.*, **28**, 227–30.

KLEMM, E. & NINNEMANN, H. (1979). Nitrate reductase – a key enzyme in blue light-promoted conidiation and absorbance change of *Neurospora*. *Photochem. Photobiol.*, **29**, 629–32.

LANG-FEULNER, J. & RAU, W. (1975). Redox dyes as artificial photoreceptors in light-dependent carotenoid synthesis. *Photochem. Photobiol.*, **21**, 179–83.

LEMBERG, R. & BARRETT, J. (1973). *Cytochromes*. London: Academic Press.

LEONG, T.-Y. & BRIGGS, W. R. (1981). Partial purification and characterization of a blue light-sensitive cytochrome-flavin complex from corn membranes. *Plant Physiol.* **67**, 1042–6.

LEONG, T.-Y., CAUBERGS, R. J. & BRIGGS, W. R. (1979). Solubilization of a photoactive, membrane-associated flavin–cytochrome complex from corn. Carnegie Institution of Washington, Yearbook 79, pp. 134–35.

LIPSON, E. & PRESTI, D. (1977). Light-induced absorbance changes in *Phycomyces* photomutants. *Photochem. Photobiol.*, **25**, 203–8.

LIPSON, E. & PRESTI, D. (1980). Graphical estimation of cross sections from fluence-response data. *Photochem. Photobiol.*, **32**, 383–91.

LORIMER, G. M., ANDREWS, T. J. & TOLBERT, N. E. (1973). Ribulose diphosphate oxygenase. II. Further proof of reaction products and mechanism of action. *Biochemistry*, **12**, 18–23.

MAYER, F. (1964). Lichtorientierte Chloroplasten-Verlagerung bei *Selaginella martensii*. *Z. Bot.*, **52**, 346–81.

MICHEL, H. & HEMMERICH, P. (1981). Substitution of the flavin chromophore with lipophilic side chains: a novel membrane redox label. *J. Membrane Biol.*, **60**, 143–53.

MICHEL, H., SCHMIDT, W., HEMMERICH, P. (1982). On the environment and the rotational motion of amphiphilic flavins in artificial membrane vesicles as studied by electron paramagnetic resonance spectroscopy. *Biophys. Chem.*, **15**, 121–30.

MIKOLAJCZYK, E. & DIEHN, B. (1975). The effect of potassium iodide on photophobic responses in *Euglena*: evidence for two photoreceptor pigments. *Photochem. Photobiol.*, **22**, 269–71.

MUÑOZ, V. & BUTLER, W. L. (1975). Photoreceptor pigment for blue light in *Neurospora crassa*. *Plant Physiol.*, **55**, 421–6.

NINNEMANN, M., STRASSER, R. J. & BUTLER, W. L. (1977). The superoxide anion as electron donor to the mitochondrial electron transport chain. *Photochem. Photobiol.*, **26**, 41–7.

NULTSCH, W. & HÄDER, D.-P. (1979). Photomovement of motile microorganisms. Review article. *Photochem. Photobiol.*, **29**, 423–37.

OTTO, M. K., JAYARAM, M., HAMILTON, R. M. & DELBRÜCK, M. (1981). Replacement of riboflavin by an analogue in the blue light photoreceptor of *Phycomyces*. *Proc. Natl. Acad. Sci., USA*, **78**, 266–9.

PAGE, R. M. & CURRY, G. M. (1966). Studies on phototropism of young sporangiophores of *Philobolus kleinii*. *Photochem. Photobiol.*, **5**, 31–40.

POFF, K. L. & BUTLER, W. L. (1974). Absorbance changes induced by blue light in *Phycomyces blakesleeanus* and *Dictyostelium discoideum*. *Nature, London*, **248**, 799–801.

PRESTI, D. & DELBRÜCK, M. (1978). Photoreceptors for biosynthesis, energy storage and vision. *Plant Cell Environ.*, **1**, 81–100.

PRESTI, D., HSU, W.-J. & DELBRÜCK, M. (1977). Phototropism in *Phycomyces* mutants lacking β-carotene. *Photochem. Photobiol.*, **26**, 403–405.

ROLDÁN, J. M. & BUTLER, W. L. (1980). Photoactivation of nitrate reductase from *Neurospora crassa*. *Photochem. Photobiol.*, **32**, 375–81.

SARGENT, M. L. & BRIGGS, W. R. (1967). The effect of light on a circadian rhythm of conidiation in *Neurospora*. *Plant Physiol.*, **42**, 1504–10.

SCHMID, G. H. & SCHWARZE, P. (1969). Blue-light enhanced respiration in a colorless *Chlorella* mutant. *Z. Physiol. Chem.*, **350**, 1513–20.

SCHMIDT, W. (1979). On the environment and the rotational motion of amphiphilic flavins in artificial vesicles as studied by fluorescence. *J. Membrane Biol.*, **47**, 1–25.

SCHMIDT, W. (1980a). Physiological bluelight reception. In: *Molecular Structure and Sensory Physiology*, ed. P. Hemmerich, pp. 1–44. Structure and Bonding, vol. 41. Heidelberg: Springer-Verlag.

SCHMIDT, W. (1980b). Artificial flavin/membrane systems: a possible model for physiological bluelight action. In: *The Blue Light Syndrome*, ed. H. Senger, pp. 212–220. Heidelberg: Springer-Verlag.

SCHMIDT, W. (1981a). Fluorescence properties of isotropically and anisotropically embedded flavins. *Photochem. Photobiol.*, **34**, 7–16.

SCHMIDT, W. (1981b). Reply to: Temperature effects on the polarity of lipid bilayers and the localization of amphiphilic flavins in artificial membrane vesicles. *J. Membrane Biol.*, **60**, 164–5.

SCHMIDT, W. & HEMMERICH, P. (1981). On the redox reactions and accessibility of amphiphilic flavins in artificial membrane vesicles. *J. Membrane Biol.*, **60**, 129–41.

SCHMIDT, W., THOMSON, K. & BUTLER, W. L. (1977*a*). Cytochrome *b* in plasma membrane enriched fractions from several photoresponsive organisms. *Photochem. Photobiol.* **26**, 407–11.

SCHRANK, A. R. (1947). Electrical and curvature responses of the *Avena* coleoptile to transversely applied direct current. *Plant Physiol.*, **23**, 188–200.

SCHROTT, E. L. (1980). Dose response and related aspects of carotenogenesis in *Neurospora crassa*. In: *The Blue Light Syndrome*, ed. H. Senger, pp. 309–18. Heidelberg: Springer-Verlag.

SENGER, H. (ed.) (1980). *The Blue Light Syndrome*. Proceedings of the first conference on Physiological Blue Light Reception. Berlin, Heidelberg & New York: Springer-Verlag.

SENGER, H. & BRIGGS, W. R. (1981). The blue light receptor(s): primary reactions and subsequent metabolic changes. In: *Photochemical and Photobiological Reviews*, vol. 6, ed. C. K. Smith, pp. 1–38. New York: Plenum.

SONG, P.-S. (1969). Electronic structures and spectra of flavins: an improved Pariser-Parr-Pople MO and semiempirical unrestricted Hartree-Fork computations. *Ann. N.Y. Acad. Sci.* **158**, 410–23.

SONG, P. -S. (1980). Spectroscopic and photochemical characterization of flavoproteins and carotenoproteins as blue light photoreceptors. In *The Blue Light Syndrome*, ed. H. Senger, pp. 157–71. Heidelberg: Springer-Verlag.

SONG, P.-S., WALKER, E. B., VIERSTRA, R. D. & POFF, K. L. (1980). Roseoflavin as a blue light receptor analog: spectroscopic characterization. *Photochem. Photobiol.* **32**, 393–8.

SPURGEON, S. L., TURNER, R. J. & HARDING, R. W. (1979). Biosynthesis of phytoene from isopentenyl phytophosphate by a *Neurospora* enzyme system. *Arch. Biochem. Biophys.* **195**, 23–9.

SUN, M., MOORE, T. A. & SONG, P. -S. (1972). Molecular luminescence studies of flavin. I. The excited states of flavin. *J. Amer. Chem. Soc.* **94**, 1730–40.

THEIMER, R. R. & RAU, W. (1970). Untersuchungen über die lichtabhängige Carotenoidsynthese. V. Aufhebung der Lichtinduktion durch Reduktionsmittel und Ersatz des Lichts durch Wasserstoffperoxid. *Planta*, **92**, 129–37.

TRUMPOWER, B. L. & LANDEEN, C. E. (1977). Properties of ubiquinone relevant to its function in cellular respiration. *The Ealing Review* Sept./Oct. 1977.

ULASZEWSKI, S., MAMOUNAS, T., SHEN, W. -K., ROSENTHAL, P. J., WOODWARD, J. R. & EDMUNDS JR., L. N. (1979). Light effects in yeast: evidence for participation of cytochromes in photoinhibition of growth and transport in *Saccharomyces cerevisiae* cultured at low temperatures. *J. Bacteriol.* **138**, 523–29.

VIERSTRA, R. & POFF, K. (1979). *Perception of light by plants*. Report 79 of the Plant Research Laboratory of Michigan State University.

VON GUTTENBERG, H. (1959). Über die Perzeption des phototrophen Reizes. *Planta*, **53**, 412–33.

WATENPAUGH, K. D., SIEKER, L. C. & JENSEN, L. M. (1973). The binding of riboflavin-5′-phosphate in a flavoprotein: flavodoxin at 2.0 Å resolution. *Proc. Natl. Acad. Sci, USA*, **70**, 3857–60.

WEBSTER JR., W. W. & SCHRANK A. R. (1953). Electrical induction of lateral transport of 3-indolacetic acid in *Avena* coleoptile. *Arch. Biochem. Biophys.* **47**, 107–18.

YOSHIDA, M. & MILLOTT, N. (1959). The shadow reaction of *Diadema Antillarum Philippi*. III. Reexamination of the spectral sensitivity. *J. Exp. Biol.* **37**, 390–7.

ZENK, M. H. (1967*a*). Untersuchungen zum Phototropismus der *Avena*-Koleoptile: I. Photo-oxidation *in vivo*. *Z. Pflanzenphysiol.* **56**, 57–69.

ZENK, M. H. (1967*b*). Untersuchungen zum Phototropismus der *Avena*-Koleoptile: II. Pigmente. *Z. Pflanzenphysiol.* **56**, 122–40.

ZURZYCKI, J. (1980). Blue light-induced intracellular movements. In *The Blue Light Syndrome*, ed. H. Senger, pp. 50–68. Heidelberg: Springer-Verlag.

THE PERCEPTION OF POLARISED LIGHT*

RÜDIGER WEHNER

Zoologisches Institut der Universität Zürich, Winterthurerstrasse 190,
CH-8057, Zürich, Switzerland

Insects and many other invertebrates are equipped with photoreceptors that are sensitive to polarised light. What use do they make of such photoreceptors?

In Nature, polarised light arises mainly from two sources: from the *scattering* of sunlight by the atmosphere (and hydrosphere), and from the *reflection* of light by water surfaces or wet surfaces of soil, rocks, and vegetation. Of the two, skylight offers the most reliable source of polarisation. Thus, this chapter sets out to discuss the perception of polarised light in terms of how the pattern of skylight polarisation is used for deriving compass information from the sky.

Besides being involved in skylight navigation, dichroic photoreceptors might mediate a number of functions. For example, they might reduce glare resulting from polarised reflections on wet surfaces, and thus improve the detection of prey through water surfaces or against any glossy background such as leaves or mud. This possibility has been discussed for two groups of predatory insects which live on or near to the water surface (water skaters: Schneider & Langer, 1969; Bohn & Taeuber, 1971; long-legged flies: Trujillo-Cenoz & Bernard, 1972).[1] The polarisation of light reflected from water surfaces might also demarcate an artificial horizon, as proposed for dragonflies flying over water (Laughlin, 1976). Furthermore, polarisation sensitivity might help to enhance contrast in underwater vision (Lythgoe & Hemmings, 1967; Leggett, 1978: 123),[2] or it might increase absolute sensitivity, as most recently suggested by Snyder (1979). What apparently have not evolved as yet in any living organism are optical devices that specifically emit or reflect polarised light and thus provide visual cues which (analogous to colour cues) might be used in intra- or interspecific communication.[3]

In most of the cases mentioned above the animal need not abstract the angle of polarisation as distinct from the intensity of light and its degree of polarisation. In skylight navigation, however, this is just what

---

* Dedicated to Prof. Dr H. Autrum, who first proposed the idea of insect photoreceptors acting as analysers for polarised light, in honour of his 75th birthday.

the animal should achieve because the celestial patterns of intensity and degree of polarisation are highly susceptible to even weak atmospheric disturbances like haze or thin clouds, but the pattern of the angles of polarisation is not.

Provided that insects are equipped with polarisation-sensitive photo-receptors (Section 1) as well as the neural machinery to determine angles of polarisation unambiguously (Section 2), the main questions to be asked in trying to understand the insect's celestial compass can be phrased as follows: what do the insects know about the pattern of polarised light in the sky, and how do they use this knowledge to steer compass courses? These questions have been raised ever since von Frisch (1949) first demonstrated that bees use polarised skylight as a compass cue. In recent years many new pieces of the puzzle have been discovered (Section 5), and one can hope that with the discovery of a few more, a pattern for fitting them together will emerge. Nevertheless, in spite of the progress made in the past decade, controversies and uncertainties persist.

What is lacking first and foremost is a definition of what the insect must accomplish in reading compass information from the sky. Thus, the conceptual part of the problem deserves more attention than it has received. For example, are insects, in any abstract way, aware of the laws of light scattering within the earth's atmosphere (see Section 3), and thus capable of performing rather abstract geometrical construc-tions in the sky? Or, instead, do they know all the possible e-vector patterns in the sky that vary with the height of the sun and thus the time of the day, which is to suggest they know exactly where a particular e-vector occurs at a particular point in the sky at a particular time of the day? However unlikely either hypothesis might appear, it is only after one has explicitly asked these questions that an answer will emerge.

As discussed in more detail below (Section 5), half the answer is now clear. What the insect uses is a generalised version of the pattern of polarised light in the sky – some type of master-image that does not vary with the time of day, but only changes its azimuth position as the sun moves across the sky. If one assumes that the insect relies on such an invariant celestial map, neat explanations of a number of familiar observations tumble out.

What we do not know yet is how the actual e-vector information from the sky is processed by the nervous system to allow for a comparison of the neural image actually experienced by the insect with the master-image already laid down in the insect's brain. A first stage in answering this question is to map skylight patterns on to arrays of photoreceptors and interneurones (Section 4), but the prospect of analysing the insect's celestial compass by proceeding neurophysiologically step by step from one visual neuropile to the next is daunting. Instead, it may well be that upstream analyses starting with behavioural outputs, rather than downstream neurophysiological approaches, will 'bring home the bacon'.

In spite of the brief space allotted to me in this chapter, I shall try to cover the main aspects of polarisation vision, starting with the dichroism of photoreceptors and ending with the insect's celestial map. Nevertheless, this chapter is not intended as a comprehensive review of the literature in this field[4] but is rather an attempt to acquaint the reader with some general principles and theoretical positions.

1. Analysers: dichroism and polarisation sensitivity

Although it has occasionally been suggested that the effective analysers for polarised light are located somewhere within the dioptric systems of compound eyes (Berger & Segal, 1952; Stephens, Fingerman & Brown, 1953; Baylor & Smith, 1953, 1957; Hazen & Baylor, 1962; Skrzipek & Skrzipek, 1971),[5] this is certainly not the case (Autrum & Stumpf, 1950; Stockhammer, 1956; Shaw, 1967; Seitz, 1969; Skrzipek & Skrzipek, 1974). Instead, it is an intrinsic property of the photoreceptors themselves to be dichroic, to differentially absorb linearly polarised light that varies in the angle of polarisation (e-vector direction; χ). As first proposed by Autrum (1949; in von Frisch, 1965: 423) and de Vries, Spoor & Jielof (1953), the sensitivity of photoreceptors to polarised light is based on the intrinsic dichroism of the visual pigment and its preferential alignment within the photoreceptor membrane. The sensitivity (S) of a rhabdomeric photoreceptor is greatest when the angle of polarisation is parallel to the microvillar axes, and smallest when it is at right angles to them. Correspondingly, polarisation sensitivity, S_p, is defined as the ratio $S(\chi_{max})/S(\chi_{max} \pm 90°)$, where S is the sensitivity of the receptor and χ_{max} the value of χ that maximises S. The actual values of S_p as recorded intracellularly from rhabdomeric photoreceptors are as large as 5–15 (cephalopods: Sugawara, Katagiri & Tomita, 1971; crustaceans: Shaw, 1969a; Waterman & Fernandez, 1970; Mote, 1974; Leggett, 1978: 92; insects: Butler & Horridge, 1973; Menzel & Snyder, 1974; Menzel, 1975; Laughlin, 1976; Mote & Wehner, 1980; Labhart, 1980). By combining both anatomical reconstructions and electrophysiological recordings it can be shown that the microvillar directions and the values of χ_{max} actually coincide (flies: Kirschfeld, 1969; Kirschfeld & Franceschini, 1977; Hardie, Franceschini & McIntyre, 1979; bees:

Sommer, 1979; Labhart, 1980). The same conclusion can be drawn from the microspectrophotometric work cited below.

In summary, it is the dichroism of photoreceptor structures (the alignment of both molecular absorption vectors and microvilli) that gives rise to the polarisation sensitivities of rhabdomeric photoreceptors. However, the direct demonstration of the dichroism of rhabdomeres had to await microspectrophotometric measurements first performed in the sixties (cephalopods: Hagins & Liebman, 1963; insects: Langer, 1965; crustaceans: Hays & Goldsmith, 1969). In these early reports as well as in many subsequent ones, the dichroic absorbance ratios $\Delta D = D(\chi_{max})/D(\chi_{max} \pm 90°)$[6] are considerably smaller than one would expect from the values of S_p measured electrophysiologically. As most of the spectroscopic work has been performed with isolated rhabdoms, misalignment of microvillar orientation relative to the e-vector direction of the measuring beam, or bending of the microvilli, might partly account for the observed mismatch between ΔD and S_p (Goldsmith, 1975). However, data taken from measurements of photo-induced dichroism correspond to dichroic ratios of about $\Delta D = 5–7$ (Goldsmith & Wehner, 1977). These values are in good agreement with the average polarisation sensitivities of crustacean photoreceptors.

In vertebrate rods, dichroic absorption can be observed readily when isolated rods are illuminated and viewed from the side (*side-on* dichroism: Schmidt, 1935; Liebman, 1962), but cannot be observed at all when the measurements are performed *end-on*. Thus, under natural conditions of illumination vertebrate photoreceptors do not show any dichroism ($\Delta D = 1$). This is due to the fact that in vertebrate photoreceptors the rhodopsin molecules undergo rotational diffusion within the disk membrane of the rod outer segment. The speed of this type of diffusion is such that any photoinduced dichroism is randomised within less than 0.1 ms (Cone, 1972).

There might be several reasons for the fact that the microvilli-type photoreceptors of cephalopods and arthropods exhibit dichroism (and thus polarisation sensitivity), but the disk-type photoreceptors of vertebrates do not. First, the geometry of the microvillar tube creates some kind of form dichroism that is in the order of $\Delta D = 2$ or less, even when the absorption vectors of rhodopsin molecules are randomly distributed within the membrane (Moody & Parriss, 1961; Laughlin, Menzel & Snyder, 1975). Second, asymmetries in the shape of the rhodopsin molecule and the incorporation of the molecule into a membrane that is bent into a narrow tube[7] favour molecular alignment parallel to the microvillar axis by restricting the movement of the rhodopsin molecules to the long axis of the tube (Laughlin *et al.*, 1975). Third, the viscosity of the photoreceptor membrane is likely to be higher in rhabdomeric photoreceptors than in rod outer segments (for example, as a consequence of the higher content of cholesterol in invertebrate photoreceptor membranes: Mason, Fager & Abrahamson, 1973). Fourth, some kind of cytoskeleton within the microvilli, as well as possible membrane linkages between adjacent microvilli, could provide some means for stabilising the arrangement of proteins within the membrane (Saibil, 1982). In evolutionary terms, it might well be that the ordered arrangement of rhodopsin molecules within densely packed stacks of microvillar membranes was designed primarily to increase photon absorption (Shaw, 1969*b*; Snyder, 1979) rather than to provide the animal with the exotic capacity to perceive polarised skylight. Indeed, in certain circumstances polarisation sensitivity is more of a handicap than an advantage. Thus, the detection of objects is improved when the insect is not dazzled by the disturbing effects of polarised light as caused by reflections on wet surfaces. Also, for identifying the hue of colour it is better to use photoreceptors that are insensitive to polarised light, so that polarised reflections do not disturb the apparent colour of an object.

In this context, it is worth mentioning that there are mechanisms for reducing the overall polarisation sensitivity of a rhabdomeric photoreceptor to a level that is considerably lower than one would expect, given the dichroism ΔD of a single microvillus. First, a rather straightforward strategy of reducing S_p is to design a photoreceptor that is equipped with *two or more rhabdomeres pointing in different* (preferentially orthogonal) *directions*, as is the case in the dorsal eighth cell of brachyuran crustaceans (Kunze & Boschek, 1968; Eguchi & Waterman, 1973). Second, S_p can be reduced by *twisting* the rhabdomeres (anatomical evidence: Menzel & Blakers, 1975; Wehner, Bernard & Geiger, 1975; Smola & Tscharntke, 1979; Wehner & Meyer, 1981; Smola & Wunderer, 1981; optical calculations: Wehner *et al.*, 1975; McIntyre & Snyder, 1978), or by other kinds of structural irregularities within the stacks of microvilli. Third, the polarisation sensitivity resulting from the alignment of microvillar membranes can be reduced by *self-screening* within the rhabdomere (Shaw, 1969*b*; Snyder, Menzel & Laughlin, 1973; Goldsmith & Bernard, 1974). In a single rhabdomere exhibiting one per cent absorption per micron for light polarised parallel to χ_{max} and an intrinsic (microvillar) dichroic ratio of $\Delta D = 10$, the effective dichroism given by the absorption ratio $\Delta A = A(\chi_{max})/A(\chi_{max} \pm 90°)^8$ is reduced to 2.7, when the rhabdomere is 200 μm long. Consider that the polarisation sensitivity S_p depends on ΔA rather than ΔD. Fourth, *electrical* *coupling* between photoreceptor cells characterised by different values of χ_{max} has been referred to as a mechanism for reducing S_p (Shaw, 1969a; Menzel & Snyder, 1974). However, it has still to be elucidated in more detail to what degree the observed electrical crosstalk among photoreceptor cells is due to naturally occurring low membrane resistances between adjacent cells, or whether artefactual coupling caused by impaling the retinular cells contributes to the effect. Fifth, one could also conceive of possibilities of *decreasing the dichroism* of the photoreceptor membrane itself, for example by incorporating into the membrane some kind of sensitising antenna pigment which has its absorption vectors aligned at an angle to the absorption vectors of the rhodopsin molecules.

As soon as rhabdomeric photoreceptors become part of navigational systems exploiting the polarised light in the sky, high polarisation sensitivities are at a premium. To achieve this goal, the photoreceptors should be endowed with high dichroic ratios, equipped with straight rhabdomeres that are isolated electrically from their nearest neighbours, and provided with some means for counterbalancing the inevitable effects of self-screening. The latter problem is solved in a number of ways which all come down to bringing the polarisation sensitivity, S_p, of the cell close to the dichroic ratio, ΔD, of the microvilli: either by designing a *short rhabdomere* in which self-screening does not severely reduce S_p, and/or by combining several rhabdomeres into a single waveguide structure, the rhabdom (optical coupling: Shaw, 1969b; Snyder, 1973a, 1979). In the latter case, adjacent rhabdomeres differing in χ_{max} provide *serial* or *lateral dichroic filters*, depending on whether they are incorporated into a tiered or fused rhabdom, respectively (Snyder, 1979). Thus, short proximal or distal photoreceptor cells found, for example, in bees (for summary see Wehner, 1976a), flies (Kirschfeld & Franceschini, 1977; Hardie *et al.*, 1979), dragonflies (Laughlin, 1975), and crayfish (Waterman & Fernandez, 1970; Naessel, 1976) – all equipped with ultraviolet- or blue-absorbing photopigments – are likely candidates for polarisation detectors. In the banded, fused rhabdoms of decapod crustaceans (Eguchi, 1965), as well as some insect species (Meyer-Rochow, 1971, 1972; Paulus, 1975; Kolb, 1977; Maida, 1977; Gordon, 1977), optical coupling by overlying dichroic filters of different χ_{max} has been carried to an extreme. The individual rhabdomeres are divided longitudinally into periodic stacks (tongues) of microvilli which are all parallel within any one rhabdomere but interdigitate orthogonally with corresponding stacks of adjacent rhabdomeres. Thus, in a typical banded rhabdom the microvilli aggregate into layers with their axes parallel in any given layer (formed, for example, by cells nos. 1, 3 and 5) and at right angles to the microvilli of the overlying and underlying layers (formed for example by cells nos. 2, 4 and 6). Finally, S_p can be enhanced by the effects of *lateral filtering within the photoreceptor membrane* itself. In one class of fly photoreceptors, cells nos. 7y, a blue-absorbing photostable pigment occurs that has its absorption vectors oriented perpendicularly to the absorption vectors of the visual pigment. As usual, the latter are aligned parallel to the microvillar axes. The dichroism of the accessory pigment explains the early finding, perplexing at first, that in this type of photoreceptor maximum absorption occurs when light is polarised perpendicularly to the microvilli (Kirschfeld, 1969; Kirschfeld & Franceschini, 1977). Note, however, that maximum sensitivity as measured electrophysiologically is obtained when light is polarised parallel to the microvilli (Hardie *et al.*, 1979).

Within the retina of an insect's eye there can be marked regional differences with respect to polarisation sensitivity. For example, in the bee's retina there is a specialised dorsal rim area where the rhabdoms are straight. Concomitantly, all ultraviolet receptors of this area exhibit high polarisation sensitivities ($S_p > 5$; Labhart, 1980). In the remainder of the bee's eye the rhabdoms are twisted. There, the long ultraviolet receptors do not respond to polarised light ($S_p < 1.5$; Menzel & Snyder, 1974; Labhart, 1980).

2. Detectors: two- and three-dimensional systems

Any visual system that is able to abstract information about angles of polarisation from the sky must be provided with several types of polarisation-sensitive photoreceptors. Just as colour vision depends on the interaction of several types of spectral analysers (the so-called colour receptors) which differ in λ_{max}, so polarisation vision requires several types of polarisation analysers which differ in χ_{max}.

The number of analysers required depends on the number of independent optical variables that characterise the light emanating from a given point in the sky. These variables are intensity (I), degree of polarisation (D), angle of polarisation (χ), and the variation of all these parameters with the wavelength of light (λ).[9] As it is now well established that polarisation vision of hymenopterans is exclusively mediated by one spectral type of receptor, the ultraviolet receptor (Duelli & Wehner, 1973; von Helversen & Edrich, 1974), only three types of measurements have to be taken from three types of polarisation-sensitive receptor to describe fully the state of polarisation at any one point in the sky. Hence, just as colour can be described as a point in a three-dimensional space where the dimensions are wavelength, saturation, and intensity, so a state of polarisation can be described by angle, degree, and intensity. And just as a wavelength detector needs inputs from three spectral types of receptor, an e-vector detector needs inputs from three polarisational types of receptor (see Bernard & Wehner, 1977 for a fuller discussion of the analogy between polarisation vision and colour vision).

Theoretically, a three-dimensional system for polarisation vision can be designed in a number of ways: by wiring up three polarisational types of receptor characterised by three different directions of maximum sensitivity ($\chi_{max\,1}$, $\chi_{max\,2}$, $\chi_{max\,3}$) (Kirschfeld, 1973), or by combining two such types of receptor ($\chi_{max\,1}$; $\chi_{max\,2} \neq \chi_{max\,1} \pm 90°$) with a receptor that is insensitive to polarised light ($S_p = 1$) (Wehner *et al.*, 1975).[10] Actually, however, three-dimensional polarisation vision has never been demonstrated directly in any species. Since the early work of von Frisch (1965: 397) it has generally been agreed that bees are able to determine χ independent of I and D; but it is not known how such a three-dimensional detector system works neurophysiologically, and there is little direct behavioural work critical enough to decide to what degree measurements of χ are really independent of D and I.

Let us first turn to the *behavioural* work. Any test for true polarisation vision must allow for independent changes of χ, D and I of the optical stimulus. It is only then that one can determine the dimensionality of the detector system by checking for neutral points and confusion states within the stimulus domain, that is to ask whether different combinations of certain values of χ, D and I are confused by the animal. Behavioural analyses of this kind have not yet been performed. It has been very fashionable instead, especially in the fifities and early sixties, to study so-called polarotactic responses. Many insects and crustacean species move preferentially at a certain angle relative to the e-vector direction of an overhead polariser. For example, daphnids (entomostracan crustaceans) prefer to swim at right angles to the e-vector direction of linearly polarised light (Baylor & Smith, 1953; Waterman, 1960) by adjusting the transverse axes of their cyclopean compound eyes parallel to the e-vector direction (Stockhammer, 1959). Such kinds of polarotactic responses have been extensively summarised by Waterman (1973, 1981), and so it would be of no service to go over that ground again. I would like to mention the difficulty of interpreting these results in terms of polarisation vision. At least in a few examples (*Daphnia*: Verkhovskaya, 1940;[11] *Ocypode:* Schoene & Schoene, 1961) it has been shown that polarotactic responses to horizontally and vertically polarised light are merely differential phototactic responses. They not only vary dramatically by changing the intensity of one of the two polarised lights displayed for comparison, but they can also be mimicked by different intensities of unpolarised light. On the other hand, bees and ants, while using skylight polarisation as a compass cue, seem to respond to χ independently of D and I (Section 5), but even in this case systematic studies are missing.

The second question posed above relates to the *neurophysiological* mechanisms of e-vector detection. As no one has yet recorded from any type of polarisation-sensitive interneurone in any insect,[12] we are forced to rely exclusively upon inferences from the distribution of microvillar (analyser) directions within the insect's retina. In bees, e-vector detection can be accomplished by rather small spots (diameter $<5°$) of natural skylight (Rossel, Wehner & Lindauer, 1978) as well as artificially polarised light (Edrich & von Helversen, 1976; Brines & Gould, 1979). Thus, at least in the most dorsal part of the bee's compound eye any small area of the retina comprising only a limited number of retinulae must contain the necessary set of polarisation-sensitive photoreceptors.

Apparently, a single retinula does not suffice, because in the species studied so far a single retinula does not contain three polarisation-sensitive ultraviolet receptors that vary in χ_{max} and/or S_p. Other spectral types of receptor can be disregarded for the moment, because it is mainly the ultraviolet type of receptor that is involved in the detection of polarised skylight (p. 331). By postulating a certain interaction between the ultraviolet receptors of adjacent ommatidia it is possible to design a three-dimensional (three-channel) e-vector detecting system (Menzel & Snyder, 1974; Wehner *et al.*, 1975), but at present such schemes can be no more than speculative.

What is most striking, however, is the widespread occurrence of 'crossed analysers', that is perpendicularly arranged sets of rhabdo-meres, within the retina of cephalopods (Zonana, 1961; Yamamoto, Tasaki, Sugawara & Tonosaki, 1965), spiders (Eakin & Brandenburger, 1971; Melamed & Trujillo-Cenoz, 1966; Schroeer, 1974), crustaceans (Eguchi & Waterman, 1966) and insects.[13] There is now increasing evidence (Sommer, 1979; Labhart, 1980, unpublished data) that in the specialised dorsal rim areas of the compound eyes of both bees and ants, each ommatidium is equipped with a set of orthogonally arranged ultraviolet receptors. Thus, it seems pertinent to ask whether the orthogonal arrangement of χ_{max}-values within this specialised region of

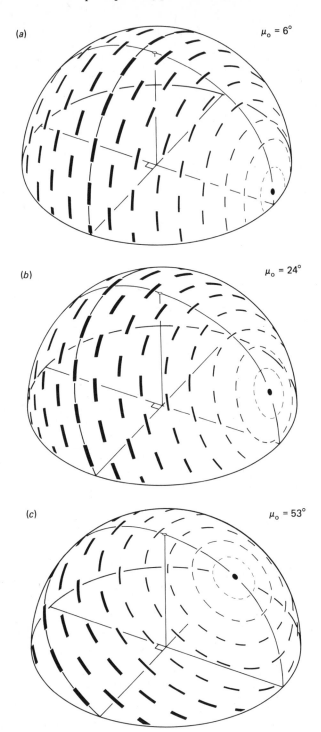

(a) $\mu_o = 6°$

(b) $\mu_o = 24°$

(c) $\mu_o = 53°$

the eye is to enhance the e-vector sensitivity of some type of second-order interneurone. In this context, it is a plausible hypothesis that the orthogonally arranged photoreceptors interact antagonistically in a way analogous to the well-studied mechanism of colour opponency (vertebrates: De Valois, 1973; insects: Menzel, 1979).[14]

As is immediately apparent from atmospherical optics, any two-dimensional e-vector detecting system must exhibit neutral points and confusion states. Nevertheless, such a system is by no means polarisation-blind, just as dichromatic systems are not colour blind (for a theoretical treatment of two-dimensional e-vector detectors see Bernard & Wehner, 1977). On the other hand, crossed analysers can be used to create polarisation-insensitive signals if their outputs are summed.

Until now we have been concerned exclusively with the way in which e-vector information might be obtained from a given *point* in the sky. However, in order to understand how insects derive compass information from the sky it is necessary to understand the *pattern* of skylight polarisation. What do these patterns look like, and how are they mapped on to the insect's array of photoreceptors? These are the questions we shall discuss in the next two sections.

3. Skylight patterns

Direct light from the sun is unpolarised. However, due to the scattering of unpolarised sunlight within the atmosphere of the earth, skylight becomes partially linearly polarised (Strutt, 1871). At any one point in the sky, the *degree* of polarisation (D) is related to the scattering angle (ϱ) through $D = \sin^2\varrho/(1 + \cos^2\varrho)$. As the light vector vibrates perpendicularly to the plane of the scattering angle, the *direction* (angle) of polarisation (χ) extends at right angles to the great circle passing throught the sun and the point in question. Thus, e-vectors form concentric circles around the sun (Fig. 1).

In general, scattered skylight is characterised by the spatial distributions of the *angle* of polarisation (χ), the *degree* of polarisation (D), the radiant *intensity* (I), and the variations of all these parameters with the *wavelength* of light (λ). In Figs. 3(a–f) the patterns of $\chi(\lambda)$, $D(\lambda)$, and $I(\lambda)$ are presented for $\lambda = 371$ nm and two positions of the sun ($\mu_o = 6°$ and 53°). For definitions of μ, μ_o, χ, and Φ see Fig. 2.

Fig. 1. The pattern of polarised light in the sky plotted for three different elevations of the sun (μ_0). The *directions* and *widths* of the black bars mark the direction (χ) and degree (D) of polarisation, respectively. The *black dot* is the sun. The two *great circles* represent the solar vertical (solar and anti-solar meridian) and the arc of highest degree of polarisation.

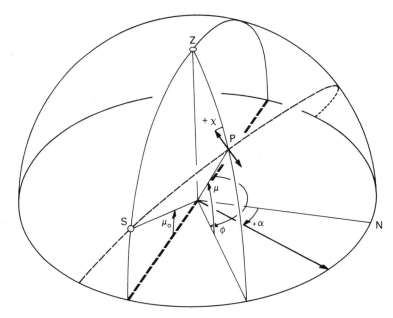

Fig. 2. Skylight polarisation: symbols and definitions. Skylight is polarised perpendicular-ly to the great circle (*dashed arc*) that runs through any one point (P) and the sun (S). Z, zenith; N, north; α, compass bearing of the animal ($\alpha = 0°$ corresponds to north); μ, elevation of P; μ_0, elevation of the sun; Φ, azimuth of P ($\Phi = 0°$ corresponds to the anti-solar meridian); χ, direction of polarisation (e-vector direction; $\chi = 0°$ corresponds to the vertical e-vector); α, Φ, and χ are counted clockwise (positive values) or anti-clockwise (negative values).

It is immediately apparent from these figures that the *solar vertical* comprising the *solar meridian* ($\Phi = 180°$) and *anti-solar meridian* ($\Phi = 0°$) forms the symmetry line of all skylight patterns. Along the solar vertical, light is polarised parallel to the horizon ($\chi = 90°$).[15] Any two points ($+\Phi_i$, μ_j and $-\Phi_i$, μ_j) that are equidistant to the solar or anti-solar meridian coincide in $D(\lambda)$ and $I(\lambda)$ and are characterised by $-\chi_k(\lambda)$ and $+\chi_k(\lambda)$, respectively. In the case of $\chi = 0°$ (vertical e-vectors) both points coincide in all skylight parameters.

In general, at a given elevation, μ_j, each e-vector orientation χ_k occurs twice (see χ/Φ functions in Fig. 4). As can be inferred from Figs. 3(a,d and Fig. 4(a), the azimuthal distance between identical e-vectors is $\Delta\Phi = 180°$ for $\chi = 90°$, and $\Delta\Phi \neq 180°$ for all other values of χ. Points coinciding in χ differ in $D(\lambda)$ (Fig. 4b) and $I(\lambda)$, the sole exception being $\chi = 0°$.

The ideal pattern of polarisation present in the deep blue sky is markedly affected by atmospheric disturbances like haze, fog or clouds.

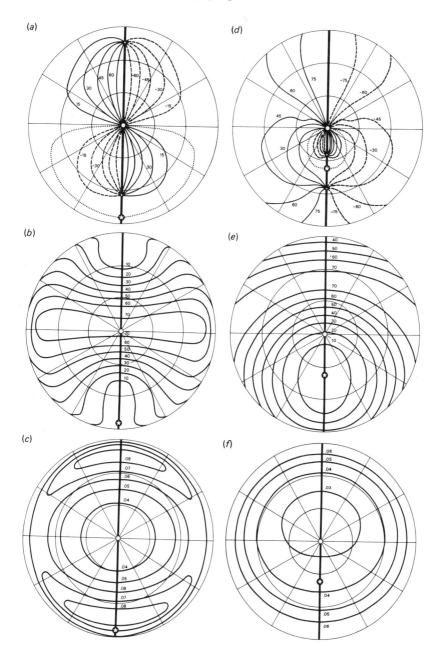

Fig. 3. Maps of skylight parameters, that is, spatial distributions of angle of polarisation, $\chi(a, d)$, degree of polarisation, D (b, e), and relative intensity, I (c, f). All data refer to the ultraviolet range of the spectrum ($\lambda = 371$ nm) and are shown for two elevations of the sun: $\mu_0 = 6°$ $(a–c)$, $\mu_o = 53°$ $(d–f)$. Small white disc, zenith; asterisk, sun; bold line, solar vertical, isoline for $\chi = 90°$; numbers in (b) and (e), percentage polarisation; numbers in (c) and (f), radiative transfer.

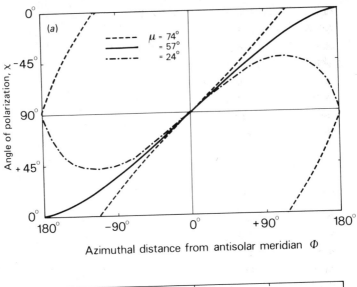

Azimuthal distance from antisolar meridian Φ

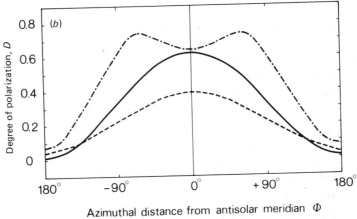

Azimuthal distance from antisolar meridian Φ

Fig. 4. Angle of polarisation (a) and degree of polarisation (b) for three different elevations (μ) above the horizon. The elevation of the sun is $\mu_o = 53°$ (see Figs. 1 and 3d, e). The three different elevations have been chosen so that one is above the sun ($\mu = 74°$), one approximately at the height of the sun ($\mu = 57°$) and one below the sun ($\mu = 24°$). In the text, the functions plotted in (a) are referred to as χ/Φ functions. For conventions of how to count χ and λ, see Fig. 2.

Under such conditions the most reliable criterion is $\chi(\lambda)$, whereas $D(\lambda)$ and $I(\lambda)$ may vary dramatically. On the other hand, if there are (high) clouds in the sky, and if the sun is not obscured, scattering of direct sunlight between the clouds and the surface of the earth generates a pattern of polarisation that continues the pattern of polarisation present in the blue parts of the sky (Stockhammer, 1959; Waterman, 1981).

4. Receptor arrays and neural response patterns

How are the skylight patterns as described in the preceding section mapped on to the insect's eye? First of all this begs the question as to how the insect's system of eye coordinates is adjusted relative to the external system of celestial coordinates.

In desert ants, there is now sufficient evidence drawn from cinematographic work that the ant stabilises its head position relative to pitch and roll movements about its transverse and longitudinal body axis, respectively, whenever it takes a compass reading (for summary see Wehner, 1982a). By exploiting the optical phenomenon of the luminous pseudopupil, Raeber (1979) in our group has further shown that the equator of the ant's eye – marked by the borderline between the areas of two structurally different types of retinula – looks straight at the horizon when the head is inclined to the horizontal, just by the pitch and roll angles most frequently observed in running ants (Fig. 5). Thus, this feature of its retina could provide the ant with a horizon detector, with which it could stabilise its head position against movements relative to the horizontal plane. The ocelli could also function as additional horizon detectors as proposed by Hesse (1908) and Wilson (1978) and demonstrated experimentally by Stange & Howard (1979), Stange (1981) and Taylor (1981). Since *Cataglyphis* is able to home correctly even after its ocelli have been occluded (Wehner, 1982a), it is clear that, if the ocellus is a horizon detector in the ant, it is not the only one. Of course, the spatial positions of skylight patterns could also be assessed when roll and pitch movements occurred, provided that the amount of roll and pitch is monitored by proprioceptive control of the position of the head (and thus the eyes) relative to the thorax.

In summary, as we know the angular position of the insect's head with respect to the horizontal (roll and pitch angles) as well as the directions of view of the individual ommatidia with respect to the head, we can map the celestial system of coordinates (meridians and parallels of altitude) on to the compound eye (Figs. 6 and 7).

Such maps are a necessary prerequisite for interpreting behavioural

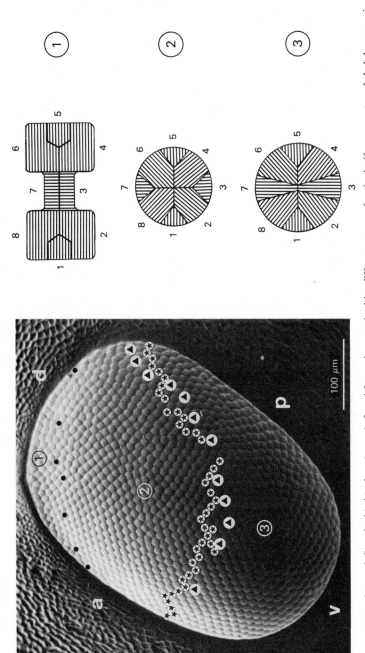

Fig. 5. The compound eye of *Cataglyphis bicolor*. Areas 1, 2 and 3 are characterised by different types of retinula (for geometry of rhabdoms see inset figures on the right). *Area 1*: dorsal rim area (dumb-bell-shaped rhabdoms, characterised by a strictly orthogonal arrangement of rhabdomeres). *Area 2*: remainder of dorsal retina ([4 + 4]-retinulae). *Area 3*: ventral retina ([6 + 2]-retinulae). The borderline between areas 2 and 3 is indicated by asterisks. Triangles mark those ommatidia that look at the horizon (see Fig. 6). Designed according to optical and anatomical measurements of Herrling (1976) and Raeber (1979).

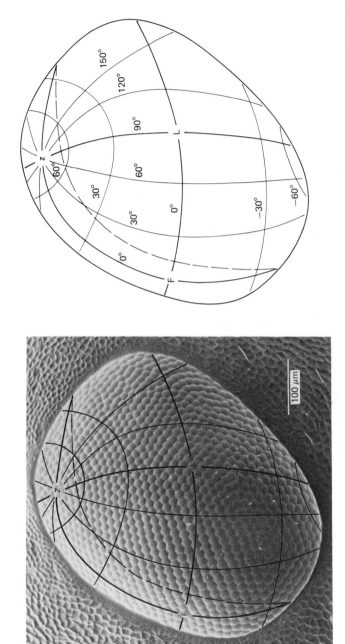

Fig. 6. Projection of the system of celestial coordinates on to the ommatidial lattice of the left eye of *Cataglyphis bicolor*. The pitch angle of the head, and thus the inclination of the eye relative to the horizontal, is taken from cinematographic observations of walking ants. The viewing directions of the ommatidia have been determined by pseudopupil measurements (antidromic illumination) performed by C. Zollikofer. F, frontal point (forward direction); L, lateral point; Z, zenith. The dashed line marks the borderline of the binocular visual field.

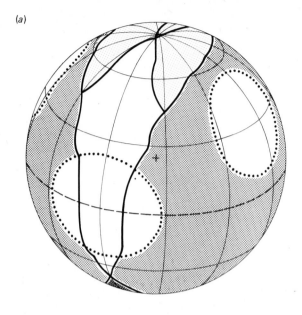

(a)

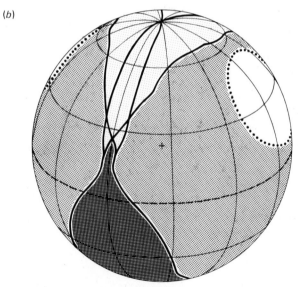

(b)

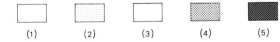

(1) (2) (3) (4) (5)

experiments in which specific regions of the eye are occluded. If, in *Cataglyphis*, the dorsal halves of both eyes are covered with light-tight paint down to the equator, the ant is no longer able to use its celestial compass. Even though it then continually turns its head and thus views the sky with the ventral ommatidia of either eye, it is not able to record compass information from the sky (Wehner, 1982a). Apparently, the neural machinery for the analysis of skylight cues is confined to the dorsal half of the eye.

Furthermore, a specialised *dorsal rim area* can be defined within the dorsal half of both the ant's and bee's eyes by a number of anatomical and physiological peculiarities. Specifically, the dorsal rim area is characterised by relatively large rhabdoms, an orthogonal arrangement of microvillar directions within the rhabdom, and a fan-like arrangement of the transverse axes of the rhabdoms within the retina (Schinz, 1975; Herrling, 1976; Raeber, 1979; Sommer, 1979). The centre of the fan is located in the dorsal apex of the eye where the ommatidia look contralaterally. The optical axis of the fan centre points about 30° away from the zenith (Fig. 8). Consequently, the microvillar directions of adjacent rhabdoms change most dramatically at an elevation of about 60° above the horizon. In *Cataglyphis*, the retinulae of the dorsal rim area are characterised by specialised dumb-bell-shaped rhabdoms. In *Apis*, they consist of nine long photoreceptor cells that are not twisted (Wehner *et al.*, 1975) and are provided with extraordinarily wide visual fields (Labhart, 1981). In the remainder of the bee's eye the retinulae are twisted and consist of one short and eight long photoreceptor cells (Wehner *et al.*, 1975; Wehner & Meyer, 1981). As can be inferred from electrophysiological recordings (T. Labhart, unpublished data), each retinula within the dorsal rim area of the eye of the bee as well as the ant is equipped with two sets of orthogonally arranged ultraviolet receptors.

In both bees and ants, the dorsal rim area seems to be sufficient for recording compass information from skylight patterns. Moreover, bees in which the dorsal rim areas of both eyes have been occluded are

Fig. 7. Visual field of the desert ant, *Cataglyphis bicolor*. The visual sphere as seen from the front (*a*) and from the rear (*b*). More specifically, the observer is coaxial with ommatidia pointing 25° to the left and 25° above the horizon (*a*), as well as 155° to the right and 25° above the horizon (*b*). The directions of view of the ommatidia have been determined by pseudopupil measurements. The pitch angle of the head has been taken from photographic pictures of walking ants. For more detailed descriptions of methods and results see Wehner (1982a). Symbols: dashed line, horizon; (1) and (2), binocular visual field; (2), dorsal rim area; (3), visual fields of the ocelli (the dotted lines represent the 50% angular sensitivity values as determined by ERG measurements); (4), monocular visual field; (5), blind zone covered by neither eye.

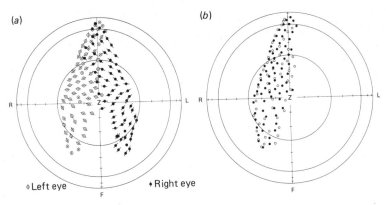

Fig. 8. The visual field of the dorsal rim area (see area 1 in Fig. 5) of *Cataglyphis bicolor*. The directions of view of all ommatidia within the dorsal rim areas of both eyes are projected on to the celestial hemisphere. The latter is shown in zenith projection: Z, zenith; F, frontal point; L, R, left and right lateral point, respectively. Parallels of altitude are drawn every 30°. In (*a*) the transverse axes of the dumb-bell-shaped rhabdoms (axes passing through rhabdomeres nos. 1 and 5, see Fig. 5) are indicated for most of the ommatidia. In (*b*), the dorsal rim areas for a large ant (●) and a small ant (○) are plotted for comparison. Although the numbers of dorsal rim ommatidia differ between the large ant ($n = 73$, that is 6.5% of in total 1118 ommatidia) and the small ant ($n = 55$, that is 6.6% of in total 836 ommatidia), the directions of view of the dorsal rim areas of both animals exactly coincide. The same holds true for the visual fields of the whole eyes. Thus, visual acuity increases with body size. Designed according to anatomical reconstructions of Dr E. Meyer and optical measurements of C. Zollikofer.

unable to dance correctly when they are allowed to see only a small area of the celestial hemisphere (diameters smaller than 10°). But they are well oriented when such a small area of the sky is viewed by the dorsal rim area, even when vision is restricted to one eye (Wehner, 1982*a*).[16] In ants, experiments that are strictly comparable have not yet been performed. The results of occluding either the dorsal rim area of the remainder of the eye show that both parts of the eye allow for correct orientation by polarised skylight (Wehner, 1982*a*). In both bees and ants, the role which the dorsal rim area plays in navigation by skylight has still to be worked out and compared with the role of the remainder of the eye.

Let us now return to the physiology of the hymenopteran retina. At least in the dorsal rim area each retinula contains a set of crossed analysers $(\chi_{\max 1} = \chi_{\max 2} \pm 90°)$ formed by ultraviolet receptors $(\lambda_{\max} \simeq 350\,\text{nm})$ that are highly sensitive to polarised light $(S_p \simeq 10)$.

Armed with this information one feels encouraged to ask what the sky looks like when viewed through an array of such sets of crossed analysers. It is possible to provide an answer because the relevant physical properties of the sky – the spatial distributions of $\chi(\lambda)$, $D(\lambda)$, and $I(\lambda)$ – and the physiological properties of the retina – the spatial distributions of $S(\lambda)$ and $S(\chi)$ – are both known (Sections 3 and 4). The false colour images shown in Wehner (1982*a*, Plates 3 and 4) demonstrate the behaviour of a model composed of a network of P-interneurones that antagonistically compare the outputs of the crossed analysers of each ommatidium. The geometrical arrangement of the retinal analysers, that is the structure of the fan array, is taken from the geometry of the dorsal rim area of the eye (Fig. 8).

What one learns from such hypothetical response patterns is how skylight patterns are pre-processed by the insect's peripheral visual system. But in trying to provide an answer to the question of how the insect reads these patterns, we must skip the higher order neuropiles of the hymenopteran visual pathway (about which we know but little) and turn to the behaviour of the whole animal.

5. Compass strategies

The basic fact which has prompted most of the theorising about skylight navigation is that insects can derive compass information equally as well from an isolated point of blue sky (or a small source of polarised light displayed artifically) as they can from the position of the sun present within a depolarised sky. Since von Frisch's early discoveries (von Frisch, 1949, 1965: 388f) the question of how the insect could use the pattern of polarised light in the sky for navigational means has usually been phrased in the following way: how does the insect infer the position of the sun from a small area of polarised skylight?[17]

One possibility is that the insect performs something analogous to a geometrical construction. From any point in the sky the sun can be located by following the great circle which passes through that point in a direction perpendicular to the local angle of polarisation (Fig. 2). Provided with this knowledge, the insect could compute the position of the sun by using at least two points of polarised light in the sky (Fig. 9), and tracing out the relevant great circles. The position of the sun is then given by the point at which the great circles intersect above the horizon.

This solution would lead to ambiguities whenever the vision is restricted to a small patch of sky. Kirschfeld, Lindauer & Martin (1975)

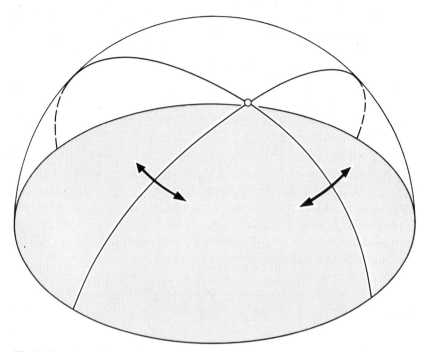

Fig. 9. How the position of the sun could be determined unambiguously by using e-vector information from (at least) two points in the sky: the sun is located where the great circles, running at right angles to the e-vectors of both points in the sky, intersect above the horizon.

have proposed that in this situation the insect could use additional information about the time of day and thus the elevation of the sun. Then the position of the sun could be inferred by computing the intersection point of the circle of celestial altitude, as defined by the elevation of the sun, and the great circle that runs at right angles to the e-vector in the sky. However, both circles usually intersect twice. In Fig. 10, the angular distance between the azimuthal positions of these two intersection points is denoted by ψ (dark shaded sector pointing towards the reader). Therefore, if the insect is allowed only to view a single e-vector in the sky, it should exhibit ambiguous orientation in confusing the correct compass direction, α, with another one separated by ψ from α. However, it does not (von Frisch, 1965: 394f; Zolotov & Frantsevich, 1973; Rossel, Wehner & Lindauer, 1978; Brines & Gould, 1979).

In attributing to a bee the ability to perform three-dimensional constructions in the sky, the investigator has perhaps been more ingenious than the bee itself. Are we really to assume that bees behave like little astronomers in tracing circles across the celestial hemisphere? Do they really come equipped with rather abstract knowledge about the physics of Rayleigh scattering? Most probably not. Let us look for more direct ways of exploiting skylight patterns.

For example, the insect could memorise a skylight pattern that it has experienced while heading towards a particular food source, and later try to match the memorised image with the current one when foraging again in the same direction.[18] It is this type of visual memory that has been assumed, more or less tacitly, in characterising the insect's celestial compass (Stockhammer, 1959; von Frisch, 1965: 398). Although image matching is an attractive hypothesis for explaining how insects steer a straight course using skylight cues, complications arise when one tries to understand how the insect's celestial compass operates during vector navigation. Consider an ant that while foraging along a devious path continually computes the mean vector pointing from the start to its current position and finally to the place where it finds food. The ant having reached a food source may never have experienced the retinal image of the celestial pattern that is associated with the direction of the mean vector, and yet it is this mean compass course, and not the sequence of courses actually steered, which the ant must remember and use during future forays. Apparently, the navigating insect continually refers to a reference direction in the sky, for example the solar or anti-solar meridian, and in addition knows how any one part of the skylight pattern is spatially related to this reference direction.

This hypothesis implies that bees and ants come provided with some kind of celestial map. In asking what this celestial map looks like, let us go back to basics. Theoretically, the insect's internal map of the sky could include all skylight parameters as outlined above ($\chi(\lambda)$, $D(\lambda)$, $I(\lambda)$; see Figs. 3a–f). But as it is the angle of polarisation, χ, that is least affected by atmospheric disturbances like haze or clouds, it would be advantageous for the insect to rely on χ and to ignore the highly variable parameters D and I. Indeed, most observations in bees confirm this expectation.

The next question to ask is whether the insect is really informed about all the possible e-vector patterns in the sky, that is about all the possible χ/Φ functions (Fig. 4a) that vary with both parallel of altitude (μ) and elevation of the sun (μ_o). In presenting a single e-vector in the sky to a dancing bee (for apparatus used see Fig. 11), we wondered whether the bee knew exactly where this e-vector was positioned relative to the solar (or anti-solar) meridian. To our surprise, the answer was No.

Bees make mistakes, and it is these mistakes that reveal the nature of

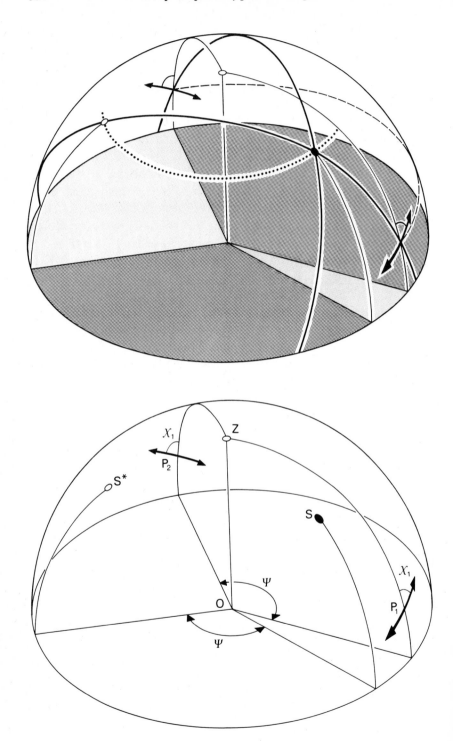

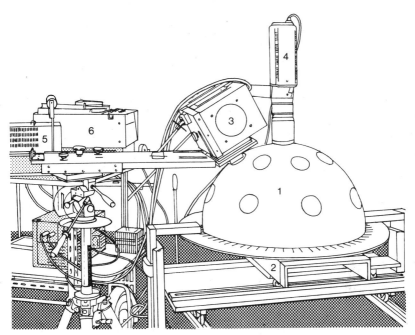

Fig. 11. Experimental device used in testing the celestial compass of bees that are dancing on a horizontal comb. 1, Translucent Plexiglass hemisphere; the apertures (diameter 10°) can be opened in order to display to the bees parts of the natural sky or artificially polarised beams of light; 2, hive consisting of a horizontally arranged comb mounted on a platform which can be moved in the x and y directions; 3, xenon bulb illuminating a set of filters (heat absorbing filters, diffusers, spectral filters, polarisers); 4, video camera; 5, monitor; 6, tape recorder.

the bee's internal representation of the sky. Suppose initially that bees viewing a single patch of sky register its elevation and angle of polarisation, and match them to an accurate internal map of these parameters over the whole sky. They should then dance in two directions because, at a given elevation (μ), each e-vector direction(χ)

Fig. 10. Theoretically, ambiguities arise when only one point of the e-vector pattern in the sky is used in navigation. This can be shown in one of two ways. (1) In general, each angle of polarisation (for example, χ_1) occurs twice (at P_1 and P_2) at a given parallel of altitude. The angular distance, ψ, between the azimuth values of P_1 and P_2 is marked by the dark sector pointing away from the reader. (2) The great circle running perpendicularly to the angle of polarisation as present in a given point in the sky (for example P_1) intersects twice with the parallel of altitude that is defined by the height of the sun. The latter can be inferred from the time of the day. The angular distance, ψ, between the azimuth values of both intersection points is marked by the dark sector pointing towards the reader. As can be shown by simple spherical geometry, both angles ψ are identical. Bees and ants, however, do not exhibit ambiguous navigation when left with only a single e-vector in the sky. How they may get rid of the ambiguities inherent in the natural pattern of skylight polarisation is described in the text.

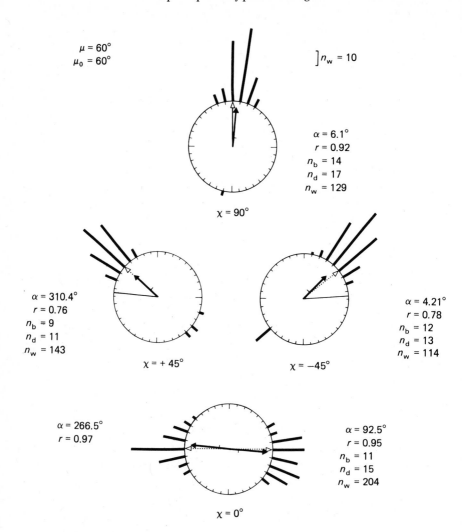

$\mu = 60°$
$\mu_0 = 60°$

$]n_w = 10$

$\alpha = 6.1°$
$r = 0.92$
$n_b = 14$
$n_d = 17$
$n_w = 129$

$\chi = 90°$

$\alpha = 310.4°$
$r = 0.76$
$n_b = 9$
$n_d = 11$
$n_w = 143$

$\chi = +45°$

$\alpha = 4.21°$
$r = 0.78$
$n_b = 12$
$n_d = 13$
$n_w = 114$

$\chi = -45°$

$\alpha = 266.5°$
$r = 0.97$

$\alpha = 92.5°$
$r = 0.95$
$n_b = 11$
$n_d = 15$
$n_w = 204$

$\chi = 0°$

Fig. 12. Navigational errors as shown by the bees which are dancing on a horizontal comb and are allowed only to view a single e-vector in the sky. The e-vector is presented by means of a xenon light source equipped with an ultraviolet filter (Schott Comp., UG 11) and an ultraviolet transmitting polariser (Polaroid Corp. HNP'B). $\chi = 90°$: no errors occur when a horizontal e-vector is presented. The bees invariably interpret the horizontal e-vector as being positioned along the anti-solar meridian. $\chi = \pm 45°$: the bees consider e-vector directions of $+45°$ and $-45°$ as being positioned nearer to the anti-solar meridian than these e-vectors actually are. $\chi = 0°$: vertical e-vectors are expected by the bees to lie at an azimuthal distance of $90°$ to the anti-solar meridian. As a consequence, bees usually exhibit bimodal orientation whenever they are confronted with a single vertical e-vector in the sky. The whole spectrum of errors made by the bees viewing arbitrary e-vectors in the sky can be read off Fig. 13. χ, e-vector direction (for conventions see Fig. 2); μ, celestial altitude of the beam of polarised light displayed to the bees; μ_0, celestial altitude of the sun; black arrow, mean vector of orientation (α, direction; r, length); white arrow (accompanied by dotted line), direction in which the bee is expected to dance if it applies

usually occurs at two positions in the sky (Φ) separated by the azimuthal distance ψ (Fig. 10, dark shaded sector pointing away from the reader). However, as mentioned earlier, bees never exhibit the ψ-ambiguity. In principle, this ambiguity could be resolved, if the bees took note of other skylight features, since the two points that coincide in the angle of polarisation vary in intensity, degree of polarisation, and hue. However, the orientations of their dances imply that bees do not resort to this possibility. Instead, they invariably interpret a small spot of the e-vector pattern as being positioned *in the more highly polarised part of the sky*, that is nearer to the anti-solar meridian, even when it is positioned in the less polarised part of the sky nearer to the solar meridian (Rossel *et al.*, 1978; Brines & Gould, 1979).

Their dancing also tells us about a second inaccuracy of their map. Bees are *not informed exactly* about the spatial distribution of e-vectors, since they make small but consistent mistakes in assigning azimuth positions to e-vector directions (Rossel *et al.*, 1978; Fig. 12). No errors occur whenever a horizontal e-vector is presented; the bees invariably interpret a horizontal e-vector as being positioned along the anti-solar meridian as is actually the case (Fig. 12, upper part). Viewing an e-vector of $\chi = +45°$ or $\chi = -45°$, they deviate to the right or left from the correct course, respectively (Fig. 12, middle part). This means no more than that, under the conditions of the experiment ($\mu = 60°$, $\mu_o = 60°$), the bees assume both e-vectors to be positioned nearer to the anti-solar meridian than the e-vectors actually are. Vertical e-vectors ($\chi = 0°$) are always considered to lie at right angles to the solar and anti-solar meridian. This holds even when, in the natural sky, vertical e-vectors do not occur at all at the elevation at which the artificial e-vector is presented (in the case of $\mu = \mu_o$; Fig. 12, lower part). When tested with a vertical e-vector in the sky, the bees usually exhibit bi-modal orientation. Since points in the sky that are characterised by vertical e-vectors are identical in all optical aspects of sky-light, this is to be expected. What is surprising, though, is that the bees always associate vertical e-vectors with meridians lying at right angles to the solar and anti-solar meridians.

In conclusion, from mistakes made by the bees, one can learn that

---

the generalised χ/Φ function as shown by the heavy lines in Fig. 13; thin black line, direction in which the bee is expected to dance if it is informed correctly about the azimuth position of the e-vector in question. In the case of $\chi = 0°$ (lower figure), the e-vector direction presented artificially is not realised in the sky (at $\mu = \mu_o = 60°$). n_b, number of bees tested; n_d, number of dances recorded; n_w, number of waggle runs performed by the bees.

bees assume a certain e-vector direction (χ) always located at a certain azimuthal distance (Φ) from the anti-solar meridian, irrespective of the parallel of altitude (μ) and the elevation of the sun (μ_o). The bees invariably apply a *generalised χ/Φ function* in which the vertical e-vectors ($\chi = 0°$) are positioned $\Phi = 90°$ off the anti-solar meridian (where $\chi = 90°$) and in which Φ is related to χ in a roughly but not exactly linear way (Rossel *et al.*, 1978). As demonstrated in Fig. 13, the bee's generalised χ/Φ function describes rather well what actually occurs in regions of the sky that are closer to the zenith than to the horizon.

It is a fascinating task to try to explain how the simplified model of the e-vector pattern in the sky has been incorporated into the insect's brain. In this respect, let us just refer to the special case of $\chi = 0°$ when the sun is at the horizon. Then, the line of maximum degree of polarisation, which is always positioned at an angular distance of 90° from the sun, runs through the zenith and comprises all vertical e-vectors in the sky. Therefore, it is under this condition that one important feature of the bee's internal model of the sky ($\chi = 0°$ at $\Phi = 90°$) is actually realised in the natural sky. However, even then only the vertical and horizontal e-vectors are located exactly at those azimuth positions where the bee expects them to occur. The whole χ/Φ function as used by the bee can be derived in a rather straightforward way from the actual skylight patterns if one pays attention exclusively to the line of maximum polarisation and how this line changes its position in the sky during the course of the day (Rossel & Wehner, 1982).

The concept of the bee's simplified version of the complex skylight patterns holds promise. However, several properties require further analysis before the concept can be fully endorsed as a general principle of the insect's celestial compass. What is needed next are experiments with more than one small point in the sky (or more than one artificial e-vector) as well as experiments in which the bees can view the natural sky through a wide-angle aperture, so that a large continuous part of the e-vector pattern (or an artificial e-vector pattern) is displayed to the bees. Do small errors occur even then? Work is proceeding apace and an answer to this question, as well as a discussion of how far

---

Fig. 13. The distributions of e-vectors (χ/Φ functions) for different elevations above the horizon ($\mu = 31°$, 66°, 78°) and different elevations of the sun ($0° \leq \mu_0 \leq 65°$). χ, e-vector direction; Φ, azimuth position; μ, elevation above horizon; μ_0, elevation of the sun. The heavy black line represents the bee's e-vector characteristic, that is its internal model of the sky. The same e-vector characteristic holds for all elevations above horizon and all elevations of the sun. This means nothing else than that the bee's internal χ/Φ function is invariant against μ and μ_0.

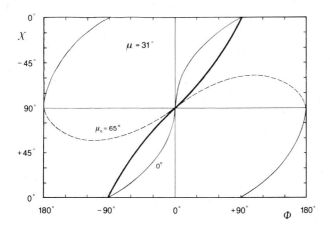

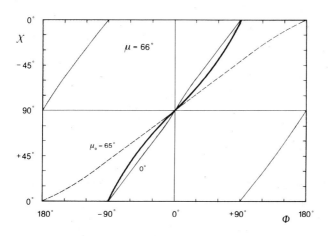

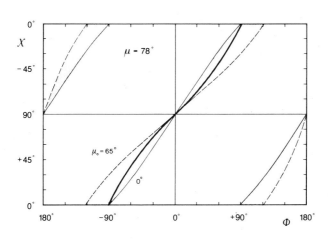

the concept of the generalised χ/Φ function might lead, will soon be available.

The bee's simplified internal representation of the e-vector pattern in the sky is not claimed to exhaust the bee's repertoire in sky light navigation. Of course, the *sun*, which is not part of the χ/Φ function, can be used as a compass cue. With respect to its compass bearing, the sun is interpreted to lie opposite to the symmetry point of the χ/Φ function, that is the meridian characterised by horizontal e-vectors. Furthermore, the sun is considered by bees and ants as a long-wavelength point in the sky (Edrich, Neumeyer & von Helversen, 1979; Brines & Gould, 1979; Wehner, 1982*a*). In addition, bees seem to be able to derive compass information from unpolarised *colour patterns* in the sky (van der Glas, 1976, 1980).[19]

I hope that I have marshalled now sufficient evidence to convince the reader that insects are not astronomers programmed to perform spherical geometry in the sky. Neither are they endowed with a complete knowledge of all possible e-vector patterns as they occur in the sky for different elevations of the sun. Instead, they rely on a generalised celestial map in which the spatial distribution of e-vectors does not vary during the day, contrary to what actually occurs in the sky. Of course, what does vary with the time of the day in both the celestial pattern and the insect's internal representation of this pattern is the compass bearing of the pattern. The insect must somehow be able to rotate its celestial map within its mind (Wehner & Lanfranconi, 1981).

Finally, one wonders how the insect can afford to use a celestial map that, though based on some principal aspects of skylight polarisation, is not a correct copy of the outside world. On closer scrutiny, the navigational errors necessarily introduced by such a strategy are usually small. They become smaller, the larger the area of the sky viewed by the insect in the more highly polarised part of the sky. They are further kept down by the additional use of some powerful backup systems drawing upon the position of the sun and non-celestial cues such as landmarks.

The ultimate reason for using a generalised rather than a correct copy of the outside world certainly is to sacrifice absolute precision for a workable neural strategy. This might also hold true for other modes of insect navigation, for example piloting by landmarks or vector navigation (integrating an indirect route).

Let us conclude by speculating about the evolutionary history of some aspects of the insect's celestial compass. First, it has been noted earlier that the detection of polarised light is dominated by the short-wavelength (ultraviolet) type of receptor. As the scattered light from the radiant sky is rich in ultraviolet components, but reflected light from terrestrial objects is not (see colour plate 5 in Wehner, 1982a), it seems likely that ultraviolet receptors have evolved in the functional context of exploiting skylight cues for one or another type of navigational purpose. Of course, in bees as well as many other insect species ultraviolet receptors are also part of a trichromatic colour vision system, used, for example, for recognising floral patterns of flowering plants. However, it might well have been that in mid-Cretaceous times when angiosperm plants evolved and certain hymenopterans became flower-visiting wasps, that is bees, the angiosperms took advantage of the ultraviolet receptors that insects already possessed and developed ultraviolet patterns on their flowers. Thus, in evolutionary terms, the ultraviolet receptors of insects could well be older than the ultraviolet patterns of flowers. Second, the evolution of e-vector-detecting (not merely polarisation-sensitive) visual systems might also have been triggered by skylight cues rather than by other stimuli present in the insect's visual surroundings. This is borne out, for example, by the fact that e-vector-detecting systems are restricted to the dorsal parts of the eye. The need for deriving compass information from skylight cues might well have been responsible for the evolution of a sensory capacity that lies beyond our ken – the perception of polarised light.

I thank Drs. G. D. Bernard (Yale University) and T. S. Collett (University of Sussex) for reading the manuscript and for valuable suggestions. My own experimental work described in this paper has been supported by Swiss NSF 3.313–0.78.

Notes

1. As light reflected from water surfaces is polarised horizontally, populations of photoreceptors which have their microvilli aligned vertically would be relatively blind to the surface glare. Such vertical analysers have been found in the eyes of gerrid bugs and dolichopodid flies (for references see above).
2. At shallow depths, underwater polarisation is due partly to skylight polarisation (within the 'aerial window') and partly to polarisation originating in the water itself (Timofeeva, 1962; Ivanoff, 1974). As most natural objects, and those under water as well, depolarise the light reflected from them, they stand out in greater contrast against the (polarised) background when viewed through a rotating polariser. This is because then the brightness of the background will fluctuate more than that of the object (Lythgoe & Hemmings, 1967). However, it has been argued that there are so many drawbacks in using polarisers as means of enhancing contrast that underwater vision with polarisers is not reliably superior to vision without them (Luria & Kinney, 1974).
3. The 'metallic' cuticles of certain scarabaeid beetles are able to reflect selectively left-circularly polarised light as can be shown by observing the beetle's cuticle with a

circular polariser comprising a quarter-wave plate and a linear polariser (Neville & Caveney, 1969; Caveney, 1971). However, as photoreceptors do not discriminate between left- and right-circularly polarised light, it is not known what this remarkable optical property of the cuticle of some insects could mean in functional terms.

4. An early review on polarisation vision which still provides an excellent introduction to the field has been given by Stockhammer (1959). In the years following this account interest has focussed mainly on dichroism and polarisation sensitivity of photoreceptors (see, for example, the review articles published in the Proceedings of the Symposium on Photoreceptor Optics, edited by Snyder & Menzel, 1975). At the same time, behaviourists became especially engaged in the study of so-called polarotactic responses (for a comprehensive list of references see Waterman, 1973). In an all-inclusive summary Waterman (1981) has dealt extensively with a wide range of topics related to 'polarisation sensitivity'. Most recently, skylight navigation in ants and bees has been reviewed and discussed by Wehner (1982a) from both the neurophysiological and behavioural points of view.

5. It has sometimes been claimed (but since refuted) that the analyser concerned must be *extra-ocular* depending on differential scattering or reflection in the insect's immediate surroundings (Baylor & Kennedy, 1958; Baylor & Smith, 1958; Kalmus, 1958).

6. D, absorbance (optical density) [μm^{-1}].

7. The diameter of a microvillus is of the order of 50 nm and thus only about ten times the diameter of the photopigment molecule.

8. The absorption A of the rhabdomere is $A = 1 - \exp(-2.3\, Dl)$ where D is the optical density [μm^{-1}] and l the length of the rhabdomere [μm].

9. As skylight polarisation is predominantly linear (Hannemann & Raschke, 1974), the phase difference between the two orthogonal components describing the general case of ellipitically polarised light – that is one of the Stokes parameters (Clarke & Grainger, 1971) – need not be considered here. For linearly polarised light this phase difference is zero.

10. Still another way by which an e-vector-detecting system could determine the angle of polarisation is to use a single analyser and rotate it about the direction of illumination. Of course, the effectiveness of such a 'successive method' (Kirschfeld, 1973) depends on how the information acquired by the rotating receptor is processed. What is needed in any case is a fine temporal tuning of the activity of higher order interneurones.

11. Following the negative results reported by Crozier & Mangelsdorf (1924), Verkhovskaya's (1940) paper provides the first description of an animal's response to linearly polarised light.

12. Although sensitivity to polarised light has once been reported for visual interneurones of beetles (Zolotov & Frantsevich, 1977), this result later turned out to be due to experimental artefacts (Frantsevich & Zolotov, 1979). Thus, the only recordings from polarisation-sensitive interneurones have been performed in crabs (extracellular recordings from the medulla of *Scylla*: Leggett, 1976, 1978: 93). These units respond to the rotation of the e-vector of linearly polarised light, but they do not respond differentially to stationary e-vectors that vary in χ.

13. The photoreceptor membranes are aligned in an orthogonal pattern even in some vertebrates (teleosts: *Anchoa*) in which certain cones are positioned side-on to the incident light (Fineran & Nicol, 1978). As summarised by Waterman (1975), and in addition reported by Delius, Perchard & Emmerton (1976), Taylor & Adler (1978) and Taylor & Auburn (1978), some species of teleosts, amphibians, reptiles and birds have been shown to respond to certain aspects of polarised light. Snyder (1973b) has proposed that teleosts could be able to detect polarised light by exploiting the small component of incident light that is perpendicular to the rod outer segments.

14. In decapod crustaceans there is some anatomical evidence that within one ommatidial unit receptors endowed with horizontally arranged rhabdomeres project on to one type of interneurone, and those endowed with vertically arranged rhabdomeres project on to another (Meyer-Rochow & Naessel, 1977; Stowe & Leggett, 1978).

15. Owing to secondary factors such as higher order scattering, diffuse reflection, molecular absorption and anisotropy, so-called *negative polarisation* occurs in regions near the sun and the anti-solar point where the degree of polarisation due to primary (Rayleigh) scattering is very low. As in these regions of the sky the intensity component within the scattering plane is greater than that perpendicular to it, vertical e-vectors occur instead of horizontal ones along those parts of the solar vertical that are close to the sun and anti-solar point. However, as in these regions of the sky the degree of polarisation is very low, it is reasonable to assume that such anomalous polarisation is ignored by the insect's skylight compass.

16. As the dorsal rim area forms part of the bee's (and ant's) binocular visual field (Fig. 7), stimulus conditions can be established in which a small area in the sky is viewed by both the dorsal rim of one eye and part of the remainder of the dorsal retina of the other eye. For *Cataglyphis*, the binocular visual field and the directions of view of the dorsal rim areas of both eyes are plotted in Figs. 7 and 8.

17. Quantitative analyses of how skylight polarisation is used as a compass have been restricted to bees and ants. In addition, some qualitative observations are available for other groups of insects (dipterans: Wellington, 1974; Wolf, Gebhardt, Gademann & Heisenberg, 1980; coleopterans: Frantsevich *et al.*, 1977), crustaceans (amphipods, isopods: Pardi, 1957; brachyurans: Daumer, Jander & Waterman, 1963; Schoene, 1963, Altevogt & von Hagen, 1964), and spiders (lycosids: Papi, 1959). These few reports about what use different species of arthropods make of skylight polarisation contrast sharply with the plethora of accounts on so-called polarotactic responses. Most of these accounts (for references see Waterman, 1973) describe the behaviour of animals which are exposed to an overhead source of linearly polarised light.

18. Note that the retinal location of a celestial cue does not change whenever the observer is moving along a straight line and refraining from rotatory movements. This is because celestial cues are effectively at infinity and thus not subject to the phenomenon of motion parallax. For this reason, celestial cues provide compass information.

19. This interpretation of van der Glas' results is not consistent with the one given by the author himself. Instead, van der Glas proposes that bees see the e-vector pattern in the sky as a colour pattern. According to this hypothesis all polarisation-sensitive receptors are plugged into a colour-coding network. Although the author is not explicit about the way the neural system is to accomplish this proposed modification of the colour vision system, he argues that 'all colour channels in the central nervous system required for colour coding' are used to convert the pattern of polarised skylight into a 'perceived polarisation-induced colour pattern' (van der Glas, 1980). Thus, rather than extracting the angle of polarisation from a given point in the sky, the insect is supposed to extract the hue of colour. This conclusion, however, is not compelling. What the author's experiments do show is that bees can deduce compass information from *unpolarised* colour patterns displayed above them. This is a necessary condition for van der Glas' hypothesis, but is by no means sufficient. The question of how bees perceive *polarised* light still remains.

Note added in proof: Recently, Brines & Gould (*J. Exp. Biol.*, **96**, 69–91, 1982) have provided elaborate data on the skylight parameters χ, D, and I measured at 350, 500 and 650 nm.

References

ALTEVOGT, R. & HAGEN, H. VON (1964). Ueber die Orientierung von *Uca tangeri* im Freiland. *Z. Morph. Oekol. Tiere*, **53**, 636–65.

AUTRUM, H. & STUMPF, H. (1950). Das Bienenauge als Analysator für polarisiertes Licht. *Z. Naturforsch.* **5b**, 116–122.

BAYLOR, E. R. & KENNEDY, D. (1958). Evidence against a polarizing analyser in the bee eye. *Anat. Rec.* **132**, 411.

BAYLOR, E. R. & SMITH, F. E. (1953). The orientation of *Cladocera* to polarized light. *Amer. Naturalist*, **87**, 97–101.

BAYLOR, E. R. & SMITH, F. E. (1957). Diurnal migration of plankton crustaceans. In *Recent Advances in Invertebrate Physiology*, eds. B. T. Scheer, T. H., Bullock, L. H. Kleinholz and A. W. Martin; pp. 21–35. Eugene: University of Oregon Publishing.

BAYLOR, E. R. & SMITH, F. E. (1958). Extraocular polarization analysis in the honeybee. *Anat. Rec.* **132**, 411–12.

BERGER, P. & SEGAL, J. (1952). La discrimination du plan de polarisation de la lumière par l'oeil de l'abeille. *Compt. Rend. Acad. Sci.*, *Paris*, *Ser D*, **234**, 1308–1310.

BERNARD, G. D. & WEHNER, R. (1977). Functional similarities between polarisation vision and colour vision. *Vision Res.* **17**, 1019–28.

BOHN, H. & TAEUBER, U. (1971). Beziehungen zwischen der Wirkung polarisierten Lichtes auf das Elektroretinogramm und der Ultrastruktur des Auges von *Gerris lacustris*. *Z. vergl. Physiol.* **72**, 32–53.

BRINES, M. L. & GOULD, J. L. (1979). Bees have rules. *Science*, **206**, 571–3.

BUTLER, R. & HORRIDGE, G. A. (1973). The electrophysiology of the retina of *Periplaneta americana*. II. Receptor sensitivity and polarized light sensitivity. *J. comp. Physiol.* **83**, 279–88.

CAVENEY, S. (1971). Cuticle reflectivity and optical activity in scarab beetles: the role of uric acid. *Proc. Roy. Soc. London*, **178B**, 205–25.

CLARKE, D. & GRAINGER, J. F. (1971). *Polarized Light and Optical Measurement*. Oxford, New York: Pergamon Press.

CONE, R. A. (1972). Rotational diffusion of rhodopsin in the visual receptor membrane. *Nature, London*, **236**, 39–43.

CROZIER, W. J. & MANGELSDORF, A. F. (1924). A note on the relative photosensory effect of polarized light. *J. Gen. Physiol.* **6**, 703–9.

DAUMER, K., JANDER, R. & WATERMAN, T. H. (1963). Orientation of the ghost-crab *Ocypode* in polarized light. *Z. Vergl. Physiol.* **47**, 56–76.

DELIUS, J. D., PERCHARD, R. J. & EMMERTON, J. (1976). Polarized light and an electroretinographic correlate. *J. Comp. Physiol. Psychol.* **90**, 560–71.

DE VALOIS, R. L. (1973). Central mechanisms of colour vision. In *Handbook of Sensory Physiology*, vol. VII/3A, ed. R. Jung, pp. 209–254. Berlin, Heidelberg, New York: Springer-Verlag.

DUELLI, P. & WEHNER, R. (1973). The spectral sensitivity of polarized light orientation in *Cataglyphis bicolor* (Formicidae, Hymenoptera). *J. Comp. Physiol.* **86**, 37–53.

EAKIN, R. M. & BRANDENBURGER, J. L. (1971). Fine structure of the eyes of jumping spiders. *J. Ultrastruct. Res.* **37**, 618–63.

EDRICH, W. & HELVERSEN, O. VON (1976). Polarized light orientation of the honey bee: the minimum visual angle. *J. Comp. Physiol.* **109**, 309–14.

EDRICH, W., NEUMEYER, C. & HELVERSEN, O. VON (1979). 'Anti-sun orientation' of bees with regard to a field of ultraviolet light. *J. Comp. Physiol.* **134**, 151–7.

EGUCHI, E. (1965). Rhabdom structure and receptor potentials in single crayfish retinular cells. *J. Cell. Comp. Physiol.* **66**, 411–30.

EGUCHI, E. & WATERMAN, T. H. (1966). Fine structure patterns in crustacean rhabdoms. In: *The Functional Organisation of the Compound Eye*, ed. C. G. Bernherd, pp. 105–24. *Wenner-Gren Centre Int. Symp. Ser.* **7**, 105–24. Oxford: Pergamon Press.

EGUCHI, E. & WATERMAN, T. H. (1973). Orthogonal microvillus pattern in the eighth rhabdomere of the rock crab *Grapsus*. *Z. Zellforsch.* **137**, 145–57.

FINERAN, B. A. & NICOL, J. J. A. C. (1978). Studies on the photoreceptors of *Anchoa mitchilli* and *Anchoa hepsetus* (Engraulidae) with particular reference to the cones. *Phil. Trans. Roy. Soc., London*, **283B**, 25–60.

FRANTSEVICH, L. I. & ZOLOTOV, V. V. (1979). Criticism of the method for identification of polarization sensitive neurones in insects. *Neurophysiol. (Kiev)*, **11**, 326–37 (in Russian).

FRANTSEVICH, L. I., GOVARDOVSKI, V., GRIBAKIN, F., NIKOLAJEV, G., PICHKA, V., POLANOVSKY, A., SHEVCHENKO, V., ZOLOTOV, V. V. (1977). Astroorientation in *Lethrus* (Coleoptera, Scarabaeidae). *J. Comp. Physiol.* **121**, 253–71.

FRISCH, K. VON (1949). Die Polarisation des Himmelslichts als orientierender Faktor bei den Tänzen der Bienen. *Experientia*, **5**, 142–8.

FRISCH, K. VON (1965). *Tanzsprache und Orientierung der Bienen*. Berlin, Heidelberg & New York: Springer-Verlag.

GLAS, H. W. VAN DER (1976). Polarization induced colour patterns: a model of the perception of the polarized skylight by insects. II. Experiments with direction trained dancing bees, *Apis mellifera*. *Neth. J. Zool.* **26**, 383–413.

GLAS, H. W. VAN DER (1980). Orientation of bees, *Apis mellifera*, to unpolarized colour patterns, simulating the polarized zenith skylight pattern. *J. Comp. Physiol.* **139**, 225–41.

GOLDSMITH, T. H. (1975). The polarization sensitivity – dichroic absorption paradox in arthropod photoreceptors. In *Photoreceptor Optics*, ed. A. W. Snyder & R. Menzel, pp. 392–409. Berlin, Heidelberg, New York: Springer-Verlag.

GOLDSMITH, T. H. & BERNARD, G. D. (1974). The visual system of insects. In *The Physiology of Insecta*, vol. 2, ed. M. Rockstein, pp. 165–272. New York, San Francisco & London: Academic Press.

GOLDSMITH, T. H. & WEHNER, R. (1977). Restrictions of rotational and translational diffusion of pigment in the membranes of a rhabdomeric photoreceptor. *J. Gen. Physiol.* **70**, 453–90.

GORDON, W. C. (1977). Microvillar orientation in the retina of the nymphalid butterfly. *Z. Naturforsch*, **32c**, 662–4.

HAGINS, W. A. & LIEBMAN, P. A. (1963). The relationship between photochemical and electrical processes in living squid photoreceptors. Abstr. Biophys. Soc. 7th Ann. Meet. New York.

HANNEMANN, D. & RASCHKE, E. (1974). Measurements of the elliptical polarization of sky radiation: preliminary results. In *Planets, Stars and Nebulae*, ed. T. Gehrels, pp. 510–13. Tucson: Arizona University Press.

HARDIE, R. C., FRANCESCHINI, N., McINTYRE, P. D. (1979). Electrophysiological analysis of fly retina. II. Spectral and polarisation sensitivity in R7 and R8. *J. Comp. Physiol.* **133**, 23–39.

HAYS, D., GOLDSMITH, T. H. (1969). Microspectrophotometry of the visual pigment of the spider crab, *Libinia emarginata*. *Z. Vergl. Physiol.* **65**, 218–32.

HAZEN, W. E. & BAYLOR, E. R. (1962). Behaviour of *Daphnia* in polarized light. *Biol. Bull.* **132**, 243–52.

HELVERSEN, O. VON & EDRICH, W. (1974). Der Polarisationsempfänger im Bienenauge: ein Ultraviolettrezeptor. *J. Comp. Physiol.* **94**, 33–47.

HERRLING, P. L. (1976). Regional distribution of three ultrastructural retinula types in the retina of *Cataglyphis bicolor* (Formicidae, Hymenoptera). *Cell Tiss. Res.* **169**, 247–66.

HESSE, R. (1908). *Das Sehen der niederen Tiere*. Jena: G. Fischer Verlag.

IVANOFF, A. (1974). Polarization measurements in the sea. In *Optical Aspects of Oceanography*, ed. N. G. Jerlov & E. S. Nielsen, pp. 151–75. London: Academic Press.

KALMUS, H. (1958). Responses of insects to polarized light in the presence of dark reflecting surfaces. *Nature, London*, **182**, 1526–7.

KIRSCHFELD, K. (1969). Optics of the compound eye. In *Processing of Optical Data by Organisms and by Machines*, ed. W. Reichardt, pp. 144–66. New York & London: Academic Press.

KIRSCHFELD, K. (1973). Vision of polarized light. Int. Biophys. Congr. Moscow **4**, 289–296; ed. by Int. Union for Pure and Applied Biophysics, Acad. Sci. USSR.

KIRSCHFELD, K. & FRANCESCHINI, N. (1977). Photostable pigments within the membrane of photoreceptors and their possible role. *Biophys. Struct. Mechan.* **3**, 191–4.

KIRSCHFELD, K., LINDAUER, M. & MARTIN, H. (1975). Problems of menotactic orientation according to the polarized light of the sky. *Z. Naturforsch.* **30c**, 88–90.

KOLB, G. (1977). The structure of the eye of *Pieris brassicae* (Lepidoptera). *Zoomorphologie*, **87**, 123–46.

KUNZE, P. & BOSCHEK, C. B. (1968). Elektronenmikroskopische Untersuchung zur Form der achten Retinulazelle bei *Ocypode*. *Z. Naturforsch.* **23b**, 568–9.

LABHART, T. (1980). Specialized photoreceptors at the dorsal rim of the honey bee's compound eye: polarizational and angular sensitivity. *J. Comp. Physiol.* **141**, 19–30.

LANGER, H. (1965). Nachweis dichroitischer Absorption des Sehfarbstoffes in den Rhabdomeren des Insektenauges. *Z. Vergl. Physiol.* **51**, 258–63.

LAUGHLIN, S. B. (1975). Receptor function in the apposition eye – an electrophysiological approach. In *Photoreceptor Optics*, ed. A. W. Snyder & R. Menzel, pp. 479–398, Berlin, Heidelberg, New York: Springer-Verlag.

LAUGHLIN, S. B. (1976). The sensitivities of dragonfly photoreceptors and the voltage gain of transduction. *J. Comp. Physiol.* **111**, 221–47.

LAUGHLIN, S. B., MENZEL, R. & SNYDER, A. W. (1975). Membranes, dichroism and receptor sensitivity. In *Photoreceptor Optics*, ed. A. W. Snyder & R. Menzel, pp. 237–59. Berlin, Heidelberg, New York: Springer-Verlag.

LEGGETT, L. M. W. (1976). Polarised light sensitive interneurones in a swimming crab. *Nature, London*, **262**, 709–11.

LEGGETT, L. M. W. (1978). Some visual specializations of a crustacean eye. PhD Thesis, Australian National University, Canberra.

LIEBMAN, P. A. (1962). *In-situ* microspectrophotometric studies on the pigments of single retinal rods. *Biophys. J.* **2**, 161–78.

LURIA, S. M. & KINNEY, J. A. S. (1974). Linear polarising filters and underwater vision. *Undersea Biomed. Res.* **1**, 371–8.

LYTHGOE, J. N. & HEMMINGS, C. C. (1967). Polarized light and underwater vision. *Nature, London*, **213**, 893–4.

MAIDA, T. M. (1977). Microvillar orientation in the retina of a pierid butterfly. *Z. Naturforsch.* **32c**, 660–1.

MASON, W. T., FAGER, R. S. & ABRAHAMSON, E. W. (1973). Characterization of the lipid composition of squid rhabdom outer segments. *Biochimm. Biophys. Acta*, **306**, 67–73.

McINTYRE, P. & SNYDER, A. W. (1978). Light propagation in twisted anisotropic media: application to photoreceptors. *J. opt. Soc. Amer.* **68**, 149–57.

MELAMED, J. & TRUJILLO-CENOZ, O. (1966). The fine structure of the visual system of *Lycosa* (Araneae: Lycosidae). I. Retina and optic nerve. *Z Zellforsch.* **74**, 12–31.

MENZEL, R. (1975). Polarization sensitivity in insect eyes with fused rhabdoms. In *Photoreceptor Optics*, ed. A. W. Snyder & R. Menzel, pp. 372–87. Berlin, Heidelberg & New York: Springer-Verlag.

MENZEL, R. (1979). Spectral sensitivity and color vision in invertebrates. In *Handbook of Sensory Physiology*, vol. VII/6A, ed. H. Autrum, pp. 503–580. Berlin, Heidelberg & New York: Springer-Verlag.

MENZEL, R. & BLAKERS, M. (1975). Functional organization of an insect ommatidium with fused rhabdom. *Cytobiologie* **11**, 279–98.

MENZEL, R. & SNYDER, A. W. (1974). Polarised light detection in the bee, *Apis mellifera*. *J. Comp. Physiol.* **88**, 247–70.

MEYER-ROCHOW, V. B. (1971). A crustacean-like organization of insect-rhabdoms. *Cytobiologie*, **4**, 241–9.

MEYER-ROCHOW, V. B. (1972). The eyes of *Creophilus erythrocephalus* and *Sartallus signatus* (Staphylinidae: Coleoptera). Light-, interference-, scanning electron-, and transmission electron microscope examinations. *Z. Zellforsch.* **133**, 59–86.

MEYER-ROCHOW, V. B. & NAESSEL, D. R. (1977). Crustacean eyes and polarization sensitivity. *Vision Res.* **17**, 1239–40.

MOODY, M. F. & PARRISS, J. R. (1961). The discrimination of polarized light by *Octopus*: a behavioural and morphological study. *Z. vergl. Physiol.* **44**, 268–91.

MOTE, M. I. (1974). Polarization sensitivity. A phenomenon independent of stimulus intensity or state of adaptation in retinular cells of the crabs *Carcinus* and *Callinectes*. *J. Comp. Physiol.* **90**, 389–403.

MOTE, M. I. & WEHNER, R. (1980). Functional characteristics of photoreceptors in the compound eye and ocellus of the desert ant, *Cataglyphis bicolor*. *J. Comp. Physiol.* **137**, 63–71.

NAESSEL, D. R. (1976). The retina and retinal projection on lamina ganglionaris of the crayfish *Pacifastacus leniusculus*. *J. Comp. Neurol.* **167**, 341–60.

NEVILLE, A. C. & CAVENEY, S. (1969). Scarabaeid beetle exocuticle as an optical analogue of cholesteric liquid crystals. *Biol. Rev.* **44**, 531–62.

PAPI, F. (1959). Sull'orientamento astronomico in specie del gen. *Arctosa* (Araneae, Lycosidae). *Z. vergl. Physiol.* **41**, 481–9.

PARDI, L. (1957). L'orientamento astronomico degli animali: risultati e problemi attuali. *Boll. Zool.* **24**, 473–523.

PAULUS, H. F. (1975). The compound eyes of apterygote insects. In *The Compound Eye and Vision of Insects*, ed. G. A. Horridge, pp. 3–17. Oxford: Clarendon Press.

RAEBER, F. (1979). Retinatopographie und Sehfeldtopologie des Komplexauges von *Cataglyphis bicolor* (Formicidae, Hymenoptera) und einiger verwandter Formiciden-Arten. Dissertation Universität Zürich.

ROSSEL, S. & WEHNER, R. (1982). The bee's map of the e-vector pattern in the sky. *Proc. Natl. Acad. Sci., USA*, **79**, 4451–55.

ROSSEL, S., WEHNER, R. & LINDAUER, M. (1978). E-vector orientation in bees. *J. Comp. Physiol.* **125**, 1–12.

SAIBIL, H. R. (1982). An ordered membrane-cytoskeleton network in squid photoreceptor microvilli. *J. Mol. Biol.* in press.

SCHINZ, R. H. (1975). Structural specialization in the dorsal retina of the bee, *Apis mellifera*. *Cell Tiss. Res.* **162**, 23–34.

SCHMIDT, W. J. (1935). Doppelbrechung, Dichroismus und Feinbau des Aussengliedes der Sehzellen vom Frosch. *Z. Zellforsch. mikroskop. Anat.* **22**, 485–522.

SCHNEIDER, L. & LANGER, H. (1969). Die Struktur des Rhabdoms im 'Doppelauge' des Wasserläufers *Gerris lacustris*. *Z. Zellforsch.* **99**, 538–59.

SCHOENE, H. (1963). Menotaktische Orientierung nach polarisiertem und unpolarisiertem Licht bei der Mangrovekrabbe *Goniopsis*. *Z. Vergl. Physiol.* **46**, 496–514.

SCHOENE, H. & SCHOENE, H. (1961). Eyestalk movements induced by polarized light in the ghost crab, *Ocypode quadrata*. *Science*, **134**, 675–76.

SCHROEER, W. D. (1974). Zum Mechanismus der Analyse polarisierten Lichtes bei *Agelena gracilens* (Araneae, Agelenidae). I. Die Morphologie der Retina der vorderen Mittelaugen (Hauptaugen). *Z. Morph. Tiere* **79**, 215–31.

SEITZ, G. (1969). Polarisationsoptische Untersuchungen am Auge von *Calliphora erythrocephala*. *Z. Zellforsch.* **93**, 525–9.

SHAW, S. R. (1967). Simultaneous recording from two cells in the locust retina. *Z. Vergl. Physiol.* **55**, 183–94.

SHAW, S. R. (1969a). Sense-cell structure and interspecies comparisons of polarized-light absorption in arthropod compound eyes. *Vision Res.* **9**, 1031–40.

SHAW, S. R. (1969b). Interreceptor coupling in ommatidia of drone honeybee and locust compound eyes. *Vision Res.* **9**, 999–1029.

SHAW, S. R. (1969c). Optics of the arthropod compound eye. *Science*, **165**, 88–90.

SKRZIPEK, K.˙H. & SKRZIPEK, H. (1971). Zur funktionellen Bedeutung der räumlichen Anordnung des Kristallkegels zum Rhabdom im Auge der Trachtbiene (*Apis mellifica*). *Experientia*, **27**, 409–11.

SKRZIPEK, K. H. & SKRZIPEK, H. (1974). Die spektrale Transmission und die optische Aktivität des dioptrischen Apparates der Honigbiene (*Apis mellifera*). *Experientia* **30**, 314–15.

SMOLA, U. & TSCHARNTKE, H. (1979). Twisted rhabdomeres in the dipteran eye. *J. Comp. Physiol.* **133**, 291–7.

SMOLA, U. & WUNDERER, H. (1981). Fly rhabdomeres twist *in vivo*. *J. Comp. Physiol.* **142**, 43–9.

SNYDER, A. W. (1973a). Polarization sensitivity of individual retinula cells. *J. Comp. Physiol.* **83**, 331–60.

SNYDER, A. W. (1973b). How fish detect polarized light. *Invest. Ophthalmol.* **12**, 78–9.

SNYDER, A. W. (1979). The physics of vision in compound eyes. In *Handbook of Sensory Physiology*, vol. VII/6A, ed. H. Autrum, pp. 225–313. Berlin, Heidelberg, New York: Springer-Verlag.

SNYDER, A. W. & MENZEL, R. (eds.) (1975). *Photoreceptor Optics*. Berlin, Heidelberg, New York: Springer-Verlag.

SNYDER, A. W., MENZEL, R. & LAUGHLIN, S. B. (1973). Structure and function of the fused rhabdom. *J. Comp. Physiol.* **87**, 99–135.

SOMMER, E. (1979). Untersuchungen zur topografischen Anatomie der Retina und zur Sehfeldtopologie im Auge der Honigbiene, *Apis mellifera* (Hymenoptera). Dissertation Universität Zürich.

STANGE, G. (1981). The ocellar component of flight equilibrium control in dragonflies. *J. Comp. Physiol.* **141**, 335–47.

STANGE, G. & HOWARD, J. (1979). An ocellar dorsal light response in a dragonfly. *J. Exp. Biol.* **83**, 351–5.

STEPHENS, G. C., FINGERMAN, M. & BROWN, F. A. (1953). Orientation of *Drosophila* to plane polarized light. *Ann. Ent. Soc. America*, **46**, 75–83.

STOCKHAMMER, K. (1956). Zur Wahrnehmung der Schwingungsrichtung linear polarisierten Lichtes bei Insekten. *Z. Vergl. Physiol.* **38**, 30–83.

STOCKHAMMER, K. (1959). Die Orientierung nach der Schwingungsrichtung linear polarisierten Lichtes und ihre sinnesphysiologischen Grundlagen. *Erg. Biol.* **21**, 23–56.

STOWE, S. & LEGGETT, M. (1978). Retina-lamina connectivity and polarization sensitivity in Crustacea. *Vision Res.* **18**, 1087.

STRUTT, J. W. (Lord Rayleigh) (1871). On the light from the sky, its polarization and colour. *Phil. Mag.* **41**, 107–20, 274–9.

SUGAWARA, K., KATAGIRI, Y. & TOMITA, T. (1971). Polarized light responses from octopus single retinular cells. *J. Fac. Sci. Hokkaido Univ. Ser. 6 (Zool.)* **17**, 581–6.

TAYLOR, C. P. (1981). Contribution of compound eyes and ocelli to steering of locusts in flight. I. Behavioural analysis. *J. Exp. Biol.* **93**, 1–18.

TAYLOR, D. H. & ADLER, K. (1978). The pineal body: site of extraocular perception of celestial cues for orientation in the tiger salamander. *J. Comp. Physiol.* **124**, 357–61.

TAYLOR, D. H. & AUBURN, J. S. (1978). Orientation of amphibians by linearly polarized light In. *Animal Migration, Navigation, and Homing*, ed. K. Schmidt-Koenig & W. T. Keeton, pp. 334–346. Berlin, Heidelberg, New York: Springer-Verlag.

TIMOFEEVA, V. A. (1962). Spatial distribution of the degree of polarization of natural light in the sea. *Bull. Acad. Sci. USSR, Geophys. Ser.* **12**, 1843–51.

TRUJILLO-CENOZ, O. & BERNARD, G. D. (1972). Some aspects of the retinal organization of *Sympychus lineatus* (Diptera, Dolichopodidae). *J. Ultrastruct. Res.* **38**, 149–60.

VERKHOVSKAYA, I. N. (1940). The influence of polarized light on the phototaxis of certain organisms. *Bull. Moscow Nat. Hist. Soc., Biol. Sect.* **49**, 101–13.

VRIES, H. DE, SPOOR, A. & JIELOF, R. (1953). Properties of the eye with respect to polarized light. *Physica* **19**, 419–32.

WATERMAN, T. H. (1960). Interaction of polarized light and turbidity in the orientation of *Daphnia* and *Mysidium*. *Z. Vergl. Physiol.* **43**, 149–72.

WATERMAN, T. H. (1973). Responses to polarized light: animals. In *Biology Data Book*, vol. 2, ed. P. L. Altman, D. S. Dittmer, pp. 1272–89. Bethesda: Fed. Amer. Soc. Exp. Biol.

WATERMAN, T. H. (1975). Natural polarized light and e-vector discrimination by vertebrates. In *Light as an Ecological Factor*, vol. II, ed. G. C. Evans, R. Bainbridge & O. Rackham, pp. 305–35. Oxford: Blackwell.

WATERMAN, T. H. (1981). Polarization sensitivity. In *Handbook of Sensory Physiology*, Vol. VII/6B, ed. H. Autrum; pp. 281–469. Berlin, Heidelberg, New York: Springer-Verlag.

WATERMAN, T. H. & FERNANDEZ, H. R. (1970). E-vector and wavelength discrimination by retinular cells of the crayfish *Procambarus*. *Z. Vergl. Physiol.* **68**, 154–74.

WEHNER, R. (1976a). Structure and function of the peripheral visual pathway in hymenopterans. In *Neural Principles in Vision*, ed. F. Zettler & R. Weiler, pp. 280–333. Berlin, Heidelberg, New York: Springer-Verlag.

WEHNER, R. (1976b). Polarized-light navigation by insects. *Scient. Amer.* **235**/1: 106–15.

WEHNER, R. (1982a). Himmelsnavigation bei Insekten. Neurophysiologie und Verhalten. *Neujahrsbl. Naturf. Ges. Zürich*, **184**, 1–132.

WEHNER, R. (1982b). The bee's celestial map. A simplified model of the outside world. In: *The Biology of Social Insects*, ed. M. D. Breed, C. D. Michener & H. E. Evans, pp. 375–9. Boulder (Colorado): Westview Press.

WEHNER, R., BERNARD, G. D. & GEIGER, E. (1975). Twisted and non-twisted rhabdoms and their significance for polarization detection in the bee. *J. Comp. Physiol.* **104**, 225–45.

WEHNER, R., LANFRANCONI, B. (1981). What do the ants know about the rotation of the sky? *Nature*, London, **293**, 731–3.

WEHNER, R. & MEYER, E. (1981). Rhabdomeric twist in bees – artefact or *in-vivo* structure? *J. Comp. Physiol.* **142**, 1–17.

WELLINGTON, W. G. (1974). Changes in mosquito flight associated with natural changes in polarized light. *Canad. Entomol.* **106**, 941–8.

WILSON, M. (1978). The functional organization of locust ocelli. *J. Comp. Physiol.* **124**, 297–316.

WOLF, R., GEBHARDT, B., GADEMANN, R. & HEISENBERG, M. (1980). Polarization sensitivity of course control in *Drosophila melanogaster*. *J. Comp. Physiol.* **139**, 177–91.

WOLSTENCROFT, R. D. (1974). The circular polarization of light reflected from certain optically active surfaces. In *Planets, Stars, and Nebulae*, ed. T. Gehrels, pp. 495–9. Tucson: University of Arizona Press.

YAMAMOTO, T., TASAKI, K., SUGAWARA, Y. & TONOSAKI, A. (1965). Fine structure of the *Octopus* retina. *J. Cell. Biol.* **25**, 345–59.

ZOLOTOV, V. V. & FRANTSEVICH, L. I. (1973). Orientation of bees by the polarized light of a limited area of the sky. *J. Comp. Physiol.* **85**, 25–36.

ZOLOTOV, V. V. & FRANTSEVICH, L. I. (1977). Polarized light sensitive interneurons in insects. *Neurophysiol. (Kiev)* **9**, 397–401. (In Russian).

ZONANA, H. V. (1961). Fine structure of the squid retina. *Hosp. Johns Hopkins Bull.* **109**, 185–205.

WAVELENGTH PERCEPTION AND COLOUR VISION

DIETRICH BURKHARDT

University of Regensburg, Department of Zoology, Universitätsstrasse 31,
D 8400 Regensburg, GFR

The intention here is not to review the extensive material and the detailed data on spectral response curves contributed by electrophysiological studies of either the peripheral or central nervous systems of vertebrates and arthropods, nor to review fully the flood of data presented by biochemical and microphotometric analysis of visual pigments – and certainly not to review the huge amount of data arising from the psychophysics of vision. The target is the following: to focus interest on the biological context of wavelength perception and colour vision in comparative animal physiology, and to give some useful definitions, even if they are redundant to those who are familiar with this field of physiology. Thus, after a few remarks regarding the light available in some habitats, the range of wavelengths perceived by animals living in these habitats, and the problem of different regions of the spectrum governing different behavioural patterns in some animals, the major part will deal with some definitions and a few examples of true colour vision that have been investigated more thoroughly. The last part will be devoted to some speculations and open questions.

Spectral densities of lights, wavelength perceived by animals and wavelength-discriminating behaviour

Spectral flux distribution curves of natural lights differ greatly, whether from radiant or reflecting sources, depending on many parameters such as time of day and weather conditions; useful surveys are given by Gates (1980), Lythgoe (1972), Menzel (1979) and Waterman (1981). Average daylight has an overall spectral flux distribution with several peaks in the ultraviolet (UV) and the blue due to the contribution of scattered skylight, and peaks in the longer wavelength region due to the incident sunlight. Animals can be expected to be adapted to their natural light habitats. An example of an animal's spectral sensitivity matched nicely

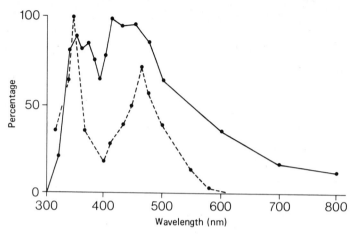

Fig. 1. Spectral flux distribution given as percentages of the maximum (solid line) for typical daylight in comparison to the spectral sensitivity as a percentage of the maximum (dotted line) for a green receptor cell (R1–6) of the blowfly *Calliphora erythrocephala* – representative of the average sensitivity of the compound eye as a whole. (Receptor sensitivity: D. Burkhardt, unpublished; daylight spectrum: adapted from D'Ans & Lax, 1949.)

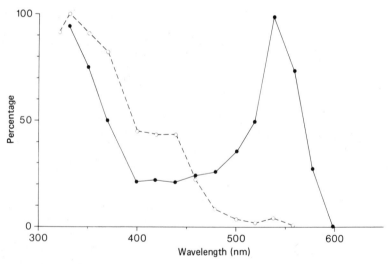

Fig. 2. Spectral sensitivity as a percentage of the maximum for a UV-receptor cell in the large dorsal part of the compound eye of a male *Bibio marci*, representative for the average sensitivity of this eye part (dotted line). For comparison the sensitivity of a green receptor cell in the small ventral part of this eye is given (solid line). (D. Burkhardt & I. de la Motte, unpublished.)

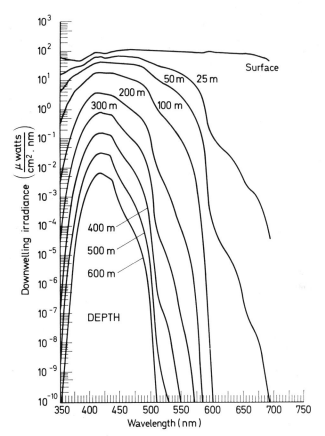

Fig. 3. Spectral flux distribution of downwelling irradiance in clear oceanic water at different depths below the surface. Note the logarithmic scale of the ordinate. (From Smith & Tyler, 1967.)

with the overall spectral flux distribution of the daylight is the compound eye of the blowfly *Calliphora* (Fig. 1).

In some insects the dorsal part of the eye is specialised for the perception of the UV, in the context of orientation with respect to the sky (possibly to the horizon or the ground, and thus also to gravity), or with the detection of objects against the sky, or finally the discrimination of the e-vector of polarised sky light (see previous chapter). When recording from the dorsal part of the compound eye of *Bibio* males with microelectrodes either the gross electroretinogram or the responses of single visual cells, we found only UV- and blue-sensitive cells (Fig. 2) (Burkhardt & de la Motte, 1972, I. de la Motte & D. Burkhardt, unpublished). *Ascalaphus* (Gogala, 1967) and the dronefly (Schatz, 1971, Bishop, 1974) are among many other examples that can be cited.

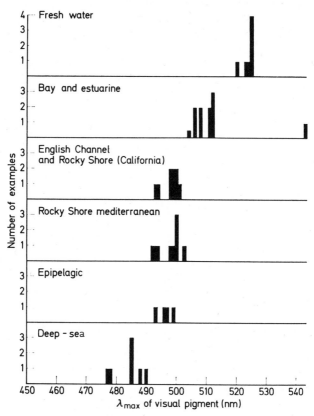

Fig. 4. Values for λ_{max} of visual pigments extracted from eyes of fishes caught in various types of natural water. (From Lythgoe, 1972.)

Animals living under water receive their daylight filtered according to depth. With increasing distance from the surface only a band of blue light peaking around 450 nm finally remains (Fig. 3). One can assemble a nice correlation between spectral sensitivity peaks and preferred depth in many fishes (Fig. 4).

As soil and sand do not reflect short wavelengths, light reflected from the ground in such surroundings has a high density in the yellow and red parts of the spectrum only. The desert ant *Cataglyphis* seems to be well adapted to these light conditions, its spectral sensitivity extending into the long wavelength range of the spectrum (Wehner & Toggweiler, 1972; Kretz, 1977).

Another habitat with special wavelength distribution is dense vegetation. As green leaves absorb nearly all the UV and blue light and in addition the orange-red part of the spectrum, animals living under the

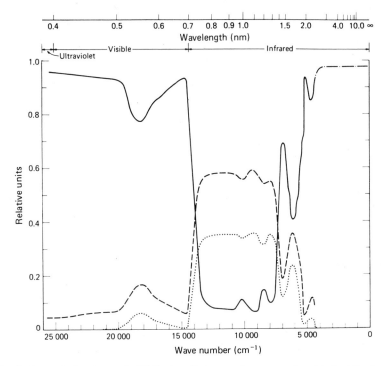

Fig. 5. Spectral absorbance (solid line), reflectance of upper surface (dashed line) and transmittance (dotted line) of green leaves of *Nerium oleander*, given in relative units. (From Gates, 1980.)

canopy of a forest receive mostly green light peaking around 520 nm and the long wavelength range above 700 nm. Some tropical insects, *Campo-notus gigas* and *Cyrtodiopsis sp.*, for example, seem to be able to use the far-red end of the spectrum (Burkhardt, 1972, D. Burkhardt & I. de la Motte, unpublished). In contrast, light reflected by green leaves contains a certain amount of UV (Fig. 5), which is of importance in hymenopteran colour vision as we shall see later.

Some animals use luminescence organs for intraspecific communication. Lall's group (for example Lall, 1981) and our group have both investigated the spectral sensitivity curves of the eyes of fire-flies and compared these with the spectral density of the light emitted. In spite of the more extensive and accurate work of Lall, I present here our unpublished results on the European species *Phausis splendidula*, because they show how mistaken it is to refrain from further considerations if one finds a neat and expected correlation in a narrow context! The spectral sensitivity of *Phausis* males shows a reasonable congruence with the bioluminescence emission curve of the female as analysed by

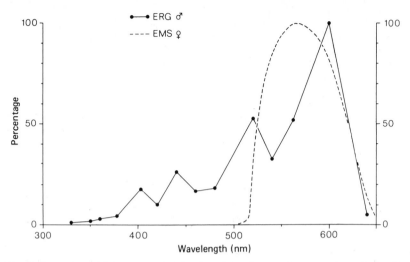

Fig. 6. Spectral sensitivity as revealed by size of the electroretinogram of a male fire-fly, *Phausis splendidula*, and spectral emission of the female's luminescence. (Spectral sensitivity: D. Burkhardt & M. Scheubeck, unpublished; luminescence: from Schwalb, 1961.)

Schwalb (1961); both curves peak between 570 and 600 nm (Fig. 6). However, the disturbing yet intriguing fact about *Phausis* is that in experiments with artificial lights of different wavelengths, performed by Schwalb, *Phausis* males were attracted much more by blue lights than by yellow ones, or even by the light of their own females.

There seems to be general agreement that, in animals equipped with only one type of receptor with regard to spectral sensitivity, it must be possible to adjust intensities at different wavelengths to elicit equal reactions. This will be true as long as the spectral response curve does not change with intensity, and if the exposure–response curves obtained at different wavelengths have the same shape and are only shifted relative to each other with respect to the x-axis (univariance principle).

While this view in general might be correct, there are possible exceptions. For example, if screening pigments interfere, their location in front or between the visual cells may alter the spectral flux distribution of the light reaching the visual pigments. If the state of the screening pigments changes with light intensity, this will have influence on the shape of the spectral response curve of the visual cells. It does not matter whether the movements of the screening pigments are activated by light stimulating directly the pigment cells or by an efferent nervous control mediated via the visual cells and the central nervous system. A

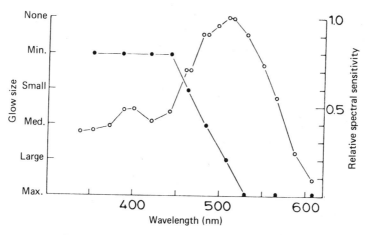

Fig. 7. Spectral efficiency of screening pigment migration (●) and spectral sensitivity of the retina (○) of the moth *Deilephila elpenor*. The left-hand scale is in arbitrary units for classes of pupil diameters. (From Hamdorf & Höglund, 1981.)

mechanism involving different spectral responses of pigment and visual cells has been described recently by Hamdorf & Höglund (1981) (Fig. 7). Formerly, positive results of selective adaptation experiments, that is a change in the spectral response curve after adapting the eye with a particular wavelength, had been considered to be proof of the existence of more than one receptor type. However, this view is incorrect since we now know that in invertebrates the metarhodopsins are thermostable and may be reconverted into the active rhodopsin pigment by absorption of specific quanta. Irradiating an eye will change the equilibrium between the active rhodopsin and the metarhodopsin, depending on the wavelength of the light used within the spectrum (for review cf. Hamdorf, 1979). Schneider, Gogala, Draslar, Langer & Schlecht (1978) have shown in *Ascalaphus*, a Neuropteran insect, that there are two types of screening pigments with different absorption properties. They form two concentric apertures. The wider aperture is formed by those screening pigments that absorb UV and most of the visible part of the spectrum, the smaller aperture by pigment grana that absorb the UV only. Thus the amount of UV reaching the visual cells and absorbed there by the UV-sensitive rhodopsin is greatly reduced, while the amount of blue light reconverting the metarhodopsin into the active visual pigment is less reduced.

Animals with more than one type of spectral sensitivity among their photoreceptors may show wavelength-specific behaviour patterns.

Furthermore, they are at least potential candidates for colour vision. Again, one should bear in mind alternative possibilities. Wavelength-specific behaviour could also be mediated by a receptor-system not obeying the univariance principle. Animals with more than one receptor-system, and capable of wavelength-specific behaviour, may still be colour blind if the two receptor-systems work independently. Animals that do have true colour vision may exhibit merely wavelength-specific behaviour or even act as though colour blind in certain behavioural contexts. Finally, the possession of different types of receptors could simply serve the purpose of enlarging the width of the usable part of the visible spectrum.

Wavelength-specific behaviour means that certain reactions are triggered within a specific (but often broad) range of wavelengths only. This is well known for organisms such as the prokaryotic halobacterians (cf. Hildebrand and Colembetti & Lenci, this volume). We observed a striking demonstration of wavelength-specific behaviour when looking for colour vision in blowflies (J. Kiepenheuer & D. Burkhardt, unpublished). Well-fed flies had to choose between two differing spectral lights in phototactic runs, then intensities were adjusted for equal choice frequencies but hungry flies suddenly preferred significantly the blue stimulus! Later on Menne & Spatz (1977) proved wavelength-specific behaviour in *Drosophila* (but not colour vision as the title of their paper suggests).

Since I started probing single visual cells in flies (Burkhardt, 1962) and found the three types necessary for colour vision, there have been several attempts to prove true colour vision in flies. Perhaps those flies tested in behavioural experiments are not the right species? Having watched last year an ototid fly in Malaysia, I was encouraged that there is still some chance to establish true colour vision. This particular animal is spectacularly coloured, with a yellow head and large yellow proboscis, and contrasting bright orange eyes. A shiny brilliant blue-green abdomen behind a dark blue thorax is surrounded by spread wings of dark brown with transparent fenestration and an ochre spot at the leading edge. The animal showed a continuous display of spread waving wings and a repetitive protrusion of its wattle-like proboscis, apparently to attract possible mates. A very good example that demonstrates the variety of possible behavioural responses is the bee: depending on light conditions and behavioural context it can exhibit true colour vision, wavelength-specific behaviour or even colour blindness. An instructive summary is given by Menzel (1979), Fig. 8, to which should be added: (i) According to Kaiser's experiments (1972) the bee's opto-motor

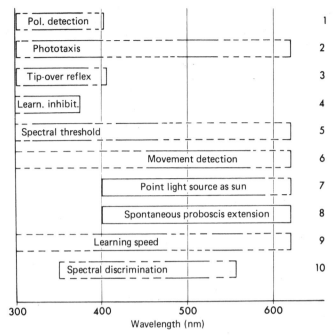

Fig. 8. Spectral ranges in which different types of wavelength-specific behaviour may be elicited in the honey bee. Ranges shown as solid lines are strong responses, those as dashed lines are weak responses. (From Menzel, 1979.)

response is independent of wavelength; its spectral sensitivity is mainly governed by the green receptor-type, yet there is also some minor contribution from the other receptors involved. (ii) Menzel (1981) proved that the colour vision of the honey bee ceases at low intensities, which closely resembles the situation in man where a switch from photopic to scotopic vision is present. All this indicates only the flexibility of information processing in the neuronal circuits of the optical pathways, and in the higher centres which finally deliver the motor output for the resulting behavioural response.

Before turning to the next section, I should like to make a few remarks about the question of wavelength-specific behaviour and colour vision in molluscs. Even in eyeless clams wavelength-specific reactions have been described. There are many molluscs having attractively coloured shells or brightly coloured bodies which certainly could be defence or camouflage colours; and there are the cephalopods, highly developed animals whose behaviour depends so much on optical inputs, for example they change their body colours to match the coloured background: yet there is no evidence of colour vision! In *Octopus* eyes only one pigment has been found and all experimental results indicate that the animal is colour blind, thus it has already been suggested that researchers stop further fruitless efforts to establish possible colour vision (Messenger, 1977, 1981).

Colour vision

Criteria of colour vision

True colour vision, in contrast to wavelength-specific behaviour, is characterised by a number of special features that are defined by physiological criteria rather than in terms of physics.

(1) Colour is to be distinguished from all shades of so-called grey stimuli, the extremes being white and black. The quality 'white' is attributed to the spectral flux distribution of daylight (in experiments it can be replaced fairly well by the emission of a xenon arc).

(2) Different regions within the visible range of wavelengths are evaluated as different colours or hues. The wavelength difference which makes the hues of two monochromatic spectral radiations distinguishable from each other changes with the position in the spectrum, and is described by the spectral discrimination function.

(3) The hues can be arranged according to their affinities in a linear manner.

(4) Mixtures of two different wavelengths may result in a hue differing from that of either components.

(5) Mixtures of the short and the long wavelength ends of the visible range are evaluated as hues not present in the spectrum. Usually they are called purple, a quality which may be species-specific.

(6) The monochromatic spectral radiations can be arranged on the perimeter of a chromaticity diagram, the purple range closing it (Fig. 9).

(7) Monochromatic spectral radiation or purple of equal quanta are not equal in brightness, but their brightness can be compared with that of a certain intensity of white light. This brightness of colours is described by the spectral brightness curve.

(8) Different hues on the perimeter of the chromaticity diagram are saturated to different degrees, that is they show differing affinities to white.

(9) Mixtures of monochromatic lights, purple included, are always less saturated than colours located on the perimeter. This means that such a mixture is equivalent to an intermediate wavelength mixed

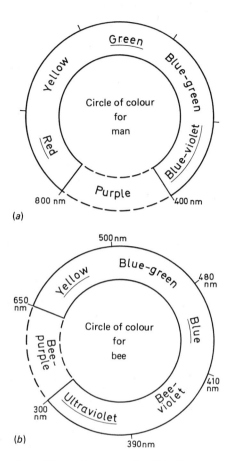

Fig. 9. Colour circle of man (*a*) and of the honey bee (*b*). (From Autrum & Thomas, 1973, based on data of Daumer, 1956.)

appropriately with white. Increasingly unsaturated hues approach a singular point within the diagram representing white.

(10) Pairs of colours lying at the intersection points of any straight line passing the white point and cutting the perimeter are called complementary colours. Mixing them in appropriate amounts results in white sensation, and filtering one of the pair from white light leaves the complementary colour.

(11) Any position within the chromaticity diagram represents a single defined hue which can be produced by a variety of appropriate mixtures of different spectral components.

(12) True and non-deficient colour vision along the whole visible spectral range (that is no part of the spectrum is mistaken for grey) can

be achieved if at least three types of photoreceptor are interacting, each of them with its own unique sensitivity function.

If we investigate colour vision, we should ask for the signal or communication value of colours in that particular species investigated. This kind of questioning was one of the inspirations in von Frisch's wonderful physiological analysis of colour-oriented behaviour in bees – he would not believe the flower colours to be without significance for their pollinators.

Colour vision in any animal can only be proved by behavioural experiments, the first point of the list given above being the crucial one. All the other approaches, whether biochemical, photochemical, electrophysiological or morphological, can only give the clue that the necessary prerequisite (a sufficient number of different receptor-types) is present, and they can help to analyse the mechanisms underlying colour vision. If an animal fails to show colour vision in spite of the necessary receptor types being present, this does not necessarily mean the animal is colour blind. At best, an accumulation of negative evidence resulting from different approaches can be taken as a measure of the unlikeliness of functioning colour vision in that particular animal.

Since the comparative study of colour vision was started at the very beginning of the century by von Frisch (1911 in fishes, 1914 in bees) some fundamental questions have been settled. Yet there are only a few cases of true colour vision that have been investigated thoroughly. The study of colour vision involves analysing an animal's behaviour in terms of responses to monochromatic lights and mixtures of them. The outcome will depend critically on the work of a patient scientist with some knowledge of physics, who is intimately familiar with an experimental animal that exhibits a stereotyped response, depending critically on the spectral distribution of stimulating light. These conditions apparently are seldom to be found.

Example for colour vision analysis

In order to analyse the picture perceived by an animal capable of colour vision, we must know both the characteristics of the animal's chromaticity diagram and the spectral characteristics of the stimulus pattern eliciting the behavioural response. A pupil of von Frisch, Karl Daumer (1956 and 1958), completed von Frisch's work by adding his sophisticated investigations with trained bees, using colourimetric analysis (on which basis a chromaticity diagram was later constructed), and measuring the reflectance of flowers within the sensitivity ranges of the postulated three receptor-types (cf. Fig. 9). One point of special interest

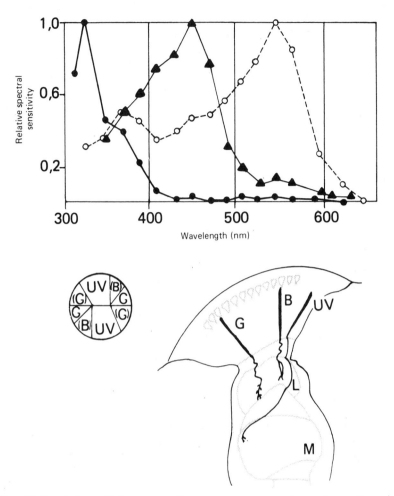

Fig. 10. Spectral sensitivity curves of three receptor types in the honey bee, and localisation of the receptor cells in the retinula cross section and their projection into the optic ganglia. L, lamina; M, medulla. (From Menzel, 1979.)

in this well-known material is the large contribution of the UV receptors, which makes the UV not only the peak of the spectral brightness curve but also the most saturated of all spectral lights. The addition of only 2% UV quanta to a monochromatic light of the yellow region of the spectrum makes this mixture distinguishable as a purple hue (called purple I) from the pure yellow. In contrast, if yellow quanta are added to monochromatic UV light about 50% must be added before this mixture is distinguishable from the pure UV as a different hue (called purple II). Thus yellow is an ineffective 'ingredient' in colour

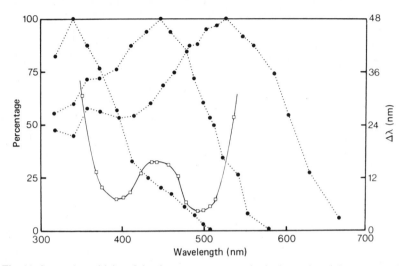

Fig. 11. Spectral sensitivity of the three receptor types in the honey bee (after Autrum & Zwehl, 1964) compared with the wavelength discrimination function (after von Helversen, 1972). Note: wavelength discrimination is best at the intersection of the flanks of neighbouring receptor curves. Spectral sensitivities (dashed lines) are given as percentages of the maximum.

mixtures and hence called highly unsaturated, while UV is powerful and hence highly saturated. The high saturation of UV explains the importance of the weak UV reflection of green leaves. As the green region is complementary to the UV in bees, and as it is a more unsaturated hue, green leaves look nearly uncoloured, that is grey to the bee. Therefore green or yellow-green flowers (for example flowers of spurge) which do not reflect UV, contrast as coloured spots against the greyish leaves; perhaps even with enhanced contrast, for bees do have simultaneous colour contrast as already shown by Kühn (1927) and recently reinvestigated by Neumeyer (1980, 1981). The results of his experiments led Daumer to postulate three types of receptors, which were indeed found at about the right wavelength positions by Autrum & Zwehl (1964). Subsequently Gribakin (1969, 1972) and Menzel & Blakers (1976) began with the investigation of receptor-type distribution within the ommatidial array (Fig. 10).

Menzel (1974), Kien & Menzel (1977a,b), Hertel (1980) and Riehle (1981) analysed the chromatic properties of interneurons. Meanwhile, in 1972, von Helversen, again using a behavioural approach, reinvestigated the wavelength-discrimination function, and found two $\Delta\lambda$ minima as low as 5 nm (Fig. 11). The lowest $\Delta\lambda$ values were found at the intersection points of the steep flanks of neighbouring receptor curves.

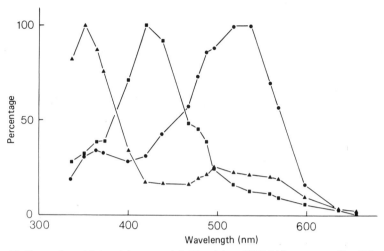

Fig. 12. Spectral sensitivity of the green (●), blue (■) and UV (▲) receptor cells of *Vespa rufa*, compound eye. Ordinate: sensitivities are given as percentages of the maximum. (D. Burkhardt & F. W. Räber, unpublished.)

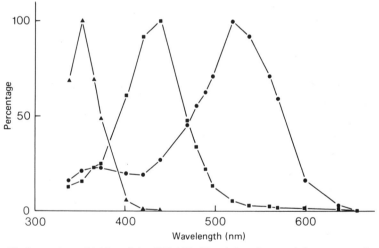

Fig. 13. Spectral sensitivities of the UV (▲), blue (■) and green (●) receptor cells of *Bombus* sp. (D. Burkhardt & F. W. Räber, unpublished.)

In other hymenopterans closely related to the bee, for example wasps and bumble-bees, our knowledge is incomplete. Behavioural work of Beier & Menzel (1972) indicates that colour vision in wasps is similar to that in bees. The existence of three receptor-types with peaks similarly positioned to those in bees was inferred by Menzel (1971), using the electroretinogram. Recently F. W. Räber in my laboratory (unpublished) started investigations with microelectrodes, and has presently demonstrated at least three types of cells with different spectral

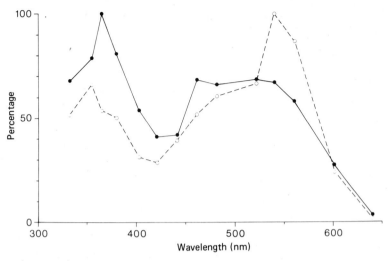

Fig. 14. Spectral sensitivity as revealed by the electroretinogram of *Bombus lucorum* in different regions of the compound eye. Sensitivities given as percentages of the maxima. Solid line: most ventral part of the eye; dashed line: dorsal to that region. (D. Burkhardt & M. Scheubeck, unpublished).

sensitivity; these peak at 360 nm, between 420 and 460 nm, and at 530 nm, respectively (Fig. 12).

Worse still is the situation in bumble-bees for which colorimetric work using behavioural responses is totally lacking. The sensitivity curves of three receptor-types so far identified in our laboratory (Fig. 13) fit very well with the photometric results of Bernard & Stavenga (1978) yet differ in some respects from those of Meyer-Rochow (1980). Electroretinogram recordings from different eye regions in *Bombus lucorum* (D. Burkhardt & M. Scheubeck, unpublished; Scheubeck, 1980) indicate that the ventral eye region is more sensitive to UV than the more equatorial regions (Fig. 14). This seems reasonable in the context of the detection of nectar guides, which in many flowers are borders between UV reflectant and non-reflectant regions of the petals. The similarity of the receptor responses in the bee, in wasps and in bumble-bees (compare Figs. 10–13) implies that colour vision in these genera of the Hymenoptera is the same.

Another group of social Hymenoptera, namely the ants, have been investigated nearly as extensively as bees. In 1968 Kiepenheuer found wavelength-specific behaviour in *Formica*, and in 1972 Wehner & Toggweiler proved true colour vision for ants. In *Cataglyphis* the wavelength discrimination threshold proved to be low, UV is a spectral

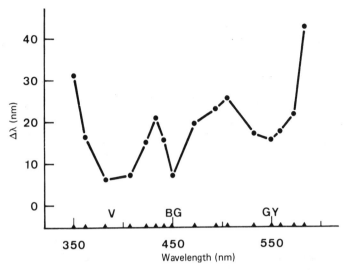

Fig. 15. Wavelength discrimination of the desert ant *Cataglyphis bicolor* for 60% positive response. (From Kretz, 1979.)

region of high effectiveness; and the visible range extends to the long wavelength region.

An extensive and thorough study of *Cataglyphis* using behavioural responses and colourimetric methods was done by Kretz (1977, 1979). In contrast to the situation in bees he found not two, but three minima in the wavelength discrimination function (Fig. 15) and not three, but four peaks in the spectral brightness function. These findings together with the results of selective adaptation were combined to form the quadrilateral chromaticity diagram of *Cataglyphis* (Fig. 16), and indicate the participation of four receptor-types peaking at 345, 430, 505 and 570 nm. Microelectrode probing by Mote & Wehner (1980) has already provided proof of two of them, namely those peaking in the UV and the green regions. In Lepidoptera, another insect group which is sensitive to red light, evidence for long-wavelength receptors is accumulating quickly: in the Noctuid *Spodoptera exemata*, Langer, Hamann & Meinicke (1979) found four visual pigments, one of them peaking roughly at 570 nm. Bernard (1979) showed for nine species of butterflies the existence of visual pigments maximally sensitive around 610 nm. Peaks this far into the red region have not very often been found in vertebrates.

However, the situation concerning behavioural studies of colour vision in butterflies is still unsatisfactory. From the work of Ilse (1928)

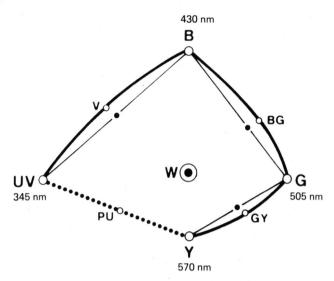

Fig. 16. Chromaticity diagram of the ant *Cataglyphis bicolor*. Tentative receptors peaks marked by their λ_{max}. Primary colours: Y (yellow), G (green), B (blue) and UV (ultraviolet). Secondary colours: GY (yellow-green region), BG (blue-green region), V (violet region), PU (purple region), W (white point). (From Kretz, 1979.)

we do know that they see colours. Magnus (1953) and many others proved that colour vision plays an important role during mating behaviour, and recently Swihart, C. A. & Swihart, S. L. (1970) and Swihart (1971) showed that they are also able to learn colours. Lutz (1924), later Crane (1954) and many others dealt with UV reflecting patterns in butterflies, which may also show a sexual dimorphism (for example: Obara & Hidaka, 1968; also Wehner, 1981). A tentative chromaticity diagram, based on the absorption curves of the contribut-ing visual pigments was constructed by Schlecht (1979) for the moth *Deilephila elpenor*. However, satisfactory behavioural studies to com-plete the picture are still missing. Nevertheless it is certain that at least some insects (for example, ants and butterflies) do extend their visible spectrum and colour vision as far into the red as do vertebrates.

During the last decade increasing evidence has accumulated that UV vision is present in some vertebrates. A work of Schiemenz published in 1924 in the very first volume of the *Zeitschrift für vergleichende Physiologie* (now the *Journal of Comparative Physiology*) evidently is almost forgotten. He trained minnows (*Phoxinus phoxinus*) to search for food at patches of spectral lights and found not only that they could visualise UV down at least to 365 nm but, in addition, were able to distinguish this as a hue differing from violet. In 1972 Huth & Burkhardt

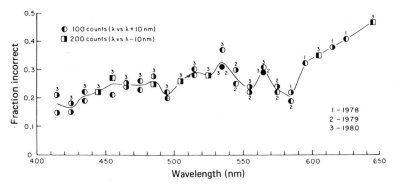

Fig. 17. Wavelength discrimination function of a hummingbird (*Archilochus alexandri*). Ordinate: fraction of incorrect choices for ±10 nm *Δλ* intervals. Note: the UV region is not shown, yet the birds are able to discriminate between UV and the short-wavelength region of the spectrum visible to man. (From Goldsmith, Collins & Perlman, 1981.)

found that hummingbirds could be trained to UV stimuli when visiting artificial feeders, while their red vision is poorer than in man.

In the same year UV sensitivity was proved in both pigeons (Wright, 1972), and toad *Bufo* (Dietz, 1972); shortly thereafter UV vision was claimed for lizards by Moehn (1974). Thus it seems likely that, with the exception of perhaps the mammals, the visible range of many vertebrates may extend nearly as far into the UV as it does in insects. Extensive work was done on birds during the seventies: Bowmaker (1977) determined the spectral properties of three visual pigments in the pigeon (peaking at 460, 515 and 567 nm) and the absorbance curves of the oil droplets which act as cut-off filters for long-wavelength light. One type of oil droplets (clear type) does not absorb visible light. Graf & Norren (1974) and Norren (1975), by means of electrophysiological methods, found two blue-sensitive systems peaking at 400 and 420 nm in the retina of the pigeon, chicken and daw (*Corvus monedula*). Fager, L. Y. & Fager, R. S. (1980) identified in the cone fraction of chicken an iodopsin-like pigment, peaking at 417 nm. In behavioural experiments, the wavelength-discrimination of pigeons was analysed by Emmerton & Delius (1980). Investigating the whole range between 360 and 660 nm they found minima at 375, 460, 530 and 595 nm. This means wavelength discrimination is especially good there with *Δλ* values of less than 10 nm. The existence of four minima indicates the participation of at least four receptor types, suggesting tetrachromatic vision.

Finally hummingbirds have been examined with respect to their wavelength discrimination function including the UV by Goldsmith, T. H. & Goldsmith, K. M. (1979), Goldsmith (1980) and Goldsmith,

Collins & Perlman (1981). This is comparable with that of pigeons, but reveals a tendency towards increasingly better discrimination at shorter wavelengths (Fig. 17).

Concluding remarks

A wide range of wavelengths from 700 nm down to ultraviolet of 300 nm is used for colour vision in some insects, as well as in birds. It is tempting to speculate about some other parallels between vertebrate and arthropod colour vision and their underlying mechanisms. For example, chromatic aberration of the lens can be compensated by different depth position of spectral receptor types in the retinae of fishes and spiders (Eberle, 1967; Eakin & Brandenburger, 1971). In vertebrates as well as in insects, colour vision may be restricted to specialised parts of the retina, in human eyes the periphery being colour blind. In recent investigations Franceschini, Hardie, Ribi & Kirschfeld (1981) and Hardie, Franceschini, Ribi & Kirschfeld (1981) showed that, in male housefly eyes, the foveal part (used for tracking females) seems to lack two receptor-types, R7y,p, which are replaced by receptors, R7r, having the green-sensitive visual pigment normally found in retinula cells R1–6, thus rendering the fovea colour blind. A regular repetitive pattern of sets of different receptors is not only an attribute of the insect compound eye, but is also a feature of the vertebrate retina apparently being correlated with increasing performance of colour vision. Examples are provided by Wagner (1972) and Kunz (1980) for teleost fishes, by Morris (1970), most beautifully, for the chicken (Fig. 18), and even for primates by Wässle & Riemann (1978). (A hexagonal array of parafoveal cones in the retina of man has been described as early as 1866 by Schultze.)

Certainly the increasing amount of knowledge is an inspiration to settle questions which remain unanswered. One would like to know the chromaticity diagrams for butterflies, birds and, of course, fishes since the colour vision of the latter was proved by von Frisch as early as 1911. It would be interesting to know whether the chromaticity diagram in fishes living in deep yet clear water, and therefore exposed to a narrowed spectral band, differ from that of fishes living close to the surface, which are exposed to normal daylight.

It would be nice to know what the world looks like for birds, since we now know about their colour vision extending into the UV. It is likely that there will be a purple quality for them which differs from our purple. It has often been claimed that birds have an excellent colour

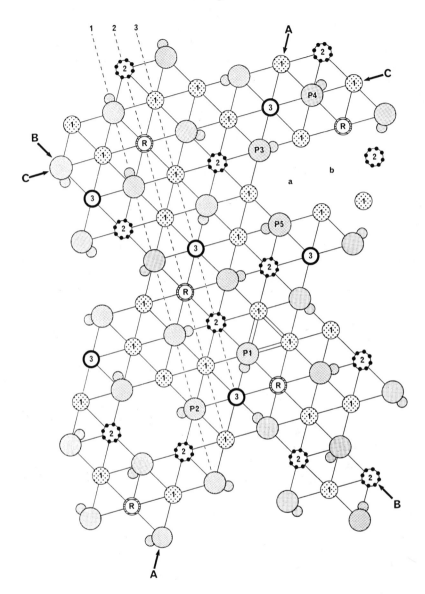

Fig. 18. Hexagonal array of different types of cones (indicated by different symbols) in the chicken retina. For details see original. (From Morris, 1970.)

Fig. 19. Photographs of a wild plum (*Prunus spinosa*) within the visible (*a*) or UV (*b*) regions (Agfapan 400). (*a*) Schott filter type KV 399, UV cut off; (*b*) Schott filter type UG 11 + BG 38, UV only. Note that the grey scale is not correct in the UV. (From Burkhardt, 1982.)

discrimination in the red yet a poor one in the blue-violet region; the many berries of orange and red colour and the red colour of many bird-pollinated flowers are two instances which can be mentioned to support that argument. Yet in some books on the ecology of pollination (for example Kugler, 1970), it is clearly stated that birds also pollinate flowers coloured blue or violet. Futhermore, a great many berries and fruits are coloured blue, violet or black. Unpublished data from our group indicates that some flowers coloured red and possibly pollinated by birds do, in addition, reflect some UV. Recently I have examined the UV reflectance of some objects which are perhaps of 'interest' to birds. Some samples of white feathers of magpies (*Pica pica*) and tits (*Parus caruleus*) do reflect UV. The brownish throat of the nuthatch (*Sitta europaea*) does not reflect UV, while most of the other feathers do (Burkhardt, 1982). Which berries reflect the UV? My preliminary results indicate that for berries of privet (*Ligustrum vulgare*), cornel (*Cornus sanguinea*), elder (*Sambucus nigra*) and guelder-rose (*Viburnum lantana*) there is not much UV reflection, yet there is in the blackberry (*Rubus caesius*) – or shall I now say UV-berry – and in the wild plum (*Prunus spinosa*) (Fig. 19) and other fruits which also have a surface bloom constituted of cuticular waxes (Burkhardt, 1982).

If one is allowed to generalise, perhaps colour vision without using the UV range is restricted to man and some mammals. Colour vision without using the 'red' range is restricted to some of the Hymenopterans and some other insects. Yet, among the insects the Lepidoptera and the ants, and among the vertebrates perhaps some fishes, Amphibians and Reptiles, and certainly the birds, make use of the whole range from UV to 'red'. For them the world must be more colourful for it includes the hues of UV and UV-purple. The significance of the UV in colour vision for intra- and interspecific communication is then still a wide field to explore.

References

ANS, D' E. & LAX, E. (1949). *Taschenbuch für Chemiker und Physiker*, 2nd edn. Berlin: Springer-Verlag.

AUTRUM, H. & THOMAS, I. (1973). Comparative physiology of colour vision in animals. In *Handbook of Sensory Physiology*, vol. VII/3, ed. R. Jung, pp. 661–92. Berlin, Heidelberg & New York: Springer-Verlag.

AUTRUM, H. & ZWEHL, V. VON (1964). Die spektrale Empfindlichkeit einzelner Sehzellen des Bienenauges. *Z. Vergl. Physiol.* **48**, 357–84.

BEIER, W. & R. MENZEL (1972). Untersuchungen über den Farbensinn der deutschen Wespe (*Paravespula germanica* F., Hymenoptera, Vespidae): Verhaltensphysiologischer Nachweis des Farbensehens. *Zool. Jahrbüch. Physiol.* **76**, 441–54.

BERNARD, G. D. (1979). Red-absorbing visual pigment of butterflies. *Science*, **203**, 1125–7.

BERNARD, G. D. & STAVENGA, D. G. (1978). Spectral sensitivities of retinular cells measured in intact, living bumblebees by an optical method. *Naturwissenschaften*, **65**, 442–3.

BISHOP, L. G. (1974). An ultraviolet photoreceptor in a dipteran compound eye. *J. Comp. Physiol.* **91**, 267–75.

BOWMAKER, J. K. (1977). The visual pigments, oil droplets and spectral sensitivity of the pigeon. *Vision Res.* **17**, 1129–38.

BURKHARDT, D. (1962). Spectral sensitivity and other response characteristics of single visual cells in the arthropod eye. *Symposia of the Society for Experimental Biology*, **16**, 86–109.

BURKHARDT, D. (1972). Electrophysiological studies on the compound eye of a stalked eye fly, *Cyrtodiopsis dalmanni* (Diopsidae, Diptera). *J. Comp. Physiol.* **81**, 203–14.

BURKHARDT, D. & DE LA MOTTE, I. (1972). Electrophysiological studies on the eyes of Diptera, Mecoptera and Hymenoptera. In *Information Processing in the Visual Systems of Arthropods*, ed. R. Wehner, pp. 147–53. Berlin, Heidelberg & New York: Springer-Verlag.

BURKHARDT, D. (1982). Birds, berries and UV, a note on some consequences of UV vision in birds. *Naturwissenschaften*, **69**, 153–7.

CRANE, J. (1954). Spectral reflectance characteristics of butterflies (Lepidoptera) from Trinidad, BWI. *Zoologica, New York*, **39**, 85–115.

DAUMER, K. (1956). Reizmetrische Untersuchung des Farbensehens der Bienen. *Z. Vergl. Physiol.* **38**, 413–78.

DAUMER, K. (1958). Blumenfarben, wie sie die Bienen sehen. *Z. Vergl. Physiol.* **41**, 49–110.

DIETZ, M. (1972). Erdkröten können UV-Licht sehen. *Naturwissenchaften*, **59**, 316.

EAKIN, R. M. & BRANDENBURGER, J. L. (1971). Fine structure of the eyes of jumping spiders. *J. Ultrastruct. Res.* **37**, 618–63.

EBERLE, H. (1967). Cone length and chromatic aberration in the eye of *Lebistes reticulatus*. *Z. Vergl. Physiol.* **57**, 172–3.

EMMERTON, J. & DELIUS, J. D. (1980). Wavelength discrimination in the 'visible' and ultraviolet spectrum by pigeons. *J. Comp. Physiol.* **141**, 47–52.

FAGER, L. Y. & FAGER, R. S. (1980). Chicken blue and Chicken violet, short wavelength sensitive visual pigments. *Vision Res.* **21**, 581–6.

FRISCH, K. VON (1911). Über den Farbensinn der Fische. *Verh. Deutsch. Zool. Ges.* 220–5.

FRISCH, K. VON (1914). Der Farbensinn und Formensinn der Bienen. *Zool. Jahrbüch. Abt. Allg. Zool. Physiol.* **35**, 1–188.

FRANCESCHINI, N., HARDIE, R., RIBI, W. & KIRSCHFELD, K. (1981). Sexual dimorphism in a photoreceptor. *Nature, London*, **291**, 241–4.

GATES, D. M. (1980). *Biophysical Ecology*. New York, Heidelberg & Berlin: Springer-Verlag.

GOGALA, M. (1967). Die spektrale Empfindlichkeit der Doppelaugen von *Ascalaphus macaronius* Scop. (Neuroptera, Ascalaphidae). *Z. Vergl. Physiol.* **57**, 232–43.

GOLDSMITH, T. H. (1980). Hummingbirds see ultraviolet light. *Science*, **207**, 786–8.

GOLDSMITH, T. H., COLLINS, J. S. & PERLMAN, D. L. (1981). A wavelength discrimination function for the hummingbird *Archilochus alexandri*. *J. Comp. Physiol.* **143**, 103–10.

GOLDSMITH, T. H. & GOLDSMITH, K. M. (1979). Discrimination of colors by the black-chinned hummingbird, *Archilochus alexandri*. *J. Comp. Physiol.* **130**, 209–20.

GRAF, V. & NORREN, D. V. (1974). A blue sensitive mechanism in the pigeon retina λ_{max}: 400 nm.

GRIBAKIN, F. G. (1969). Cellular basis of colour vision in the honey bee. *Nature, London*, **233**, 639–41.

GRIBAKIN, F. G. (1972). The distribution of the long wave photoreceptors in the compound eye of the honey bee as revealed by selective osmic staining. *Vision Res.* **12**, 1225–30.

HAMDORF, K. (1979). The physiology of invertebrate visual pigments. In *Handbook of Sensory Physiology*, vol. VII/6A, ed. H. Autrum, pp. 145–224. Berlin, Heidelberg & New York: Springer-Verlag.

HAMDORF, K. & HÖGLUND, G. (1981). Light-induced retinal screening pigment migration independent of visual cell activity. *J. Comp. Physiol.* **143**, 305–9.

HARDIE, R. C. FRANCESCHINI, N., RIBI, W. & KIRSCHFELD, K. (1981). Distribution and properties of sex-specific photoreceptors in the fly *Musca domestica. J. Comp. Physiol.* **145**, 139–52.

HELVERSEN, O. VON (1972). Zur spektralen Unterschiedsempfindlichkeit der Honigbiene. *J. Comp. Physiol.* **80**, 439–72.

HERTEL, H. (1980). Chromatic properties of identified interneurons in the optic lobes of the bee. *J. Comp. Physiol.* **137**, 215–31.

HUTH, H. H. & BURKHARDT, D. (1972). Der spektrale Sehbereich eines Violettohr-Kolibris. *Naturwissenschaften*, **59**, 650.

ILSE, D. (1928). Über den Farbensinn der Tagfalter. *Z. Vergl. Physiol.* **8**, 658–92.

KAISER, W. (1972). A preliminary report on the analysis of the optomotor system of the honey bee. In *Information Processing in the Visual System of the Arthropods*, ed. R. Wehner, pp. 167–70. Berlin, Heidelberg & New York: Springer-Verlag.

KIEN, J. & MENZEL, R. (1977*a*). Chromatic properties of interneurones in the optic lobes of the bee. I. Broad band neurons. *J. Comp. Physiol.* **113**, 17–34.

KIEN, J. & MENZEL, R. (1977*b*). Chromatic properties of interneurons in the optic lobes of the bee. II. Narrow band and colour opponent neurons. *J. Comp. Physiol.* **113**, 35–53.

KIEPENHEUER, J. (1968). Farbunterscheidungsvermögen bei der roten Waldameise *Formica polyctena* Förster. *Z. Vergl. Physiol.* **57**, 409–11.

KRETZ, R. (1977). Verhaltensphysiologische Analyse des Farbensehens der Ameise *Cataglyphis bicolor* (Formicidae, Hymenoptera). Thesis, Universität Zürich.

KRETZ, R. (1979). A behavioural analysis of colour vision in the ant *Cataglyphis bicolor* (Formicidae, Hymenoptera). *J. Comp. Physiol.* **131**, 217–33.

KÜHN, A. (1927). Über den Farbensinn der Bienen. *Z. Vergl. Physiol.* **5**, 762–800.

KUGLER, H. (1970). *Blütenökologie*, 2nd edn. Stuttgart: Gustav Fischer Verlag.

KUNZ, Y. W. (1980). Cone mosaics in a teleost retina: changes during light and dark adaptation. *Experientia*, **36**, 1371–4.

LALL, A. B. (1981). Electroretinogram and the spectral sensitivity of the compound eyes in the firefly *Photuris versicolor* (Coleoptera-Lampyridae): a correspondence between green sensitivity and species bioluminescence emission. *J. Insect Physiol.* **27**, 461–8.

LANGER, H., HAMANN, B. & MEINECKE, C. C. (1979). Tetrachromatic visual system in the moth *Spodoptera exempta* (Insecta: Noctuidae): preliminary note. *J. Comp. Physiol.* **129**, 235–9.

LUTZ, F. E. (1924). Apparently non-selective characters and combination of characters including a study of ultraviolet in relation to the flower-visiting habits of insects. *Ann. N.Y. Acad. Sci.* **29**, 181–283.

LYTHOGOE, J. N. (1972). The adaptation of visual pigments to the photic environment. In *Handbook of Sensory Physiology*, vol. VII/1, ed. H. I. A. Dartnall, pp. 566–603. New York, Heidelberg & Berlin: Springer-Verlag.

MAGNUS, D. B. (1953). Über optische 'Schlüsselreize' beim Paarungsverhalten des Kaisermantels *Argynnis paphia* (Lepidoptera, Nymphalidae). *Naturwissenschaften*, **40**, 610–11.

MENNE, D. & SPATZ, H.-C. (1977). Colour vision in *Drosophila melanogaster. J. Comp. Physiol.* **114**, 301–12.

MENZEL, R. (1971). Über den Farbensinn von *Paravespula germanica* F. (Hymenoptera): ERG und selektive Adaptation. *Z. Vergl. Physiol.* **75**, 86–104.

MENZEL, R. (1974). Spectral sensitivity of monopolar cells in the bee lamina. *J. Comp. Physiol.* **93**, 337–46.

MENZEL, R. (1979). Spectral sensitivity and color vision in invertebrates. In *Handbook of Sensory Physiology*, vol. VII/6A, ed. H. Autrum, pp. 503–80. New York, Heidelberg & Berlin: Springer-Verlag.

MENZEL, R. (1981). Achromatic vision in the honeybee at low light intensities. *J. Comp. Physiol.* **141**, 389–93.

MENZEL, R. & BLAKERS, M. (1976). Colour receptors in the bee eye – morphology and spectral sensitivity. *J. Comp. Physiol.* **108**, 11–33.

MESSENGER, J. B. (1977). Evidence that octopus is colour blind. *J. Exp. Biol.* **70**, 49–55.

MESSENGER, J. B. (1981). Comparative physiology of vision in molluscs. In *Handbook of Sensory Physiology*, vol. VII/6C, ed. H. Autrum, pp. 93–200. Berlin, Heidelberg & New York: Springer-Verlag.

MEYER-ROCHOW, V. B. (1980). Electrophysiologically determined spectral efficiencies of the compound eye and median ocellus in the bumblebee *Bombus hortorum tarhakimalainen* (Hymenoptera, Insecta). *J. Comp. Physiol.* **139**, 261–6.

MOEHN, L. D. (1974). The effect of quality of light on agonistic behavior of iguanid and agamid lizards. *J. Herpetol.* **8**, 175–83.

MORRIS, V. B. (1970). Symmetry in a receptor mosaic demonstrated in the chick from the frequencies, spacing and arrangement of the types of retinal receptor. *J. Comp. Neurol.* **140**, 359–98.

MOTE, M. I. & WEHNER, R. (1980). Functional characteristics of photoreceptors in the compound eye and ocellus of the desert ant, *Cataglyphis bicolor*. *J. Comp. Physiol.* **137**, 63–71.

NEUMEYER, CH. (1980). Simultaneous color contrast in the honeybee. *J. Comp. Physiol.* **139**, 165–76.

NEUMEYER, CH. (1981). Chromatic adaptation in the honeybee: successive color contrast and color constancy. *J. Comp. Physiol.* **144**, 543–53.

NORREN, D. V. (1975). Research Note: two short wavelength sensitive cone systems in pigeon, chicken and daw. *Vision Res.* **15**, 1164–6.

OBARA, Y. & HIDAKA, T. (1968). Recognition of the female by the male, on the basis of ultra-violet reflection, in the white cabbage butterfly, *Pieris rapae crucivora* Boisduval. *Proc. Japan Acad.* **44**, 829–32.

RIEHLE, A. (1981). Color opponent neurons of the honeybee in a heterochromatic flicker test. *J. Comp. Physiol.* **142**, 81–8.

SCHATZ, B. (1971). Über die spektrale Empfindlichkeit des Schwebfliegenauges. Elektrophysiologische Untersuchungen an *Myiatropa florea*. Thesis, Universität Frankfurt (Main).

SCHEUBECK, M. (1980). Aufbau einer Spektralapparatur mit selbsttätiger Einstellung und Regelung des quantengleichen Lichtes. Über die spektrale Empfindlichkeit des Komplexauges bei *Bombus lucorum*. Thesis, Universität Regensburg.

SCHIEMENZ, F. (1924). Über den Farbensinn der Fische. *Z. Vergl. Physiol.* **1**, 175–220.

SCHLECHT, P. (1979). Colour discrimination in dim light: an analysis of the photoreceptor arrangement in the moth *Deilephila*. *J. Comp. Physiol.* **129**, 257–67.

SCHNEIDER, L., GOGALA, M., DRASLAR, K., LANGER, H. & SCHLECHT, P. (1978). Feinstruktur und Schirmpigmenteigenschaften der Ommatidien des Doppelauges von *Ascalaphus* (Insecta, Neuroptera). *Cytobiologie*, **16**, 274–307.

SCHULTZ, M. (1866). Zur Anatomie und Physiologie der Retina. *Arch. Mikrosk. Anat.* **2**, 175–286.

SCHWALB, H. H. (1961). Beiträge zur Biologie der einheimischen Lampyriden *Lampyris noctiluca* GEOFFR. und *Phausis splendidula* LEC. und experimentelle

Analyse ihres Beutefang- und Sexualverhaltens. *Zool. Jahrbüch.*, *Abt. Systematik*, **88**, H.4, 399–550.

SMITH, R. S. & TYLER, J. E. (1967). Optical properties of clear natural water. *J. Opt. Soc. Amer.* **57**, 589–95.

SWIHART, C. A. (1971). Colour discrimination by the butterfly, *Heliconius charitonius* Linn. *Anim. Behav.* **19**, 156–64.

SWIHART, C. A. & SWIHART, S. L. (1970). Colour selection and learned feeding preferences in the butterfly, *Heliconius charitonius* Linn. *Anim. Behav.* **18**, 60–4.

WÄSSLE, H. & RIEMANN, H. J. (1978). The mosaic of nerve cells in the mammalian retina. *Proc. Roy. Soc. London, Ser. B*, **200**, 441–61.

WAGNER, H.-J. (1972). Vergleichende Untersuchungen über das Muster der Sehzellen und Horizontalen in der Teleostier-Retina (Pisces). *Z. Morphol. Tiere*, **72**, 77–130.

WATERMAN, T. H. (1981). Polarization Sensitivity. In *Handbook of Sensory Physiology*, vol. VII/6B, ed. H. Autrum, pp. 281–469. New York, Heidelberg & Berlin: Springer-Verlag.

WEHNER, R. (1981). Spatial vision in arthropods. In *Handbook of Sensory Physiology*, vol. VII/6C, ed. H. Autrum, pp. 287–616. Berlin, Heidelberg & New York: Springer-Verlag.

WEHNER, R. & TOGGWEILER, F. (1972). Verhaltensphysiologischer Nachweis des Farbensehens bei *Cataglyphis bicolor* (Formicidae, Hymenoptera). *J. Comp. Physiol.* **77**, 239–55.

WRIGHT, A. A. (1972). The influence of ultraviolet radiation on the pigeons's color discrimination. *J. Exp. Anal. Behav.* **17**, 325–37.

PHOTORECEPTION AND PHOTOMOVEMENTS IN MICRO-ORGANISMS

GIULIANO COLOMBETTI and *FRANCESCO LENCI*

Istituto Biofisica CNR – Via S. Lorenzo, 26 – 56100 Pisa, Italy

Introduction

The ability to collect and process light energy is important in the life of many organisms: while some use light directly as an *energy* source for driving metabolic processes – for example, photosynthesis – others are able to detect the *information* contained in light coming from the external world. In the latter case, light usually acts as a trigger that initiates a series of dark reactions driven by previously stored metabolic energy. A typical example is that of vision, where the absorption of a single photon can give rise to a detectable current flowing through the visual cell.

Higher organisms have developed highly sophisticated optical and neural apparatuses that enable them to obtain as much information as possible about the external environment. However, the capability of detecting changes in various parameters of incident light, such as intensity, direction, spectral composition and sometimes also polarisation, is not restricted to higher organisms; many unicellular micro-organisms are able to detect light, even though they do not possess complex optical apparatuses and neural networks. It may be of some interest to note that all the processes of light detection, signal processing and final response (we will limit ourselves to photomotile responses of micro-organisms) take place in systems whose linear dimensions can be as small as a few micrometers (Lenci & Colombetti, 1978).

In this chapter we will briefly describe the fascinating world of light-sensitive micro-organisms, starting with the different types of photoresponses and the way they can be measured; next we will consider in some detail the possible orientation mechanisms to light, and finally we will discuss briefly some of the well known examples in the field.

Photomotile responses

The capability shown by many micro-organisms of responding to light stimuli is most probably connected to the need for regulating their

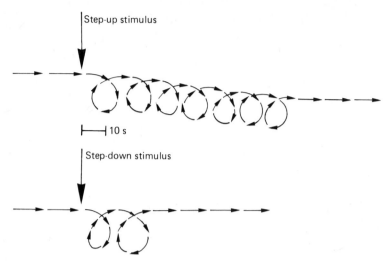

Fig. 1. Schematic representation of step-down and step-up photophobic responses. The arrows represent the cell trajectories before and during stimulation.

exposure to natural illumination (most of the light-responding cells are indeed photosynthetic). Freely swimming micro-organisms, such as bacteria (*Halobacterium halobium*), flagellated algae (*Euglena gracilis, Chlamydomonas reinhardtii* and many more) and some Protozoa (*Stentor coeruleus*) can simply achieve this goal by changing their motion as a function of the external illumination conditions. Such a behaviour is usually referred to as a photomotile response.

Micro-organisms basically show three different types of photomotile response:

(1) *Photophobic reactions*: where the stimulus is a sudden variation in light intensity, and the cell response is characterised by a rapid change in the beating pattern of the locomotory organelle(s). For example, in *Euglena* the response to a sudden increase of light intensity above a threshold (step-up photophobic response; for terminology see Diehn *et al.*, 1977) has a latency of between 100 and 400 ms, as determined by high speed cinematography (Diehn, Fonseca & Jahn, 1975). The behaviour initiated by this response is a tumbling type of swimming, with possible rotations on the spot. The response is transient; cells adapt to the new illumination conditions and resume their previous state of motion in a new swimming direction, which is unrelated to the pre-stimulus direction. Adaptation times in *Euglena* depend on the type and strength of the stimulus; responses to decreasing light intensities (step-down photophobic responses) show adaptation within 10–20 s

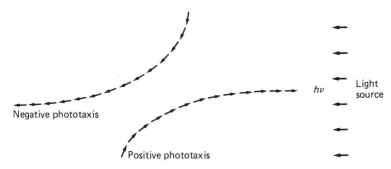

Fig. 2. Schematic illustration of positive and negative phototaxis, indicating the oriented trajectories of the cells.

whereas adaptation to step-up stimuli may take as long as 90 s (see Fig. 1).

(2) *Phototaxis*: where light direction is the external stimulus and the macroscopic response of cells is an oriented movement (Fig. 2) toward (positive phototaxis) or away from (negative phototaxis) the light source. Though phototaxis is probably the most fascinating of all the photomotile responses, very little is known either about the strategy of orientation or about the molecular pathways involved. In some cases phototaxis may not be a response '*per se*', but rather the result of a series of phobic reactions. This has been suggested for *Euglena* (see, for instance, Colombetti, Lenci & Diehn, 1981), but in other micro-organisms, such as *Anabaena variabilis*, phototaxis does not seem to be linked to phobic responses (Nultsch, Schuchart & Höhl, 1979). We will come back to phototaxis in more detail later.

(3) *Photokinesis*: where cell speed depends on light intensity. Photokinesis is usually considered as a special case among photomotile responses, for it has been shown to be a photocoupling response where light energy is directly used via the photosynthetic system to drive extra ATP synthesis (Nultsch, 1980). Moreover, there is no adaptation to the new light conditions, the photokinetic reaction remaining as long as the light level is kept higher or lower than the previous one (Fig. 3).

To summarise, photomotile responses can be divided into two main groups, according to whether light is utilised as a trigger (photosensory responses) or as an energy source (photocoupling responses). In the former case, the whole process leading from photon absorption to

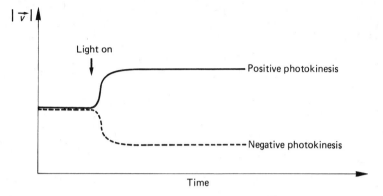

Fig. 3. Diagrammatic scheme of a photokinetic reaction. $|\vec{v}|$, speed of movement.

motile response is usually referred to as a photosensory transduction chain, as shown in the scheme of Fig. 4.

Experimental techniques

How do we know that certain micro-organisms are able to perceive light? The most immediate answer to this question comes from the observation of cell behaviour in response to light stimulation: if cells alter their motion parameters, such as speed and/or direction, we can infer that they are able to sense light stimuli. In some cases, however, this direct cause–effect relationship may be only apparent and the response, for instance, may be due to photochemical modifications of the external medium (Harayama & Ino, 1977).

Behavioural responses to external stimuli, and in particular to photic ones, can be studied by both population methods and single cell tracking. Other experimental techniques have in a few cases been employed: for example, flagellar apparatus isolation (Hyams & Borisy, 1978), electrophysiology (Häder, 1978*ab*; Ristori, Ascoli, Banchetti, Parrini & Petracchi, 1981) and micro-spectroscopy '*in vivo*' (Benedetti & Lenci, 1977; Colombetti, Ghetti, Lenci, Polacco & Quaglia, 1981). Each of these techniques can allow a better understanding of parts of the photosensory transduction chain, while a comprehensive picture of the whole process necessarily requires an integrated view of all the results obtained with the different approaches.

Population studies

The net result of photomotile responses is very often observed as a clustering or scattering of cells in an illuminated region; cell concentra-

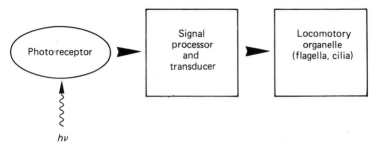

Fig. 4. Block diagram of a photosensory transduction chain.

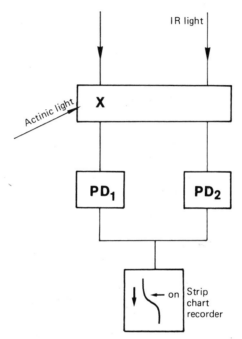

Fig. 5. Schematic drawing of a differential turbidimeter 'phototaxigraph' ('photoaccumuli-graph'). PD_1 and PD_2, infrared-sensitive photodiodes which differentially measure the cell density in the illuminated region of the cuvette (X) and in the reference region.

tion changes are most easily measured by means of optical techniques, and indeed the first automatic devices used to monitor cell responses to light were essentially differential turbidimeters such as the one shown schematically in Fig. 5. This type of experimental set-up has the advantage of good statistical sampling, since the measurement is proportional to the average response of millions of cells; however, the interpretation of the results may often be complicated by spurious effects, for example chemoresponses and photokinetic effects.

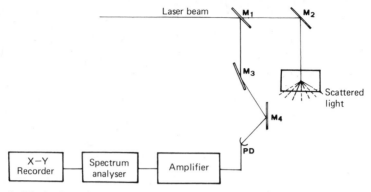

Fig. 6. Block diagram of a Doppler spectrometer used for measurements of the parameters of motion of cells by means of spectral analysis of scattered laser light. M_{1-4}, mirrors; PD, photodiode.

More recently laser light-scattering techniques have been applied to study the photomotile properties of micro-organisms (Ascoli, Barbi, Frediani & Mure, 1978). This kind of approach allows in principle a thorough analysis of the motion parameters of the cells, such as speed and direction, but the results obtained are again often difficult to interpret because of insufficient knowledge of parameters, for instance the light-scattering properties of the individual cells. What is measured is the Doppler shift of the laser light which is scattered by the micro-organisms; these shifts, that are of the order of 100 Hz, can be revealed using a heterodyne technique. A schematic drawing of a Doppler spectrometer is shown in Fig. 6.

Studies of single cells

The most reliable approach to investigate the motile responses of micro-organisms is the tracking of individual cells. All the motion parameters can be determined and a response can be unambiguously related to the stimulus. Unfortunately this method is also quite tedious, time-consuming, and not very easy to automate. Excluding the unaided human observation of cells under the microscope, which can give only very rough and qualitative information on the behaviour of the system, four different techniques have been employed for this study: dark-field photography, high-speed cinematography, tracking microscopes and video-recording systems. Dark-field photography was most frequently used in the past, but it is now being replaced by video-recordings. High-speed cinematography can give unique information about short-time responses and the primary changes in flagellar or ciliary beating;

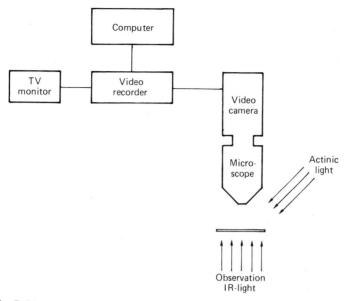

Fig. 7. Block diagram of a video-recording system with an on-line computer.

but it usually requires high levels of illumination which can severely perturb the system under investigation. The tracking microscope, which can follow the three-dimensional motion of cells (Berg, 1971), is a unique machine in the strict sense, as there is only one in the world. The most widely used technique nowadays is that of video-recording, which has all the advantages of photography and offers some of the possibilities of high-speed cinematography. A schematic diagram of a standard set-up is shown in Fig. 7. Cell trajectories are recorded and subsequently analysed; presently the analysis has been manual but the introduction of automatic or semi-automatic computerised analysis of video-recordings (Häder, Colombetti, Lenci & Quaglia, 1981) will certainly allow a much easier and faster collection of information. It should be kept in mind that these methods, tracking microscope excluded, are limited to the analysis of two-dimensional trajectories.

Orientation mechanisms

Introduction

Phototaxis, which is the capability of certain micro-organisms to direct their motion along a beam of light, is probably the most fascinating among photomotile responses, and it has been investigated in several laboratories for many years, starting from the work of Treviranus in 1817 and continuing up to the present day. Unfortunately our knowledge of the complex molecular machinery underlying this response is still quite poor. Some models have been proposed that can be of help towards a better understanding of the orientation strategies, and we will discuss some of them later, but the molecular details of light perception, signal processing and signal transduction are largely unknown in most cases.

From a behavioural viewpoint, the question has been raised as to whether phototaxis is a photomotile response '*per se*' or rather the result of a series of phobic responses. The answer to this depends very much on the organism under study; in some cases (for example, *Anabaena variabilis*, a blue-green alga investigated by Nultsch *et al.*, 1979), phototaxis is not brought about through phobic reactions, whereas other systems can use the phobic response as the means of regulating cell motion with respect to light direction, as suggested for *Euglena gracilis*. Recent findings seem to indicate that there are micro-organisms in which negative phototaxis is not necessarily related to the step-up photophobic response. We will come back later to this point.

Cells and light direction

How can a cell sense the position of a light source? The easiest way to accomplish this is to possess subcellular structures that are sensitive to the anisotropy of the external light field. Two possible solutions have been proposed (Feinleib, 1980). In one the cell possesses at least two spatially separated light detectors (Fig. 8A): the signal to be processed is then proportional to the difference in light absorbed at the same time at the two photoreceptor sites. This is known as the one-instant-mechanism; it implies the creation of a spatial gradient within the cell, and there is no requirement for any special kind of motion of the cell. In the other a single light sensor is asymmetrically located in the cell body, and the signal to be processed is proportional to the light absorbed by it at (two) different instants of time (Fig. 8B). This is the two-instant-mechanism, where the cell is able to measure a temporal gradient, and it demands that the cell rotates as it swims. The first type of solution has been chosen by Nature mostly in some gliding organisms (Häder, 1979) or in phototropic systems (Hertel, 1980) – that is slow-moving or not-freely moving micro-organisms, where the time comparison would probably require too long a memory time. The second type of solution requires that the cell body should move along a helical path during swimming (this is true for all flagellates) and that the photoreceptor

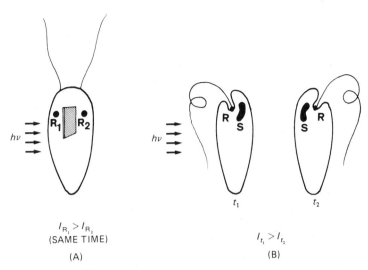

Fig. 8. Mechanisms of light direction perception: (A) one-instant; (two photoreceptors measuring light intensity at the same time); (B) two-instant (one photoreceptor measuring light intensity at two different instants of time). R, receptor; S, stigma.

should possess a certain degree of directionality. This can be achieved through a large variety of subcellular designs, employing optical phenomena such as light absorption, scattering, dichroism and possibly reflection interference, diffraction and wave-guide properties, to increase the efficiency of the sensor directionality (Foster & Smyth, 1980).

The simplest method of achieving directionality using absorption is by the presence of a screening device such as that shown in Fig. 8B. This is indeed more than a hypothetical scheme for it is well known that most of the phototactic flagellates possess an eyespot (or stigma), a deeply pigmented structure closely associated with the photoreceptor, which can act as a screen. In such a situation, light coming from one side is absorbed by the photoreceptor at time t_1, but at a subsequent time t_2, because of the cell body rotation, the screening organelle absorbs part of the incident radiation and therefore less light is absorbed by the photoreceptor. The cell will thus perceive different signals at the two different times, t_1 and t_2 and will use this information to 'deduce' which side the light is coming from.

A dichroic photoreceptor can also act as a directional detector, since oriented receptor molecules show a preferential direction for photon absorption, both in polarised and non-polarised light, the effect being stronger in the former case (Fig. 9). The fact that the directionality of a dichroic photoreceptor is enhanced with polarised light gives the investigator a very useful tool for studying the possible dichroic nature

NON-POLARISED LIGHT

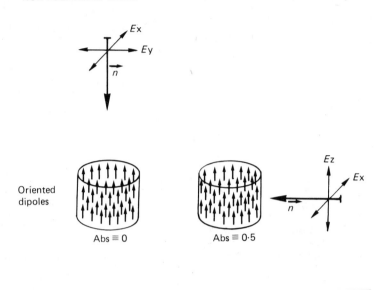

POLARISED LIGHT

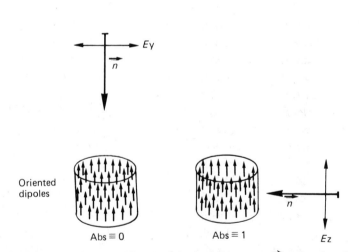

Fig. 9. Light absorption properties of a dichroic photoreceptor. $\vec{n}$ = direction of propagation of the light beam; Ex, Ey, Ez = components of the electric field of the light wave.

of photoreceptive structures (see, for example, Haupt, 1980, and this volume), but it might also be of importance for the orientation of phototactic micro-organisms in the natural environment. Direct sunlight is unpolarised, but a large amount of natural light scattered and reflected underwater is partly linearly polarised (Waterman, 1975). Both the orientation of the electrical vector and the degree of polarisation are determined by the line of sight to the sun, so that polarisation patterns underwater might indicate the sun's position in the sky. It should, however, be kept in mind that in general a dichroic photoreceptor can detect the axis of stimulus propagation, but not its direction (that is, it can distinguish between top and lateral illumination, but not between left and right).

Recently, Foster & Smyth (1980) have proposed that a number of other optical properties, such as multilayer reflection and interference of wave-guide optics can play an important role in increasing the directionality of flagellate photoreceptors. Although interesting from a theoretical point of view, this hypothesis goes, in our opinion, beyond the experimental evidence which indicates only a fair amount of reflectivity from the eyespots of some flagellated algae, for example *Chlamydomonas*. Wave-guide properties seem to be even more questionable, given the dimensions and refractive indexes of the structures involved. The amount of light channelled by wave-guide effects into the photoreceptor depends on the characteristic wave-guide parameter (Snyder, 1975):

$$V = \frac{\pi d}{\lambda} \sqrt{n_1^2 - n_2^2}, \tag{1}$$

where d is the linear dimension of the photoreceptor, n_1 its refractive index, n_2 the refractive index of the surrounding medium and λ the wavelength. In our cases, assuming $d = 0.1\,\mu$m, $n_1 = 1.5$ (probably an overestimate) and $n_2 = 1.35$ (probably an underestimate), using blue stimulating light (472 nm):

$$V = \frac{3.14 \times 0.1}{0.472} \sqrt{(1.5)^2 - (1.35)^2} = 0.4. \tag{1'}$$

V must be higher than 1.0 to have at least 20% of energy channelled. This low V value implies that, in algal flagellates, wave-guide properties play a very minor role, if any, in channelling incident light energy within the photoreceptors.

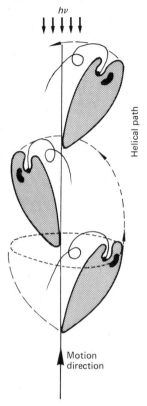

Fig. 10. Schematic representation of a flagellated cell trajectory.

Cell orientation to light

In the previous sections we have described briefly some of the mechanisms that could be the basis for a directional light detector (an antenna), which a cell can use during its motion. Now, let us assume that our cells do have such a detector: what is the best way to use it in order to be able to direct their motion towards light?

It is useful to think in terms of models to help in understanding how the system could actually operate. For example, Foster & Smyth (1980) have recently proposed that flagellates might be able to track light direction using the same principles as used in target detection by radar. Briefly, the radar rotates around the line of sight to the target (tracking direction) with its antenna slightly tilted with respect to the tracking direction. The detector thus receives a fairly constant signal and the tracked object is in the centre of the scanned circle. As soon as the target moves, the signal received by the antenna changes, becoming

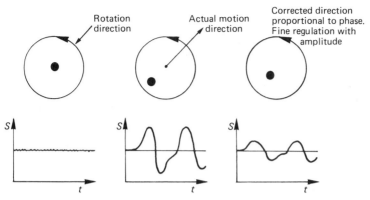

Fig. 11. Schematic description of a scanning cell pointing to the light source. The signal is stronger when the detector and the light source are on the same side with respect to the motion direction. (Redrawn after Foster & Smyth, 1980.)

greater when the antenna direction and the target are on the same side relative to the tracking direction, and smaller when they are on opposite sides. This error signal, which contains both phase and amplitude information, is fed into a feedback control that re-aligns the tracking direction until the error signal is nullified again. Flagellated algae rotate as they swim at frequencies in the range 0.2–2.0 Hz; the resulting trajectory is therefore a uniform helical path, with the cell body kept at a fixed orientation relative to the helix axis, as shown in Fig. 10. Their directional detector executes, therefore, a conical scanning of the environment, as does radar. When the cell moves towards light its transduction apparatus receives an almost constant signal, but when the cell deviates from this path the signal pattern changes and the processor elicits a cell body re-alignment until the signal is constant again. The situation is sketched in Fig. 11, projected on a plane perpendicular to the axis of motion. As the antenna moves towards the light source the signal gets stronger and reaches a maximum when the tracking direction, light source and antenna are aligned; the signal then declines to a minimum as the cell continues swimming. Now, suppose that the cell processor is programmed to elicit a cell body rotation in the direction from which the maximum signal comes. This will cause a re-alignment of the cell body until the error signal is nullified, as shown in the last part of Fig. 11. The amplitude error is not really necessary to perform a trajectory correction, but it allows a smoother re-alignment (the turning drive is then proportional to the amplitude error). This mechanism would allow a good orientation of the cell towards light. Moreover, according to Foster & Smyth (1980), the temporal pattern received by the cell processor could be used to improve the signal-to-noise ratio and to extend the useful range of light intensities.

What is the minimum amount of light that is necessary to orient the movement of a micro-organism such as a flagellated alga? The answer depends on many variables (for example, the ability of the cell processor to extract the signal from the noise), but the lowest limit is certainly set by the amount of receptor pigments available for light absorption and by the efficiency of the screen. The important parameters are the amount of light modulation and its noise during a rotation cycle. Let N_L be the photon count rate of the photoreceptor during direct illumination and N_D represent the photon count rate during the screening phase. Also let t_L and t_D represent the respective count durations. $N_L t_L$ and $N_D t_D$ are the total counts under full and screened light, respectively. The variances in each of these counts are $N_L t_L$ and $N_D t_D$ assuming a Poisson distribution. The signal is determined by the difference in the count rates:

$$N_L - N_D. \tag{2}$$

The variance for this quantity is given by:

$$\frac{N_L}{t_L} + \frac{N_D}{t_D}. \tag{3}$$

Accordingly, this counting process has a signal-to-noise ratio given by (RCA Electro-Optics Handbook, 1974):

$$\frac{S}{N} = \frac{N_L - N_d}{\sqrt{\left(\dfrac{N_L}{t_L} + \dfrac{N_D}{t_D}\right)}}. \tag{4}$$

The lowest detectable signal has $S/N = 1$, therefore:

$$N_L - N_D = \sqrt{\left(\frac{N_L}{t_L} + \frac{N_D}{t_D}\right)}. \tag{4'}$$

Now, let M be the number of receptor molecules in the photoreceptor, $ø$ be the quantum yield of the process, σ the photon absorption cross-section and I_L the number of quanta ($cm^{-2} s^{-1}$). Let us moreover assume that our screen has a transmittance T, so that the screened number of quanta is $I_D = T \cdot I_L$. Hence:

$$N_L = M \cdot ø \cdot \sigma \cdot I_L, \tag{5}$$

and

$$N_D = M \cdot T \cdot \phi \cdot \sigma \cdot I_L. \tag{6}$$

Substituting (5) and (6) in (4′) and rearranging, we obtain:

$$M \cdot \sigma \cdot \phi \cdot I_L = \frac{t_D + T t_L}{t_L t_D} \cdot \frac{1}{(1 - T)^2}. \tag{7}$$

The threshold value

$$I_L = \frac{1}{M \cdot \phi \cdot \sigma} \cdot \frac{t_D + T t_L}{t_L t_D} \cdot \frac{1}{(1 - T)^2}. \tag{8}$$

goes to a minimum when $t_D = 0.4t$, where t is the time for a full turn.

Let us now see what this means in the case of *Euglena*. In this alga, $t = 0.5$ s and $t_D = 0.2$ s; T, the transmittance of the screening organelle, the stigma, has been determined to be about 0.46 (Benedetti *et al.*, 1976). The photoreceptor of *Euglena* is probably a flavin (Colombetti & Lenci, 1980), the absorption cross-section of which, is about 3.8×10^{-17} cm^2. M, as determined by microspectrophotometric measurements, is of the order of 10^6 (Colombetti & Lenci, 1980) and we assume $\phi = 0.5$ for our process. Substituting these values in equation (8) we have for our threshold:

$$I = \frac{19.3}{10^6 \cdot 0.5 \cdot 3.8 \cdot 10^{-17}} = 10^{12} \text{ quanta cm}^{-2} \text{ s}^{-1}. \tag{8′}$$

It may be interesting to compare this value with the recent findings of Häder *et al.* (1981), who describe detectable phototaxis in *Euglena* down to about 1 lux (about 4.2×10^{-3} Wm^{-2}) or 9.5×10^{11} quanta cm^{-2} s^{-1} at 450 nm. The similarity between these two values is indeed remarkable. We would like to stress that they have been determined completely independently of each other, the former being a theoretical estimate and the latter an independent experimental measurement. This could, therefore, be a strong indication of the validity of the proposed model, at least in the case of *Euglena*.

Models such as the one we have described above assume that the cells orient to the light in an active manner. In other words, the correction angle is assumed to depend on the signal received by the processing apparatus.

It is also possible, however, to think in terms of models in which no such active orientation exists. For instance, bacterial chemotaxis is by no means an active orientation of cells along the external gradient, but rather the result of a biassed random walk (Berg & Brown, 1972). The

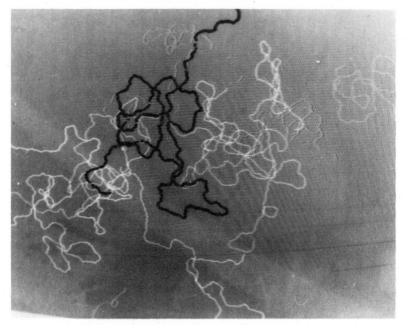

Fig. 12. Computer-simulated unbiassed cell tracks.

distribution of direction angles remains spatially isotropic as the organisms move, and it is the frequency of directional changes that varies,
decreasing when the cell moves from a lower to a higher concentration
of chemo-attractant. This eventually results in an average motion
toward the stimulus source.

A mechanism of this kind could well operate also in phototactic
responses, and we have very recently started an investigation based on a
computer simulation of cell trajectories. Our assumption is that cells
move on a surface, executing a two-dimensional random walk when no
stimulation is present. The mean free path (the path between two
random directional changes) is given as input parameter, whereas the
angle of deviation is randomly chosen by the computer. Cell tracks
obtained in these conditions are shown in Fig. 12. When the organisms
are presented with a stimulus, they are characterised by a variable mean
free path, which is maximum (l_{max}) when the cell moves towards light
and goes to a minimum (l_{min}) when the cell moves away from the light
source. The trajectories obtained with such a bias are shown in Fig. 13.
The high degree of orientation of the cell tracks depends on the values
chosen for l_{max} and l_{min}, but there is no bias on the new angular direction
chosen by a cell. A quite interesting piece of information given by such a

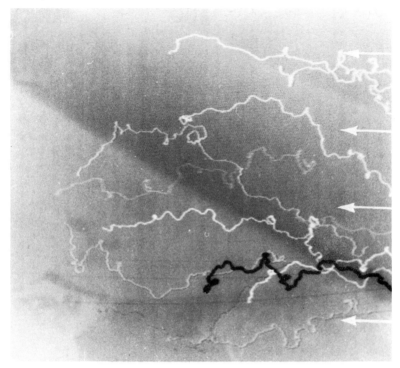

Fig. 13. Computer-simulated biassed cell tracks (the mean free path of cells moving toward the light source is ten times higher than the mean free path of cells swimming away from the light source). Arrows indicate light direction.

model is that the possibility of detecting directness in cell paths may depend on the ratio of the dimensions of the field observed under the microscope to the maximum free path. For instance, Fig. 14 shows a series of apparently non-oriented trajectories, but this picture is just an enlargement of Fig. 13 that has reduced by a factor of ten the ratio of field-dimension to mean-free-path. This means that one has to check the experimental conditions very carefully before excluding the possibility of an oriented movement of cells. Probably, however, statistical orientation analysis could help in extracting the correct information from situations like the one in Fig. 14.

It is difficult to say whether 'active' or 'passive' orientation mechanisms are operating in micro-organisms. It might well be that some cells use the former and some the latter. Moreover, the two mechanisms are not mutually exclusive: for instance a biassed random (passive) walk where the bias is on the direction angle is very hard to discriminate from an active phototactic orientation.

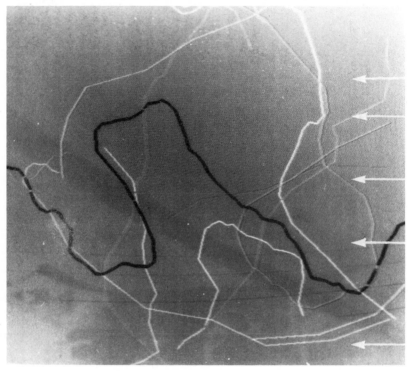

Fig. 14. Enlargement of a trajectory portion of Fig. 13. Arrows indicate light direction.

In the following section we will briefly discuss some particular cases of phototactic orientation in micro-organisms.

Phototaxis in Euglena, Ochromonas *and* Stentor

If you open an undergraduate textbook of biology and look for the term 'phototaxis' you will very likely be informed that phototaxis is an oriented movement of cells toward light and that a good example of such a photobehaviour is offered by *Euglena*. Then, if you go to modern literature covering the last 20 years, you will find many citations of phototaxis and the phototactic model by Jennings & Mast but there are no up-to-date experiments giving a clear indication of phototactic behaviour (with the exception of Creutz & Diehn's paper of 1976 in which a polarotactic response of *Euglena* is described). Indeed some authors have begun to wonder whether phototactic behaviour actually exists in *Euglena* (see, for instance, Feinleib, 1980). We have therefore decided to face this problem using a quantitative assay developed by Häder (1981), which gives measurable and reproducible parameters of

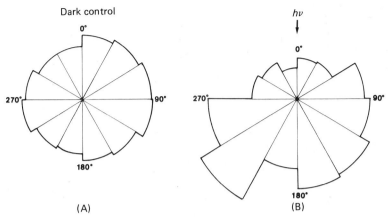

Fig. 15. Polar-wedge diagrams of unstimulated (A) and stimulated (B) cells. The length of the ith angular sector is proportional to the number of cells, the trajectories of which form, with the light direction, an angle between $30° \times (i-1)$ and $30° \times i$.

the motile behaviour of single cells (Häder *et al.*, 1981). Video-recordings, obtained as described in the section devoted to experimental techniques, were analysed semi-automatically with the help of a computer, thus allowing the storage and manipulation of a very high number of cell tracks. These tracks were apparently randomly oriented; in other words, from a visual examination of the cell trajectories one would have concluded that *Euglena* was indeed not phototactic. We have, however, analysed the cell tracks using the methods of orientation statistics described by Batschelet (1972). We have measured the angle that each track forms with the direction of actinic light and have plotted these angles, grouped into one of twelve 30° sectors in a polar wedge diagram (0° corresponding to the light direction).

Without stimulation (dark control) an isotropic distribution of angles is expected, which means that all angular sectors will have the same length as shown in Fig. 15A. If a preferred orientation exists, the circular histogram will deviate from a random distribution (Fig. 15B) and the statistical methods make it possible to quantify this deviation, giving both the meaningfulness of deviation and the preferred angular direction.

We have used three different statistical tests, the R-test, the V-test and the χ^2 test, which are described in Batschelet (1972). All statistical treatments applied show a definite orientation with respect to the light direction, while the distribution in darkness does not differ significantly from a random one. There is a clear positive phototaxis at about 0.005 W m^{-2}, 0.05 W m^{-2} and 0.2 W m^{-2}; at 1.00 W m^{-2}, however, the distribution becomes bimodal, indicating that at this light intensity the population is split into two groups, one positively and the other negatively phototactic. At 2 W m^{-2} there is a clear negative phototaxis. Two facts seem to be established by these experiments: (i) statistical analysis is necessary to answer definitely the question whether or not phototaxis exists, and, (ii) the fraction of phototactically active *Euglena* does not seem to be high. Only about 50% of the population we examined moved towards light at 0.2 W m^{-2} in a sector covering $\pm$ 60°, compared to an expected fraction of 33% in an isotropic distribution. This could indicate either that the population examined is not homogeneous with respect to phototaxis, or that the mechanism of phototactic orientation in *Euglena* is not very efficient.

We have also used this kind of approach to study the phototactic orientation of *Ochromonas danica*. This alga, shown in Fig. 16, is interesting because it has a photoreceptor apparatus consisting of a stigma and a paraflagellar swelling (PFB). Unlike *Euglena*, the stigma is inside the chloroplast. It is formed by a single layer of osmiophilic granules and does not seem to be birefringent. The PFB is contained within the membrane of the short flagellum and looks like an amorphous structure under the electron microscope. When excited by light of 450 nm, the PFB emits a greenish fluorescent light (Colombetti & Dipasquale, 1980), as does the *Euglena* PFB. *Ochromonas* shows photomotile reactions characterised by an increase in the number of directional changes per unit time when subjected to step-up or step-down blue-light stimuli. The elementary photomotile reaction seems to be a sudden bending of the long trailing flagellum towards the cell body. Nothing was known about its phototactic capability. Our results indicate that the cells show positive phototaxis at the lowest light intensity tested ($\sim$ 1 lux), at all other intensities phototaxis is negative. Once again there is no chance of detecting an oriented motion of cells unless a statistical analysis of the angular distribution of the cell trajectories is carried out. It may be of interest to note that *Ochromonas* shows negative phototaxis at light intensities well below the threshold for inducing the step-up photophobic response. Typically, a step-up photophobic reaction is elicited at irradiances of about 10 W m^{-2}, whereas negative phototaxis is brought about by irradiances of the order of only 10^{-2} W m^{-2}. This finding seems to indicate that negative phototaxis in this cell is not mediated by repetitive step-up photophobic responses as proposed for negative phototaxis in *Euglena* (Diehn *et al.*, 1975).

To conclude we will now discuss briefly the case of a blue-green coloured ciliate, *Stentor coeruleus*, whose photomotile reactivity was

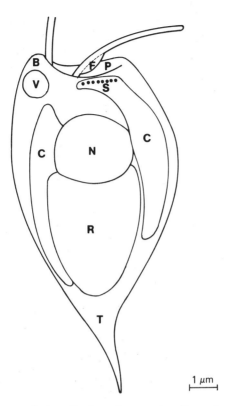

Fig. 16. *Ochromonas danica.* P, leucosin vacuole; N, nucleus; C, chloroplast; V, contractile vacuole; S, stigma; F, paraflagellar body; T, tail; B, beak.

already known at the beginning of the century (see Song & Tapley, this volume). When subjected to unilateral illumination, *Stentor* at first exhibits a typical step-up photophobic response that is characterised by a delayed stop reaction, followed by turning between 30° and 180° to the side. The delay time seems to decrease with increasing irradiance. Cells finally swim off in a new direction that is not related to the direction of the stimulating light (Song, Häder & Poff, 1980a). Once this response is completed (usually within about 1.5 s), *Stentor* smoothly swims away from the light source. This photomotile reaction of *Stentor* seems to be a true phototaxis, the actual photic stimulus being light *direction*. Strong support for this conclusion is provided by the observation that *Stentor* orients away from the light source in the converging as well as in the diverging regions of a focused light beam. The initial stop and turning reactions do not seem to be responsible for the orientation of *Stentor* away from the light source and negative phototaxis occurs without multiple step-up photophobic responses.

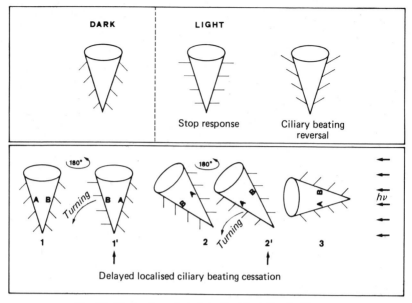

Fig. 17. Mechanism of phototactic orientation in *Stentor*.

In summary, *Stentor* is probably capable of orienting away from the light direction by a strategy which is not based on repetitive step-up photophobic responses. Thus both *Ochromonas* and *Stentor* provide further evidence that negative phototaxis and step-up responses are not necessarily related. However, Song *et al.* (1980a,*b*) and Walker, Yoon & Song (1980) show that a unique photoreceptor pigment, stentorin, is responsible for triggering both photomotile responses and that a light-induced release of protons from the photoreceptor system acts as the sensory signal for photophobic as well as phototactic responses in *Stentor*. It seems likely that, in this microorganism, two different photomotile responses are controlled by the same photoreceptor and elicited through the same sensory transduction chain.

Finally, what is the elementary motile mechanism responsible for negative phototaxis? Taking the mechanism of phototactic orientation of the colonial flagellate *Volvox* (Hand & Haupt, 1971) as a starting point, Song *et al.* (1980a) suggest that light could induce in *Stentor* a cessation of ciliary beating in the row of cilia adjacent to the row of illuminated photoreceptor pigment vesicles. This localised cessation of ciliary beating would take place with a time delay sufficient for a 180° rotation of *Stentor*. In other words, ciliary beating would stop on the side of the cell away from the light source and the continued beating of cilia in the row facing the light would cause a net turning of the cell body, resulting in orientation of the cell trajectory away from the light source (Fig. 17). An interesting feature of this hypothesis is that photophobic and phototactic responses of *Stentor* are elicited through the same locomotory apparatus (cilia) utilising measurement of the change in light intensity with time (two-instant-mechanism) for photophobic, and a measurement of the gradient of light intensity in space (one-instant-mechanism) for phototactic responses.

Part of this work has been supported by a grant from Consiglio Nazionale Ricerche (CNR, Decreti 8102190 and 8102191) for a Collaborative Bilateral Project Italia–USA.

References

ASCOLI, C., BARBI, M., FREDIANI, C. & MURE, A. (1978). Measurements of *Euglena* motion parameters by laser light scattering techniques. *Biophys. J.* **24**, 585–99.

BATSCHELET, E., (1972). Statistical methods in the analysis of problems in animal orientation and certain biological rythms. In *Animal Orientation and Navigation*, ed. S. R. Galles, K. Schmidt-Koenig, G. J. Jacobs & R. F. Belleville, pp. 61–91. Washington: NASA.

BENEDETTI, P. A., BIANCHINI, G., CHECCUCCI, A., FERRARA, R., GRASSI, S. & PERCIVAL, D. (1976). Spectroscopic properties and related functions of the stigma measured in living cells of *Euglena gracilis*. *Arch. Microbiol.* **111**, 73–6.

BENEDETTI, P. A. & LENCI, F. (1977). *In vivo* microspectrofluorometry of photoreceptor pigments in *Euglena gracilis*. *Photochem. Photobiol.* **26**, 315–18.

BERG, H. C. (1971). How to track bacteria. *Rev. Scient. Instr.* **42**, 868–71.

BERG, H. C. & BROWN, D. A. (1972). Chemotaxis in *E. coli* analysed by three-dimensional tracking. *Nature, London*, **239**, 500–4.

COLOMBETTI, G. & DIPASQUALE, M. A. (1980). Photomovement in *Ochromonas danica*. Strasbourg, Proc. VIII Inter. Congr. Photobiol. P 64.

COLOMBETTI, G., GHETTI, F., LENCI, F., POLACCO, E. & QUAGLIA, M. (1981). *In vivo* microspectrofluorometry of photoreceptor pigments. Int. Conference Photochem. Crete, Sept. 1981. *J. Photochem.* **17**, 36.

COLOMBETTI, G. & LENCI, F. (1980). Characterization of photoreceptor pigments. In *Photoreception and Sensory Transduction in Aneural Organisms*, ed F. Lenci & G. Colombetti, pp. 173–88. London: Plenum.

COLOMBETTI, G., LENCI, F. & DIEHN, B. (1982). Responses to photic, chemical and mechanical stimuli. In *The biology of Euglena*. vol. 3, ed. D. E. Buetow, pp. 169–95. New York: Academic Press.

CREUTZ, C. & DIEHN, B. (1976). Motor responses to polarized light and gravity sensing in *Euglena gracilis*. *J. Protozool.* **23**, 552–6.

DIEHN, B., FEINLEIB, M. E., HAUPT, W., HILDEBRAND, E., LENCI, F. & NULTSCH, W. (1977). Terminology of behavioral responses of motile microorganisms. *Photochem. Photobiol.* **26**, 559–60.

DIEHN, B., FONSECA J. R. & JAHN, T. L. (1975). High speed cinemicrography of the direct photophobic response of *Euglena* and the mechanism of negative phototaxis. *J. Protozool.* **22**, 492–4.

FEINLEIB, M. E. (1980). Photomotile responses in flagellates. In *Photoreception and Sensory Transduction in Aneural organisms*, ed. F. Lenci & G. Colombetti, Plenum. pp. 45–68. New York & London: Plenum.

FOSTER, K. W. & SMYTH, R. D. (1980). Light antennas in phototactic algae. *Microbiol. Rev.* **44**, 572–630.

HÄDER, D. P. (1978a). Evidence of electrical potential changes in photophobically reacting blue-green algae. *Arch. Microbiol.* **118**, 115–19.

HÄDER, D. P. (1978b). Extracellular and intracellular determination of light-induced potential changes during photophobic reaction in blue-green algae. *Arch. Microbiol.* **119**, 75–9.

HÄDER, D. P. (1979). Photomovement. In *Encyclopedia of Plant Physiology*, New Series, ed. W. Haupt & M. E. Feinleib, vol. 7, pp. 268–309. Berlin: Springer-Verlag.

HÄDER, D. P. (1981). Computer-based evaluation of phototactic orientation in micro-organisms. *EDV in Med. Biol.* **12**, 27–30.

HÄDER, D. P., COLOMBETTI, G., LENCI, F. & QUAGLIA, M. (1981). Phototaxis in the flagellates *Euglena gracilis* and *Ochromonas danica*. *Arch. Microbiol.* **130**, 78–82.

HAND, W. G. & HAUPT, W. (1971). Flagellar activity of the colony members of *Volvox aureus* Ehrbg. during light stimulation. *J. Protozool.* **18**, 361–4.

HARAYAMA, S. & INO, T. (1977). Ferric ion as photoreceptor of photophobotaxis in non-pigmented *Rhodospirillum rubrum*. *Photochem. Photobiol.* **25**, 571–8.

HAUPT, W. (1980). Localization and orientation of photoreceptor pigments. In *Photoreception and Sensory Transduction in Aneural Organisms*, ed. F. Lenci & G. Colombetti, pp. 155–72. London: Plenum.

HERTEL, R. (1980) Phototropism of lower plants. In *Photoreception and Sensory Transduction in Aneural Organisms*, ed. F. Lenci & G. Colombetti, pp. 89–105. London: Plenum.

HYAMS, S. S. & BORISY, G. G. (1978). Isolated flagellar apparatus of *Chlamydomonas*: characterization of forward swimming and alteration of waveform and reversal of motion by calcium ions *in vitro*. *J. Cell Sci.* **33**, 235–53.

LENCI, F. & COLOMBETTI, G. (1978). Photobehavior of microooorganisms. A biophysical approach. *Ann. Rev. Biophys. Bioeng.* **7**, 341–61.

NULTSCH, W. (1980). Photomotile responses in gliding organisms and bacteria. In *Photoreception and Sensory Transduction in Aneural Organisms*, ed. F. Lenci & G. Colombetti, pp. 69–87. New York & London: Plenum.

NULTSCH, W., SCHUCHART, H. & HÖHL, M. (1979). Investigations on the phototactic orientation of *Anabaena variabilis*. *Arch. Microbiol.* **122**, 85–91.

RCA Electro-Optics Handbook EOH–11 (1974).

RISTORI, T., ASCOLI, C., BANCHETTI, R., PARRINI, P. & PETRACCHI, D. (1981). Localization of photoreceptor and active membrane in the green alga *Haematoccoccus pluvialis*. VI Internat. Congr. Protozool., Warszawa July 1981, Abstr. P. 314.

SNYDER, A. W. (1975). Photoreceptor optics – theoretical principles. In *Photoreceptor Optics* ed. A. W. Snyder & R. Menzel, pp. 38–55. Berlin: Springer-Verlag.

SONG. P.-S., HÄDER, D. P. & POFF, K. L. (1980a). Phototactic orientation by the ciliate, *Stentor coeruleus*. *Photochem. Photobiol.* **32**, 781–6.

SONG, P.-S., HÄDER, D. P. & POFF, K. L. (1980b). Step-up photophobic response in the ciliate, *Stentor coeruleus*. *Arch. Microbiol.* **126**, 181–6.

WALKER, E. B., YOON, M. & SONG P.-S. (1980). The pH dependence of photosensory responses in *Stentor coeruleus* and model system. *Biochim. Biophys. Acta,* **634**, 289–308.

WATERMAN, T. H. (1975). The optics of polarization sensitivity. In *Photoreceptor Optics*, ed. A. W. Snyder & R. Menzel, pp. 339–71. Berlin: Springer-Verlag.

THE PERCEPTION OF LIGHT DIRECTION AND ORIENTATION RESPONSES IN CHLOROPLASTS

WOLFGANG HAUPT

Institut für Botanik und Pharmazeutische Biologie der Universität Erlangen-Nürnberg, Schlossgarten 4, D-8520 Erlangen, GFR

Introduction

Besides the polarisation and wavelength, parameters which have been treated elsewhere in this volume, the directionality of light may also be important in an organism's responses. The perception of this parameter by organisms is the topic of this chapter.

In many systems the photoresponse depends only on the overall number of quanta being absorbed by the photoreceptor, and the light direction is without any relevance. Thus, most of the photomorphogenetic phytochrome responses are as independent of the light direction as are the light-on and light-off photophobic responses in *Halobacterium*. There are, however, a few photoresponses in plants that are oriented with respect to the light direction. This implies perception of the light direction as the important cue. To introduce the problem, a few types of orientation responses will be referred to briefly, restricted to non-motile organisms and cells. A well-known example is the phototropism of seedlings, which show curvature towards the light, for example in oats (Dennison, 1979). Here, growth on the two opposite flanks is different and results in the observed curvature; this differential growth is under the control of light. Thus, light direction is transformed into different growth responses over a multi-cellular organ. Phototropism is observed also in the sporangiophore of the mould *Phycomyces*, but here the differential growth response concerns two flanks of a single cell-like system.

Oriented responses can even be restricted to rearrangement of cell contents without any change in the outer shape of the cell. The spore of the horsetail, *Equisetum*, germinates with an unequal cell division, the polarity of which is oriented by light (Weisenseel, 1979). One of the earliest steps in this light response is a rearrangement of chloroplasts, which assemble towards the light source (Fig. 1). While this system appears to be well suited for investigating some aspects of directional light effects, its usefulness is limited because the photo-orientation of

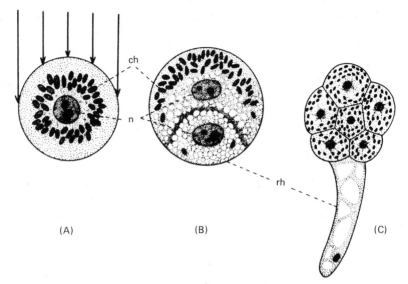

Fig. 1. Polarity induction of the *Equisetum* spore. (A) Short unilateral light pulse at the beginning of germination. (B) First cell division after rearrangement of chloroplasts. (C) Later stage of gametophyte development. ch = chloroplasts, n = nucleus, rh = rhizoid. (After Nultsch, 1964.)

Equisetum chloroplasts is restricted to a rather short period during its development, and the response is irreversible. There are, however, many organisms which display reversible and repeatable chloroplast orientation to light intensity and light direction and these have proved to be valuable tools for analysing the perception of light direction and its transduction into an oriented response (Britz, 1979).

The main feature of this kind of response, which is widely found in the plant kingdom, is shown by the duckweed *Lemna* (Fig. 2). In low-intensity light, the chloroplasts are mainly found in the regions facing the light (front) and opposite to it (rear), but the flanks are empty. In high-intensity light the reverse pattern is observed. The function of these responses appears obvious: to optimise photosynthesis in low-intensity light, and to protect chloroplasts against damage in high intensity light. However, this conclusion is not yet unequivocal, and is of little relevance to the purely physiological problems to be dealt with here. Important in this respect and particularly useful for the photo-biologist, are modifications of this overall pattern in some systems.

Response to light requires light absorption. In consequence, a differential response requires differential light absorption in the cell. The basic question, then, is how light direction is transformed into differential light absorption. It can be shown that several mechanisms

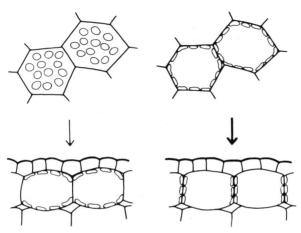

Fig. 2. Chloroplast arrangement in mesophyll cells of *Lemna* as seen from above and in cross section, in low- (left) or high-intensity light (right). (After Haupt, 1963.)

exist; however, full understanding of these mechanisms requires knowledge about the nature and localisation of the photoreceptor. I will deal with these questions mainly with reference to three green algae: *Vaucheria, Hormidium* and *Mougeotia*:

Refraction as a mechanism for perception of light direction

Vaucheria is a coenocytic alga in which the multinucleate cytoplasm is not compartmented by cross walls. The pattern of chloroplasts corresponds to that in *Lemna* (cf. Fig. 5, below). The action spectra for low-intensity and high-intensity responses are very similar to the absorption spectrum of riboflavin (Fig. 3), and therefore a flavin is suggested as the photoreceptor pigment (Haupt & Schönbohm, 1970).

If a small part of *Vaucheria* is exposed to a microbeam, chloroplasts accumulate in this region. It can be shown experimentally that this response is independent of whether the chloroplasts themselves are exposed to light or only the cytoplasm (Fig. 4), and hence localisation of the photoreceptor in the cortical cytoplasm has been demonstrated (Fischer-Arnold, 1963).

By computing the path of the light rays in a *Vaucheria* surrounded by air (Senn, 1908), a region can be identified just behind the flanks which is by-passed by the light and hence is the least-absorbing region (Fig. 5, centre). Upon close inspection it can be seen that the patterns of chloroplast distribution reflect this calculated pattern of light absorption: those regions which are emptied in low-intensity light and occupied

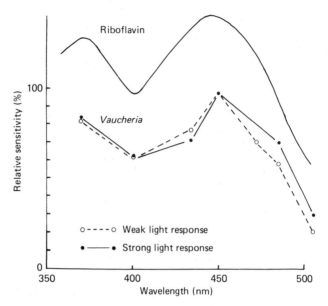

Fig. 3. Action spectra of chloroplast orientation of *Vaucheria* in low- or high-intensity light as compared with the absorption spectrum of riboflavin. The ordinate relates to the chloroplast orientation only: the riboflavin spectrum is drawn adjacent for comparison. (After Haupt, 1963.)

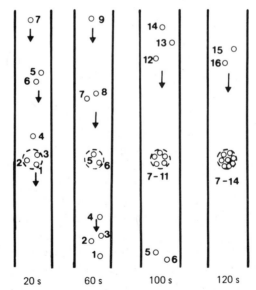

Fig. 4. Accumulation of chloroplasts in *Vaucheria* irradiated with a microbeam (dashed circle) after the times indicated. Chloroplasts are numbered. Only those chloroplasts which cross the field after it has been exposed to light for a threshold period are trapped. (After Fischer-Arnold, 1963, and Haupt & Schönbohm, 1970.)

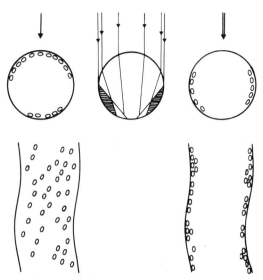

Fig. 5. Cross-section of *Vaucheria* in air with the light path upon unilateral irradiation (centre). The resulting chloroplast arrangements in low- or high-intensity light are given on the left and right, respectively. (After Haupt, 1973.)

in high-intensity light, are not exactly the flanks, but are slightly shifted towards the rear (Fig. 5, left and right). Moreover, if the organism is surrounded by water, the by-passed region becomes rather small, and concomitantly the pattern of chloroplast distribution approaches homogeneity. Thus, in *Vaucheria*, light refraction is the important factor that transforms light direction into an intracellular absorption gradient (a front-to-rear versus behind-the-flanks gradient).

Light refraction acts in a different way in *Hormidium* (Scholz, 1976a): the cylindrical cells of this alga contain a single chloroplast which occupies about half the cell circumference, and can slide along the wall. When the cells are surrounded by air in low-intensity light, the chloroplast is usually found at the rear of each cell, at the side farthest from the light source (Fig. 6a). Since, in low-intensity light, other plant chloroplasts usually gather in those regions which have the highest light absorption, the rear region is assumed to be the brightest in *Hormidium*; this points to the cell acting as a collecting lens with its focus near the back of the cell. Indeed, the pattern of chloroplast arrangement depends on the surrounding medium: in air the chloroplast goes to the rear in nearly all cells, but in paraffin oil in which the cell becomes a diverging lens, the chloroplast normally moves to the front (Fig. 6b). In a series of technical oils, the percentage of front versus rear positioning depends linearly on the refractive index of the oil (Fig. 7).

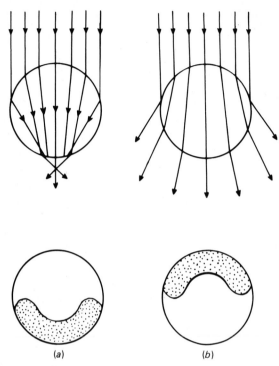

Fig. 6. Cross-section of *Hormidium* in air (*a*) and in paraffin oil (*b*), showing the light paths and chloroplast position in unilateral light. (After Scholz, 1976*a*.)

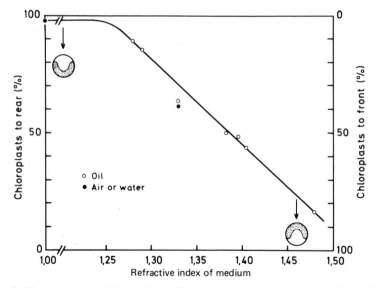

Fig. 7. The percentage of *Hormidium* chloroplasts at front and rear in unilateral light depends on the refractive index of the surrounding medium. (After Scholz, 1976*a*.)

Thus, in *Hormidium* the contribution of the lens-effect to perception of light direction has been proven equally well as in the phototropism of *Phycomyces* (Dennison, 1979).

In both *Vaucheria* and *Hormidium*, the interface between medium and cell wall is the main source of light refraction. It works well, therefore, with air as the medium, but little effect is obtained with cells in water. Thus, light refraction is not a good mechanism for perception of light direction in water. Nevertheless, there are algae living exclusively in water which perform very precise chloroplast orientations. It might be assumed that light traversing the cell is attenuated sufficiently to establish an effective absorption gradient. However, light attenuation cannot explain all the observations, as will be shown in our next system, *Mougeotia*, where a completely different mechanism for perception of light direction has been found.

Absorption dichroism as a mechanism for perception of light direction

The cylindrical cell of *Mougeotia* contains a single ribbon-shaped chloroplast which can rotate in the cell to expose its face to low-intensity light, or its edge to high-intensity light (Fig. 8); only the former will be referred to in this section. This response is obtained equally well with unidirectional light or with light from two opposite directions; thus, a front-to-rear gradient based on attenuation cannot be the decisive factor (Haupt & Bock, 1962).

The chloroplast orientation is controlled mainly by red light, with phytochrome as the photoreceptor (Haupt, 1972, 1973). This is in contrast to all responses referred to before, which are mediated by a blue-light photoreceptor, probably a flavin. But this difference has no bearing on the general conclusions to be drawn.

Phytochrome is a photochromic pigment, which exists in two photo-convertible forms, Pr and Pfr, absorbing red light ($\lambda_{max} = 667$ nm) and far-red light ($\lambda_{max} = 730$ nm), respectively; Pfr is the active form of the pigment. In *Mougeotia*, phytochrome is located in the peripheral cytoplasm (see below) comparable with flavin localisation in the systems mentioned before. Whenever, by differential photoconversion, a gradient of Pfr is established, the chloroplast reorients in the cell; its edges, merging into the peripheral cytoplasm, approach those regions with the lowest level of Pfr, as shown in Fig. 9 (Haupt, 1972). Thus, the question arises: How does unilateral light cause a differential photo-conversion of phytochrome and hence a gradient of Pfr? Neither light

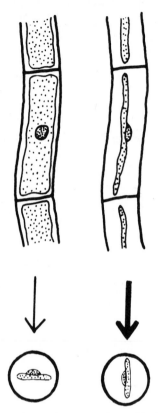

Fig. 8. *Mougeotia* in surface view and cross section, with the chloroplast orienting to low-intensity light (left) and high-intensity light (right). (After Haupt, 1963.)

attenuation in the cell nor light refraction are sufficient to explain such a gradient in *Mougeotia*.

An important observation is the action dichroism displayed by *Mougeotia*: in linearly polarised light, maximal chloroplast orientation is displayed by cells with their long axes approximately at a right angle to the electrical vector of the light; no response at all is found in cells parallel to the electrical vector (Fig. 10A). This action dichroism implies a highly ordered pattern of the transition moments of the photoreceptor molecules. By irradiating small areas of the cell with monochromatic polarised microbeams, such an array has been demonstrated (Fig. 10B), with the Pr transition moments oriented parallel to the surface and along helical lines. The resulting absorption dichroism is the basis of perception of light direction in *Mougeotia* (Haupt & Schönbohm, 1970). For a better understanding, we replace, theoretically, the helically oriented transition moments by their longitudinal and azimuthal compo-

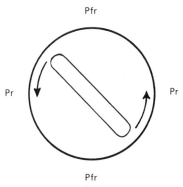

Fig. 9. Schematic cross-section through a *Mougeotia* cell, with phytochrome locally being mainly in the Pr or Pfr form, respectively. Orientation of the chloroplast in this Pfr gradient. (After Haupt, 1972.)

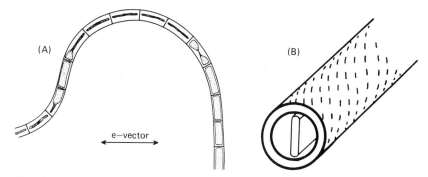

Fig. 10. Action dichroism in *Mougeotia* and its interpretation by absorption dichroism. (A) Chloroplast orientation in polarised red light, impinging normal to the plane of the paper, with the electrical vector along the two-headed arrow. Before the light exposure, the chloroplast was in profile position in all cells. (B) Surface-parallel orientation of the transition moments of phytochrome (Pr) along helical lines. (After Haupt & Bock, 1962; Haupt, 1972.)

nents, thus having two populations of absorption vectors (Fig. 11). Let us irradiate with polarised light, vibrating parallel to the cell axis. Since all the transition moments of the 'longitudinal family' will absorb light, Pr is photoconverted uniformly all around the cell and no gradient is formed (Fig. 11B). No response is observed under those conditions (cf. Fig. 10A). Next, we irradiate with polarised light, vibrating normal to the cell axis. Now the transition moments of the 'azimuthal family' will absorb, but this is possible only in the cytoplasmic layer at front and rear, whereas at the flanks the absorption vector has the wrong orientation. Thus, Pr is differentially transformed to Pfr and a gradient, front-and-rear versus flanks, is established. The chloroplast orients

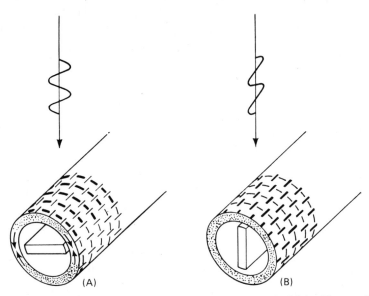

Fig. 11. Absorption characteristics of a *Mougeotia* cell in polarised light. The transition moments (Fig. 10B) are decomposed into longitudinal and perpendicular surface-parallel components. Those which can absorb are drawn in bold; the resulting Pfr formation is indicated by the dots. If the electrical vector is oriented perpendicular (A) or parallel (B) to the cell axis, a Pfr gradient or uniform Pfr results, respectively. (After Haupt & Schönbohm, 1970.)

accordingly (Fig. 11A). Unpolarised light, containing all directions of the electrical vector, can be considered as equivalent to a mixture of longitudinally and perpendicularly vibrating light; thus, differential absorption has to result, with more Pfr being formed at front and rear than at the flanks.

Therefore, whenever photoreceptor molecules are oriented, as in *Mougeotia*, both unilateral light and light coming from opposite directions, has to result in an absorption gradient. However, there are other patterns of surface-parallel photoreceptor molecules, which for theoretical reasons have to lead to an absorption gradient also; the only requirement is absorption components perpendicular to the long axis of the cell. This could be realised if all, or part, of the transition moments were oriented strictly perpendicularly, by a random distribution of orientations within the surface, or by photoreceptor molecules with disk-shaped distribution of transition moments. Figure 12 summarises some of these possibilities. The more the perpendicular component of the transition moments prevails, the stronger the absorption gradient in unpolarised light; whereas the longitudinal component acts to level the gradient.

Fig. 12. Theoretical patterns of surface-parallel orientation of transition moments of photoreceptor molecules. By proper decomposition into components, they all yield the pattern shown in Fig. 11.

These considerations can also be applied to spherical cells. In this case a front-and-rear versus equatorial-region gradient is established by unilateral light if the photoreceptor molecules are randomly distributed in surface-parallel orientation. This is found, for example, in the spores of *Equisetum*, and in the zygotes of *Fucus* where the rhizoid germinates from the darkest regions. Under certain conditions, polarised light results in germination at particular sites near the equatorial girdle (Weisenseel, 1979). Gradients will also develop in cells which have their photoreceptors oriented other than parallel to the surface: if they are oriented normal to the surface, front and rear absorb least and thus a gradient, equatorial-region versus front-and-rear, is established. Such a pattern has been reported for the conidia of the mould *Botrytis* where the germ tube originates from the brightest regions (Weisenseel, 1979). It is apparent that an absorption gradient cannot result if the photoreceptor molecules are *randomly* distributed in all three dimensions. But any deviation from a random distribution dictates that orientation of the transition moment has to contribute to perception of the light direction.

There is good reason to assume such a contribution in several examples of chloroplast orientation (Haupt, 1982).

Perception of light direction under saturation conditions

From these generalisations, we return to *Mougeotia*, in which phytochrome is the photoreceptor pigment. Because of its photochromic properties, any saturating irradiation results in a photostable state between Pr and Pfr, that is a well-defined fraction of $Pfr/(Pr + Pfr)$, which depends on the wavelength. An absorption gradient, then, would mean only that this photostable state is reached sooner at one point and later at another, but eventually the same photostable Pfr state would be reached at any point of the periphery. Since phytochrome photoconversion saturates at low irradiances and within a few minutes, no effective

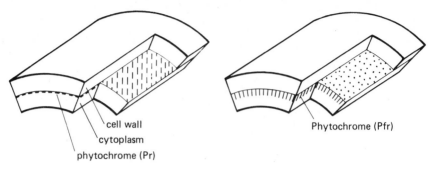

cell wall
cytoplasm
phytochrome (Pr)

Phytochrome (Pfr)

Fig. 13. Part of cell wall and cytoplasm with phytochrome molecules at the cytoplasmic membrane (dashes). Orientation of transition moments parallel (Pr) or normal (Pfr) to the cell surface. (After Haupt, 1972.)

Pfr gradient could be expected to be reached in continuous and hence saturating light. Nevertheless, in Nature, where light is acting continuously, we do observe chloroplast orientation equally as well as after pulse-irradiation in the laboratory. This can be explained only if a permanent Pfr gradient is obtained even in saturating light. We have to investigate how this is possible.

It has been mentioned above that the phytochrome molecules are oriented parallel to the cell surface and along helical lines. Strictly speaking, this has been shown only for the red-absorbing form of phytochrome, Pr. In contrast, the far-red absorbing form, Pfr, is oriented normal to the cell surface (Fig. 13). This can be demonstrated by polarised microbeams of red and far-red light, directed at various regions of the cell. To understand these experiments, we have to remember that the edge of the chloroplast always moves away from the region with the highest Pfr concentration. Thus, local movement of the chloroplast indicates local increase in Pfr concentration.

In Fig. 14 one of these experiments is shown, with microbeams placed at the cell flank (Haupt, Mörtel & Winkelnkemper, 1969). Obviously a red microbeam can establish Pfr only if vibrating parallel to the surface, and this is consistent with the surface-parallel orientation of the transition moment of Pr. If, however, we try to reverse this induction by photoconverting Pfr back to Pr with a far-red microbeam, its electrical vector has to be oriented normal to the surface, demonstrating the surface-normal orientation of the transition moment of Pfr. Thus, upon each photoconversion of phytochrome, the transition moment flips to a new position by about 90°. Recent studies to investigate this reaction at the molecular level have led to the suggestion that the chromophore

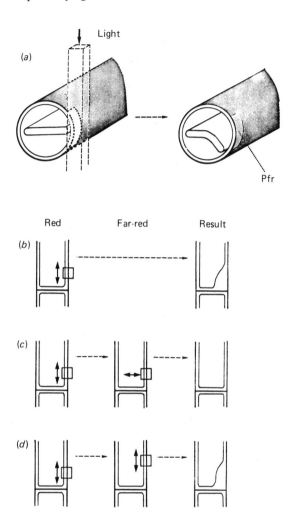

Fig. 14. Local irradiation of a *Mougeotia* cell with a microbeam (*left*) and response of the chloroplast (*right*). (*a*) General view; (*b–d*) the individual irradiation protocols. Two-headed arrows indicate the electrical vector of the light in the microbeam. Induction by parallel vibrating red light indicates surface-parallel Pr, its reversion by perpendicularly vibrating far-red light indicates surface-normal Pfr. (After Haupt, Mörtel & Winkelnkemper, 1969.)

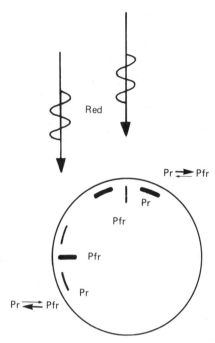

Fig. 15. Dependence of the photoequilibrium Pr ⇌ Pfr on the dichroic orientation of Pr and Pfr in the *Mougeotia* cell. Irradiation with polarised red light, vibrating normal to the cell axis. Molecules with an orientation favourable for absorption are drawn in bold. The corresponding shifts of the photoequilibrium are indicated. (After Haupt, 1972.)

swings into and out of a crevice in the protein moiety of the pigment (Hahn, Kang & Song, 1980; Song, this volume).

Whatever the molecular mechanism, the flip-flop dichroism of phytochrome can easily explain a gradient of Pfr even in saturating light (Haupt, 1972). If we assume unilateral monochromatic light (which in a randomly distributed phytochrome population would establish a given Pfr/(Pr + Pfr) level) and a *Mougeotia* cell with phytochrome partly in the surface-parallel Pr form, partly in the surface-normal Pfr form, and consider only the electrical vector of light which vibrates perpendicular to the cell axis, we can predict what will happen (Fig. 15). At front and rear, the Pr molecules are in a favourable geometric position to absorb the light, but the Pfr molecules are out of direction. Accordingly, the photoequilibrium is shifted towards a higher level of Pfr. In contrast, at the flanks, the Pfr molecules have optimal geometric orientation for absorption, but the Pr molecules do not. Thus, at the flanks the photoequilibrium is shifted towards Pr. As a result, the photoequilibrium becomes dependent on the azimuth and a permanent Pfr gradient results. It is true that the photostable state of Pfr depends on the wavelength, but the azimuthal-dependent shifts to the right and to the left always occur, and thus a gradient in saturating light is ensured as long as the flip-flop dichroism is working.

This latter statement contains an interesting problem, which is still unsolved. In *Mougeotia*, chloroplast orientation can be controlled by blue light also, although with less efficiency. We do not know yet for certain the photoreceptor for this blue-light effect (Haupt, 1982). However, if we seek to attribute the orientation in continuous blue light to phytochrome, we must check whether or not the flip-flop dichroism occurs in blue light.

Interaction of several mechanisms for perception of light direction

Having become familiar with the main mechanisms to transform light direction into an absorption gradient, we return to the alga *Hormidium* with the aim of demonstrating an interaction of different mechanisms in one system (Scholz, 1976*b*). Remember the *Hormidium* cell acts as a collecting lens if surrounded by air and therefore the chloroplast moves to the rear. This is true both in white and blue light and the final proof is the reversion (movement to front) when paraffin oil is the surrounding medium.

There is good evidence that the blue-light response is mediated by a flavin acting as the photoreceptor, which is localised in the cytoplasm. However, there is also a small orienting effect of red light and here the chloroplast always goes to the front irrespective of whether air or oil is the medium. The same behaviour can be found with blue light provided the cells have been kept in darkness for a few days, which results in a strong reduction of the cell's flavin content. This change in behaviour is reversed as soon as the cells are provided with flavin again. The changed response type is also found if the flavin triplet state is quenched by potassium iodide. It is assumed, therefore, that there is a second photoreceptor system which absorbs in the blue and red regions and detects the light direction other than via light refraction. This is possible if such a system is localised in a different compartment from the flavin, and a reasonable candidate is the chloroplast with chlorophyll as the photoreceptor. The absorption gradient front-to-rear would then be simply due to attenuation of the light when traversing the chloroplast.

In blue light under normal conditions, the light refraction mechanism strongly predominates over the attenuation mechanism, and the latter can be detected only under specific conditions. The effect of light

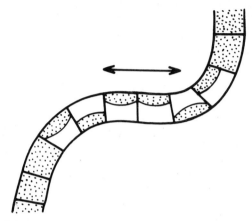

Fig. 16. Action dichroism in *Hormidium*. Chloroplast orientation in polarised blue light, impinging normal to the plane of the paper, with the electrical vector along the two-headed arrow. Before the light exposure, the chloroplast was in profile position in all cells. (After Haupt, 1972.)

refraction, however, can be additionally modified by the effect of absorption dichroism. If technical oils are mixed so that their resulting refractive index exactly matches that of the cell, a random distribution of chloroplasts would be expected. However, in most of the cells the chloroplast goes to either the front or the rear with equal probability, but the flanks are empty (Scholz, 1976*a*). Thus, even with no light refraction, a front-and-rear versus flanks gradient must have been established. This is most easily explained by a surface-parallel dichroic orientation of the photoreceptor molecules; and indeed, as can be seen in Fig. 16, *Hormidium* can exhibit a pronounced action dichroism in polarised light (Scholz, 1976*b*).

Thus, in *Hormidium* all three mechanisms – attenuation, refraction and dichroism – interact in transforming light directionality into an absorption gradient.

Localised transduction processes as a requirement for intracellular orientation movement

Intracellular orientation movement in response to light direction not only requires an absorption gradient, that is local differences of light absorption, but also local transduction into the response: the transduction processes must not spread over the whole cell. It is therefore interesting to review briefly what is known about these processes in two of our examples.

First we consider *Vaucheria* and describe the microbeam experiment more thoroughly (Fig. 4). In darkness or red light all chloroplasts move in an approximately longitudinal direction, with frequent autonomous reversal of direction. If a small area is illuminated, all chloroplasts entering this field stop moving and hence are captured there, which results in the accumulation mentioned earlier (Fischer-Arnold, 1963).

How does light stop the autonomous movement? By combined ultrastructural and inhibitor studies it has been shown that actin microfilaments are responsible for generating the motive force for the longitudinal movements, and light acts by inactivating these microfilaments (Blatt & Briggs, 1980; Blatt, Wessels & Briggs, 1980). Moreover, an outward electrical current has been discovered at the illuminated surface, which has been found to be an electrogenic proton efflux causing the cell to hyperpolarise. This effect slightly precedes the inactivation of the microfilaments and the chloroplast accumulation. After switching off the light, the former cell-potential is re-established. All three light effects – proton efflux, microfilament inactivation and chloroplast accumulation – have the same dependence on light intensity and wavelength (Blatt, Weisenseel & Haupt, 1981).

It is tempting, therefore, to suggest the following transduction chain (Blatt, Weisenseel & Haupt, 1981; Haupt, 1982): light via a flavin activates a proton pump; the resulting change in potential causes redistribution of other ions, which finally control actin–myosin interaction so as to stop movement. Obviously the modified ion redistribution is a strictly local function of light absorption without any signal transmission.

A similar approach has been made in *Mougeotia* (Haupt, 1982). Here, too, actin microfilaments have been demonstrated to exist. Their localisation, their behaviour during movement and the results of inhibitor experiments strongly point to actin–myosin interaction as the motive-force generator (Schönbohm, 1975; Wagner & Klein, 1978; Klein, Wagner & Blatt, 1980). It is well known that calcium ions control this interaction in muscles. Accordingly the role of calcium in the transduction chain has been investigated. Calcium uptake by the cell is increased by red light, and this is reversible by far-red light (Dreyer & Weisenseel, 1979). Thus, calcium uptake is under Pfr control. Moreover, calcium accumulates in vesicles of a particular type, which are most abundant close to the chloroplast edge where actin is supposed to act as the chemomechanical transducer (Wagner & Rossbacher, 1980). Finally, upon depleting the cells of calcium the chloroplast ceases to

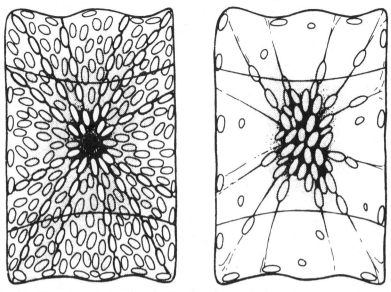

Fig. 17. Chloroplast distribution in *Biddulphia* in low intensity light (left) and in high intensity light or darkness (right), respectively. (After Haupt, 1977.)

orient to light; and it regains its ability to orient soon after calcium is added again (Wagner & Klein, 1978).

Although we have still incomplete knowledge of the whole transduction chain, calcium seems to be an important messenger connecting light absorption by phytochrome with actin–myosin interaction. Again, this messenger acts strictly locally, thus ensuring that local differences in light absorption are transduced into local activity of the motive force. The well-known slowness of the diffusion rate of calcium makes this ion particularly well suited for this function.

Finally, an exception provides further support for our thesis. There are some diatoms which perform a different type of chloroplast redistribution (Senn, 1919): in low-intensity light, the chloroplasts are spread over the whole cell or over its surface, but in darkness and in strong light they assemble in the centre around the nucleus (Fig. 17). No orientation to the light direction is observed. Full displacement can be obtained even by illuminating part of the cell, provided the total number of quanta absorbed by the cell corresponds to the requirement for that response. Obviously, the signal generated by differential light absorption does not remain localised, but spreads over the whole cell. Thus, lack of orientation does not necessarily mean lack of perception of light direction: it could be due to lack of localised transduction leading to the uniform activation/inactivation of the motor apparatus in the cell – even if light as the primary signal has been absorbed in a strong gradient. This exceptional transduction process awaits analysis.

Summary

We may summarise what is known, in stationary cells, about perception of light direction. Cells have evolved three different mechanisms to transform light direction into an absorption gradient: light attenuation, light refraction, and dichroic orientation of the photoreceptors. Furthermore, these mechanisms can be combined and can cooperate in the same cell. With the photochromic pigment phytochrome, a regular change in dichroism upon photoconversion is an important improvement for effectively establishing an absorption gradient.

However, perception of the direction of light, while a necessary condition, is not solely sufficient for a directional response. Equally important is a strongly localised transduction chain effecting localised control of the intracellular motor apparatus. Slowly-diffusing ions seem to fulfil this requirement, but further work needs to be done to elucidate the mechanisms completely.

References

BLATT, M. R. & BRIGGS, W. R. (1980). Blue-light induced cortical fiber reticulation concomitant with chloroplast aggregation in the alga *Vaucheria sessilis*. *Planta*, **147**, 355–62.

BLATT, M. R., WEISENSEEL, M. H. & HAUPT, W. (1981). A light-dependent current associated with chloroplast aggregation in the alga *Vaucheria sessilis*. *Planta*, **152**, 513–26.

BLATT, M. R., WESSELLS, N. K. & BRIGGS, W. R. (1980). Actin and cortical fiber reticulation in the siphonaceous alga *Vaucheria sessilis*. *Planta*, **147**, 363–75.

BRITZ, S. J. (1979). Chloroplast and nuclear migration. In *Encyclopedia of Plant Physiology*, New Series, vol. 7, ed. W. Haupt & M. E. Feinleib, pp. 170–205. Berlin, Heidelberg & New York: Springer-Verlag.

DENNISON, D. S. (1979). Phototropism. In *Encyclopedia of Plant Physiology*, New Series, vol. 7, ed. W. Haupt & M. E. Feinleib, pp. 506–66. Berlin, Heidelberg & New York: Springer-Verlag.

DREYER, E. M. & WEISENSEEL, M. H. (1979). Phytochrome-mediated uptake of calcium in *Mougeotia* cells. *Planta* **146**, 31–9.

FISCHER-ARNOLD, G. (1963). Untersuchungen über die Chloroplastenbewegung bei *Vaucheria sessilis*. *Protoplasma*, **56**, 495–520.

HAHN, T. R., KANG, S. S. & SONG, P. S. (1980). Difference in the degree of exposure of chromophores in the Pr and Pfr forms of phytochrome. *Biochem. Biophys. Res. Commun.* **97**, 1317–23.

HAUPT, W. (1963). Photoreceptorprobleme der Chloroplastenbewegung. *Ber. Deutsch. Bot. Ges.* **76**, 313–22.

HAUPT, W. (1972). Perception of light direction in oriented displacements of cell organelles. *Acta Protozool.* **11,** 179–88.

HAUPT, W. (1973). Role of light in chloroplast movement. *Bio Science,* **23,** 289–96.

HAUPT, W. (1977). *Bewegungsphysiologie der Pflanzen.* Stuttgart: Thieme.

HAUPT, W. (1982). Light-mediated movement of chloroplasts. *Ann. Rev. Plant Physiol.* **33,** 205–33.

HAUPT, W. & BOCK, G. (1962). Die Chloroplastendrehung bei Mougeotia. IV. Die Orientierung der Phytochrom-Moleküle im Cytoplasma. *Planta,* **59,** 38–48.

HAUPT, W., MÖRTEL, G. & WINKELNKEMPER, I. (1969). Demonstration of different dichroic orientation of phytochrome P_R and P_{FR}. *Planta,* **88,** 183–6.

HAUPT, W. & SCHÖNBOHM, E. (1970). Light-oriented chloroplast movements. In *Photobiology of Microorganisms,* ed. P. Halldal, pp. 283–307. London: Wiley.

KLEIN, K., WAGNER, G. & BLATT, M. R. (1980). Heavy-meromyosin decoration of microfilaments from *Mougeotia* protoplasts. *Planta,* **150,** 354–6.

NULTSCH, W. (1964). *Allgemeine Botanik.* Stuttgart: Thieme.

SCHOLZ, A. (1976a). Lichtorientierte Chloroplastenbewegung bei *Hormidium flaccidum*: Perception der Lichtrichtung mittels Sammellinseneffekt. *Z. Pflanzenphysiol.* **77,** 406–21.

SCHOLZ, A. (1976b). Lichtorientierte Chloroplastenbewegung bei *Hormidium flaccidum*: Verschiedene Methoden der Lichtrichtungsperception und die wirksamen Pigmente. *Z. Pflanzenphysiol.* **77,** 422–36.

SCHÖNBOHM, E. (1975). Der Einfluss von Colchicin sowie von Cytochalasin B auf fädige Plasmastrukturen, auf die Verankerung der Chloroplasten sowie auf die orientierte Chloroplastenbewegung. *Ber. Deutsch. Bot. Ges.* **88,** 211–24.

SENN, G. (1908). *Die Gestalts- und Lageveränderung der Pflanzen-Chromatophoren.* Stuttgart: Engelmann.

SENN, G. (1919). Weitere Untersuchungen über Gestalts- und Lageveränderungen der Chromatophoren. IV. und V. *Z. Bot.* **11,** 81–139.

WAGNER, G. & KLEIN, K. (1978). Differential effect of calcium on chloroplast movement in *Mougeotia. Photochem. Photobiol.* **27,** 137–40.

WAGNER, G. & ROSSBACHER, R. (1980). X-ray microanalysis and chlorotetracycline staining of calcium vesicles in the green alga *Mougeotia. Planta,* **149,** 298–305.

WEISENSEEL, M. H. (1979). Induction of polarity. In *Encyclopedia of Plant Physiology,* New Series, vol. 7, ed. W. Haupt & M. E. Feinleib, pp. 485–505. Berlin, Heidelberg & New York: Springer-Verlag.

THE BIOPHYSICS OF INTERMEDIATE PROCESSES IN PHOTORECEPTOR TRANSDUCTION: 'SILENT' STAGES, NON-LOCALITIES, SINGLE-PHOTON RESPONSES AND MODELS

PETER HILLMAN

The Institute of Life Sciences, The Hebrew University of Jerusalem

Introduction

The nature of the transduction process in photoreceptors remains a fascinating and unsolved problem. Considerable progress has been made recently in biochemical studies of the transduction process in vertebrate photoreceptors (Bownds, 1981), and to a lesser extent in those of invertebrates (Fein & Corson, 1981); but in developing biochemical models of transduction only very limited use has been made of the strong constraints provided by biophysical observations. The biophysical requirements that have been taken into account are: that the relevant biochemical pathway must be fast enough not to exceed the physiological response latency; in the vertebrate rod some agent must travel from the disks to the plasma membrane; and that perhaps only a 'positive' transmitter can explain the remarkably large signal-to-noise ratio for weak stimuli (Kilbridge & Ebrey, 1979). However, no attempt has yet been made to adapt any model based on known biochemical pathways to account for a variety of other biophysical observations. It is the purpose of this chapter to review, in a form suitable for confrontation with the biochemistry, some of the most relevant of these biophysical observations.

The observations to which I refer are concerned with the existence of 'silent' stages of the transduction process and states of the cell; with evidence that various processes related to transduction spread out within the cell from the point of light absorption; and with the individual and stochastic properties of single-photon responses.

Certain agents have been found which suspend the transduction process following light absorption but are prior to the appearance of a conductance change. This clearly shows the existence of at least two stages to the process: the first immune to the agents and the second suspended by them. Following the disappearance of a light-induced conductance change, the cell remains for some time in a changed state which affects the responses to further stimuli. This shows the existence of an additional process *parallel* to that inducing the conductance change and outlasting it.

Ranges and speeds of spread of the various processes within the cell have been determined. Comparison of these with measurements of translational diffusion of visual pigment in the photoreceptor membrane, and with calculations of diffusion rates of other materials in the membrane and cytoplasm, demonstrated that pigment diffusion in the membrane and even Ca^{2+} ion diffusion in the cytoplasm are excluded as carriers of at least some of the spreads.

Observations on the stoichiometry and the individual and stochastic properties of single-photon responses (at least in invertebrates) point to separate successive stages in the transduction process underlying the latency and time course of the response; and to contributions to both the latency and the time course from non-diffusional (presumably chemical and probably non-linear) processes.

The biophysical processes discussed below relate to two types of photoreceptor response: the late receptor potential (LRP) and the prolonged depolarising afterpotential (PDA). The PDA has been seen only in invertebrates and only in certain circumstances, and its exact relation to the LRP (which is the normal receptor response) is unclear, yet they probably share at least part of their underlying processes (Hochstein, 1979). The PDA appears following transfer, by an appropriate light stimulus, of a substantial fraction of the visual pigment from the stable rhodopsin (R) to the long-lived meta-rhodopsin (M) state. A further stimulus (usually of different wavelength) transforming a comparable amount of pigment from the M to the R state depresses the PDA or, when presented prior to the PDA-evoking stimulus, in some cases impedes its induction for a limited time. Depression of the PDA and its impedance have very similar parameters (absorption spectrum, absolute sensitivity), so they may be due to the same process (but see Hamdorf & Razmjoo, 1977). The PDA has been seen in all invertebrate preparations in which it has been properly sought, but not yet in a vertebrate.

I begin this review with a survey of several experiments which relate to 'silent' processes, intermediate between absorption of photons and changes in ionic conductance of the plasma membrane, or parallel to the direct process. The primary pigment processes have recently been reviewed by Birge (1981) and the final conductance changes dealt with by Fain & Lisman (1981). The experimental observations prove the existence of intermediate processes and, also characterise them in various ways. In the case of the LRP, the observations we shall deal with relate to the modulation of cell sensitivity by light: light-adaptation and facilitation.

The long dark-persistence of the PDA (up to hours) makes possible experiments on intermediate processes which could not be performed on the stimulus-coincident LRP. For example, it has been shown that there exists a stage or process underlying the PDA even when its expression as a conductance change is suppressed: when the suppression is relieved, a PDA appears *without further light stimulation.*

Intermediate processes in both the LRP and the PDA are shown also by observations on the spread of various processes within the photo-receptor cell, together with demonstrations that translocation of the pigment molecule itself cannot explain all of these spreads, or the non-localities. Pigment translocation is limited in vertebrate rods by the disk–plasma-membrane gap, and in invertebrates by restricted diffusion rates.

The second part of the review is devoted to a survey of determinations of the ranges and speeds of pigment translocation and of the spreads of the transduction processes.

I have calculated corresponding diffusion coefficients assuming free diffusion of a stable material within these ranges. In principle, however, conversion of these coefficients into 'free' diffusion coefficients requires consideration of anatomical barriers (Lamb, McNaughton & Yau, 1981), itinerant adsorption (McLaughlin & Brown, 1981), and the possibility that the material has a finite lifetime. The processes whose spreads are examined are: for the LRP, excitation (activation), light adaptation (desensitisation), and facilitation (enhancement); and for the PDA, excitation, facilitation, and the interaction of the PDA with the PDA-depressing process and with the impedance of PDA.

Measurements and calculations of the diffusion of pigment in the membrane and of other materials in the membrane or cytoplasm are also considered. In the vertebrate rod, for example, diffusion of the pigment cannot be the only spreading agent, as the pigment cannot cross the gap between the disks and the plasma membrane, but pigment diffusion throughout a disk would apparently take a time short compared with the response latency and so could be responsible for part of the spread and amplification (Liebman & Pugh, 1979). In the invertebrate the upper limit on pigment diffusion is low enough to exclude this as the mechanism of spread of LRP excitation, adaptation, and facilitation; but not as the mechanism of spread of PDA facilitation or the PDA–PDA impedance interaction.

In the third part of the review I deal with the properties of the responses of photoreceptors to single photons because a fundamental understanding of the mechanism of generation of single-photon responses is necessary, and perhaps sufficient, for the understanding of the whole transduction process. This assertion is based on the idea that all responses are sums of independent single-photon responses with interactions among them being only perturbations on the basic process. I survey observations on: the stoichiometry; the individual time-course; the stochastic or ensemble properties of single-photon responses; and finally, I present a short critical analysis of extant mathematical models of their time-courses and stochastic properties. These models, in so far as they fit the observations, can serve as a convenient intermediary between the biochemical and biophysical pictures. Most models to date have been linear. Early models predicted the time-course of responses and – very elegantly – the simultaneously diminished amplitude and duration resulting from light adaptation. Recently some attempts have been made to include the stochastic properties of invertebrate single-photon responses – amplitude (or 'area' = amplitude × duration) and latency distributions – but a suggestion has been made that 'ordinary' linear models can produce *only* area distributions declining monotonically, in contrast to the experimentally peaked distributions (N. Grzywacz, personal communication). The number of physiologically reasonable linear models is relatively restricted, and it has been worthwhile seeking out useful specific models, since many can be excluded by the strong constraints provided by the biophysical observations. If it turns out that linear models are untenable and non-linearities are dominant, the same approach may not be worth pursuing. Although one or two non-linear models have indeed been attempted, the possible variety of non-linear models is so great that it is perhaps only through a closer collaboration with biochemistry in the course of development of a model that useful results can be obtained.

Silent stages and states

I define a 'silent stage of the transduction process' as one following light absorption by the visual pigment but preceding any detectable membrane conductance change; and a 'silent state of the photoreceptor' as one in which the cell's response to light is modified even though its membrane potential and conductance are 'baseline'. The silent states to which I will refer follow light stimulation of various kinds in otherwise unmolested cells; the stages are revealed by application of external agents. Such agents suspend the final manifestation of the transduction process as a conductance change but are shown not to have suppressed some earlier stage in the process by the appearance of a conductance change without further stimulation when the agent is removed. The existence of silent stages implies that the transduction process comprises at least two *successive* stages with different characteristics; silent states imply *parallel* processes of different duration.

The late receptor potential

The most common silent state is light adaptation. Bright light reduces the time scale, and particularly the amplitude, of the cell's response to test stimuli long after conductance has returned to the resting level. For a very bright adapting light pigment depletion by bleaching is an element in the vertebrate adapting process, but for weaker adaptation depletion or enhancement of some other material in the cell is probably involved. In the invertebrate, light adaptation is probably due to a rise in Ca^{2+} concentration (Lisman & Brown, 1975); in the vertebrate the material is unknown. A similar phenomenon is facilitation (invertebrate only), in which a cell's sensitivity may be enhanced in certain circumstances following an 'adapting' light. Facilitation is probably *not* due to a fall in Ca^{2+} (Dahl, 1978; M. Hanani, personal communication), and its basis is unknown.

The prolonged depolarising afterpotential (invertebrate only)

A light-induced silent state of facilitation occurs also with respect to the PDA, but it appears only following an adapting light that has induced a PDA (Hillman, Hochstein & Minke, 1976). That is, decline of an unsaturated PDA is always followed by a period during which further PDAs have enhanced amplitudes and durations. The length of this period appears to be comparable with the duration of a saturated PDA in that cell. (A *reduction* in the PDA ('adaptation') by prior stimulation which does not alter the pigment state is never seen.) The facilitation is substantial – the PDA duration can be greatly increased.

Another silent state which influences PDA responses is the impedance of a PDA response by appropriate prior stimulation. Absorption of light by the long-lived invertebrate meta-rhodopsin state of the visual pigment does not appear to lead directly to any cell conductance changes (Minke, Hochstein & Hillman, 1973). However, a stimulus which transfers a substantial net fraction of the pigment from the meta-rhodopsin to the rhodopsin state (not during a PDA) leaves the cell for some time in such a state that PDA induction by a further stimulus is impeded or suppressed. Proposed mechanisms for the impedance include a hypothetical 'low-energy' state of the pigment (Hamdorf & Razmjoo, 1977) and a hypothetical inhibitory transmitter (Hochstein, Minke & Hillman, 1973) (see below).

The most strikingly direct evidence for an intermediate state in the transduction process, however, is the observation that when the response to a stimulus which would otherwise induce a PDA (and LRP) is entirely suppressed by various agents; *the PDA then appears without further stimulation when the agent is removed.* The agents used were anoxia (Wong, Wu, Mauro & Pak, 1976), carbon dioxide (Wong *et al.*, 1976; Atzmon, 1978), background light in the trp^{CM} mutant of *Drosophila*, and temperature in the norp A^{H52} mutant (Minke, 1979). This observation clearly divides the transduction process into two stages; the first is long-lived and unaffected by the agents used, and the second suppressed. There is as yet no experimental hint as to the mechanisms of these two stages, or even whether all the agents act on the same late stage.

Non-localities

A useful and potentially definitive characterisation of any cellular process is provided by a determination of whether, how far, and how quickly the process spreads out from its point of origin. While in the photoreceptor certain undetermined parameters limit the definitiveness of the characterisation, in some cases limits can be obtained which exclude putative mechanisms. In particular, if certain processes are shown to spread out in the photoreceptor, and if translational diffusion of the visual pigment can be ruled out as the mechanism of the spread, an additional substance must be involved in the transduction chain. Furthermore, the observations place limits on the nature of that substance and the medium through which it diffuses.

The late receptor potential: excitation

I shall deal here with the spatio-temporal distribution of ionic conductance change in the photoreceptor plasma membrane with respect to the pigment molecule absorbing the photon. Both indirect and direct evidence argues strongly against the conductance changes being confined to the pigment molecule or its immediate neighbourhood.

Of the indirect arguments, the most fundamental is that based on the very high sensitivity of the dark-adapted photoreceptor. A single photon induces a conductance change which is 1/500 (*Limulus*: Brown & Coles, 1979), 1/20 (toad rod: Baylor, Lamb & Yau, 1979; Bastian & Fain, 1979), or 1/10 (frog rod: Donner & Hemilä, 1978) of the saturated light-dependent conductance change. If one makes the reasonable assumption that the pigment and/or the channels are homogenously distributed through the rhabdomere or along the rod, and notes the high quantum efficiency for inducing responses (Lillywhite, 1977; Baylor *et al.*, 1979), these observations imply a spread of the excitatory influence over at least the corresponding fraction of the cell surface – a distance in all three cases of the order of a few micrometres, in a time of the order of a second. This corresponds to a diffusion constant, D, of about $10^{-7}\,\mathrm{cm^2\,s^{-1}}$. This is of course a lower limit, since either the spread could be impeded (see below) or the excitation could be spreading to much larger areas and, for each photon, failing to saturate the conductance change locally.

In *Limulus*, a further striking argument based on the high sensitivity suggests the necessity for some spread of excitation, though it places a slightly smaller lower-limit on the spread. This is the calculation that if a photon is absorbed at the distal end of a microvillus, no matter how much conductance it opens locally (even if it blows the end off the microvillus), the electrical impedance of the length of the microvillus (1 μm long, 0.1 μm diameter) will prevent a current as large as 5 nA (the observed maximum single-photon current, Behbehani & Srebro, 1974) from flowing (Brown & Coles, 1979). Brown & Coles even argue that cutting off a single microvillus *at its base* is unlikely to supply 5 nA, because of local *extra*cellular impedance, which ultrastructure suggests is comparable with the internal impedance. Again making the assumption of pigment homogeneity, and noting the high quantum efficiency, one can conclude that the excitation process spreads at least a micrometre in less than a second.

Two other indirect observations in invertebrates suggest that some aspect of the excitatory process, but not necessarily the conductance increase itself, spreads out within the cell. The first is the supralinearity of the stimulus–response curve in *Limulus* (Brown & Coles, 1979): when more than about ten photons are simultaneously absorbed in a cell, the response *per photon* begins to increase substantially. (This phenomenon may be related to the 'facilitation' mentioned above.) This observation implies an interaction at a distance of at least several micrometres, and in a time of about a fifth of a second, corresponding to a diffusion constant of at least several times 10^{-6}.

An intriguing recent observation is that of Hamdorf & Kirschfeld (1980), in a fly. They found that the latency of the onset of the response to flashes of light was independent of light intensity up to the point where there began to be an appreciable probability of absorption of more than one photon per microvillus. At this point a quantum reduction in latency was observed in the appropriate fraction of the responses. This suggests that some influence spreads throughout each microvillus in which a photon is absorbed (they could not exclude the possibility that the unit involved comprised up to three microvilli), and that this influence modulates the transduction process which otherwise occurs independently in each microvillus. That this spead is so much more limited in distance than those demonstrated above suggests that different mechanisms are involved, presumably related to different stages of the transduction process.

The observation that has long been recognised as directly requiring a spread of excitation in the vertebrate rod is that of the separateness of the disk and plasma membranes. If one assumes radial diffusion in the cytoplasm from the point of photon absorption to the plasma membrane, a spread of the order of a couple of micrometres in a couple of seconds $(D \sim 10^{-7})$ is implied. However, Baylor *et al.* (1979) note that they see a remarkably small variation in the latency and time-course of single-photon responses (see below). They point out that if these times are determined by diffusion from the point of photon absorption to the plasma membrane, there ought to be a considerable variation in these times according to where in the disk the photon is absorbed. They suggest, therefore, that the cytoplasmic diffusion is considerably faster $(D \gg 10^{-7})$ and that the time-courses are determined by other stages in the transduction process (chemical reaction times). An alternative (McLaughlin & Brown, 1981) might be a primary rapid spread in or on the disk, with a diffusion constant much greater than 10^{-7}, preceding a slower cytoplasmic spread with $D \gtrsim 10^{-7}$, or 10^{-8} if the transmitter has only to bridge the gap between the edge of the disk and the plasma membrane.

Aside from this anatomical argument the most direct measurements of excitation spread are those in which a small portion of the cell is illuminated and the locus of the associated conductance change determined. This must be done by extracellular recording, as the interior of the cell is almost an isopotential and integrates all sources of current. Such recordings have been made using pairs of extracellular microelectrodes (Hagins, Zonana & Adams, 1962; Hagins, Penn & Yoshikami, 1970) or suction electrodes (Fein & Charlton, 1975*a*; Jagger,

1979; Lamb *et al.*, 1981). Stimulation with various intensities was used, and in one case (Fein & Charlton, 1975*a*) the localisation of the membrane current was confirmed for single-photon responses.

In all the measurements the spread of excitation was found to be so small that only an upper limit could be placed on it: 'a few' micrometres in the squid (Hagins *et al.*, 1962), perhaps 50 μm in *Limulus* (Fein & Charlton, 1975*a*), 3 μm in toad rods (Lamb *et al.*, 1981), about 12 μm in rat rods (Hagins *et al.*, 1970), and less than 40 μm (the rod length) in frog rods (Jagger, 1979). Upper limits for the corresponding diffusion constants are about 10^{-6}, 4×10^{-4}, 3×10^{-7}, 4×10^{-6} and 4×10^{-5}. However, the actual 'free' diffusion constants may be considerably higher than these values because of anatomical and itinerant adsorption factors (see below). Furthermore, if the spreads are limited by re-uptake or destruction of the transmitter, rather than by diffusion rates, as is not unlikely, these limits have no direct significance. It may be possible to decide if the spreads are diffusion-limited by comparing the current distribution for flashes to that for weak (in order not to saturate the mechanism) *continuous* stimuli, but this has not been done.

In summary, the diffusion constant of whatever is responsible for the spread of excitation in photoreceptors is likely to be in the range 10^{-6}–10^{-7} cm^2 s^{-1}.

The late receptor potential: adaptation and facilitation

I have already referred to the 'silent' states of adaptation and facilitation, and here review observations on their spatial spreads.

Indirect evidence again is based on the relatively weak light found sufficient to modify substantially the entire cell's sensitivity. In *Limulus*, I calculate from the data of Stieve & Bruns (1980) that a single absorbed photon is sufficient to *increase* sensitivity of an appreciable portion of the cell within a few seconds,[*] although other authors suggest that much higher intensities are required (Hanani & Hillman, 1976; Fein & Charlton, 1977). *Adaptation* begins to set in for 100 photons in *Limulus* (Brown & Coles, 1979; Stieve & Bruns, 1980) but for only about one photon in toad rods (Bastian & Fain, 1979; Baylor, Matthews & Yau, 1980). These results place lower limits on the diffusion constant of about 10^{-6} (*Limulus* facilitation), and 10^{-7} and 10^{-6} (*Limulus* and rat rod adaptation).

---

[*] The results presented are averages only, and in principle could be explained by a very large facilitation of a very small fraction of the cell. However, single-photon responses were examined, and the rare very large responses implied by this hypothesis would probably have been noticed in the raw data.

Direct tests of the spread of adaptation and facilitation require localised illumination of a portion of the cell by both the adapting and the test stimuli (the recording technique is irrelevant). Stimulation techniques were straightforward, except for that of Hemilä & Reuter (1981) who illuminated a frog retina obliquely from the distal side and relied on self-screening by the tightly-packed outer segments to confine the illumination to the distal third or so of the outer segment. In *Limulus* only upper limits to the spread of adaptation (Fein & Charlton, 1975*b*) and facilitation (Fein & Charlton, 1977) were established: some $50\,\mu$m in 30 s and 0.1 s respectively, corresponding to diffusion constants of about 4×10^{-7} and $10^{-4}\,\mathrm{cm^2\,s^{-1}}$. *Values* for the adaptation spread were measured of 5–$20\,\mu$m in frog rods (Hemilä & Reuter, 1981) and $6\,\mu$m in toad rods (Lamb *et al.*, 1981) in a few seconds. These figures correspond to a diffusion constant of $\sim10^{-7}\,\mathrm{cm^2\,s^{-1}}$. Because of the same reservations expressed about the excitation spread, the vertebrate value must be considered a lower limit and the invertebrate upper limits subject to the absence of anatomical and adsorption effects.

Thus a diffusion constant of $10^{-7}\,\mathrm{cm^2\,s^{-1}}$ is consistent with all the facilitation and adaptation observations.

The prolonged depolarising afterpotential: excitation

No direct measurement has yet been made of the spread of conductance corresponding to a PDA from the point of excitation. One observation, however, suggests that at least one component of the PDA mechanism is not localised to its pigment molecules of origin: this is the non-linearity of the stimulus–response relationship (Hillman *et al.*, 1976). This non-linearity manifests itself in two ways: a supralinearity of the amplitude dependence and an increase in the duration of the PDA. The non-linearity begins to appear when the stimulus hits more than a few percent of the pigment molecules, indicating a range of at least a fraction of a micrometre, in a time of the order of a second. (Since each microvillus contains about a thousand pigment molecules, no spread beyond a microvillus is required.) This non-linearity could be related to the PDA facilitation described above, and the long duration of the facilitation period means (if one assumes an intermediate substance as the source) that an initially fast and then very slow (non-exponential) decline of the substance would be required.

A separate experiment places an *upper* limit on what is presumably (but not necessarily) the same spread: if two separate portions of a cell are simultaneously illuminated, the observed PDA will be the sum of the two separate PDAs if no spatial overlap of the processes occurs, while it will be higher and longer if there is overlap because of the non-linearity described above. For technical reasons the experiment was carried out in a slightly different way (Almagor, 1981): the PDA induced in a small spot was compared with that induced by general illumination of the cell when the same number of photons was absorbed by the pigment in both cases. (The total absorption was monitored by using the early receptor potential to measure the ratio of the pigment populations in the rhodopsin and meta-rhodopsin states, Hillman *et al.*, 1976.) The non-saturating, diffuse illumination resulted in a high-amplitude but short-lived PDA, while the saturated concentrated stimulus induced a relatively low but much longer-lived PDA – the lifetime in fact being comparable with that seen following saturating diffuse illumination of the cell. This indicates that the transduction mechanism did not spread out substantially from the spot during the PDA lifetime (a minute or so); or at least its early stages did not, if the 'substance' whose concentration determines the PDA lifetime is an early intermediate (perhaps the pigment itself?).

If the spread of the PDA excitation is limited only by free diffusion, lower and upper limits of about 10^{-11} and $4 \times 10^{-8} \mathrm{cm^2 s^{-1}}$ can be placed on the diffusion constant from these experiments. No observations have been made on the spatial spread of the PDA facilitation.

Interaction of the PDA with the PDA-depressing and impedance processes

Lower and upper limits have been placed on the ranges of these interactions in separate experiments.

If each of two successive diffuse stimuli hits a fairly small fraction, n, of the pigment molecules in a cell, only a sub-fraction of those hit by the first stimulus will also be hit by the second. Thus if the PDA, PDA-depression and PDA-impedance processes were all confined to the pigment molecules in which they were induced, PDA-depressing or PDA-impeding stimuli would have (if n is small) only a small effect on the PDA induced by a similar stimulus since each treatment would involve mainly different pigment molecules. In fact, a much larger effect is observed: the PDA induced by a 2% stimulus (in an exceptionally sensitive cell) is *fully* suppressed by a following 2% PDA-depressing stimulus (Hillman *et al.*, 1976), and a much-reduced PDA is seen following a 20% PDA-inducing stimulus when it is closely preceded by a 20% PDA-impeding stimulus (Almagor, 1981). (This last experiment was done only on one cell and is much less definitive than the PDA-depression experiment.) This result implies a spread of some component of the transduction process (underlying either the PDA or the PDA-depression and PDA-impedance) to at least some tens of neighbouring pigment molecules, or a distance of some tens of nanometres in a few seconds. This result corresponds to a diffusion constant of only 10^{-11} cm^2 s^{-1}.

An upper limit if placed on these interactions by local-illumination experiments (Almagor, 1981). A PDA induced in one half of the cell is almost unaffected by PDA-impeding or PDA-depressing stimuli in the other half. The residual effect is consistent with estimates of light scattering. It is of course desirable to reduce this limit below a 'half-cell', but the illumination can be reduced in size only to a limited extent because of light scattering, and because below about 10% of the cell area (30 μm radius), the PDA, due to its supralinear stimulus–response characteristic, generally becomes too small to monitor. Almagor has used an elegant technique to overcome this problem: she noted in electron-microscopic examination of barnacle photoreceptors that the microvilli were grouped in 'blocks' of parallel microvilli about 1 μm in diameter separated by about 6 μm on average, with a tendency for neighbouring blocks to have orthogonal alignments. Invertebrate photoreceptors, because of the geometry of microvilli and the fact that the chromophore in all photoreceptors remains within the plane of the membrane, have dichroic ratios of at least 2 for illumination perpendicular to the microvillar axis (Moody & Parriss, 1961). Thus exposure of this photoreceptor to two polarised stimuli with mutually orthogonal directions of polarisation should activate two populations of pigment molecules with only a partial overlap and with an average distance between their elements of something like 6 μm. Again, only a limited effect of the PDA-impedance process on the PDA, attributable to population overlap and to light scattering, was found in this experiment, with a 30 s interval between stimuli. By careful analysis of the data, Almagor was able to set an upper limit on the range of the interaction between the PDA and its impedance process which is very close to the lower limit found by Hillman *et al.* (1976). If the value really does coincidentally fall on these limits, the corresponding diffusion constant would be about 10^{-11} cm^2 s^{-1}, which is much less than the lower limits for LRP spread. It seems more likely, however, that the PDA spread is limited not by diffusion rate but by a barrier, either between microvilli or blocks of microvilli.

Mechanisms of spread

It has been suggested (Srebro & Behbehani, 1971) that the mechanism of amplification, and therefore of spread, of the excitation process in *Limulus* might be electrical. However, the observation of comparable amplifications under voltage-clamp (Behbehani & Srebro, 1974), together with the demonstration that the membrane potential is fairly homogeneous except at the highest light levels (Brown & Coles, 1979), appear to have negated this hypothesis. The most likely mechanism of spread therefore appears to be diffusion of some substance *in* the membrane, *on* the membrane, or in the cytoplasm (in the case of the vertebrate rod, necessarily at least partly in the cytoplasm).

Are the measured or calculated diffusion constants of various substances in the membrane or cytoplasm sufficiently small that these materials can be excluded as transmitters for the distances and times of spread described above? Arguments against diffusion being the *determinant* of the response time are relevant to the present question, and will be cited here, only when the calculated diffusion time is comparable with the observed response time. General arguments for and against diffusion being the response time determinant will be mentioned in the next section.

Diffusion in the membrane

The one substance known to be involved in all photoreceptor transduction is of course the visual pigment itself. Could diffusion of this pigment be responsible for any of the spreads described above?

The translational diffusion of the pigment in both vertebrate and invertebrate has been measured by 'local illumination' techniques. In unfixed vertebrate rods of *Necturus* (Poo & Cone, 1973) and frog (Liebman & Entine, 1974) a small area of pigment was bleached, and it was found photometrically that the bleached molecules diffused away and were replaced by unbleached molecules with a radial diffusion constant of about $4 \times 10^{-9}\,\mathrm{cm^2\,s^{-1}}$. There was, as expected, no diffusion along the axis of the rod.

In invertebrates, the pigment does not 'bleach', but interconverts between the rhodopsin and meta-rhodopsin states. This interconversion can be monitored either photometrically or by the early receptor potential (ERP) (Hillman *et al.*, 1976). Using the first technique, Goldsmith & Wehner (1977) placed an upper limit of less than $10^{-9}\,\mathrm{cm^2\,s^{-1}}$ on the diffusion constant of rhodopsin *within* microvilli in the formaldehyde-fixed crayfish photoreceptor. Almagor, Hillman & Minke (1979) used the ERP to place an upper limit of $3 \times 10^{-10}\,\mathrm{cm^2\,s^{-1}}$ on the diffusion constant in unfixed, intact barnacle photoreceptors, but only for diffusion *between* microvilli.

It is clear that invertebrate membranes are more 'rigid' than verte-brate ones. This can be ascribed either to the different lipid composi-tions of the membranes (cholesterol has a higher relative concentration in invertebrate membranes that is known to decrease their fluidity), or to the existence in invertebrates of a cytoskeletal structure associated with microvilli that is not present in vertebrates (H. Saibil, personal communication).

If one assumes universality for the observed value of the diffusion constant in vertebrates, a rhodopsin molecule could diffuse to the periphery of a frog or rat disk in 1.0 or 0.1 s respectively. Therefore, at least in the latter case diffusion of the activated rhodopsin molecule to the disk periphery is not excluded as the mechanism for the initial (radial) part of the spread of excitation and adaptation in the vertebrate rod. However, since the calculated time is a substantial fraction of the observed response time-to-peak, the arguments cited in the next section against diffusion being the major determinant of the response time-course argue against this mechanism of spread. In addition, the crossing to the plasma membrane cannot be explained by pigment diffusion.

In the invertebrate (again assuming universality), the upper limits on pigment diffusion of 3×10^{-10} or even 10^{-9} cm^2 s^{-1} clearly exclude this as the mechanism of spread of LRP adaptation, facilitation, or excita-tion, all of which require diffusion constants in the range 10^{-6}–10^{-7} cm^2 s^{-1}. The PDA spreads, however, which are of the order of 10^{-11} cm^2 s^{-1} (*if* there is no barrier) are *not* excluded as arising from pigment diffusion.

What about diffusion of other substances within the membrane? In the vertebrate, any molecule smaller than rhodopsin, if it experiences the same viscosity, is a candidate for the radial portion of the spread (and in principle also the longitudinal spread, but presumably not for the cytoplasmic gap). In the invertebrate, even the smallest molecule *is* excluded for the LRP spreads if the membrane is as rigid for such molecules as it is for rhodopsin – which is of course uncertain. However, even if the effective viscosity for small molecules were that of the vertebrate membrane, diffusion within the membrane would probably be excluded for the LRP spreads, since a diffusion constant of 10^{-8} cm^2 s^{-1} would be predicted.

Diffusion in the cytoplasm

No direct measurements have been made on diffusion of any substance in photoreceptor cytoplasm, but some conclusions are nevertheless possible. Ca^{2+}, as the putative candidate for both the adaptation spread in the invertebrate and the excitation spread in the vertebrate, is presently the most interesting substance to be considered.

The smallest molecules and ions have a diffusion constant in water of about $10^{-5}\,\text{cm}^2\,\text{s}^{-1}$ (Ca^{2+} about $6 \times 10^{-6}\,\text{cm}^2\,\text{s}^{-1}$), and none of the spreads discussed above require diffusion constants close to this. However, the effective diffusion constant for all substances will be slowed by 'anatomical' factors and for some, additionally, by 'itinerant adsorption'. By anatomical factors I mean restrictions and elongations of the cytoplasmic diffusion pathway by internal organelles and membrane foldings; by 'itinerant adsorption' I mean the transient binding and release of the transmitter on sites in the cytoplasm or on the membrane or organelles. The two factors are of course not actually separable, and indeed are distinguished only by the 'reflectance' of the objects encountered by the transmitter. However, if they *are* separately calculated, the final effect would be the product of the two factors only if they have the same geometrical distribution along the diffusion pathway. This is, for instance, not true for the rod, where the radial and longitudinal diffusion paths are quite different, so that one must at least calculate for each of the two paths separately.

Radial diffusion

If electron-microscopic hints at some inter-disk structure are ignored, this diffusion may be assumed anatomically unimpeded and, in the absence of itinerant adsorption, would require only a few milliseconds for small ions (Lamb *et al.*, 1981) and only a few tens of milliseconds for larger proteins – both short compared with the dim-light response latency. Itinerant adsorption will multiply these values by some factor, which McLaughlin & Brown (1981) calculate as at least 10 to 100 for Ca^{2+}. While the resulting calculated diffusion time of at least a few tens of milliseconds is still small compared to the time-to-peak of the dim-light response of (0.5–2 s) (Lamb *et al.*, 1981), it may be large compared with the very small *variation* in time-course and amplitude of toad rod quantal responses seen by Baylor *et al.* (1979). On the basis of this small variation, Baylor *et al.* concluded that radial diffusion could not be a major source of response delay since if it were, photons absorbed at the centre and at the periphery of the disks should give very different delays. If more quantitative measurements and calculations of quantal response variation and Ca^{2+} diffusion support these considerations, Ca^{2+} would seem to be precluded as the primary excitation transmitter in rods. Diffusion of other small molecules or ions with smaller itinerant adsorptions might of course not be excluded.

Lamb *et al.* (1981) have calculated the anatomical effect of the disk structure of rods on the *longitudinal* diffusion. They conclude that the effective diffusion constant is reduced by a factor which is the ratio of the area reduction factor to the volume reduction factor. They estimate the area reduction factor, which is the ratio of the cross-sectional area of the rod to the area of the disk–plasma membrane gap, as about 100, and the volume reduction factor as about 2, leaving a diffusion reduction factor of about 50. This makes the effective longitudinal diffusion constant expected for small ions or molecules about $2 \times 10^{-7} \, \text{cm}^2 \, \text{s}^{-1}$, compared with their observed value of $10^{-7} \, \text{cm}^2 \, \text{s}^{-1}$ for activation spread. The adsorption factor for this geometry has not been calculated but seems likely to be substantial for Ca^{2+}, at least the factor 10 to 100 calculated for radial diffusion, giving an effective calculated longitudinal diffusion constant for Ca^{2+} of less than $2 \times 10^{-8} \, \text{cm}^2 \, \text{s}^{-1}$. This is incompatible with the diffusion constant for desensitisation given above, so Ca^{2+} seems excluded as a desensitisation transmitter. The possibility that Ca^{2+} is the excitation transmitter is *not* excluded by the results of Lamb *et al.*, since they only present an upper limit (although in one cell they appeared to observe a *value* of $3 \times 10^{-7} \, \text{cm}^2 \, \text{s}^{-1}$). However, the argument given above that the excitation associated with a single-photon response must spread over at least 1/20 of the toad rod in 1–2 s sets a *lower* limit on the diffusion constant which is close to $10^{-7} \, \text{cm}^2 \, \text{s}^{-1}$. This is sufficiently high to make it unlikely that Ca^{2+} is the excitatory transmitter. However, the assumptions and approximations contributing to this conclusion need much more careful analysis before this conclusion can be considered sound.

The situation in invertebrates is less well worked out. The preparations in which lower limits on the LRP adaptation and excitation spreads have been established, *Limulus* and barnacle, have such irregular rhabdomeric geometries that it would be very difficult to calculate the anatomical and itinerant adsorption factors. It does not seem possible at present to exclude diffusion of Ca^{2+} or of even large protein molecules in the cytoplasm as bases for the LRP spreads (but see the general argument below about the temperature-dependence of the single-photon response time-course).

Single-photon responses and models

Stieve (this volume) shows examples of responses of fully dark-adapted photoreceptors to very weak light flashes. The justification for calling these 'single-photon responses' is outlined below.

I stated in the Introduction the importance of understanding single-photon responses, based on the idea that all photoreceptor responses are the sums of these independent events. The following considerations support this idea.

At very low light intensities, when less than a few photons are absorbed per second per cell, there are strong indications in both vertebrates (Baylor *et al.*, 1979) and invertebrates (Lillywhite, 1977) that each photon induces (with some, probably high, excitation efficiency) an independent, and similar, response. In those invertebrate preparations in which such responses have been carefully sought and not found, one may suspect that they are simply too small, or too spread out in time, to be seen above background electrical noise. The alternative that more than one photon may be required to induce a response (the extreme case of non-independence), would lead to an *initially* (at very low intensities) supralinear dependence of bump rate or response amplitude on stimulus intensity, which has so far never been seen. It is true that supralinear dependences *have* been noted in both vertebrates (McNaughton, personal communication) and invertebrates (Brown & Coles, 1979), but in the former case so far only at relatively high intensities and in zero Ca^{2+}, and in the latter only following an initially linear dependence. Of course, any model must ultimately explain the interactions underlying these supralinearities, but I think them still, for the moment, best considered as perturbations on the basic process.

At higher light intensities, the best evidence in favour of the concept of a perturbed summation of single-photon responses is the highly successful 'adapting-bump model' developed mainly by the Rockefeller group (Wong & Knight, 1980; Wong, Knight & Dodge, 1982). (Also see article by Stieve, this volume.) These authors have shown that, with certain simple assumptions about the *average* effect of steady stimulation on the single-photon responses and only (at most) a small additional violation of independence, it is possible to predict quantitatively the response of *Limulus* photoreceptors to steady, or not-too-strongly and rapidly modulated light, entirely from the time-course and stochastic properties of the single-photon responses. (The model cannot yet handle the transient responses to sudden large changes of intensity (but see Wong, Knight & Dodge, 1974).)

The PDA in invertebrates also seems to be made up of single-photon responses, though the evidence is weaker: Minke, Wu & Pak (1975) showed by noise analysis in the trp[CM] mutant of *Drosophila* that the PDA is probably made up of the same single-photon responses as the LRP. An analysis similar to that applied to the invertebrate LRP, though more restricted, has been successfully applied to a vertebrate (Baylor *et al.*, 1979; Baylor, Matthews & Yau, 1980), and it seems not unreasonable tentatively to set as the essential initial task of a bio-chemical model for both vertebrates and invertebrates the prediction of the time-course and stochastic properties of single-photon responses. The 'average' effects of light adaptation on these properties must then be included, but the problem of any other perturbations resulting from bright illumination can be postponed until later. Accordingly, I shall discuss here mainly the time course and stochastic properties of single-photon responses in the dark-adapted cell. I shall also briefly cover the effects of light adaptation on these properties.

It is convenient to divide a description of the characteristics of single-photon responses into their *individual* and their *stochastic* prop-erties. By individual I mean the full time-course of a 'typical' single-photon response following light absorption, and by stochastic I mean the ensemble distribution of the properties of the individual single-photon responses, as well as the stoichiometry of the relation between them and photon absorptions. The ensemble properties discussed are the latency, amplitude and duration distributions. Since not all single-photon responses have the same form of time-course (Goldring, 1980), the latency, amplitude and duration do not fully characterise the individual response, thus this stochastic description is incomplete; but it is all that the present data allow, and all that present modelling can handle.

A full physiological description also covers what happens to all these properties during and following a light exposure.

The stoichiometry

What happens when a single photon is absorbed by a visual pigment molecule in a photoreceptor? Does each event arise from absorption of a single photon; does absorption of a photon give rise to at most one event, and if so with what probability (excitation efficiency)? While these questions have been most carefully investigated in an invertebrate (Lillywhite, 1977), the answers appear to be similar in a vertebrate (Baylor *et al.*, 1979). As already noted, absorption of a single photon is followed, with a high probability, by a single response event.

A reservation must be expressed with respect to the PDA. It is probable that this too is made up of apparently normal light-adapted single-photon responses (Minke *et al.*, 1975), but whether each absorbed photon induces a string of individual responses whose frequency dies away during the course of the PDA (Horridge &

Tsukahara, 1978), or a single response with very long latency, is undetermined. Hamdorf & Razmjoo (1977) calculate that the 'area' (duration × amplitude) of the PDA in *Calliphora* does not exceed the single-photon response area times the number of photons absorbed; but we have come to the opposite conclusion in *Drosophila*, where the PDA is more than six hours long. The main problem is the establishment of the effective area of the presumably light-adapted single-photon response during the PDA. This should be derivable, with certain assumptions, from noise analysis, but this has not yet been done.

The excitation efficiency remains high in *Limulus* up to fairly high light intensities, but then begins to drop substantially in the lateral eye of *Limulus* (Wong & Knight, 1980) but not in the ventral eye (Wong *et al.*, 1982). In the *Drosophila* trp^{CM} mutant, even relatively moderate lights appear greatly to reduce the excitation efficiency (Minke, 1982). Stieve & Bruns (1980) find a small light-induced increase in excitation efficiency in *Limulus* at very low light levels.

The amplitude and time course

In both the vertebrate and invertebrate, the large amplitude of the single-photon response appears to require involvement of many particles at some stage of the process. In the toad rod the conductance decrease is about 30 pS, which is only slightly larger than the squid Na$^+$ channel (Baylor *et al.*, 1979). However, Baylor *et al.* (1979) argue that the smooth fixed shape and fixed size of the single-photon response (±20% amplitude variation) make it unlikely that the number of channel-blocking particles is less than 100. In *Limulus* the observed 2 nS maximum single-photon response conductance would correspond to a single channel with an unlikely 10 nm diameter, and so is also almost certainly made up of at least 100 channels, presumably opened by at least 100 'transmitter' particles.

In both vertebrate (Baylor *et al.*, 1979) and invertebrate (Wong, Knight & Dodge, 1980) preparations, the single-photon response time-course is characterised by a latent period followed by a rapid rise and then a slower decay, and by a very large amplitude. The time-course is generally slower in the vertebrate. The most difficult to model, and therefore the most restrictive and interesting, of these properties are the large amplitude and, in the invertebrate in particular, the shortness of the rise time with respect to the latency. The latency-to-rise-time ratio is small in toad rods (Baylor *et al.*, 1979) but up to 5–10 or more in invertebrates (Wong *et al.*, 1980; Lillywhite, 1977).

In *Limulus* a close fit to the average single-photon response time-course in the ventral eye is given by a Γ function $B(t) = \tau^{-1}(t/\tau)^n e^{-t/\tau}$ with $n = 2$ and τ about 12 ms (Wong et al., 1982), but *offset* from zero time by some 80–300 ms. However, Goldring (1980) suggests that the detailed time-course of the single-photon response is much more complex.

During, and for some time following exposure to bright light, the time-course is slightly shortened and the amplitude greatly reduced in invertebrates (Wong et al., 1982) and apparently also in vertebrates (Baylor et al., 1979). The Γ-function description of the *Limulus* single-photon response time-course remains valid, but with n declining to zero (Wong, 1978) and the offset to 40–180 ms (Wong et al., 1982) at high irradiances. The single photon response amplitude decreases with about the 0.4 and 0.7 powers of the irradiance in *Limulus* lateral and ventral eyes, respectively. Since the amplitude recovers on a time scale of seconds to minutes following an adapting flash, one would expect the amplitude of the response to depend at any time on the declining effect of the adapting irradiance or response rate, backwards in time, with an integration constant of seconds to minutes. In the *Limulus* lateral eye, an integration time of about a second was seen (Wong & Knight, 1980), while in the ventral eye the time was longer than the resolution of the experiment, which was about two and a half seconds (Wong et al., 1982).

The single-photon response duration has a Q_{10} of about 2.5 in *Limulus* ventral eye (Wong et al., 1982), and the time-course is unaffected by lowering external Ca^{2+} concentration (Wong et al., 1982; Stieve, this volume).

The stochastic properties

In preparations from both vertebrates and invertebrates, the amplitude distributions show maxima at non-zero values and relatively narrow distributions around these maxima (Baylor et al., 1979; Stieve & Bruns, 1980).

The latency distribution in *Limulus* ventral eye is fitted well by a Γ-distribution with $n = 3$, τ about 30 ms, and an offset of some 100 ms (Wong et al., 1982). This distribution is considerably broader than the time-course of a single-photon response, so that the impulse response is much more spread out in time than the latter (Wong et al., 1980). The same applies in locusts (Lillywhite, 1977); but in contrast, in the fly *Calliphora* (Hamdorf & Kirschfeld, 1980) and in toad rods (Baylor et al., 1979), there is a remarkably narrow single-photon response latency distribution, as judged from the similarity in the time-courses of the latter and the weak-light impulse responses.

No quantitative data on the distribution of single-photon response durations (or of τ in the Γ-distribution) have appeared, but again the similarity of the time-courses of the single-photon responses and of the impulse responses in toad rods suggests that the durations are narrowly distributed. In the invertebrate, the duration distributions also appear narrow. Presently, fruitful experiments have been designed to determine the *correlation* among the variations of the three major properties of the single-photon response: amplitude, latency, and duration. Stieve (this volume) reports the very important result that there is *no* appreciable correlation between the latency on the one hand and the amplitude and duration on the other.

Under light adaptation, the amplitude distribution in *Limulus* tends to lose its non-zero maximum as the maximum moves towards zero (Stieve & Bruns, 1980). The latency distribution retains its Γ character with n unchanged, but with both the offset and apparently τ reduced by about 40% for an intensity change of 10^5 (Wong *et al.*, 1982).

The Q_{10} of both the offset and the width of the latency distribution is about 4–5 (Wong *et al.*, 1980). Zero-Ca^{2+} increased the latency dispersion by 15–30% (Wong *et al.*, 1982).

Thus the latency offset and dispersion are affected similarly by light adaptation and temperature. Their dependences on light adaptation, temperature and external Ca^{2+} concentration are, however, different from those of the single-photon response time-course. Wong *et al.* (1980) conclude from this that the processes determining the latency and the time-course of the response are separate and different. They support the argument by referring to the results of Pak, Ostroy, Deland & Wu (1976) on the *Drosophila* mutant norp A^{H52}, in which, for temperatures above 17 °C, a very much broadened latency distribution is observed, with no change in the response time course.

The hypothesis of separate latency and time-course processes is strongly supported by the results of the experiments mentioned above on the correlation between single-photon response latency, and amplitude and duration, for the hypothesis predicts no correlation between latency and the other variables – exactly as observed.

Models

The most difficult problems for a model of single-photon responses are the large amplification, the amplitude distribution with a maximum at a non-zero amplitude, the long latency and short rise time in the invertebrate, the large invertebrate latency spread (in *Limulus* and locust but not in the fly) and the simultaneous effect of light adaptation on the amplitude and (at least in the invertebrate) time-course of the responses. I shall refer to each of these separately in the context of intermediate process hypotheses.

The amplitude and amplitude distribution

I have already pointed out that the large single-photon response amplitudes appear to require an intermediate, presumably enzymatic, amplifying process. However, a single-stage first-order process of this kind would result in an amplitude distribution declining monotonically, in fact exponentially, from a peak amplitude. This is because an ensemble of single molecules decaying in one step from an active to an inactive state has an exponential distribution of active lifetimes and therefore, for a fixed enzymatic rate, an exponential distribution of quantities of the product.

An obvious first hypothesis to explain the peaked amplitude distribution is anatomical, by analogy with the vesicular basis of miniature endplate potentials in the neuromuscular junction; that is, some means of ensuring release of a package of material of relatively fixed size. The disks in vertebrate rods and the microvilli of invertebrate photoreceptors form natural vesicles; these could store either the final channel-closing or -opening transmitter or an intermediate enzyme. (In vertebrate cones there is no comparable structure, unless the infoldings of the plasma membrane serve in some way, but there is no information about single-photon responses in cones.)

A related 'anatomical' idea is that a fixed number, n, of enzyme molecules is directly attached to each pigment molecule and activated by a photon. However, the narrowness of the amplitude distribution calls for an n of at least 10 to 15 (Grzywacz, personal communication), an unlikely figure for the number of directly-attached molecules – especially since this large quantity of protein would probably have been seen biochemically (the visual pigment is by far the most dominant protein in all photoreceptors examined until now).

An alternative to the release of a fixed quantity of transmitter is that amplitudes could be defined by limiting the area to which the transmitter diffuses, with the conductance change being saturated in this area. Since this area is much bigger than a disk or a microvillus, no appropriate barriers appear to exist in rods or most invertebrates although, in barnacle, blocks of microvilli of perhaps the appropriate size are indeed seen (see above; but single-photon responses have *not* been seen in barnacle).

Within the framework of the vesicular model, the transmitter or enzyme would have to do two things: open or close channels over a large area (a few square micrometres in the dark-adapted cell) and desensitise a comparable area with comparable onset and much slower decay times. Two separate transmitters could be involved, but a single transmitter could do both jobs with the slower adaptation recovery time-course being due to local reactions or to the local release of a secondary transmitter. This last possibility is the suspected mechanism in invertebrates, where the unknown primary transmitter is presumed to open Na^+ and Ca^{2+}, or Na^+/Ca^{2+} channels, and Ca^{2+} inhibits the primary transmitter or competes locally with it at channel receptor sites.

An alternative explanation for adaptation is that transmitter release is inhibited. Exhaustion or inhibition at the *primary* source is excluded by the observation of inhibition spread, but the excitation transmitter, or another transmitter, could inhibit release of excitation transmitter at other vesicles.

In summary, the idea that the peaked amplitude distribution is explained by the release or activation by a single photon of the entire contents of a single disk or microvillus is compatible with present observations. In the invertebrate at least, it seems possible that light adaptation is due to competition with the primary transmitter at channel-opening sites by a locally released secondary transmitter (Ca^{2+}).

An entirely distinct approach to the problem of peaked amplitude distributions is that based on the types of chemical reaction cascade used to explain the single-photon response time-course without anatomical limitation. Tiedge (1981) has shown that a linear, non-saturating chain of first-order reactions with two stages of amplification can lead to a peaked amplitude distribution close to that observed in *Limulus*. However, the calculations are difficult, and there is a suspicion that the model implies a correlation between amplitude and duration, and that the *area* distribution (amplitude × duration) in this model (in fact in all linear models with a single active stage), is necessarily monotonically declining. Further calculations as well as experiments are clearly called for.

There is, however, at least one linear chain model which apparently will give a peaked amplitude distribution. The idea is to turn the pigment molecule, or some successor, into an enzyme with a relatively fixed, non-exponentially distributed, lifetime. This would be the case if the molecule cascaded thermally through a series of stages, and were enzymatically active not at one stage only, but at several successive stages. Such a situation would occur, for instance, if the pigment molecule cascaded thermally through a series of states, all of which, up to the nth state, were active; or if, as has been suggested, successive phosphorylations of the active rhodopsin molecule occurred, which left it active until the nth phosphorylation. The lifetime distribution of the enzyme, and therefore the final amplitude distribution can be made indefinitely narrow by increasing the number of active stages. The model has not been quantified, but the number of active stages needed may be large.

It is clear that a wide variety of non-linear models involving feedback can predict peaked amplitude distributions. Although some are quite simple, only one or two have been investigated (Srebro & Behbehani, 1971; Kramer, 1975; also see Levinson, 1972; Baylor, Hodgkin & Lamb, 1974b; and the 'threshold' model below), probably because there are so many possibilities and it is difficult to know where to begin. However, a start must now be made in the context of the growing body of biochemical data, which are known to involve feedback pathways (Bownds, 1981). Perhaps the simplest feedback model which would predict a peaked amplitude distribution is one in which the feedback is used to give an enzyme a relatively defined lifetime, as in the model of the preceding paragraph. This would occur if in a chain of enzymatic reactions the product of a late reaction 'switched off' an early one. Such a reaction, for example by a Ca^{2+}- or GMP-controlled rhodopsin phosphorylation, is biochemically feasible.

The individual time-course

The long latency and relatively rapid rise of the invertebrate single-photon response are particularly striking, and historically these were the first characteristics modelled theoretically (Borsellino & Fuortes, 1968). The first model, based on a filter-chain which could be interpreted as chemical, succeeded brilliantly not only in achieving a good fit to the flash-elicited response and, later, the single-photon response time-courses, but also in explaining the amplitude- and time-scale-reducing influences of light adaptation based on a single 'mechanism'. Before dealing with chemical-chain models in more detail, I will refer to two alternative models for the single-photon response time-course: one is based on diffusion and the second assumes a threshold be-haviour.

Early support for the idea of diffusion as the determinant of the response time-course was Cone's (1973) observation that the time-to-peak of the responses to weak stimuli (though single-photon responses were not considered) increases monotonically with rod radius in a variety of vertebrate preparations. If the single-photon response time-course were determined by diffusion of a transmitter from the point of photon absorption over adjacent homogeneous plasma membrane, there would clearly be no latency in the response. A latency could be introduced either by anatomical distance between the source and plasma membrane (for which a clear basis is visible in vertebrate rods but not in cones nor in invertebrates), or by assuming a chemical delay either before the emission of the transmitter or between the arrival of the transmitter at the channels and their opening or closing. The relatively broad latency distribution in the invertebrate must arise from the portion of the chemical chain that precedes amplification, suggesting that in this case the delay in fact occurs *before* transmitter emission. Support for the idea of two separate mechanisms for the single-photon response latency and intrinsic time-course is provided by Wong *et al.* (1980), who showed that the temperature dependences of the latency and intrinsic time-course of the *Limulus* single-photon response are very different, with Q_{10} values of 4–5 and 2.5, respectively. (A relatively high Q_{10} for the time-to-peak values of responses to weak stimuli was also observed by Baylor, Hodgkin & Lamb (1974*a*) in toad rods.) Although both values are too high for unrestricted diffusion, itinerant adsorption of the transmitter can, while decreasing the diffusion coefficient, in principle increase the Q_{10} to any arbitrary value, depending on the value and Q_{10} of the adsorption coefficient.* Nevertheless, the observed values of the Q_{10} coefficients, in so far as they do not argue against the whole idea of a diffusion contribution, give some further support for this mixed model with a high Q_{10} for the chemically-determined latency and a much lower Q_{10} for the diffusion-determined intrinsic time-course.

Another observation which argues against a diffusion determination of the latency but is consistent with the mixed model has been made in a vertebrate. Baylor *et al.* (1979) found in toad rod that the latency spread is much smaller than would be predicted by the variation in distance from various points on a rod disk to the plasma membrane. However, McLaughlin & Brown (1981) suggest that diffusion could still be the latency determinant if its slowest component is the collection of some material from all over the disk (interior?) for release at the point of photon absorption. This idea predicts a small, but non-zero, spread of single-photon response time-courses, and a careful experimental and theoretical study should determine if the mechanism is feasible. The weakest point of any simple diffusion or mixed model is the absence of any intrinsic, natural, or simple mechanism for light adaptation;

---

* However, McLaughlin & Brown (1981) observe a Q_{10} close to 1 for Ca^{2+} adsorption on artificial lipid membranes.

especially one that speeds up the time-course (both the latency and the intrinsic time-course) as it depresses the amplitude. For instance, a factor which decreases the itinerant adsorption of the transmitter would reduce the latency, but presumably also increase the amplitude. The overall amplitude and intrinsic time scale reductions could come from the chemical chain (see below) but light would then have to act by two entirely different mechanisms on the chemical and diffusion components.

An alternative model based on thresholds (and therefore intrinsically non-linear) has recently been proposed by Payne & Howard (1981) for the time-course of responses to light flashes in the locust. The model assumes a logarithmic rise of a transmitter and a normal (Gaussian) distribution of channel-opening thresholds as a function of transmitter concentration. With only two parameters the authors obtain a good fit to the response time-course and, by varying only one of these parameters, to the effects of light adaptation and temperature on that time-course. The authors do not discuss mechanisms in detail but, when applied to single-photon responses, the model would presumably read mechanistically as follows: a photon activates an enzyme which in turn begins releasing a transmitter whose concentration at first rises linearly but then saturates. Presumably the transmitter then decays in a time that is short compared with some channel 'dead-time'. Channels open for short times when the concentration exceeds certain thresholds whose values are distributed normally about a level approximately half-way up the concentration curve. These 'thresholds' are presumably determined by a cooperativity requirement. Adaptation and temperature vary the rate of transmitter production only.

The model is useful mainly as a remarkably succinct descriptive framework for comparing preparations. The succinctness promises a mechanistic basis, but in its present form its physiological significance is limited by its incompleteness: (i) the degree of cooperativity necessary to explain the strikingly large latency:rise-time ratio has not been calculated, and may be unreasonably high; (ii) the model has no explanation for the most striking effect of light adaptation, which is amplitude reduction or desensitisation; (iii) a homogeneous concentration of transmitter over the channel-loaded area of the membrane is assumed, and no account is taken of the temporal spread of transmitter.

Most other models in the literature are based on chemical reaction chains in one form or another – without specifying at all the particular chemicals involved. It is not surprising that so much attention has been paid to such chains since, on the one hand, chemical chains are found throughout biochemistry, and on the other, they are a natural and simple way of modelling a process with a latency and an amplification.

In general, each stage of such a model is considered to be an enzymatic amplification process with the enzyme having a finite (exponential) lifetime and the product appearing at a constant, non-saturating rate and serving as the enzyme of the next stage. (Each stage can be considered instead as an electrical filter (Fuortes & Hodgkin, 1964), but no physiological correlate to such a filter appears likely.) Each stage is characterised by two parameters, a time constant and a gain or rate constant, and the whole process in addition by the number of stages. This number is determined mainly by the ratio of single-photon response latency to rise time: this is because the latency, which is clearly zero for a single stage, increases more rapidly with number of stages than does the rise time. In fact, the minimum number of stages required turns out to be something like twice the latency : rise-time ratio. This number works out at a reasonable 4 to 6 in vertebrates (Baylor *et al.*, 1974*b*, 1979, 1980), but an uncomfortable 10 to 20 or more in various invertebrates (Payne & Howard, 1981). Both the large number and the variability of this number from species to species argue against the model. One can easily imagine time and rate constants varying with species and even with specific circumstances, but a variation in *number* of stages is much less acceptable. (I owe this argument to R. Payne, personal communication.)

The early models all assumed, for simplicity, equal time and rate constants for all stages, and the values given above for the minimum number of stages are based on this assumption. The choice of equal time constants is also dictated by the fact that, *for equal rate constants*, the least number of stages required to fit a given latency : rise-time ratio is given by a model with equal time constants. On the other hand, if the amplification is concentrated in one or a few stages (see below), the latency : rise-time ratio can be large for a single-photon response even for a single pre-amplification stage. However, for too few pre-amplification stages, the *spread* of latencies becomes very large and the *flash* response develops a low latency : rise-time ratio (see below).

With the assumption of ten stages with equal time and rate constants Fuortes & Hodgkin (1964) were able to get a good fit to the time-course of responses to flashes in *Limulus*. The particular power of the model, however, is in its automatic prediction that varying the single universal time constant results in *both* of the major effects of light adaptation: the reduction in amplification and the speeding-up of the time-course. (Reducing the time constant of course speeds up the whole process and, by reducing the duration of the enzymatic processes, also decreases the amplification.) Furthermore, the fits obtained to experimental data were remarkably good. In the vertebrate, Baylor *et al.* (1974*b*, 1979) found they could get a better fit to the time courses of the flash responses in turtle rods, and of the single-photon responses and flash responses in toad, with a slightly modified reaction-chain model in which the time constants increased along the chain. Six stages were needed for the turtle and four for the toad.

In summary, the distributed-amplification reaction-chain model appears to be able to fit the single-photon and flash responses in vertebrates but perhaps requires too large and too variable a number of stages to do so in invertebrates. Since it is clear from stochastic considerations that these models need modification, they should probably not, despite their triumphs, be pursued in their present form. Nevertheless, it seems likely that single-photon response time-courses in both vertebrates and invertebrates are determined by chemical reaction chains, probably modified by non-linearities and feedbacks.

The latency distribution

The relatively fixed latency of the vertebrate single-photon response (Baylor *et al.*, 1979) puts constraints on the models discussed above for the individual time-course. In particular, the diffusion model has difficulty in explaining the apparent failure of the location of photon absorption (periphery versus centre of disk) to influence the response time-course.

In the reaction-chain model all that is required by the narrow latency spread is that substantial amplification occurs at an early stage of the chain. To illustrate this let us consider a chain consisting of a single particle cascading through several stages and triggering a single amplifying stage. In such a system, the time-course of the response is determined only by the characteristics of the amplifying stage; while the time of arrival of the single particle at that stage, and so the latency and latency distribution are determined solely by the number of stages preceding the amplification and their time constants. (This separation is just what is required by the Wong *et al.* (1980) analysis described above.)

Such a split system has not been considered for the vertebrate, and may not be necessary. Even if one assumes no contribution to the latency spread from diffusion (and there must be *some*), the relatively small overall amplification required may allow a latency spread matching the (as yet unmeasured) experimental value even if the amplification is equally distributed over the stages; and if the calculated value is too large, the model amplification can always be concentrated in the first stage(s).

In the invertebrate, the relatively large latency spread constitutes a more serious constraint. If the spread comes from diffusion, a corresponding variation in the source-target distance would have to be postulated. The only obvious anatomical candidate might be the distance from the point of absorption of the photon to the base of the microvillus. However, as with the latency itself, the high Q_{10} of the latency distribution width argues against the diffusion explanation. In the threshold model, variability of response latency could arise from a variability in the rate of rise of the 'transmitter'. However, one would then expect a close correlation between the single-photon response duration and latency, and Stieve (this volume) reports that there is *no* such correlation.

In the reaction-chain model, the large experimental latency spread means that at least the several initial stages must have little or no amplification. If the overall amplification is distributed equally over the stages, it is clear that this overall amplification must be small. In fact, Borsellino & Fuortes (1968) showed that the ten-stage model which fits the *Limulus* flash response time-course (and the effect of adaptation on it), would also predict the single-photon response latency spread if the overall amplification were assumed to be 25 for 'small' responses and 1 for 'large' responses. The true amplification is far larger than this so, within the reaction-chain model, one is forced to split the chain into non-amplifying and amplifying portions, with the non-amplifying prior portion comprising about ten stages in *Limulus*. Using additional data (Stieve & Bruns, 1980), Tiedge (1981) was able to fit the latency distributions with only six non-amplifying stages, which were followed by two amplifying stages to give the single-photon response time-course.

The source of the non-amplifying stages which make up the latency process is problematical. If the analysis done by Atzmon, Hillman & Hochstein (1978) for the barnacle is correct and applicable to *Limulus*, the transduction process must leave the pigment molecule within a few milliseconds, so the pigment cascade would not be involved. On the other hand, a long enzymatic cascade with unit gain at each stage would have a large percentage of failures – which is inconsistent with the observed high excitation efficiency. A careful analysis of these and related possibilities could be very fruitful. The effects of non-linearities and feedbacks on the latency spread predicted by chemical chains have not been investigated, and it is difficult to see how they fit in.

Conclusions

I have reviewed a variety of biophysical observations relevant to the characterisation of intermediate stages in the transduction process in photoreceptors. I have tried to do this in such a way as to encourage the interaction between biophysicists and biochemists which will surely be the focus of further progress in understanding this process. The main conclusions I have reached are as follows.

The existence of 'silent states of the photoreceptor' suggests that there may be several transduction processes going on in parallel. In addition to the direct process effecting the late receptor potential (LRP) and the prolonged depolarising afterpotential (PDA) there are apparently separate long-lived processes underlying LRP adaptation, LRP facilitation, PDA facilitation and PDA impedance.

Observation of 'silent stages of the transduction process' shows that the direct process comprises at least two successive stages, the first of which is relatively immune to external agents and the second very susceptible.

In both vertebrate and invertebrate photoreceptors, the spread of the various LRP processes within the photoreceptor is greater than can be explained by diffusion of the visual pigment in the membrane; an additional 'transmitter' must therefore be involved in the transduction process. In the vertebrate, this transmitter is probably *not* Ca^{2+} for either the excitation or the adaptation, but other small ions or molecules with less 'itinerant adsorption' are not excluded. In the invertebrate, Ca^{2+} may well be the LRP adaptation transmitter, while pigment diffusion cannot be excluded as the source of the PDA spread.

The study of the characteristics of single-photon responses is most advanced in invertebrates and leads to certain ideas which can only very tentatively be applied to vertebrates. The transduction process appears to be clearly divisible into two stages; the first 'silent', non-amplifying, and with a time scale not primarily determined by diffusion; and the second strongly amplifying, due either to the release of the contents of an anatomical 'vesicle' or more probably to a feedback-controlled enzymatic reaction, and with a time scale which could be determined either by diffusion or by the chemical reactions involved.

In summary, a skeletal model for the transduction process consistent with all the data could be this: following absorption of a photon by a visual pigment molecule, a primary process is initiated as well as several parallel processes, some of which may arise from the primary process. The primary process proceeds through several non-amplifying stages at least the first of which is insensitive to external molestation and the last of which initiates an amplifying process. This amplifying process, which is a feedback-limited enzymatic reaction, releases a 'transmitter' (ion or molecule) into the cytoplasm which diffuses to membrane sites and there modulates ionic conductance. The parallel light-adaptation process also involves a transmitter which reduces the amplification either by competing with the excitation transmitter for the membrane sites or by impeding the amplification process.

I am grateful to Drs Shaul Hochstein and Baruch Minke for critical readings of the manuscript from which I, and it, greatly benefited, and to Karen Handford for remarkably fast and accurate typing.

References

ALMAGOR, E. (1981). Electrophysiological study of spatial processes related to the transduction process in invertebrate photoreceptors. PhD Thesis, Hebrew University of Jerusalem.

ALMAGOR, E., HILLMAN, P. & MINKE, B. (1979). Upper limit on translational diffusion of visual pigment in intact unfixed barnacle photoreceptors. *Biophys. Struct. Mechanism*, **5**, 243–8.

ATZMON, Z. (1978). Does CO_2 abolish light response of barnacle photoreceptor by anoxia? *Israel. J. Med. Sci.* **14**, 1087a.

ATZMON, Z., HILLMAN, P. & HOCHSTEIN, S. (1978). Visual response in barnacle photoreceptors is not initiated by transitions to and from metarhodopsin. *Nature, London*, **274**, 74–5.

BASTIAN, B. L. & FAIN, G. L. (1979). Light adaptation in toad rods: requirement for an internal messenger which is not calcium. *J. Physiol.* **297**, 493–520.

BAYLOR, D. A., HODGKIN, A. L. & LAMB, T. D. (1974a). The electrical response of turtle cones to flashes and steps of light. *J. Physiol.* **242**, 685–727.

BAYLOR, D. A., HODGKIN, A. L. & LAMB, T. D. (1974b). Reconstruction of the electrical responses of turtle cones to flashes and steps of light. *J. Physiol.* **242**, 759–91.

BAYLOR, D. A., LAMB, T. D. & YAU, K.-W. (1979). Responses of retinal rods to single photons. *J. Physiol.* **288**, 613–34.

BAYLOR, D. A., MATTHEWS, G. & YAU, K.-W. (1980). Two components of electrical dark noise in toad retinal rod outer segments. *J. Physiol.* **309**, 591–621.

BEHBEHANI, M. & SREBRO, R. (1974). Discrete waves and phototransduction in voltage-clamped ventral photoreceptors. *J. Gen. Physiol.* **64**, 186–200.

BIRGE, R. R. (1981). Photophysics of light transduction in rhodopsin and bacteriorhodopsin. *Ann. Rev. Biophys. Bioeng.* **10**, 315–54.

BORSELLINO, A. & FUORTES, M. G. F. (1968). Responses to single photons in visual cells of *Limulus*. *J. Physiol.* **196**, 507–39.

BOWNDS, M. D. (1981). Biochemical pathways regulating transduction in frog photoreceptor membranes. *Curr. Top. Memb. Transp.* **15**, 203–15.

BROWN, J. E. & COLES, J. A. (1979). Saturation of the response to light in *Limulus* ventral photoreceptor. *J. Physiol.* **296**, 373–92.

CONE, R. A. (1973). The internal transmitter model for visual excitation: some quantitative implications. In *Biochemistry and Physiology of Visual Pigments*, ed. H. Langer, pp. 275–82. Berlin: Springer-Verlag.

DAHL, R. D. (1978). Facilitation in arthropod photoreceptors. *J. Gen. Physiol.* **71**, 221–2.

DONNER, K. O. & HEMILÄ, S. (1978). Excitation and adaptation in the vertebrate rod photoreceptor. *Med. Biol.* **56**, 52–63.

FAIN, G. L. & LISMAN, J. E. (1981). Membrane conductances of photoreceptors. *Prog. Biophys. Mol. Biol.* **37**, 91–147.

FEIN, A. & CHARLTON, J. S. (1975a). Local membrane current in *Limulus* photoreceptors. *Nature, London*, **258**, 250–2.

FEIN, A. & CHARLTON, J. S. (1975b). Local adaptation in the ventral photoreceptor of *Limulus*. *J. Gen. Physiol.* **66**, 823–36.

FEIN, A. & CHARLTON, J. S. (1977). Enhancement and phototransduction in the ventral eye of *Limulus*. *J. Gen. Physiol.* **69**, 553–69.

FEIN, A. & CORSON, D. W. (1981). Excitation of *Limulus* photoreceptors by vanadate and by a hydrolysis-resistant analog of guanosine triphosphate. *Science*, **212**, 555–7.

FUORTES, M. G. F. & HODGKIN, A. L. (1964). Changes in time scale and sensitivity in the ommatidia of *Limulus*. *J. Physiol.* **172**, 239–63.

GOLDRING, M. A. (1980). Quantum bumps in *Limulus* ventral photoreceptors have complex shapes. *Fed. Proc.* **39**, 2063a.

GOLDSMITH, T. H. & WEHNER, R. (1977). Restrictions on rotational and translational diffusion of pigment in membranes of a rhabdomeric photoreceptor. *J. Gen. Physiol.* **70**, 453–90.

HAGINS, W. A., PENN, R. D. & YOSHIKAMI, S. (1970). Dark current and photocurrent in retinal rods. *Biophys. J.* **10**, 380–412.

HAGINS, W. A., ZONANA, H. V. & ADAMS, R. G. (1962). Local membrane current in the outer segments of squid photoreceptors. *Nature, London*, **194**, 844–7.

HAMDORF, K. & KIRSCHFELD, K. (1980). 'Prebumps': evidence for double-hits at functional subunits in a rhabdomeric photoreceptor. *Z. Naturforsch.* **35c**, 173–4.

HAMDORF, K. & RAZMJOO, S. (1977). The prolonged depolarization after potential and its contribution to the understanding of photoreceptor function. *Biophys. Struct. Mechanism*, **3**, 163–70.

HANANI, M. & HILLMAN, P. (1976). Adaptation and facilitation in the barnacle photoreceptor. *J. Gen. Physiol.* **67**, 235–49.

HEMILÄ, S. & REUTER, T. (1981). Longitudinal spread of adaptation in the rods of the frog's retina. *J. Physiol.* **310**, 501–28.

HILLMAN, P., HOCHSTEIN, S. & MINKE, B. (1976). Nonlocal interactions in the photoreceptor transduction process. *J. Gen. Physiol.* **68**, 227–45.

HOCHSTEIN, S. (1979). On the implications of bistability of visual pigment systems. *Biophys. Struct. Mechanism*, **5**, 129–36.

HOCHSTEIN, S., MINKE, B. & HILLMAN, P. (1973). Antagonistic components of the late receptor potential in the barnacle photoreceptor arising from different stages of the pigment process. *J. Gen. Physiol.* **62**, 105–28.

HORRIDGE, G. A. & TSUKAHARA, Y. (1978). The distribution of bumps in the tail of the locust photoreceptor afterpotential. *J. Exp. Biol.* **73**, 1–14.

JAGGER, W. S. (1979). Local stimulation and local adaptation of single isolated frog rod outer segments. *Vision Res.* **19**, 381–4.

KILBRIDGE, P. & EBREY, T. G. (1979). Light-initiated changes of cGMP levels in the frog retina measured with quick freezing techniques. *J. Gen. Physiol.* **14**, 415–26.

KRAMER, L. (1975). Interpretation of invertebrate photoreceptor potentials in terms of a quantitative model. *Biophys. Struct. Mechanism*, **1**, 239–57.

LAMB, T. D., MCNAUGHTON, P. A. YAU, K.-W. (1981). Spatial spread of activation and background desensitization in toad rod outer segments. *J. Physiol.* **319**, 463–96.

LEVINSON, J. Z. (1972). Interpretation of generator potentials. In *Physiology of Photoreceptor Organs, Handbook of Sensory Physiology*, Vol. VII/2, ed. M. G. F. Fuortes, pp. 339-56. Berlin: Springer-Verlag.

LIEBMAN, P. A. & ENTINE, G. (1974). Lateral diffusion of visual pigment in photoreceptor disk membranes. *Science*, **185**, 457–8.

LIEBMAN, P. A. & PUGH, E. N. JR. (1979). The control of phosphodiesterase in rod disk membranes: kinetics, possible mechanisms and significance for vision. *Vision Res.* **19**, 375–80.

LILLYWHITE, P. G. (1977). Single photon signals and transduction in an insect eye. *J. Comp. Physiol.* **122**, 189–200.

LISMAN, J. E. & BROWN, J. E. (1975). Effects of intracellular injection of calcium buffers on light adaptation in *Limulus* ventral photoreceptors. *J. Gen. Physiol.* **66**, 489–506.

McLAUGHLIN, S. & BROWN, J. (1981). Diffusion of calcium ions in retinal rods. *J. Gen. Physiol.* **77**, 475–87.

MINKE, B. (1979). Transduction in photoreceptors with bistable pigments: intermediate processes. *Biophys. Struct. Mechanism*, **5**, 163–74.

MINKE, B. (1982). Light-induced reduction in excitation efficiency in the *trp* mutant of *Drosophila*. *J. Gen. Physiol.* **79**, 361–85.

MINKE, B., HOCHSTEIN, S. & HILLMAN, P. (1973). Antagonistic process as source of visible-light suppression of afterpotential in *Limulus* UV photoreceptors. *J. Gen. Physiol.* **62**, 787–91.

MINKE, B., WU, C.-F. & PAK, W. L. (1975). Induction of photoreceptor voltage noise in the dark in *Drosophila* mutant. *Nature, London*, **258**, 84–7.

MOODY, M. F. & PARRISS, J. R. (1961). The discrimination of polarized light by *Octopus*: a behavioural and morphological study. *Z. Vergl. Physiol.* **44**, 268–91.

PAK, W. L., OSTROY, S. E., DELAND, M. C. & WU, C.-F. (1976). Photoreceptor mutant of *Drosophila*: is protein involved in intermediate steps of phototransduction? *Science*, **194**, 956–9.

PAYNE, R. & HOWARD, J. (1981). Response of an insect photoreceptor: a simple log-normal model. *Nature, London*, **290**, 415–16.

POO, M.-M. & CONE, R. A. (1973). Lateral diffusion of rhodopsin in *Necturus* rods. *Exp. Eye Res.* **17**, 503–10.

SREBRO, R. & BEHBEHANI, M. (1971). A stochastic model for discrete waves in the *Limulus* photoreceptor. *J. Gen. Physiol.* **58**, 267–86.

STIEVE, H. & BRUNS, M. (1980). Dependence of bump rate and bump size in *Limulus* ventral nerve photoreceptor on light adaptation and calcium concentration. *Biophys. Struct. Mechanism*, **6**, 271–85.

TIEDGE, J. (1981). Deterministische und stochastische Modelltheorien zur Photorezeption des Pfeilschwanzkrebses *Limulus polyphemus*. PhD Thesis, Aachen University.

WONG, F. (1978). Nature of light-induced conductance changes in ventral photoreceptors of *Limulus*. *Nature, London*, **276**, 76–8.

WONG, F. & KNIGHT, B. W. (1980). Adapting-bump model for eccentric cells of *Limulus*. *J. Gen. Physiol.* **76**, 539–57.

WONG, F. F., KNIGHT, B. W. & DODGE, F. A. (1974). Transient behaviour of the adapting-bump model parameters in the eccentric cell of *Limulus*. In *Electrophysiology: Photoreceptors*, ARVO, Florida.

WONG, F., KNIGHT, B. W. & DODGE, F. A. (1980). Dispersion of latencies in photoreceptors of *Limulus* and the adapting-bump model. *J. Gen. Physiol.* **76**, 517–37.

WONG, F., KNIGHT, B. W. & DODGE, F. A. (1982). Adapting-bump model for ventral photoreceptors of *Limulus*. *J. Gen. Physiol.* **79**, 1089–113.

WONG, F., WU, C.-F., MAURO, A. & PAK, W. L. (1976). Persistence of prolonged light-induced conductance change in arthropod photoreceptors on recovery from anoxia. *Nature, London*, **264**, 661–4.

DROSOPHILA MUTANTS WITH REDUCED RHODOPSIN CONTENT

R. S. STEPHENSON, J. O'TOUSA, N. J. SCAVARDA, L. L. RANDALL and W. L. PAK

Department of Biological Sciences, Purdue University, West Lafayette, Indiana
47907, USA

Introduction

One of the objectives of our research in the past few years has been to identify the genes involved in rhodopsin synthesis and/or function, and to isolate mutations in these genes. The rationale for this work is that these mutations are likely to provide invaluable clues to the role(s) that rhodopsin plays in photoreceptor function. Identification of the structural gene for opsin (the gene that codes for the amino acid sequence of opsin) would be particularly interesting, because some mutant alleles of this gene are expected to produce molecular variants of opsin. Such variants would make it possible to investigate the role of opsin in the structure and function of the photoreceptor. Moreover, because of the recent advances in recombinant DNA technology (see Abelson, 1980, and references cited therein), once the opsin gene is precisely localized, it would be possible to explore the molecular organization of the wild-type opsin gene, to determine the nucleotide sequence of the gene, or to prepare large amounts of rhodopsin for biochemical work.

We describe in this review the current status of our effort to identify the genes that are important in synthesizing rhodopsin or maintaining its content at a normal level. Our main approach has been to isolate and examine mutants which have significantly reduced rhodopsin levels compared to wild type. The decrease in rhodopsin level in the mutants was initially detected on the basis of a defect in the prolonged depolarizing afterpotential (PDA).

The PDA is induced by an intense, colored (blue for flies) stimulus that photoconverts a substantial ($\geq 20\%$) net amount of rhodopsin to metarhodopsin. It consists of a depolarization that persists for many seconds to many hours after the stimulus is turned off. Depending on the intensity of the PDA-inducing stimulus, the photoreceptor is inactivated (or desensitized) to a varying degree during the PDA, so that it is incapable of responding with a full amplitude receptor potential to a stimulus applied during the PDA*. These effects of the PDA are reversed by photoconverting a net amount of metarhodopsin back to rhodopsin. Although the physiological significance of the PDA is still obscure, it proved to be valuable as a relatively simple assay for rhodopsin-deficient mutants because the integrity of the PDA process is dependent on the amount of rhodopsin present (Stark & Zitzmann, 1976; Larrivee, Conrad, Stephenson & Pak, 1981).

The *Drosophila* compound eye contains three anatomically distinct classes of photoreceptors, R1–6, R7, and R8. All photoreceptors of the majority class, R1–6, contain the same rhodopsin, R_{480}, which absorbs maximally at ~480 nm and photointerconverts with a thermally stable metarhodopsin, M_{580}, absorbing maximally at ~580 nm (Ostroy, Wilson & Pak, 1974; Harris, Stark & Walker, 1976), while the photoreceptors of the other classes contain classes of rhodopsin that are spectrally different from R_{480} (Harris *et al.*, 1976; for more recent information on larger flies: Kirschfeld, 1979; Hardie, Franceschini & McIntyre, 1979). In this chapter we report the characterization of mutants in which the concentration of the major class of rhodopsin contained in the R1–6 class of photoreceptors (R1–6 rhodopsin, R_{480}) is reduced.

Methods

Mutagenesis

The mutagenesis schemes were designed to isolate mutations from the second and third chromosomes, which comprise ~80% of the *Drosophila* genome. Details of the schemes are described elsewhere (Pak, 1979). Briefly, wild-type male flies of the Oregon-R strain were treated with the chemical mutagen, ethyl methanesulfonate (EMS). Using a suitable balancer stock for either the second or the third chromosome, stocks containing a single EMS-treated second or third chromosome were constructed. Each stock, containing flies homozygous for a single EMS-treated autosome, was then tested for the PDA phenotype.

Almost all flies used in this work, including the rhodopsin-deficient mutants, had their eye color pigments removed genetically. This was achieved by introducing the mutation white (w) into the wild-type stock used for mutagenesis. The elimination of the eye-color pigments facilitated the induction of the PDA and the M-potential, and removed

---

* Two separate mechanisms appear to be responsible for the afterpotential and inactivation processes of the PDA. Thus, there are mutants that lack the sustained depolarizing afterpotential but still display the inactivation phenomenon (Pak, 1979).

sources of extraneous absorbance in spectrophotometric and spectral sensitivity measurements.

Electroretinogram (ERG) recording

The ERGs were recorded from intact, live animals with glass micropipettes filled with Hoyle's saline (for details of the recording method see Larrivee *et al.*, 1981). Voltage signals were amplified with a high-impedance amplifier and displayed on both an oscilloscope and a strip chart recorder. The light source was a Bausch and Lomb tungsten lamp, filtered by Corning filters (CS 5–59 for blue and CS 2–73 for orange stimuli). The light intensities at the plane of the eye were 5.9×10^{14} photons $cm^{-2} s^{-1}$ for blue and 1.5×10^{16} photons $cm^{-2} s^{-1}$ for orange.

Rhodopsin measurements

The amount of rhodopsin in R1–6 photoreceptors was determined by means of the following three methods:

(a) In vivo *spectrophotometry*
In this technique, absorbance changes of R1–6 rhodopsin due to photoconversions between the rhodopsin and metarhodopsin states were determined by performing spectrophotometry on the deep pseudo-pupil (Franceschini & Kirschfeld, 1971) of the living fly (see Stavenga, Zantema & Kuiper, 1973; Lo, 1977; Schinz, Lo, Larrivee & Pak, 1982). The deep pseudopupil consists of the superposition of the virtual images of the corresponding rhabdomere tips of many neighboring ommatidia. The technique, thus, allows spectrophotometry of the rhabdomeres to be performed on the living fly.

(b) In vitro *spectrophotometry*
In this technique, absorbance measurements were carried out on digitonin extracts of fly heads. To obtain an extract, 1000 flies were adapted in the dark and frozen. Their heads were then removed, homogenized, and extracted into a 2% digitonin solution. Absorption spectra were obtained from the extract using a Cary 14 spectrophoto-meter. A more detailed description of the procedure may be found in Ostroy (1978), Larrivee (1979), and Larrivee *et al.* (1981).

(c) Early receptor potential (ERP)
Pak & Lidington (1974) showed that a biphasic response is seen in the initial portion of the fly ERG evoked by a bright orange flash. The

biphasic response, which they termed the M-potential, consists of a small, corneal-negative M_1 followed by a much larger, corneal-positive M_2. Stephenson & Pak (1980) showed that M_1 is a true ERP arising from R1–6 metarhodopsin. Thus, its amplitude is proportional to the amount of R1–6 metarhodopsin activated by a flash and, therefore, may be used as a measure of the amount of visual pigment present in R1–6 photoreceptors.

The ERG setup was used to record the M-potential except that precautions were taken to prevent photoartifacts. The light source was a photographic strobe lamp (0.5 ms duration) filtered by a broadband orange filter (Corning CS 3–67). The quantum flux of the orange flash was 1.5×10^{16} photons cm^{-2} at the plane of the eye.

Spectral sensitivity

Spectral sensitivity was obtained by determining the quantum flux needed to elicit a criterion ERG response of 1.5 mV at ten different wavelengths. To obtain stimuli of different wavelengths, Baird-Atomic interference filters having the following peak transmission wavelengths were used: 376, 387, 401, 431.5, 460, 480, 499, 544.5, 581, and 621.5 nm. Before each set of measurements, animals were adapted in the dark overnight, mounted on the ERG setup in dim red light, and allowed to adapt further for 15 min in the dark. Test flashes were 1.5 s in duration and repeated at intervals of 45 s or longer at any given wavelength.

Freeze-fracture electron microscopy

Details of procedures for freeze-fracture electron microscopy are found in Schinz *et al.* (1982). Briefly, the eyes were frozen in liquid nitrogen-cooled Freon 22 and fractured at a specimen stage temperature of -116 to -106 °C in a vacuum of $<2 \times 10^{-6}$ Torr. The fracture replicas were viewed with a Philips EM 300 electron microscope.

Retinoid supplementation of the media

In this experiment, the Storr's cornmeal–yeast–agar medium was supplemented with the following compounds: β-carotene, retinol, retinal, or retinoic acid. For this purpose, the compounds were dissolved in a suitable solvent (EtOH or EtOEt) and added to the surface of the medium. The solvent was then allowed to evaporate overnight before transferring adult flies to the enriched medium. The quantity added was about 10 μM per vial. β-carotene did not dissolve in the medium but crystallized in patches on the surface. The other retinoids penetrated somewhat into the surface of the medium, to a depth of less than 1 mm.

The compounds added were in the all-*trans* isomeric form. However, no precaution was taken to prevent their photoisomerization. The flies were on the supplemented media for 3 to 15 days before PDA or ERP measurements were performed on them.

Optical neutralization

Individual rhabdomere tips were visualized in the intact head using the optical neutralization technique of Franceschini & Kirschfeld (1971). Heads of red eyed (w^+) flies were removed, mounted on a plexiglass slide, covered with immersion oil, and examined under a compound microscope.

Deficiency stocks

A deficiency for the region containing *ninaE* was constructed by utilizing the stocks and techniques described in Lindsley *et al.* (1972). The stocks used were *T(Y;3)A155* and *T(Y;3)B189*, which have breakpoints at 92A and 92D, respectively, on the third chromosome. The synthesized deficiency, therefore, lacked the region 92A through to 92D.

Results

PDA-defective mutants: general characteristics

We have isolated in the past few years several dozen autosomal mutants that are defective in the PDA. These mutants fall into two classes: those in which the afterpotential is small or absent but the inactivation process is intact, and those in which both the afterpotential and inactivation processes are defective (see footnote on p. 472). We have named the first class of mutants *ina* (*i*nactivation but *n*o *a*fterpotential) and the second class *nina* (*n*either *i*nactivation *n*or *a*fterpotential) (Pak, 1979). We consider only the second class of mutants, *nina*, in this paper because in these mutants the R1–6 rhodopsin level is consistently reduced by a significant amount.

(a) Complementation groups*

To date complementation tests have been carried out on 22 independently isolated *nina* mutants. On the basis of these tests the 22 mutants have been assigned to five complementation groups* shown in Table 1.

---

* A complementation test consists of testing the phenotype of a fly heterozygous for two independently isolated mutations. If the fly displays the mutant phenotype, the mutations are said not to 'complement'. A group of noncomplementing mutations define a complementation group. All of the mutations within a given complementation group are regarded as affecting the same gene.

Table 1. nina *mutants*

Gene	No. of alleles isolated	Map position	Cytogenetic localization
ninaA	2	2–1.4[a]	21D2–22A1[b]
ninaB	2	3–53.5[c]	87E5–87F12[c]
ninaC	9	2nd chrom. (unmapped)	—
ninaD	3	2–57[d]	36C3–37B11[d]
ninaE	6	3–66.1[e]	92A–B[f]

Sources: Data obtained by: [a]N. J. Scavarda & F. Wong; [b]N. E. Kremer & F. Wong; [c]N. E. Kremer; [d]Q. Pye; [e]J. O'Tousa & N. J. Scavarda; [f]S. Davis & J. O'Tousa.

Thus, there appear to be at least five genes on the second and third chromosomes of *Drosophila* which, when defective, yield the *nina* phenotype, to be described below in some detail. These five genes will be designated by *ninaA*, . . . *E*, and alleles of each gene by superscripts over the gene symbol. Thus, for example, *ninaE^{P332}* and *ninaE^{P334}* are two different alleles of the gene *ninaE*. Table 1 also shows the number of alleles isolated in each complementation group and our current knowledge of the chromosome location for each locus.

(b) ERG phenotypes

Fig. 1 compares the ERG phenotype of one representative allele from each of the five *nina* complementation groups with the wild-type ERG phenotype (top row). The stimulus protocol consisted of three unattenuated blue stimuli of 4s duration presented at ~30s intervals followed by three unattenuated orange stimuli of same duration. This protocol was chosen to insure that the R1–6 PDA shows up clearly in the ERG. The mutant ERGs are arranged approximately in the order of their departure from the wild-type ERG.

In wild-type flies, a single blue stimulus produces a fully developed PDA in R1–6 photoreceptors and inactivates them completely so that the second and third blue stimuli elicit only the small responses originating from the central photoreceptors R7 and R8 (Cosens & Wright, 1975; Minke, Wu & Pak, 1975). The first orange stimulus repolarizes and reactivates the R1–6 photoreceptors. In the case of *ninaC^{P238}*, the relative amplitude of the PDA (measured as a fraction of the peak amplitude of the light-coincident response to the first blue

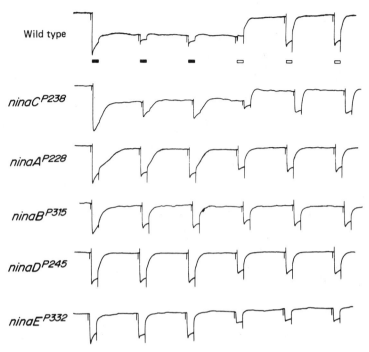

Fig. 1. ERG records from wild type and from one representative allele of each of the five *nina* complementation groups. Records are arranged in order of departure from wild type. Stimulus protocol was three unattenuated blue stimuli (filled bars under wild-type record) followed by three unattenuated orange stimuli (open bars) at 30 s intervals. Stimulus duration was 4 s and calibration pulse shown before each response was 5 mV.

stimulus) is smaller than in wild type. More importantly, the second and third blue stimuli elicit relatively large responses, suggesting that the first blue stimulus has not completely inactivated the R1–6 photoreceptors and that these photoreceptors contribute to the responses elicited by the subsequent blue stimuli. In the case of the other mutants, little or no evidence of either afterpotential or inactivation of R1–6 photoreceptors by the first blue stimulus is discernable by the time the succeeding blue stimulus is delivered, as judged by the fact that the response has returned to the baseline and that the amplitudes of the responses to the second and third blue stimuli are nearly as big as the response to the first blue stimulus. In the case of the *ninaA^P228* and *ninaB^P315* mutants, however, there is a suggestion of an afterpotential that rapidly decays following the first blue stimulus. Even this effect is absent in the case of the other two mutants, *ninaD^P245* and *ninaE^P332*.

Table 2 displays the results of measurements of amount of rhodopsin

Table 2. *Rhodopsin content as percentage wild-type content*[a]

Genotype	In vivo spectrophotometry[b]		Digitonin extracts[c]	ERP (M_1) amplitude[d]
Wild type	100 ± 15 (12)	100 ± 16 (11)	100 ± 9 (5)	100 ± 18 (17)
$ninaC^{P238}$		32 ± 6 (4)	35 ± 3 (3)	
$ninaB^{P315}$	20 ± 8 (17)	18 ± 11 (2)		24 ± 12 (5)
$ninaA^{P228}$	15 ± 4 (10)	13 ± 4 (3)	12 ± 4 (5)	14 ± 7 (10)
$ninaD^{P245}$		3 ± 1 (4)		10 ± 6 (5)
$ninaE^{P318}$		-2 ± 4 (3)		
$ninaE^{P332}$		1 ± 1 (3)		3 ± 8 (4)
$ninaE^{P334}$		-2 ± 5 (3)		

[a] Given as mean $\pm$ SD (n); n = number of extracts for digitonin extracts and number of flies for others. Each digitonin extract was obtained from 1000 heads (see Methods).
Sources: [b]Schinz et al. (1982) (*left*), N. J. Scavarda (*right*); [c]Larrivee et al. (1981); [d]R. S. Stephenson.

contained in R1–6 cells of various *nina* mutants. Three different techniques were used for this purpose: spectrophotometry of the deep pseudopupil of the living fly, spectrophotometry of digitonin extracts of heads, and electrophysiological measurements of the ERP (M_1 potential) amplitude (see Methods). In the case of deep pseudopupil spectrophotometry, two sets of measurements are shown: one obtained by Schinz et al. (1982) before the *ninaD* or *ninaE* mutants were isolated and a more recent set of data obtained by N. J. Scavarda. All values are shown normalized to the corresponding wild-type values. It may be seen that agreement among different sets of measurements is quite reasonable. It is clear from the table that the rhodopsin content in R1–6 photoreceptors is significantly reduced in all *nina* mutants examined. Moreover, the amount of reduction in R1–6 rhodopsin appears to parallel approximately the extent to which the mutant PDA phenotype differs from the wild-type PDA phenotype (Fig. 1). Results such as these suggested that the *nina* mutations affect the amount of rhodopsin in the R1–6 photoreceptors and that the PDA defect arises as a consequence of the decrease in the amount of rhodopsin (cf. Larrivee et al., 1981). This interpretation is consistent with the findings of previous workers (Stark & Zitzmann, 1976; Larrivee et al., 1981) that depletion of rhodopsin by vitamin A deprivation results in the elimination of the PDA in wild-type flies.

(c) Spectral sensitivity

If all five classes of *nina* mutations reduce the rhodopsin content by essentially the same mechanism, one would expect the shape of ERG spectral sensitivities of all five classes of mutants to be similar. Spectral sensitivity measurements were, therefore, carried out, using the ERG, on wild type and on one allelic representative from each of the following four *nina* complementation groups: *ninaA*, *B*, *D*, and *E*. The resulting spectral sensitivity curves are shown in Fig. 2. Five flies were used to obtain the data for each curve.

An obvious feature of these curves is that the sensitivity is reduced in all four mutants at all wavelengths tested. The reduction is particularly striking in the mutants *ninaD* and *ninaE*. At 480 nm the sensitivity of these mutants is nearly two logarithm units lower than that of wild type. In the order of decreasing sensitivity at 480 nm the four mutants can be arranged as follows: $ninaA^{P228} \approx ninaB^{P315} > ninaD^{P246} \approx ninaE^{P332}$. It may be noted that the above order of sensitivity at 480 nm approximately parallels the order of decreasing R1–6 rhodopsin content (Table 2).

In addition to the overall sensitivity decrease at all wavelengths, the shape of spectral sensitivity curves differs markedly from mutant to mutant. Thus, for example, the UV peak is practically absent in $ninaB^{P315}$ and $ninaD^{P246}$, whereas it is very prominent in $ninaA^{P228}$ and $ninaE^{P332}$, even though at 480 nm $ninaA^{P228}$ has nearly the same sensitivity as $ninaB^{P315}$, and $ninaE^{P332}$ nearly the same sensitivity as $ninaD^{P246}$ (Fig. 2). In other words, two mutants that have similar sensitivities at 480 nm have very different sensitivities at 380 nm. The amount of decrease in sensitivity in long wavelength regions (>550 nm) also differs in some mutants. In the case of $ninaE^{P332}$, for example, the sensitivity decrease in the long wavelength region is significantly less than that at 480 nm, while in $ninaD^{P246}$ the amount of sensitivity decrease in the two regions is nearly the same. Thus, the results of spectral sensitivity measurements suggest that different mechanisms are responsible for rhodopsin reduction in different classes of *nina* mutants.

nina mutations affecting only R1–6 rhodopsin

There are two general ways by which a mutation could reduce the amount of rhodopsin in the organism: (i) to interfere with the synthesis, or to alter the structure, of the protein (opsin) portion of rhodopsin and (ii) to interfere with utilization of carotenoids in the synthesis of the chromophore. Since different classes of rhodopsin differ in opsin but not in the chromophore, mutations that affect the opsin portion of rhodop-

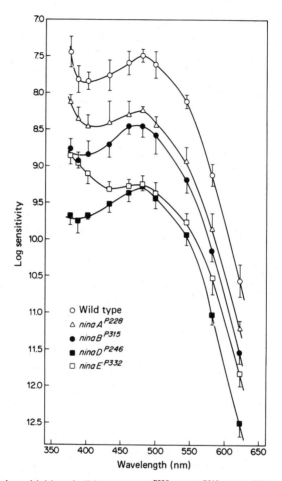

Fig. 2. Spectral sensitivities of wild type, *ninaA*^{P228}, *ninaB*^{P315}, *ninaD*^{P246}, and *ninaE*^{P332}. Sensitivity was measured as the quantum flux necessary to elicit a 1.5 mV criterion response at each wavelength. Each curve represents data from five flies. Within each type of fly, curves from the five individual flies were normalized to the average sensitivity at the 480 nm peak. Error bars are standard deviations, shown in one direction only for the mutants for clarity.

sin would be specific for one particular class of rhodopsin, whereas mutations that interfere with carotenoid utilization would affect all classes of rhodopsin. Therefore, the following two methods were used to determine if the effects of mutations in any of the five *nina* genes might be specific for R1–6 rhodopsin: (*a*) freeze-fracture electron microscopy, and (*b*) measurements of the PDA arising from R7 photoreceptors. In addition, various *nina* mutants were tested for the effect of supplementing their diet with excess amounts of retinoids to

Table 3. *Comparison of rhabdomeric membrane particle density* (particles μm^{-2}) *in two classes of retinula cells[a]*

Genotype	Retinula cells R1–6	Retinula cell R7
wild type	2870 ± 750 (15)	3160 ± 870 (5)
ninaB^{P315}	1270 ± 610 (10)	770 ± 260 (5)
ninaA^{P228}	1020 ± 270 (22)	3030 ± 980 (5)

[a]Results are given as mean $\pm$ SD of the number of cells given in parentheses. *Source*: Schinz *et al.* (1982).

see more directly if any of the *nina* mutations affect carotenoid utilization.

(a) Freeze-fracture electron microscopy

Several investigators have shown by freeze-fracture electron microscopy that the protoplasmic face (PF) of the rhabdomeric membrane of fly photoreceptors is packed with a large number of membrane particles (Boschek & Hamdorf, 1976; Brown & Schwemer, 1977; Harris, Ready, Lipson, Hudspeth & Stark, 1977; Chi & Carlson, 1979; Larrivee *et al.*, 1981; Schinz *et al.*, 1982). Most of these membrane particles apparently are associated with rhodopsin because various experimental procedures that reduce the rhodopsin content also decrease the membrane particle density (Boschek & Hamdorf, 1976; Harris *et al.*, 1977; Larrivee *et al.*, 1981; Schinz *et al.*, 1982).

Shown in Fig. 3 are freeze-fracture replicas of cross-fractured ommatidia and enlarged views of portions of rhabdomeric membranes of R1–6 and R7 photoreceptors of wild type, *ninaA^{P228}*, and *ninaB^{P315}* mutants. The amount of interrhabdomeric space seen in the figure varies among the three cross-fracture replicas of ommatidia primarily because of the difference in fracture levels. In the case of wild-type flies, the PF of the microvillar membranes of both the R1–6 and R7 rhabdomeres are packed with membrane particles. In the case of *ninaA^{P228}*, on the other hand, the PF of the R7 rhabdomeric membrane is filled with membrane particles, but that of the R1–6 rhabdomeric membrane has a substantially reduced particle content (see also Larrivee *et al.*, 1981). In contrast, in *ninaB^{P315}* the rhabdomeric membrane particle density is reduced in both R1–6 and R7 photoreceptors. These qualitative observations are supported by quantitative measurements of particle density shown in Table 3 (Schinz *et al.*, 1982). The results shown in Fig.

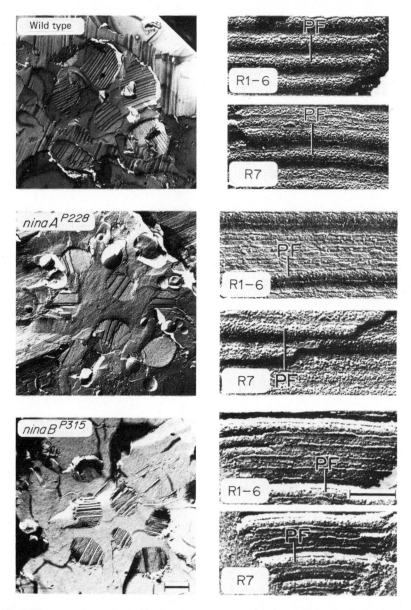

Fig. 3. Freeze-fracture replicas of cross-fractured ommatidia of wild type (top), *ninaA*[P228] (middle), and *ninaB*[P315] (bottom). At right are enlarged views of portions of replicas of R1–6 and R7 rhabdomeres. PF, protoplasmic face of rhabdomeric microvillus. Calibration bars: 1 nm for cross-fractured ommatidia on the left and 0.2 nm for replicas of microvilli on the right. Modified from Schinz *et al.* (1982) with permission from the *Journal of Cell Biology*.

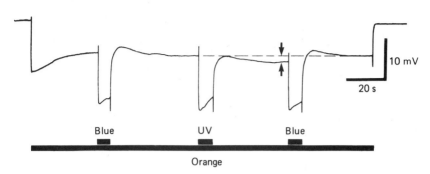

Blue   UV   Blue

Orange

Fig. 4. ERG record obtained from *ninaA*^{P228}, illustrating protocol used to elicit R7 PDA. An orange background light (long bar at the bottom) was left on throughout to cancel R1–6 PDA. UV and blue stimuli (short bars) superimposed on the orange background light alternately elicit (UV) and cancel (blue) the R7 PDA (arrows). The orange light was obtained by filtering a tungsten–halogen lamp with a 600 nm interference filter, and the blue and UV lights were obtained by filtering a xenon lamp with a 480 nm interference filter and a Corning broadband filter, respectively. Reproduced from Larrivee *et al.* (1981) with permission from the *Journal of General Physiology*.

3 and Table 3 lead to the interpretation that the mutation *ninaB*^{P315} reduces the rhabdomeric membrane particle density (hence, the rhodopsin concentration) in both R1–6 and R7 classes of photoreceptors, whereas *ninaA*^{P228} reduces it only in R1–6 photoreceptors.

To determine whether the mutation *ninaA* spares both classes of central photoreceptors, R7 and R8, freeze-fracture electron microscopy was carried out on the double mutant carrying both *ninaA*^{P228} and *sev*^{LY3} by Schinz *et al.* (1982). Since the mutation *sev*^{LY3} has been shown to eliminate R7 photoreceptors (Campos-Ortega, Jürgens & Hofbauer, 1979; see also Harris *et al.*, 1976), the only central cell rhabdomeres remaining in the double mutant are those belonging to the R8 photoreceptors. Thus, the use of the double mutant obviates the difficulty of distinguishing R8 rhabdomeres from R7 rhabdomeres. The results of this study showed clearly that the density of rhabdomeric membrane particles is normal in the R8 photoreceptors of the double mutant.

(b) R7 PDA

In as much as all *nina* mutants were isolated because of their defects in the R1–6 PDA, an examination of the PDA originating from R7 photoreceptors would indicate whether the mutations also affect R7 photoreceptors. Fig. 4 illustrates the stimulus protocol used to elicit the R7 PDA and the ERG responses obtained from a *ninaA* mutant fly using this protocol (see Larrivee *et al.*, 1981). Because the R7 photoreceptors are UV photoreceptors (Harris *et al.*, 1976; Hardie *et al.*,

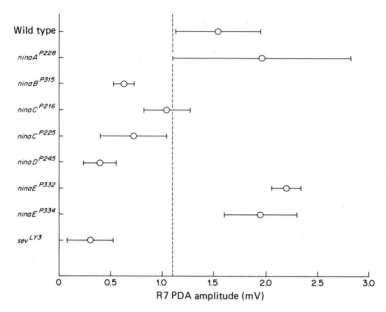

Fig. 5. Amplitude (mean ± SD) of R7 PDA in wild type, the mutant sevenless (sev^{LY3}) and representative mutants (each carrying a different allele) from each of the five *nina* complementation groups. Note *ninaA* and *ninaE* have R7 PDAs in the same range of amplitude as wild type (right of dashed line) while the others have smaller R7 PDAs (left of dashed line), almost as small as the mutant sevenless.

1979; Kirschfeld, 1979), an intense UV stimulus was used to elicit a PDA from these photoreceptors. Since the UV stimulus induces the PDA also from R1–6 photoreceptors, a bright orange light was turned on throughout the duration of the protocol to suppress the generation of the PDA from R1–6 photoreceptors by the UV stimulus. As may be seen in Fig. 4, a blue stimulus, which would normally induce a PDA in R1–6 photoreceptors, elicits no trace of afterpotential under these experimental conditions. A UV stimulus, on the other hand, evokes a small afterpotential indicated by the arrows in Fig. 4. This increment was taken as the R7 PDA, although since it is recorded extracellularly, one cannot exclude the possibility that cells other than R7 may also make some contributions to it.

Figure 5 displays the results of R7 PDA measurements obtained from the *nina* mutants, wild-type flies, and the *sev* mutant. One allele each of the *ninaA*, *B*, and *D* genes and two alleles each of the *ninaC* and *E* genes were examined for this purpose. As seen in Fig. 4, the amplitude of the R7 PDA is as big as or bigger than that of wild type in *ninaA* or *ninaE*. The R7 PDA, however, is considerably smaller in *ninaB*, *C*, or

D. In fact, in the case of *ninaB* or *D* the R7 PDA is nearly as small as in *sev*, which lacks R7 photoreceptors. The R7 PDAs of *ninaC*, on the other hand, tend to be larger than that of *ninaB*, *ninaD*, or *sev*, though not as large as that of wild type. This result probably reflects the fact that the effect of *ninaC* mutations is often milder than that of mutations in the other *nina* genes, as is seen also in the results of R1–6 rhodopsin measurements (Table 2).

Thus, results of R7 PDA measurements lead to the conclusion that the *ninaB*, *C*, and *D* genes, when defective, reduce the amplitudes of PDAs in both R1–6 and R7 classes of photoreceptors whereas defects in the *ninaA* and *E* genes reduce the R1–6 PDA but not the R7 PDA.

(c) Retinoid enriched media
In this series of experiments the fly medium was supplemented with various retinoids (β-carotene, retinal, retinol, and retinoic acid) to see if any of the *nina* mutants could be rescued when maintained on retinoid enriched media. Any indication that a mutant can be so rescued would establish that the mutation affects a step in the uptake and/or utilization of retinoids. If the mutation affects the protein portion of rhodopsin, on the other hand, the retinoid supplement should have no effect on the mutant.

Figure 6 shows the results of these experiments on *ninaB*^{P315} and *ninaD*^{P245}, respectively. The mutant phenotypes were assayed using the R1–6 PDA and the ERP (M_1 potential). The amplitudes of these potentials are shown normalized to the respective wild-type value. As may be seen in Fig. 6(*a*), retinal significantly increased the amplitudes of the R1–6 PDA and ERP of *ninaB*^{P135}. Retinol and retinoic acid, on the other hand, caused only a slight increase of questionable significance, and β-carotene seemed clearly without effect. In the case of *ninaD*^{P245}, all retinoids used had a significant effect on the values of R1–6 PDA and ERP (Fig. 6*b*). In fact, the *ninaD* flies maintained on the β-carotene or retinal supplemented media had R1–6 PDA or ERP values indistinguishable from those of wild type (Fig. 6*b*). Supplementation of the media with retinoids, however, had no effect whatever on either *ninaA* or *ninaE*. In the case of *ninaC* the results were equivocal, and no definitive conclusions could be drawn.

Table 4 summarizes the results of the above three sets of experiments: membrane particle measurements, R7 PDA measurements, and the effects of retinoid supplementation. The results are internally consistent. Membrane particle and PDA measurements indicate that the *ninaA* and *ninaE* mutations affect the R1–6 photoreceptors but not the

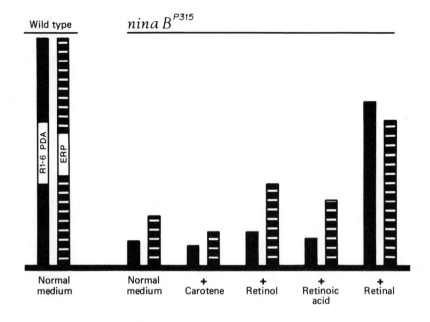

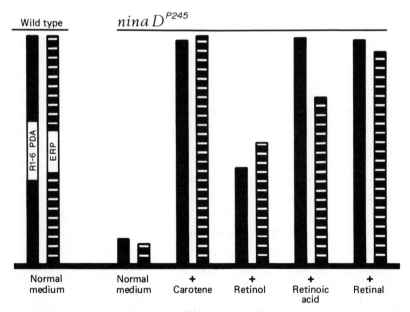

Fig. 6. The effects of raising *ninaB^P315* and *ninaD^P245* mutants on normal medium supplemented with various retinoids (see Methods) on amplitudes of R1–6 PDA (solid bars) and ERP (striped bars). The R1–6 PDA was measured 30s after onset of the PDA-inducing stimulus of 5s duration. The mutant values are shown normalized to wild-type values. Adult flies were on enriched media for 3–15 days before measurement.

Table 4. *Summary of specificity of* nina *mutations*

Mutant	Membrane particles		PDA		Rescued by retinoid supplement
	R1–6	R7	R1–6	R7	
ninaA	−	+	−	+	no
ninaB	−	−	−	−	yes
ninaC			−	−	—
ninaD			−	−	yes
ninaE			−	+	no
vitamin A deprived	−	−	−	−	yes

R7 photoreceptors. These mutants are also the very ones that are unaffected by maintenance on retinoid enriched media.

Gene dosage effect

The results summarized in Table 4 indicate that if any of the five *nina* genes identified to date is the structural gene for R1–6 opsin, it would have to be either *ninaA* or *ninaE*. The structural gene for a protein usually displays a gene dosage effect, that is the amount of protein produced depends on the number of copies of the structural gene for the protein present in the organism (O'Brien & MacIntyre, 1978, and references contained therein). We, therefore, sought to determine if either the *ninaA* or the *ninaE* gene displays this effect.

The *ninaA* gene was tested for gene dosage effect by Kremer & Wong (see Pak *et al.*, 1980). They assayed the amount of R1–6 rhodopsin by microspectrophotometry of the deep pseudopupil (Methods) and varied the dose of the wild-type allele of *ninaA* (*ninaA*+) using the deficiency Df(2L)S3 (Lindsley & Grell, 1968), which uncovers *ninaA*, and a duplication of the *ninaA* region constructed from T(Y;2)P51 (Lindsley *et al.*, 1972). Their results showed that the amount of R1–6 rhodopsin does not depend on the dose of *ninaA* (Pak *et al.*, 1980).

Similar methods were used to vary the dosage of the functional *ninaE* gene. Thus, flies heterozygous for a deficiency uncovering the *ninaE* locus were compared with wild-type flies for their rhodopsin content in R1–6 photoreceptors, and also flies homozygous or heterozygous for a number of *ninaE* alleles were compared with wild-type flies for R1–6 rhodopsin content. These methods reduce the number of copies of the functional *ninaE* gene from two (in wild type) to either one (deficiency

Table 5. *Dependence of rhodopsin content on* ninaE+ *dosage*

Genotype	Age (days)	% of wild type[a]
ninaE^{P332}/ninaE^{P332}		1 ± 1 (3)
ninaE^{P332}/+	0–1	21 ± 2 (5)
	1–2	29 ± 10 (5)
	3–5	43 ± 8 (4)
	9–16	55 ± 10 (6)
+/Def		57 ± 17 (9)
+/+		100 ± 16 (11)

[a]Mean ± SD (*n*).

or *ninaE* heterozygotes) or zero (*ninaE* homozygotes). Experiments utilizing a duplicated third-chromosome segment that would increase the number of copies of the wild-type *ninaE* allele have not yet been carried out.

The results of the *ninaE* gene dosage experiments are shown in Table 5. As may be seen, deficiency heterozygotes have approximately one-half as much R1–6 rhodopsin as wild-type flies. Also, while the amount of R1–6 rhodopsin in *ninaE^{P332}* homozygotes is very small (see also Table 2), the amount of R1–6 rhodopsin in *ninaE^{P332}* heterozygotes is considerably larger, and it gradually increases during the first few days after eclosion, stabilizing at about 50% about 7 days after eclosion. Similar experiments were also carried out using *ninaE^{P318}* and *ninaE^{P334}*. Both these *ninaE* alleles yielded results similar to those obtained with *ninaE^{P332}* (data not shown).

Relationship between ninaE and ora

Koenig (1975), Koenig & Merriam (1977) reported previously on a third chromosome mutation that maps at 65.3 ± 0.4 and causes a defect in the ERG. The mutation was named *ora^{JK84}* (outer *r*habdomeres *a*bsent) by Harris *et al.* (1977), because it either eliminates the R1–6 rhabdomeres or reduces them to vestigial remains without affecting R7 or R8 rhabdomeres. Because the map positions of *ninaE* and *ora* are, within experimental errors, identical (see Table 1 for *ninaE* map position), we carried out complementation tests to determine whether *ninaE* and *ora^{JK84}* are allelic with each other. Flies carrying *ora^{JK84}* and *ninaE^{P332}* in heterozygous combination displayed a very low R1–6 rhodopsin content, a very small ERP, and no R1–6 PDA. Similar results were also obtained with several other *ninaE* alleles, suggesting that *ora^JK84* and *ninaE* are members of the same complementation group.*

Unlike *ora^JK84*, however, none of the *ninaE* alleles we have isolated seems to eliminate the R1–6 rhabdomeres. When the optical neutralization technique (see Methods) is applied to intact heads of wild-type flies, seven bright spots arranged in a trapezoidal pattern are seen in each ommatidium, corresponding to the six rhabdomeres of R1–6 photoreceptors plus the R7 and R8 rhabdomeres acting as a single optic waveguide (Fig. 7). Harris *et al.* (1977) showed that in the case of *ora^JK84* only one spot is present in each ommatidium when examined by this technique (Fig. 7), because R1–6 rhabdomeres have been eliminated as waveguides. The *ninaE* mutants, when examined by this technique, showed a pattern of light spots indistinguishable from that of wild type (Fig. 7), suggesting that R1–6 rhabdomeres are sufficiently intact to act as waveguides. Detailed histology has not yet been performed on any of the *ninaE* mutants, however. Therefore, the possibility that the R1–6 rhabdomeres of *ninaE* may be abnormal at the microscopic level has not yet been excluded.

Discussion

The results of the freeze-fracture and R7 PDA studies strongly support the idea that the effect of mutations in the *ninaA* or *ninaE* genes is to reduce the amount of rhodopsin only in R1–6 photoreceptors. In as much as the R1–6 PDA is reduced or absent in all five classes of *nina* mutants (Fig. 1), the R7 PDA results (Fig. 5; Table 4) demonstrate that the *ninaA* and *ninaE* mutations affect R1–6 photoreceptors but not R7 photoreceptors. Moreover, in so far as the PDA amplitude is closely linked to the amount of visual pigment (see Results, ERG phenotypes; Stark & Zitzmann, 1976; Larrivee *et al.*, 1981), the above results suggest that in *ninaA* or *E* mutants the amount of rhodopsin is reduced in R1–6 photoreceptors but not in R7 photoreceptors.

The effect of the *ninaA* and *ninaE* mutations on the other central photoreceptor, R8, cannot be inferred from R7 PDA results. In the case of *ninaA*, however, freeze-fracture studies have shown that the membrane particle density is normal in the rhabdomeres of both R7 and R8 photoreceptors but is significantly reduced in R1–6 rhabdomeres (Fig. 3; Table 3; Larrivee *et al.*, 1981; Schinz *et al.*, 1982). Since most of the

---

* Further studies are currently being carried out to establish the relationship between *ora^JK84* and *ninaE* more clearly. Until these studies are complete, we will continue to refer to the mutations in question as *ninaE*, even though they may be alleles of *ora^JK84*.

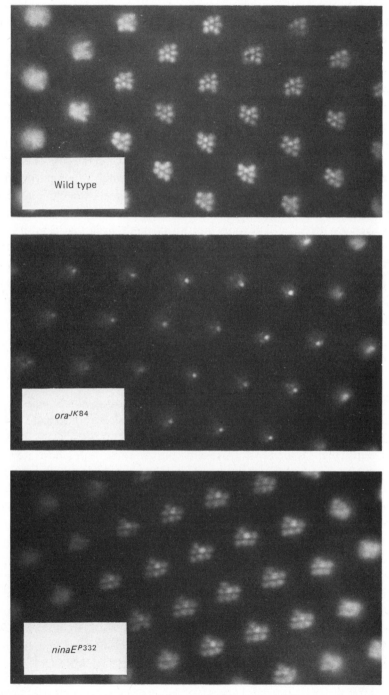

Fig. 7. Rhabdomere structure of wild type, ora^{JK84} and $ninaE^{P332}$ as visualized by the technique of optical neutralization (see Methods). While the R1–6 rhabdomeres are absent in ora^{JK84}, they are present in $ninaE^{P332}$.

rhabdomeric membrane particles are thought to be associated with rhodopsin (Boschek & Hamdorf, 1976; Harris *et al.*, 1977; Larrivee *et al.*, 1981; Schinz *et al.*, 1982), the freeze-fracture results suggest that the concentration of rhodopsin is reduced only in R1–6 photoreceptors, and not in R7 or R8 photoreceptors, in the *ninaA* mutant.

Similar results of freeze-fracture studies are not yet available on any of the *ninaE* alleles. We have obtained evidence, however, that our *ninaE* alleles belong to the same complementation group as ora^{JK84}, previously shown to be specific for R1–6 photoreceptors in eliminating the rhabdomeres (Koenig, 1975; Koenig & Merriam, 1977; and Harris *et al.*, 1976). Moreover, Schinz *et al.* (1982) have shown by freeze-fracture electron microscopy that the membrane particle density is greatly reduced in the vestigial rhabdomeres of R1–6 photoreceptors of ora^{JK84} but is normal in the rhabdomeres of the central photoreceptors. (They did not distinguish R7 from R8 rhabdomeres in this study.) These results suggest that the effect of the ora^{JK84} mutation, and *ninaE* alleles by implication, is specific for R1–6 photoreceptors, and that the main effect of mutations in this gene (*ninaE*) is most likely to reduce the amount of R1–6 rhodopsin. Results of the R7 PDA studies on *ninaE* (Fig. 5) are consistent with this view.

Since R1–6 rhodopsin presumably differs from R7 or R8 rhodopsin in the opsin portion of the molecule, the specificity of *ninaA* and *ninaE* mutations for R1–6 rhodopsin suggests that *ninaA* and *ninaE* mutations affect the opsin portion of R1–6 rhodopsin. The fact that *ninaA* and *ninaE* mutants cannot be rescued by retinoid enrichment of the diet is consistent with this interpretation.

Spectral sensitivities of various *nina* mutants (Fig. 2) are also consistent with the above interpretation. The sensitivity at 480 nm is nearly the same in *ninaA* and *ninaB*, as it is also in *ninaD* and *ninaE*. One member of each pair of mutants with nearly the same 480 nm sensitivity, however, has a very prominent UV peak, while the other shows a virtual absence of the UV peak (Fig. 2). The mutants with a prominent UV peak (*ninaA* and *E*) are also the same ones with a rhodopsin reduction only in R1–6 photoreceptors. Thus, the reduced UV peak in *ninaB* and *D* apparently comes about, in part, through a reduction in the rhodopsin content in the UV photoreceptors, R7, which are unaffected by the mutations *ninaA* or *E*. In addition, Kirschfeld, Franceschini & Minke, (1977) have suggested that there is a UV-sensitizing pigment of carotenoid origin in R1–6 photoreceptors and that the observed absence of UV peak in flies deprived of vitamin A (Stark, Ivanyshyn & Greenberg, 1977) is due to a reduction in the UV-sensitizing pigment as well as R1–6 rhodopsin. Similarly, the mutations *ninaB* and *D* may also reduce the hypothetical UV-sensitizing pigment. The mutations *ninaA* and *E*, which appear to affect the opsin portion of R1–6 rhodopsin, on the other hand, should have no effect on the sensitizing pigment and, hence, on the UV sensitivity.

Although there are several different ways in which a mutation could affect the opsin portion of R1–6 rhodopsin, one obvious way is for the mutation to occur in the gene that codes for the amino acid sequence of R1–6 opsin. Indeed, the most likely consequence of a mutation in the structural gene for opsin is to reduce the amount of the corresponding rhodopsin because of a reduction in the amount of opsin synthesized or the amount of rhodopsin incorporated into the membrane. Our autosomal mutagenesis program of the past few years, which has led to the identification of the five *nina* genes, has yielded at least two alleles in all five *nina* genes identified to date (see Table 1), suggesting that the autosomes may be nearing saturation with respect to R1–6 rhodopsin-deficient mutations. Thus, it appears likely that, if the R1–6 opsin gene is located on the second or third chromosome, it would have been identified as one of the *nina* genes.

The structural gene for R1–6 opsin, however, could be on the X chromosome, which comprises ~20% of the genome, or the very small fourth chromosome which holds 1% of the genome. If such were the case, we would not have detected it in the present study. Extensive mutagenesis of the X chromosome, however, has been carried out previously by three different laboratories for the purpose of isolating visually defective mutants (see Pak, 1975). This early work did not yield any X-chromosome *nina* mutants. On the other hand, because these early X-chromosome mutagenesis programs were carried out before the PDA and its relationship to the amount of rhodopsin were understood, definitive statements regarding the existence of '*nina* genes' on the X chromosome cannot be made without further mutagenesis work on the X chromosome. There is still another way in which we could have failed to isolate mutations in the opsin structural gene. The *Drosophila* genome is known to contain families of genes consisting of repeated elements (Rubin, Finegan & Hogness, 1976; Finnegan, Rubin, Young & Hogness, 1977). If the R1–6 opsin gene consists of a family of such repeated elements, one would not expect any single mutational event to cause a detectable defect in R1–6 opsin.

Of the five autosomal *nina* genes that have been identified to date, only *ninaA* and *E*, when mutated, give rise to unequivocally specific effects on R1–6 opsin and thus are likely candidates for the structural gene for R1–6 opsin. Of these two, *ninaE* is the more attractive possibility because it, unlike *ninaA*, displays a gene dosage effect with respect to the amount of R1–6 rhodopsin (Table 5). The homozygous phenotypic expression of various *ninaE* alleles is also consistent with this view. The phenotypic expression varies from a nearly complete absence of R1–6 rhodopsin for more extreme *ninaE* alleles to a nearly normal amount of R1–6 rhodopsin for milder alleles (data for milder alleles not shown). Although none of the above observations constitutes proof that *ninaE*⁺ is the structural gene for R1–6 opsin, they do suggest that it is the most likely candidate that we have identified to date. At least, *ninaE* is an important locus for R1–6 opsin synthesis and function.

We thank Quentin Pye, Sherry Conrad, Jessica Miller, and Frances Piaszynski for their technical assistance with mutagenesis during various phases of this work; and Steven Davis, Norbert Kremer, Quentin Pye, and Fulton Wong for releasing their mapping data on various *nina* mutants. This work was supported by grants from the National Eye Institute (EY 00033 and EY 03381) and National Science Foundation (BNS 80 15599). J. O. was supported by a National Eye Institute postdoctoral traineeship (EY 07008).

References

ABELSON, J. (1980). A revolution in biology. *Science,* **209,** 1319–21.

BOSCHEK, C. B. & HAMDORF, K. (1976). Rhodopsin particles in the photoreceptor membrane of an insect. *Z. Naturforsch.* **31c,** 763.

BROWN, P. K. & SCHWEMER, J. (1977). Insect visual pigments. *Invest. Ophthalmol. Visual Sci.* **16**(Suppl.), 93–4(abstr.).

CAMPOS-ORTEGA, J. A., JÜRGENS, G. & HOFBAUER, A. (1979). Cell clones and pattern formation: studies on sevenless, a mutant of *Drosophila melanogaster. Wilhelm Roux's Arch.* **186,** 27–50.

CHI, C. & CARLSON, S. D. (1979). Ordered membrane particles in rhabdomeric microvilli of the housefly (*Musca domestica L.*). *J. Morphol.* **161,** 309–21.

COSENS, D. & WRIGHT R. (1975). Light elicited isolation of the complementary visual input systems in white-eye *Drosophila. J. Insect Physiol.* **21,** 1111–20.

FINNEGAN, D. J., RUBIN, G. M., YOUNG, M. & HOGNESS, D. S. (1977). Repeated gene families in *Drosophila melanogaster. Cold Spring Harbor Symp. Quant. Biol.* **42,** 1053.

FRANCESCHINI, N. & KIRSCHFELD, K. (1971). Les phénomènes de pseudopupille dans l'oeil composé de *Drosphila. Kybernetik,* **9,** 159–82.

HARDIE, R. C., FRANCESCHINI, N. & MCINTYRE, P. D. (1979). Electrophysiological analysis of fly retina. II. Spectral and polarization sensitivity in R7 and R8. *J. Comp. Physiol.* **133,** 23–39.

HARRIS, W. A., STARK W. S. & WALKER, J. A. (1976). Genetic dissection of the photoreceptor system in the compound eye of *Drosophila melanogaster. J. Physiol.* **256,** 415–39.

HARRIS, W. A., READY, D. F., LIPSON, E. D., HUDSPETH A. J., & STARK, W. S. (1977). Vitamin A deprivation and *Drosophila* photopigments. *Nature, London,* **266,** 648–50.

KIRSCHFELD, K (1979). The function of photostable pigments in fly photoreceptors. *Biophys. Struct. Mechanism,* **5,** 117–28.

KIRSCHFELD, K., FRANCESCHINI, N. & MINKE, B. (1977). Evidence for a sensitising pigment in fly photoreceptors. *Nature, London,* **269,** 386–90.

KOENIG, J. (1975). The isolation and preliminary characterization of autosomal electroretinogram defective mutants in *Drosophila melanogaster*. PhD thesis, University of California at Los Angeles.

KOENIG, J. & MERRIAM, J. R. (1977). Autosomal ERG mutants. *Drosophila Information Service*, **52**, 50–1.

LARRIVEE, D. C. (1979). A biochemical analysis of the *Drosophila* rhabdomere and its extracellular environment. PhD thesis, Purdue University.

LARRIVEE, D. C., CONRAD, S. K., STEPHENSON, R. S. & PAK, W. L. (1981). Mutation that selectively affects rhodopsin concentration in the peripheral photoreceptors of *Drosophila melanogaster*. *J. Gen. Physiol.* **78**, 521–45.

LINDSLEY, D. L. & GRELL, E. H. (1968). *Genetic variations of* Drosophila melanogaster. Washington DC: Carnegie Institution of Washington.

LINDSLEY, D. L., SANDLER, L., BAKER, B. S., CARPENTER, A. T. C., DENELL, R. E., HALL, J. C., JACOBS, P. A., GABOR MIKLOS, G. L., DAVIS, B. K., GETHMANN, R. C., HARDY, R. W., HESSLER, A., MILLER, S. M., NOZAWA, H., PARRY, D. M. & GOULD-SOMERO, M. (1972). Segmental aneuploidy and the genetic gross structure of the *Drosophila* genome. *Genetics*, **71**, 157–84.

LO, M. -V. C. (1977). Darkening of deep pseudopupil in *Drosophila*: an optical indication of inactivation of the peripheral photoreceptors. PhD thesis, Purdue University.

MINKE, B., WU, C. -F. & PAK, W. L. (1975). Isolation of light-induced response of the central retinula cells from the electroretinogram of *Drosophila*. *J. Comp. Physiol.* **98**, 345–55.

O'BRIEN, S. J. & MacINTYRE, R. J. (1978). Genetics and biochemistry of enzymes and specific proteins of *Drosophila*. In *The Genetics and Biology of* Drosophila, vol. 2a, ed. M. Ashburner & T. F. R. Wright, pp. 396–551. New York: Academic Press.

OSTROY, S. E., WILSON, M. & PAK, W. L. (1974). *Drosophila* rhodopsin: photochemistry, extraction and differences in the $norpA^{P12}$ phototransduction mutant. *Biochem. Biophys. Res. Commun.* **59**, 960–6.

OSTROY, S. E. (1978). Characteristics of *Drosophila* rhodopsin in wild type and *norpA* vision transduction mutants. *J. Gen. Physiol.* **72**, 717–32.

PAK, W. L. (1975). Mutations affecting the vision of *Drosophila melanogaster*. In *Handbook of Genetics*, vol. 3, ed. R. C. King, pp. 703–33. New York: Plenum Press.

PAK, W. L. (1979). Study of photoreceptor function using *Drosophila* mutants. In *Neurogenetics: Genetic Approaches to the Nervous System*, ed. X. Breakefield, pp. 67–99. Amsterdam & New York: Elsevier/North-Holland.

PAK, W. L. & LIDINGTON K. J. (1974). Fast electrical potential from a long-lived, long-wavelenth photoproduct of fly visual pigment. *J. Gen. Physiol.* **63**, 740–56.

PAK, W. L., CONRAD, S. K., KREMER, N. E., LARRIVEE, D. C., SCHINZ, R. H. & WONG, F. (1980). Photoreceptor function. In *Development and Neurobiology of* Drosophila, ed. O. Siddiqi, P. Babu, L. M. Hall & J. C. Hall, pp. 331–46. New York: Plenum.

RUBIN, G. M., FINNEGAN, D. J. & HOGNESS, D. S., (1976). The chromosomal arrangement of coding sequences in a family of repeated genes. *Prog. Nucleic Acid Res. Mol. Biol.* **19**, 221–6.

SCHINZ, R. H., LO, M.-V. C., LARRIVEE, D. C. & PAK, W. L. (1982). Freeze-fracture study of the *Drosophila* photoreceptor membrane: mutations affecting membrane particle density. *J. Cell Biol.* **93**, 961–69.

STARK, W. S. & ZITZMANN, W. G. (1976). Isolation of adaptation mechanisms and photopigment spectra by vitamin A deprivation in *Drosophila*. *J. Comp. Physiol.* **105**, 15–27.

STARK, W. S., IVANYSHYN, A. M. & GREENBERG, R. M. (1977). Sensitivity and photopigments of R1–6, a two-peaked photoreceptor, in *Drosophila*, *Calliphora*, and *Musca*. *J. Comp. Physiol.* **121**, 289–305.

STAVENGA, D. G., ZANTEMA, A. & KUIPER, J. W. (1973). Rhodopsin processes and the function of the pupil mechanism in flies. In *Biochemistry and Physiology of the Visual Pigments*, ed. H. Langer, pp. 175–80. New York: Springer-Verlag.

STEPHENSON, R. S. & PAK, W. L. (1980). Heterogenic components of a fast electrical potential in *Drosophila* compound eye and their relation to visual pigment photo-conversion. *J. Gen. Physiol.* **75**, 353–79.

THE PHOTORECEPTOR IN
STENTOR COERULEUS

PILL-SOON SONG,* KENNETH J. TAPLEY, JR.*
and JERRY D. BERLIN†

* Department of Chemistry, Texas Tech University, Lubbock, TX 79409, USA
† Department of Biological Sciences, Texas Tech University,
Lubbock, TX 79409, USA

Introduction

Stentor coeruleus is a blue-green protozoan ciliate which exhibits a photophobic and negative phototactic response when exposed to visible light. The photoresponse of *Stentor* has been attributed to the protein-bound chromophore, hypericin,‡ which gives the organism its character-istic blue-green color (Walker, Lee & Song, 1979). The photoreceptor protein is located in membraneous vesicles or pigment granules which reside just beneath the outer pellicle, and close observation of *Stentor* reveals that the pigment granules are arranged in rows which run longitudinally along the organism, giving the appearance of green stripes (Fig. 1). Running parallel to and between the stripes are rows of cilia which are used for locomotion. The frontal field is surrounded by a membranelle band of cilia, which is used to rotate the cell body directionally in the swimming motion of the cell, and is also partially responsible for changing its swimming direction.

Stentor exhibits two types of photoresponse: a photophobic or light avoidance response, and a negative phototactic response. The photo-phobic response is actually a step-up response, which occurs when the organism encounters a sudden increase in light intensity. This behavior has been observed and recorded on videotape, as shown in Fig. 2. When the organism encounters a region of higher light intensity (light trap), it stops momentarily and then turns to the side while reversing its direction slightly and swims away from the light trap. The time required for this entire process to take place is dependent on the intensity of the light in the trap. If the light is of sufficiently high intensity, the *Stentor* stops instantly without even moving completely into the light trap. Conversely, at lower light intensities the *Stentor* will sometimes swim completely through the light trap without any response. The stop

‡ Hypericin is the chromophore of stentorin, which is the photo-receptor of photoresponses in *Stentor coeruleus*.

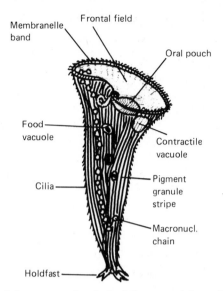

Fig. 1. A drawing of *Stentor coeruleus* in its fully matured form. The overall length is approximately 300–500 µm. (Based on Tartar, 1961.)

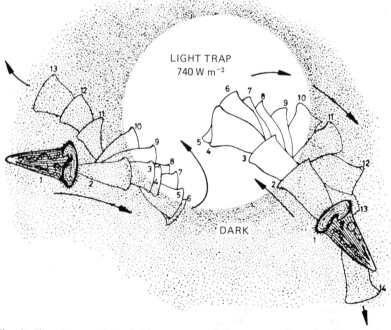

Fig. 2. The step-up photophobic response of *Stentor coeruleus*. As the organisms encounter the light trap, they stop, reverse direction, and swim away from the lit area. The positions of the cells have been traced every 0.17 s from a video recording. (Redrawn from Song, Häder & Poff, 1980*b*.)

Focusing lens

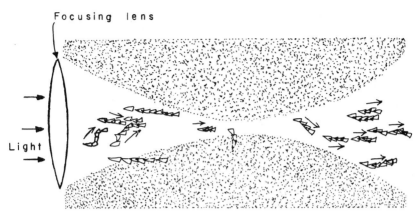

Light

Fig. 3. Tracks of *Stentor coeruleus* in a focused light beam of 94 W m⁻². The light area represents the focused beam. Note that the organisms swim away from the source in both the converging and diverging beams. The positions of the cells were traced at 0.17 s intervals from a video recording. (Redrawn from Song *et al.*, 1980*a*).

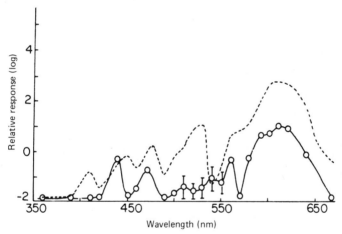

Fig. 4. The action spectra for the step-up photophobic response (- - -) and the negative phototactic response (○—○) of *Stentor coeruleus*. The action spectra were determined from sets of dose–response curves measured at various wavelengths, as described by Song *et al.* (1980*a*).

response of *Stentor* is an integral part of the light avoidance response. As can be observed under a light microscope, the sudden increase in light intensity causes the direction of the ciliary beating to reverse. It is this ciliary reversal which causes the *Stentor* to stop and then change swimming direction. It is interesting to note that the stop response can be observed even if only the anterior portion of the *Stentor* is illuminated with high intensity light, this would indicate that the frontal field may be more sensitive to light. Electron microscopic examination,

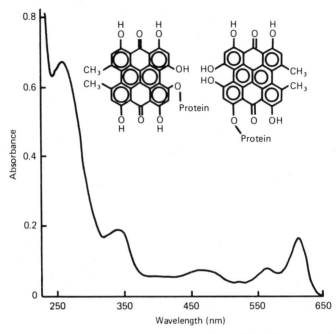

Fig. 5. The absorption spectrum of the photoreceptor isolated from *Stentor coeruleus*. The photoreceptor was isolated by sonication in phosphate buffer, followed by sucrose density gradient centrifugation and preparative isoelectric focusing. The sample was in 10 mM phosphate buffer, pH 7.4. Also shown are two possible structures for the stentorin chromophore. Other structures having the peptide linkage elsewhere are also possible. The most likely linkage, which must be readily hydrolyzable with acid, is the ester linkage with a carboxyl group of the protein. (Absorption spectrum redrawn from Walker, 1980).

however, reveals that the concentration of pigment granules in this region is quite low in comparison to other areas of the organism.

The negative phototactic response of *Stentor* has also been observed and recorded on videotape, as depicted in Fig. 3. When irradiated with unilateral light focused on the center of the cuvette, the *Stentor* can be seen swimming in a direction parallel to the direction of light propagation in both the converging and diverging beams of focused light (Song, Häder & Poff, 1980*a*). The action spectra for the photoresponses and the absorption spectrum of the chromophore are shown in Figs. 4 and 5, respectively. It can be seen that both action spectra closely resemble the absorption spectrum of stentorin, implying that the stentorin chromophore does indeed mediate both photoresponses.

Photosensory transduction

Previous studies have shown the phototactic response of *Stentor* to be

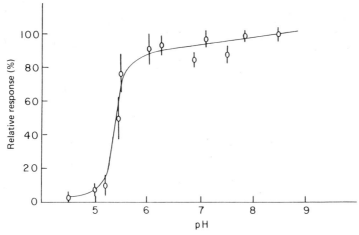

Fig. 6. pH-dependent, negative phototaxis in *Stentor coeruleus* constructed from a series of dose–response curves for red light obtained at various pH values. The light intensity was 1.6 W m^{-2}. (Redrawn from Walker *et al.*, 1981.)

dependent on the pH of the surrounding medium (Walker, Yoon & Song, 1981). As shown in Fig. 6, the response of *Stentor* to red light decreases rapidly as the pH of the medium is lowered below pH 6. This implies that a pH difference across the *Stentor* membrane is important for the photoresponse. This idea is supported further by the observation that the addition of nigericin, which exchanges H$^+$ for K$^+$, also results in a significant decrease in the photoresponse (Walker *et al.*, 1981). A similar result was obtained using carbonyl cyanide *p*-trifluoromethoxy-phenylhydrazone (FCCP) to dissipate the pH gradient across the *Stentor* membrane (Song *et al.*, 1980*b*; Walker *et al.*, 1981). The formation of a pH gradient can be further supported by examination of the excited-state properties of the stentorin chromophore. Spectroscopic studies have shown that the acidity of hypericin hydroxyl groups is markedly increased upon photoexcitation (Song, Chin, Yamazaki & Baba, 1977; Walker *et al.*, 1979). In fact, the pKa decreases from approximately 6 in the ground state to roughly 2 in the excited state. This results in a rapid proton release from the excited-state stentorin to form the excited-state anion, which has its fluorescence emission maximum at 660 nm as opposed to 610 nm for the neutral stentorin molecule. Further experimental evidence on the isolated *Stentor* chromoprotein indicates that protons are being released upon photoexcitation, thus accounting for the observed pH dependence of the photomovement.

More recent studies using whole *Stentor* organisms, have implicated calcium in the photosensory transduction of *Stentor*. Studies of the

Stentor photoresponse in the presence of Ruthenium Red, a Ca^{2+}-blocking agent, showed that the photoresponse was greatly inhibited at micromolar concentrations of the calcium-blocking agent (Chang, Walker & Song, 1981). Other studies using the Ca^{2+} ionophore A23187 showed strong inhibition of the photoresponse in the absence of Mg^{2+} but showed no effect in the presence of Mg^{2+} (Chang *et al.*, 1981). This result may be due to a reduction in the effective A23187 concentration, since A23187 may form a complex with Mg^{2+}.

Based on the limited results available, we have postulated a tentative working model for the photoresponse of *Stentor coeruleus*.

(*a*)   The initial step involves absorption of actinic light by the chromophores of the photoreceptor protein, which is located in the pigment granules.

(*b*)   The excited-state chromophores generate the primary signal in the form of a proton gradient by ejecting protons into the cytoplasm *via* a proton translocation network located in the photoreceptor protein.

(*c*)   The ΔpH produced in step (*b*) triggers depolarization of the ciliary membranes and activates voltage-sensitive Ca^{2+} channels in the ciliary membranes.

(*d*) Ca^{2+} influx results from the opening of the Ca^{2+} channel, which further depolarizes the ciliary membrane.

(*e*)   The ciliary beating is reversed as a result of the depolarization in step (*d*) causing the organism to turn and swim in the opposite direction.

The remainder of this chapter will deal with the photoreceptor protein, the pigment granules, and the experimental evidence for their role in this photosensory transduction mechanism.

The *Stentor* photoreceptor

As previously mentioned, most of the *Stentor* chromoprotein is located in the pigment granules, which reside just beneath the outer pellicle of the organism. The fact that the protein is only slightly soluble in water and can only be solubilized in detergent or organic solvents indicates that it is a membrane protein. This is further supported by the unusually high content of hydrophobic amino acids (Walker, 1980). The location of the chromoprotein in the pigment granules is not known at this time. It seems reasonable to assume that the protein is embedded in the outer membrane of the pigment granule, based on results from liposome

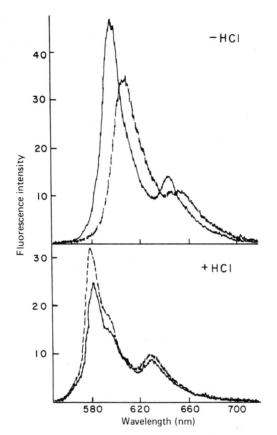

Fig. 7. Fluorescence emission spectra of hypericin (——) and the acetone-extracted photoreceptor pigment (- - -) in ethanol. The upper traces are without acid and the lower traces are in 10 mM HCl. The excitation bandpass was 10 nm and the emission bandpass was 3 nm in all cases. (Taken from Walker, 1980.)

experiments which show that the chromoprotein does function as a proton source when entrapped in a bilayer lipid membrane (Song, Walker & Yoon, 1980c). However, recent electron microscopic examination has revealed a peculiar internal structure in the pigment granules that may provide clues as to the location and function of the photoreceptor protein. (These results are discussed in greater detail in the next section.)

Numerous attempts have been made to purify the photoreceptor protein. Due to its low solubility in aqueous buffer solutions, the protein must be isolated by less conventional means such as organic solvent extractions and solutions of anionic detergents.

Some purification has been achieved by sonication of whole *Stentor* in

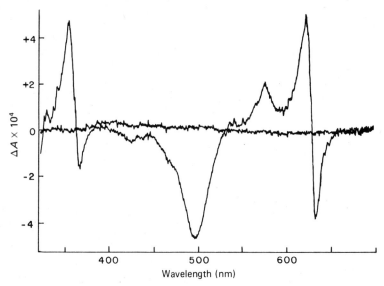

Fig. 8. The circular dichroic spectrum of the acetone-extracted photoreceptor from *Stentor coeruleus*. The sample was dissolved in acetone with $A_{610} = 1.0$. The recorded baseline is also shown. (Taken from Walker, 1980.)

aqueous phosphate buffer followed by sucrose density gradient centrifugation and preparative isoelectric focusing (Song *et al.*, 1980*c*). The isoelectric point of the protein isolated in this manner was 5.1 ± 0.1. SDS polyacrylamide gel electrophoresis (SDS PAGE) of this preparation shows four major bands which emit pink fluorescence and correspond reasonably well with the protein bands detected by Coomassie blue staining. The apparent molecular weights (M_r) of these bands are roughly 130, 65, 16 and 13×10^3, respectively (Walker *et al.*, 1979). Prolonged destaining (three to four weeks) of these gels, however, results in the disappearance of the 13, 65 and 130×10^3 M_r bands leaving the 16 000 M_r protein as the only chromophore-containing protein band (Walker, 1980).

Acetone extraction of whole *Stentor* has also been used to solubilize the chromoprotein (Walker, 1980). In contrast to the buffer-extracted pigment, the acetone extract exhibits a bright red fluorescence which resembles hypericin. Subsequent studies of this preparation have revealed substantial differences between the acetone extract and hypericin. The fluorescence emission spectra of the acetone-extracted pigment is noticeably red-shifted when compared to a similar solution of hypericin (Fig. 7). However, upon treatment with 10 mM HCl, both solutions exhibit similar emission spectra, indicating that the chromophore is easily hydrolyzed by mild acid treatment (Walker *et al.*, 1979).

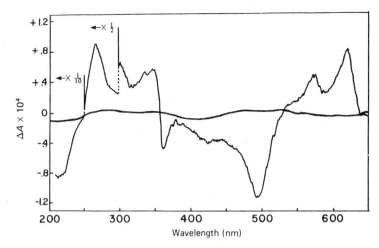

Fig. 9. The circular dichroic spectrum of the *Stentor* photoreceptor prepared by sucrose gradient centrifugation and isoelectric focusing. The sample was dissolved in 10 mM phosphate buffer, pH 7.4, $A_{610} = 0.17$. (Taken from Walker, 1980.)

Thin layer chromatography of the acid-treated solution yields identical R_f values for both hypericin and the acetone extract, indicating that the chromophore is indeed hypericin.

SDS PAGE of the acetone-extracted pigment yields the same four fluorescent bands as the buffer-solubilized preparation with only a small band at 16 000 M_r remaining after Coomassie blue staining (Walker, 1980). Isoelectric focusing of the acetone-extracted pigment shows an isoelectric point of 4.0 ± 0.1, indicating that the acetone-extracted pigment may be in a different conformation from that in the buffer-extracted preparation. Figures 8 and 9 show the circular dichroic spectra of the acetone-solubilized and buffer-extracted pigments, respectively. Although similar in overall appearance, the presence of a strong negative band at 630 nm in the spectrum of the acetone extract reflects some difference in the chromophore environment in the two preparations. This difference may account for the different pIs and appearance of the two preparations. It is also probable that the buffer-extracted preparation contains some residual lipids, which may affect the protein conformation.

Numerous detergents have also been employed in an attempt to purify the protein. While not entirely successful, some conclusions have been drawn regarding the protein's physical properties. Non-ionic detergents such as Triton X-100 and Nonidet P-40 appear to be ineffective in solubilizing the pigment. Although some chromophore appears to be solubilized, column chromotography indicates an M_r in

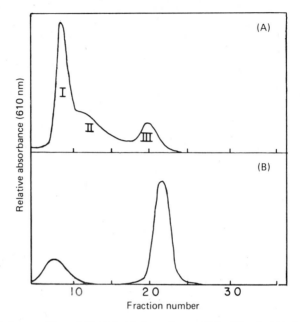

Fig. 10. Elution profiles of BioGel A1.5 m columns (1.5 × 75 cm). (A) Sample was prepared by sonication in 0.1% sodium cholate buffered with 10 mMTris, pH 7.4. Eluted with the same buffer and detergent. (B) Sample prepared the same as in (A) above, except incubated in 1% sodium cholate for 24 h prior to elution with the same detergent and buffer.

excess of one million, which is probably the result of aggregation or incomplete solubilization from the membrane.

Sodium cholate, a relatively mild anionic detergent, produces somewhat better results, as three chromophore peaks are eluted from a BioGel A1.5 m column equilibrated with a solution of 0.1% sodium cholate (Walker, 1980). Figure 10A shows the elution profile of a BioGel A1.5 m column equilibrated with 0.1% cholate in 10 mM Tris buffer, pH 7.4. The sample was prepared by sonicating whole *Stentor* in the elution buffer and centrifuging to remove insoluble debris. Fraction I elutes with the void volume, giving an apparent molecular weight of more than 2.0×10^6. Fractions II and III have apparent M_r values of $500–1000 \times 10^3$ and $250–350 \times 10^3$ respectively. Both fractions II and III are composed mainly of the 16 000 M_r protein, with fraction I containing the 16 000 M_r protein along with some other contaminants. Fractions I and II both exhibit circular dichroic (CD) spectra which are similar to that obtained with the sonicated buffer extract (Fig. 9), indicating retention of their native conformation. Fraction III, however, exhibits a very weak, almost nonexistent CD characteristic of a dena-

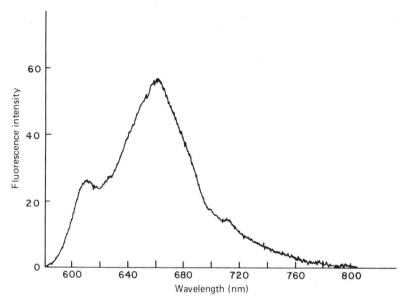

Fig. 11. Fluorescence emission spectrum of whole, living *Stentor* as measured by surface excitation in a triangular cuvette. The excitation wavelength was 560 nm. The bandpasses for excitation and emission were 5 nm and 4 nm, respectively. The emission was filtered with a Corning CS2–63 longpass filter. (Taken from Walker, 1980.)

tured protein. Fractions I and II also possess red-shifted fluorescence emission spectra at higher concentrations, which is indicative of the native protein, as will be discussed later.

If the concentration of sodium cholate is increased from 0.1%, as in the previously described work, to 1.0%, an interesting result is obtained. Fractions I and II collapse together to form one small peak, and Fraction III increases substantially. The elution profile of the BioGel column run in 1.0% cholate is shown in Fig. 10B. The noticeable increase in the lower molecular weight peak indicates that the higher detergent concentration disrupts the interactions between the 16 000 M_r proteins, resulting in a lower apparent molecular weight. It should be noted however, that the apparent M_r of the resulting peak is still relatively large ($100–200 \times 10^3$), probably due to residual interactions which remain even at higher detergent concentrations. SDS PAGE shows the major component to be the 16 000 M_r chromoprotein. Circular dichroic spectra show a very weak signal which resembles that obtained with Fraction III at lower detergent concentrations. Likewise, there is no concentration-dependent 660 nm fluorescence emission, indicating that the high detergent concentration severely disrupts the conformation of the native protein complex.

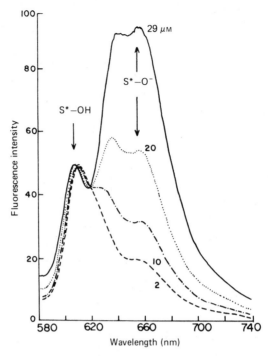

Fig. 12. Fluorescence emission spectra of the *Stentor* photoreceptor in 10 mM phosphate buffer, pH 7.4, as a function of concentration. All spectra were normalized with respect to the emission at 610 nm. S*–OH and S*–O⁻ represent emission occurring from the neutral and anionic forms of the chromophore, respectively. Excitation wavelength was 560 nm. (Redrawn from Walker *et al.*, 1981.)

As mentioned earlier, the 660 nm fluorescence emission seems to be a characteristic of the native protein. Figure 11 shows the fluorescence emission spectrum of whole, living *Stentor* as measured by surface excitation from a triangular cuvette. A similar emission spectrum is obtained with hypericin in basic aqueous solution indicating that the hypericin anion is actually responsible for the observed longer-wavelength fluorescence emission (Walker, 1980).

The 660 nm fluorescence emission can be reproduced in some photo-receptor preparations only at high concentrations. Figure 12 shows the fluorescence emission of a crude extract prepared by sonicating whole cells of *Stentor* in 10 mM phosphate buffer at pH 7.4, followed by low speed contrifugation to remove insoluble debris. The dramatic increase in 660 nm fluorescence only at high photoreceptor concentrations leads us to believe that the native photoreceptor complex is a necessary requirement for the formation of the chromophore anion. It also seems reasonable to assume that the native photoreceptor complex is com-

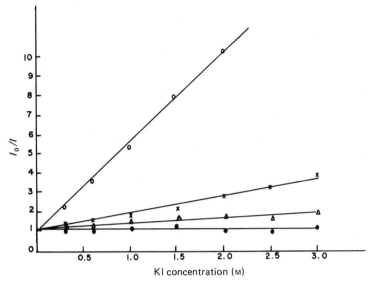

Fig. 13. Stern–Volmer quenching plots of the photoreceptor protein prepared by different methods. Potassium iodide was used as a quencher. The salt concentration was maintained at 3.0 M for all data points by the addition of sodium chloride. Excitation wavelength was 560 nm and emission was monitored at 610 nm for all data points. ●, Isoelectrically focused protein in 10 mM Tris buffer, pH 7.4. No quenching was observed, $k_q = 0 \, (\text{M}^{-1}\text{s}^{-1})$. △, Fraction II from the BioGel A1.5 m column equilibrated with 0.1% sodium cholate detergent in 10 mM Tris buffer, pH 7.4, $k_q = 8.7 \times 10^7 \, \text{M}^{-1}\text{s}^{-1}$. ×, Fraction III from the same column, $k_q = 2.2 \times 10^8 \, \text{M}^{-1}\text{s}^{-1}$. ○, Synthetic hypericin in 0.1% sodium cholate detergent buffered with 10 mM Tris, pH 7.4. $k_q \times 9.1 = 10^8$. (Taken from Walker, 1980.)

posed of a large number of 16 000 M_r proteins, as was observed in Fractions I and II from the 0.1% sodium cholate column.

In an attempt to elucidate further the mechanism of formation of the chromophore anion, Stern–Volmer quenching of the fluorescence was examined (Song, Walker, Auerbach & Robinson, 1981). No fluorescence quenching was observed in the native photoreceptor at KI concentrations up to 3 M. Fractions II and III in 0.1% sodium cholate were quenched by KI, as was hypericin in 0.1% sodium cholate (Fig. 13). Attempts to quench the native photoreceptor fluorescence with CsCl and acrylamide also failed, using quencher concentrations of up to 4 M. Thus, it is apparent from these quenching data that the chromophore must be deeply buried within the protein and not accessible to quenching ions and molecules.

The fluorescence lifetimes of the anionic species of the chromophore were also measured in both live *Stentor* and the native photoreceptor complex (Song *et al.*, 1981). Live *Stentor*, which emitted predominantly anionic fluorescence, exhibited fluorescence lifetimes of 1.21 ns (90% anionic component) and 4.95 ns (10% neutral component). A 20 μM solution of *Stentor* photoreceptor showed similar two component lifetimes, 1.47 ns (82% anionic) and 4.48 ns (18% neutral). Picosecond excitation at 527 nm was also used to follow the rise and decay of the anionic component in *Stentor* photoreceptor preparations. Surprisingly, the risetime profile very closely approximated the 10 ps risetime of the excitation pulse. This suggests that the chromophore anion is formed at a rate considerably faster than the diffusion-controlled rate. This observation can be explained by postulating the existence of an acid–base network inside the photoreceptor protein, which facilitates efficient removal of hydroxyl protons from the excited-state chromophore. Some evidence for this exists in the fact that the photoreceptor has been shown to contain relatively large amounts of aspartate and glutamate, both of which are known for their general acid–base catalytic roles in many enzymes.

Further evidence supporting this proton translocation hypothesis has been obtained using photoreceptor-entrapped and photoreceptor-embedded liposomes (Walker, 1980; Song *et al.*, 1980c). In order to ascertain whether or not the photoreceptor actually releases protons into the surrounding medium, the *Stentor* photoreceptor protein was entrapped inside artificial liposomes and 9-aminoacridine was added as a pH probe. Upon irradiation with actinic light of suitable wavelength, the fluorescence of 9-aminoacridine was quenched, indicating proton release from the photoreceptor protein.

The photoreceptor protein was also embedded into the lipid bilayer membrane. Using umbelliferone as an external pH probe, it was found that protons were indeed released into the surrounding medium. Light-induced pH changes in suspensions of whole, living cells of *Stentor* have also been observed using 9-aminoacridine as a pH probe (Walker *et al.*, 1980; Yoon, 1981). In these experiments, the fluorescence quenching of 9-aminoacridine by protons released from the photoreceptor was abolished by the addition of FCCP or nigericin, suggesting that the cytoplasmic pH does decrease upon irradiation with actinic light. Additional work has been done using a micro-pH electrode to measure intracellular pH changes in living *Stentor* (Walker, 1980). As shown in Table 1, the internal pH of *Stentor* decreased from 6.8 with no actinic light, to 6.21 after five minutes irradiation with a filtered mercury light source. Although the time required for this pH drop is too slow to be correlated directly with the ciliary reversal, which actually occurs on a subsecond time scale, it is certainly qualitatively consistent with the

Table 1. *Light-induced pH changes in the cytoplasm of* Stentor coeruleus *as measured by a micro-pH electrode[a]*

Irradiation time (min)	pH[b]	Number of cells measured
0	6.80 ± 0.11	14
1	6.62 ± 0.13	10
2	6.28 ± 0.15	13
5	6.21 ± 0.10	12

[a] The electrode was inserted through the frontal field or through the side of the cell by means of a micromanipulator. An Ag/AgCl reference electrode with a 3M KCl salt bridge was in the extacellular medium. Broad-band illumination was provided by a high pressure mercury lamp equipped with CS3-74 and IR-absorbing filters, transmitting visible radiation only, that is 400 nm < λ < 700 nm. (Taken from Walker, 1980.)
[b] In these pH recordings, correction for membrane potential changes was not made.

proposed excited-state proton release as an initial step in photosensory transduction.

Stentor pigment granules

From the previous discussion, it is clear that a more complete characterization of the photoreceptor is necessary in order to fully understand phototransduction and its role in the photoresponses of *Stentor*. Also lacking is a full understanding of the pigment granules and how they function as an integral part of the photosensory transduction chain.

Presently, it is not known whether the photoreceptor is embedded in the membrane or entrapped inside the pigment granule. Recent electron micrographs clearly show an outer membrane surrounding an inner, more dense material, which could contain the photoreceptor protein (Fig. 14). These studies failed to show any direct physical connection between the pigment granules and other organelles such as the cilia and mitochondria, indicating that the granules function independently in the initial photosensory transduction step.

It is interesting to note that the colorless species, *S. polymorphus*, possesses similar granules which, of course, contain no pigment. Even some of the granules in *S. coeruleus* appear empty and contain no pigment (Kennedy, 1965).

The pigment granules of *S. coeruleus* were previously thought to be mitochondrial in nature, based on the observation that they are stainable with Janus Green and contain labile cytochrome oxidase

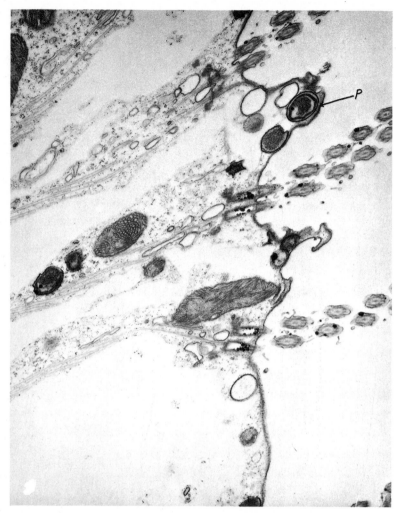

Fig. 14. Transverse section of *Stentor coeruleus* stained with uranyl acetate and lead citrate. The outer membrane of a pigment granule (P) is heavily stained and appears as an intact membrane. Some internal structure is also visible. Magnification: × 28 000 approx.

(Weisz, 1949; Weisz, 1950). Later work revealed that *Stentor* possesses mitochondria which are distinctly different from the pigment granules, indeed the granules have been shown to contain some type of fine structure which does not resemble that of the mitochondria. The granules are also substantially smaller than mitochondria, being only 0.3–0.7 μm in diameter.

The granules are normally located near the outer pellicle of *Stentor*, between the rows of cilia, and are sometimes clustered in groups of three to five. There are, however, significant numbers of granules located well beneath the outer pellicle, sometimes as deep as 10 μm inside the outer membrane. The pigment granules of *Stentor* can be readily shed along with the pellicle upon treatment with ice-cold buffer or mild chemicals (for example, 15% sucrose or 2% urea). The integrity of the granules prepared in this manner is questionable, however, as electron microscopic examination failed to reveal any granules resembling those prepared by *in-situ* fixation. Further work is now being conducted in an attempt to isolate intact pigment granules that can be characterized by both chemical and electron microscopic studies.

Concluding remarks

Although work on the *Stentor* photoreceptor and pigment granules is by no means complete, existing data do permit us to draw some conclusions about the mechanism by which photosensory transduction occurs in *Stentor*.

The photoreceptor appears to be composed of a large number of relatively small proteins, which are associated with the pigment granule. In the native photoreceptor, the chromophores are buried deep within the complex in a manner which permits efficient anion formation from the excited state. The protons released from the excited-state chromophore are conducted to the cytoplasm via a proton translocation network located within the protein. This 'proton flux' creates a pH gradient across the outer membrane of *Stentor coeruleus* with the net result being ciliary reversal.

Preliminary work suggests that calcium plays a role in the ciliary reversal, possibly via a calcium influx through a voltage-sensitive calcium channel. Further studies are necessary to clarify this process, especially with respect to the calcium influx.

The work described in this review has been supported in part by the Robert A. Welch Foundation (D-182) and USPHS NIH grants (NS15426).

References

CHANG, T.-H., WALKER, E. B. & SONG, P. S. (1981). Effects of ionophores and ruthenium red on the phototaxis of *Stentor coeruleus* as measured by simple devices. *Photochem. Photobiol.*, **33**, 933–6.

KENNEDY, J. R. (1965). The morphology of *Blepharisma undulans* Stein. *J. Protozool.*, **12**, 542–61.

SONG, P. S., CHIN, C. A., YAMAZAKI, I. & BABA, H. (1977). Excited states of photobiological receptors. II. Chlorophylls, phytochrome, and stentorin. *Int. J. Quantum Chem: Quantum Biol. Symp.*, **4**, 305–15.

SONG, P. S., HÄDER, D. P. & POFF, K. L. (1980a). Phototactic orientation by the ciliate, *Stentor coeruleus. Photochem. Photobiol.*, **32**, 781–6.

SONG, P. S., HÄDER, D. P. & POFF, K. L. (1980b). Step-up photophobic response in the ciliate, *Stentor coeruleus. Arch. Microbiol.*, **126**, 181–6.

SONG, P. S., WALKER, E. B., AUERBACH, R. A. & ROBINSON, G. W. (1981). Proton release from *Stentor* photoreceptors in the excited state. *Biophys. J.*, **24**, 551–5.

SONG, P. S., WALKER, E. B. & YOON, M. J. (1980c). Molecular aspects of photoreceptor function in *Stentor coeruleus*. In: *Photoreception and Sensory Transduction in Aneural Organisms*, ed. F. Lenci & G. Colombetti, pp. 241–52. New York: Plenum Press.

TARTAR, V. (1961). *The Biology of* Stentor. New York: Plenum Press.

WALKER, E. B. (1980). Photosensory Transduction in *Stentor coeruleus*. PhD dissertation, Texas Tech University.

WALKER, E. B., LEE, T. Y. & SONG, P. S. (1979). Spectroscopic characterization of the *Stentor* photoreceptor. *Biochim. Biophys. Acta*, **587**, 129–44.

WALKER, E. B., YOON, M. & SONG, P. S. (1981). The pH dependence of photosensory responses in *Stentor coeruleus* and model system. *Biochim. Biophys. Acta*, **634**, 289–308.

WEISZ, P. B. (1949). A cytochemical and cytological study of differentiation in normal and reorganizational stages of *Stentor coeruleus. J. Morphol.*, **84**, 335–59.

WEISZ, P. B. (1950). On the mitochondrial nature of the pigmented granules in *Stentor* and *Blepharisma. J. Morphol.*, **86**, 177–84.

YOON, M. J. (1981). Photosensory transduction in *Stentor coeruleus*: liposome model systems. PhD dissertation, Texas Tech. University.

THE PHOTOCONTROL OF MOVEMENT OF
CHLAMYDOMONAS

WILHELM NULTSCH

Lehrstuhl für Botanik, Fachbereich Biologie, Philipps-Universität,
355 Marburg a.d. Lahn, Lahnberge, GFR

Introduction

Three different types of photomovement occur in microorganisms:
photokinesis, phototaxis and photophobic responses (Nultsch, 1975;
Nultsch & Häder, 1979). Since the main energy source for movement in
Chlamydomonas is oxidative phosphorylation, photokinesis can be
observed only after long periods of darkness, exceeding three days
(Nultsch & Throm, 1975). Therefore, photokinesis of *Chlamydomonas*
has not yet been studied in detail and may be disregarded here.
Photophobic responses can be observed frequently in *Chlamydomonas*,
but as yet have been scarcely investigated. Therefore, this chapter deals
almost exclusively with phototaxis, that is the orientation of movement
with respect to light. Depending on the species or strain and the external
conditions (Nultsch, 1977), *Chlamydomonas* reacts either positively or
negatively, by swimming towards the light source or away from it.

The organism

Chlamydomonas is a unicellular, photosynthetic flagellate with two
flagella at the anterior end. The cell is asymmetrical to its long axis with
respect to the stigma, the pyrenoid and the cup-shaped chloroplast.
Schötz (1972) who ultrasectioned a whole (+) gamete of *Chlamydomonas reinhardtii* and constructed a model of the cell found that the
chloroplast shows one or more perforations.

The flagella show the typical eukaryotic 9 + 2 microtubuli pattern
(Fig. 1A). The axonemes are at least in part delimitated from the cell
lumen by a diaphragm (Fig. 1B). In swarmers of the *Ulvaphyceae*
(Melkonian, 1980), which are also biflagellate and react phototactically,
the proximal ends of the basal bodies are closed by a so-called terminal
cap. However, whether or not a separate ionic compartment in the
flagellum or axoneme is established by these structures cannot yet be
decided; although it is important for the discussion of the phototactic
reaction chain (see below).

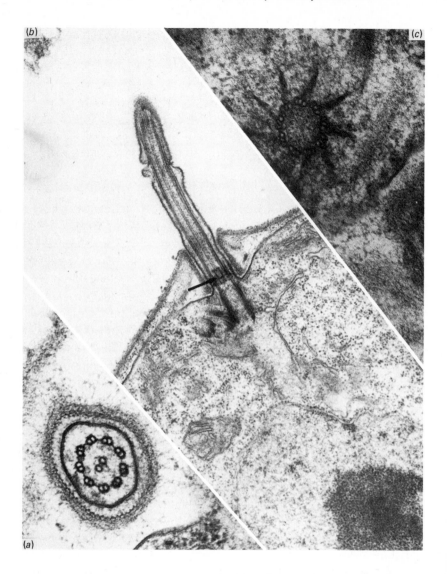

Fig. 1. Flagellar apparatus of *Chlamydomonas reinhardtii*. (*a*) Cross section through the flagellum (× 100 000); (*b*) Longitudinal section through the basal body and the proximal part of the flagellum. The arrow points to the diaphragm in the cylinder-like structure (× 35 000); (*c*) Cross section through the distal end of the basal body (× 100 000). Fixation: glutaraldehyde/OsO_4. (Courtesy of Dr D. G. Robinson.)

The two basal bodies are coupled by a distal striated connecting fibre which attaches to their distal part below the transitional region (Ringo, 1967; Melkonian, 1980); its function is not yet understood. However, the investigations on isolated and reactivated flagellar apparatus of *Chlamydomonas reinhardtii* by Hyams & Borisy (1975*a*, 1978) indicate that this connecting fibre probably plays a role in the coordination of the beat of both flagella. We will come back to this problem later. In addition two proximal connecting fibres with a similar striation pattern occur in *Chlamydomonas*.

From the basal bodies a flagellar root system originates, consisting of 4–4–4–4 microtubules according to Ringo (1967), but 4–2–4–2 according to Melkonian (1977). It has been suggested that it functions either as a cytoskeleton, or to distribute the stresses caused by the flagellar beat over the whole cell. According to a more recent hypothesis its function may be to determine the exact positions of cell organelles, for example, the stigma, with respect to the flagellar apparatus (Melkonian, 1978). For the discussion of movement photocontrol it is important to mention that in *Chlamydomonas*, as in many other green algae, a structural contact of root microtubules with the outer chloroplast membrane in the region of the stigma has been demonstrated (Melkonian, 1978).

Swimming

The average speed of swimming of *Chlamydomonas* is about $100 \mu m\,s^{-1}$, and the maximum speed is $200 \mu m\,s^{-1}$ (Nultsch, 1974). In normal forward swimming, cells move with the anterior flagella-bearing end toward the direction of motion. *Chlamydomonas* swims in the so-called breast-stroke style (Ringo, 1967), consisting of a 'power stroke' and a 'return stroke', also called 'effective stroke' and 'recovery stroke' (Sleigh, 1974). According to Ringo (1967), who investigated cells swimming in a viscous medium and, in addition, took flash micrographs of cells swimming in a non-viscous medium, the flagella are extended forward at the beginning of the effective stroke. They sweep backward, bending near the base. During the recovery stroke the wave of bending passes along the flagella from the base to the tip, until the flagella have reached the initial position. However, high-speed cinematographic studies which we carried out in cooperation with the Institut für den Wissenschaftlichen Film, Göttingen, revealed that in a normal, non-viscous medium the flagellar beat differs from Ringo's description in so far as, due to the high beat frequency (40–50 Hz), the effective strokes are never completed; nor are the flagella really extended forward during

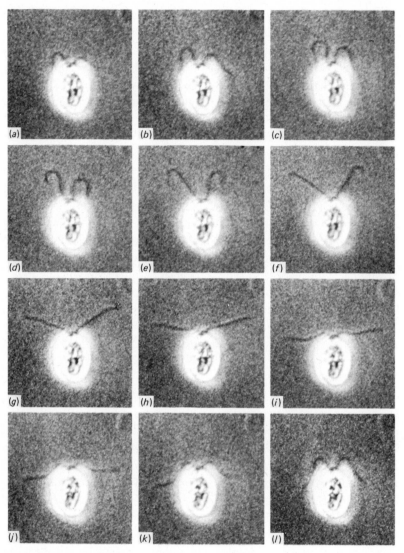

Fig. 2. Flagellar beat of *Chlamydomonas reinhardtii* strain mutant 622 E according to phase contrast, high-speed microcinematographic analysis (500 frames s^{-1}). Duration of one flagellar beat is 1/45 s. (*f–j*) effective stroke. (*k, l*) and (*a–e*) recovery stroke. Magnification about × 2000.

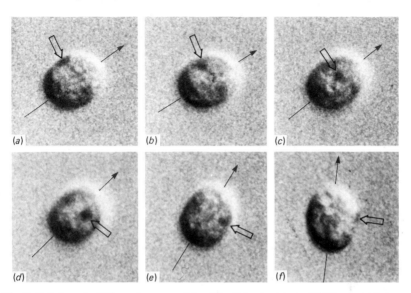

Fig. 3. Rotation of a *Chlamydomonas reinhardtii* cell (mutant strain 622 E) around its long axis (indicated by the black arrow). The changing position of the stigma (light arrow) was followed with interference contrast high-speed microcinematography (300 frames s^{-1}). Duration of the half rotation (*a–f*) about 1/3 s. Only every 15th frame is reproduced. Magnification about × 2500.

the recovery strokes (Fig. 2). Moreover, the flagellar beat is not planar, so that the cells normally rotate around their long axis 1–2 times per second (Fig. 3). This is consistent with the report of Boscov & Feinleib (1979) who observed rotation of *Chlamydomonas* cells under the microscope. Thus the flagellar beat of *Chlamydomonas* is not a breaststroke in the strict sense because of the rotational component which is essential for the phototactic orientation mechanism. Frequently the path of swimming is helical.

Chlamydomonas cells can also swim backwards, for example, in response to very high light intensities, with the non-flagellated end towards the direction of movement. In this case the flagella are extended backwards close to each other, and waves of bending run from the base to the tip, exerting a pushing force. Also during backward swimming rotation can be observed and demonstrated by cinematographic analysis.

Methods

Static cultures are of little use in phototactic investigations with *Chlamydomonas* since the phototactic sensitivity alters considerably with the age of the culture and also with the age of the cells. Therefore, even in

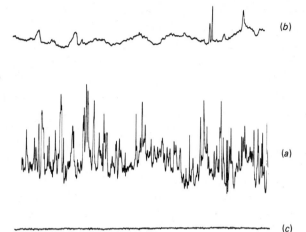

Fig. 4. Motility 'noise' diagram. (*a*) Untreated culture with 2×10^6 cells ml^{-1}. (*b*) Same culture, treated with 3×10^{-5} mol l^{-1} fluphenazine. (*c*) Nutrient medium as a control.

synchronous cultures in which all cells are in the same developmental stage, the phototactic sensitivity of the population changes. Moreover, circadian rhythms in phototactic reactivity are initiated by a light–dark regime (Nultsch & Throm, 1975). For these reasons we use continuous cultures in our phototaxis experiments (Nultsch, Throm & Rimscha, 1971).

Two principal methods are available for the quantitative analysis of phototaxis: single cell observations and population analysis. Since *Chlamydomonas* cells move very fast, video-recording (Boscov, 1974; Boscov & Feinleib, 1979) or cinematography is necessary for single cell observations. A disadvantage of these methods is that many cells have to be studied to obtain statistically significant results.

Using a video-recording system, Boscov & Feinleib (1979) observed that about 50% of the cells showed a turn in response to a single flash as well as to repetitive flashes; whereas about 10% responded with a stop. In our micro-cinematographic studies in which only the light direction, but not the irradiance, was changed, stop responses rarely occurred. Thus the stopping is not a prerequisite for phototactic orientation (Boscov & Feinleib, 1979), and does not occur during a normal phototactic reorientation caused solely by a change of light direction.

In the population method, the behaviour of a cell population under unilateral illumination is analysed by measuring changes in the optical density at both ends of a cuvette resulting from mass movement (Feinleib & Curry, 1967; Nultsch *et al.*, 1971). The advantages of this method are that it integrates over the whole cell population and can be fully automated (Nultsch & Throm, 1975). However, since the behaviour of the individual cells cannot be watched, population and single cell studies complement each other.

Another source of error in the population method is that the signal obtained also depends on the motility, that is on the number and the speed of moving organisms. Therefore, a set-up for the quantitative determination of motility has been devised in which the number of organisms passing a square red light field ($25 \times 25\,\mu$m) per unit time is counted by monitoring the absorption changes caused by the passing organisms. This method yields a motility 'noise' diagram which is shown in Fig. 4 for a normal culture (2×10^6 cells ml^{-1}) and a fluphenazine-inhibited one. This method, still in development, can be fully automated and computerised.

Photoreceptor pigment

In order to determine the chemical nature of the photoreceptor pigment(s) the phototactic action spectrum of *Chlamydomonas reinhardtii* wild-type, strain Göttingen 11–32 (+) was determined with the aid of intensity–response curves by Nultsch *et al.* (1971). According to the derived action spectrum (Fig. 5), phototaxis is a typical blue-light response (Nultsch, 1980). The main maximum lies between 500 and 510 nm and a second minor maximum occurs at 440 nm. No activity was found above 540 nm. In the UV, the activity decreases gradually without any peak or shoulder around 370 nm that would be indicative of riboflavin. This action spectrum, which resembles the action spectra measured by Halldal (1958) with *Platymonas, Dunaliella*, and *Stephanoptera*, points to a yellow pigment as the photoreceptor molecule but gives no unequivocal information about its chemical nature.

(1) Absorption spectra of carotenoids (Fig. 5) lack a UV peak around 370 nm and sometimes extend to 550 nm (for example, hydroxy-echinenone). However, the action spectrum does not resemble a typical three-peaked carotenoid absorption spectrum, and the main absorption maximum of carotenoids is found at about 450 nm rather than at 510 nm.

(2) Riboflavin and related flavin compounds are also not promising candidates for the photoreceptor pigment because of the lack of 370 nm peak in the action spectrum (Fig. 5). Moreover, potassium iodide, a quencher of flavin-excited states, inhibits phototaxis of *Chlamydomonas* only at very high concentrations (10^{-2} to 2×10^{-2} mol l^{-1}). As shown in Fig. 6, even potassium chloride inhibits phototaxis at these concentrations by about the same degree. Thus, iodide cannot be regarded as a

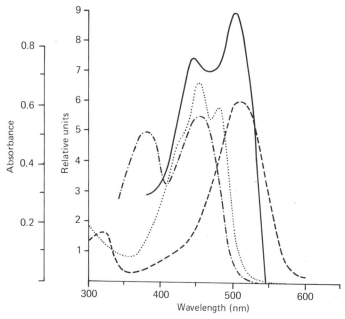

Fig. 5. Phototactic action spectrum of *Chlamydomonas reinhardtii* (——) (after Nultsch *et al.*, 1971). For comparison the absorption spectra of β-carotene (· · ·) roseoflavin (- - -) and riboflavin (— · —) are shown.

specific inhibitor of *Chlamydomonas* phototaxis, and this experiment does not support the idea that flavins could be the photoreceptor pigments in this organism.

Recently, Foster & Smyth (1980) calculated the intercept intensities I_0 from the logarithmic intensity–response curves measured by Nultsch *et al.* (1971) and plotted their reciprocals against wavelength. From these data they suggested rhodopsin or a rhodopsin-like compound to be the photoreceptor pigment. This is an interesting speculation, although the occurrence of rhodopsin in *Chlamydomonas* has not yet been demonstrated.

In general the similarity of an absorption spectrum of a pigment to an action spectrum does not unequivocally prove that it is the photoreceptive molecule of the photoresponse in question. This may be demonstrated by the following example: the maxima of the absorption spectrum of roseoflavin (=7-methyl-8-dimethyl-amino-10-(1′-D-ribityl)-isoalloxazin (Kasai, Miura & Matsui, 1975) and the phototactic action spectrum of *Chlamydomonas* agree fairly well (Fig. 5). However, roseoflavin is obviously not the phototaxis photoreceptor pigment for the following reasons: (i) its occurrence in *Chlamydomonas* has not been demon-

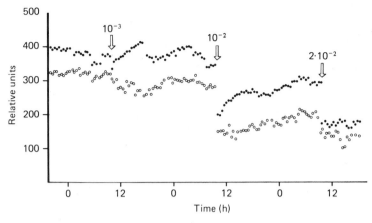

Fig. 6. Effects of potassium iodide (○) and potassium chloride (●) on phototaxis of *Chlamydomonas reinhardtii* strain Göttingen. The arrows indicate the times of the addition of both salts; the numbers are the final concentrations in mol l^{-1}. At the beginning of both plots no substance was added. Phototactic reaction is given in relative units, the scale for potassium iodide being slightly shifted downwards for clarity.

strated, and (ii) according to Song, Walker, Vierstra & Poff (1980), the fluorescence quantum yield and lifetime of roseoflavin are substantially lower than those of other flavins, thus making it kinetically less efficient as a photoreceptor. In summary, the chemical nature of the photoreceptor pigment in *Chlamydomonas* is still an open question.

The photoreceptor site

In the past the stigmata of unicellular algae were often suspected of being the photoreceptive structures, and therefore were called eyespots. In *Chlamydomonas* the stigma constitutes a part of the chloroplast. It consists of carotenoid-containing lipid granules which are arranged in one or more parallel layers (Walne & Arnott, 1967; Feinleib & Curry, 1971; Foster & Smyth, 1980). Because of these structural properties the stigma is not an appropriate candidate for the photoreceptor site and might function rather as a screen.

Walne & Arnott (1967) cross-sectioned *C. eugametos* cells and found that the plasmalemma and the outer chloroplast envelope membrane adjacent to the stigma differ from the other areas of both membranes. Using the freeze-fracture technique Melkonian & Robenek (1980) have shown that, in the plasmalemma and in the outer chloroplast-envelope membrane of the cell-wall-less mutant CW 15 of *C. reinhardtii*, the number of intramembraneous particles is significantly higher in the

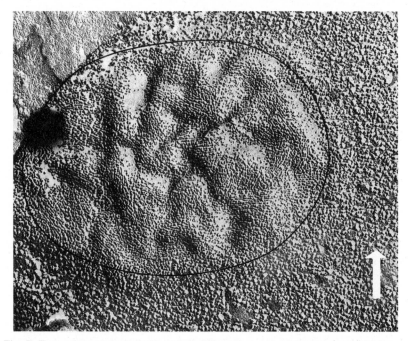

Fig. 7. Freeze-fractured plasmalemma of *Chlamydomonas reinhardtii* strain Göttingen revealing the stigma specialization on the protoplasmic fracture face. A high density and quasicrystalline arrangement of intramembrane particles is seen in the stigma adjacent area. Intramembrane particles are considerably smaller in the stigma region (encircled) compared with plasmalemma areas outside the stigma. The arrow indicates the direction of shadowing. Fixation: glutaraldehyde/OsO_4 (× 100 000). For details of method see Melkonian & Robenek, 1980. (Courtesy of Dr M. Melkonian.)

region overlying the stigma lipid globules than in the area outside the stigma. Most recently these authors obtained the same results (personal communication) with the *C. reinhardtii* strain Göttingen 11–32(+) used by Nultsch and coworkers (Fig. 7). The diameters of the intramembraneous particles in the stigma area are 8–12 nm in the plasmalemma and 4–6 nm in the outer chloroplast-envelope membrane, whereas large size particles of 16–20 nm diameter, typical for these membranes, seem to be excluded from the stigma area. For comparison, the size of rhodopsin molecules is 4–5 nm. These findings suggest that the stigma-adjacent area of either the plasmalemma or the outer chloroplast envelope membrane, or both, are the photoreceptor sites in *Chlamydomonas*. It is noteworthy that the distance of 25 nm between the plasmalemma and the chloroplast membrane overlying the stigma is constant (Melkonian & Robenek, 1980).

If the above assumption is right, the photoreceptive structure is not located between the stigma and the long axis of the cell, as suggested by Buder (1917) not only for *Euglena* but also for *Chlamydomonas*. Consequently, the phototactic orientation is the result of a 'periodic irradiation' rather than of a 'periodic shading', a possibility which was also discussed by Buder (1917).

In this connection the 'quarter–wave–stack' hypothesis proposed by Foster & Smyth (1980) is of interest. From the spacing of the stigma granule layers, these authors suggested that the stigma acts as a quarter–wave interference reflector, consisting of a stack of alternating layers of high and low refractive index, the optical distance of each layer being 1/4λ. In such a quarter–wave–stack, light reflected at low- to high-refractive-index interface changes phase by π radians, whereas light reflected at high- to low-refractive-index interface does not change the phase. Foster & Smyth calculated the pattern of irradiance within and near the stigma at different angles of incident light of 480 nm. If the photoreceptor were located between the stigma and the cell surface, as suggested by Melkonian & Robenek (1980), then light shining on this surface would produce at the receptor an irradiance that is the sum of both the incident and reflected irradiances. This would modulate the light by causing strong fluctuations in irradiance during rotation and in this way produce a directional antenna (for further information see Colombetti & Lenci, this volume). This hypothesis is supported by the observation that mutants with defective stigmata also react phototactically, but display less straight swimming paths and less well directed orientation than the wild type (M. E. Feinleib, personal communication).

Another explanation of the stigma function is that it acts as a screen protecting the inner surface of the photoreceptor membrane from 'wrong' light signals, for example those caused by a change of irradiance from inside when a 'hole' in the chloroplast (see above) passes between the light source and the stigma. Although there is no experimental evidence for such a screening function it cannot be excluded.

Sensory transduction

Since the assumed photoreceptor site and the effector, that is the flagellar apparatus, are located at some distance from each other, the stimulus received by the photoreceptor (a quantum absorbed) has to be transformed into a transmittable signal which acts at the flagellar bases where it triggers a specific motor response (Nultsch & Häder, 1979, 1980). This so-called sensory transduction chain is not yet known for

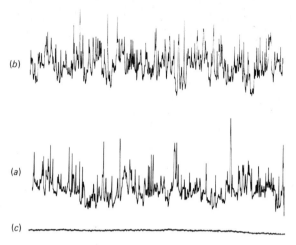

Fig. 8. Motility 'noise' diagrams of *Chlamydomonas*. (*a*) Without Ca^{2+}, (*b*) with Ca^{2+} (final concentration 10^{-3} mol l^{-1}), (*c*) nutrient medium as a control.

Chlamydomonas, but there is some evidence that calcium ions and electrical potential changes are involved.

The important role of calcium in the control of swimming and some photoresponses has been shown by several authors in different ways. Stavis & Hirschberg (1973) and Stavis (1974) demonstrated that photo-accumulations occur only if calcium and either potassium or ammonium ions are present in the medium. Schmidt & Eckert (1976) observed that, in *Chlamydomonas*, a light stimulus of 300 ms caused different responses at different Ca^{2+} concentrations. They concluded that the photostimulated flagellar reversal, the so-called stop-response, depends on the extracellular Ca^{2+} concentration and that the light stimulus results in an influx of Ca^{2+} through the cell membrane, producing a transient increase in Ca^{2+}. Thus it seems that Ca^{2+} acts as the agent that couples flagellar responses to photostimulation. Hyams & Borisy (1975*b*, 1978) working with isolated, reactivated flagellar apparatus of a cell-wall-less mutant of *Chlamydomonas reinhardtii* have shown that $Ca^{2+} < 10^{-6}$ mol l^{-1} causes forward swimming, $Ca^{2+} > 10^{-6} < 10^{-3}$ mol l^{-1} causes backward swimming, and $Ca^{2+} \gtrsim 10^{-3}$ mol l^{-1} causes a stop of flagellar beat. They concluded that depolarisation of the flagellar membranes in response to a stimulus allows the influx of Ca^{2+}, thus inducing flagellar reversal. Once the swimming direction has changed, the resting potential of the membrane is restored, calcium is removed from the axonemes by the action of a membrane pump and forward swimming is resumed.

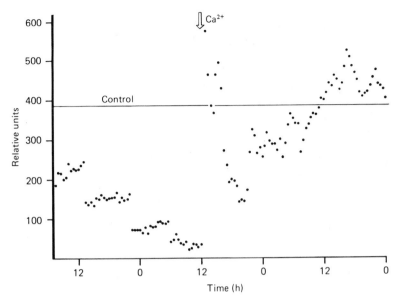

Fig. 9. Effect of the addition (arrow) of Ca^{2+} ions (final concentration 10^{-3} mol l^{-1}) on phototaxis of a culture grown in a Ca^{2+} free medium. At the beginning of the experiment Ca^{2+} ions were removed by centrifugation and the cells were resuspended in a Ca^{2+}-free medium. The horizontal line indicates the phototactic reaction values of the culture before Ca^{2+} was removed. (Modified after Nultsch, 1979.)

The calcium requirement of phototaxis was demonstrated by Nultsch (1979). Any increase of the Ca^{2+} concentration of the culture medium up to 2×10^{-3} mol l^{-1} causes a transient stimulation of phototaxis. If the culture was centrifuged and the cells washed and resuspended in a medium in which Ca^{2+} was substituted by Mg^{2+} (so-called Ca^{2+}-free medium which, however, contained traces of Ca^{2+} as impurities of the other components of the nutrient medium), the motility was of the same order of magnitude as in the calcium-containing medium, as shown by the motility 'noise' diagram (Fig. 8). In contrast to the motility, the phototactic reactivity of *Chlamydomonas* was reduced to about 20–50% of the control by the substitution of calcium (Fig. 9). If *Chlamydomonas* is grown in the absence of calcium in a continuous culture with repeated dilution, the phototactic reactivity is decreased in a stepwise fashion and finally reaches zero. Obviously, the photosensitivity is markedly decreased by the removal of the extracellular calcium, but the intracellular Ca^{2+} concentration is still sufficient to couple the flagellar responses to photostimulation. This suggests that phototactic reactivity depends on both the extracellular and the intracellular levels of calcium, and hence that Ca^{2+} acts at two sites. Addition of Ca^{2+} ions (final concentration

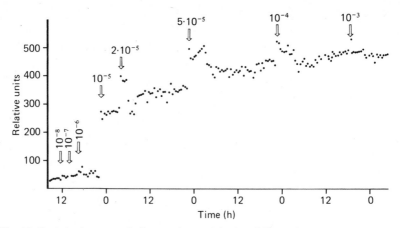

Fig. 10. Stepwise increase of phototactic reactivity by addition of increasing amounts of calcium. The arrows indicate the times of addition; the numbers are the concentrations in mol l^{-1} after addition of Ca^{2+}.

10^{-3} mol l^{-1}) restores the phototactic reactivity immediately. After a transient stimulation to about 150% of the control, the initiation of a circadian rhythm in phototactic reactivity is observed (Fig. 9).

In another experiment the Ca^{2+} concentration of an initially calcium-free culture was increased in steps (Fig. 10): a slight but significant effect was observed after increasing the Ca^{2+} concentration to 10^{-6} mol l^{-1} and the Ca^{2+} requirement of phototaxis was saturated at 10^{-4} mol l^{-1}.

Marbach & Mayer (1971) have reported that phototaxis of *C. reinhardtii* is completely suppressed by external electrical fields of $10-20$ V cm^{-1}. Recently, light-induced changes in membrane potential, which were dependent on the presence of extracellular Ca^{2+}, were measured using extracellular electrodes by Litvin, Sineschekov & Sineschekov (1978) with *Haematococcus pluvialis*.* The action spectrum for these changes in potential agrees essentially with the action spectrum for the photoaccumulation response. More recently, Ascoli et al. (personal communication) have measured light-induced potential changes in the same species, using intracellular electrodes. These experiments give at least a hint that membrane depolarisation might be involved in the photic sensory transduction chain of flagellates.

Discussion

Although the phototactic reaction chain of *Chlamydomonas* is not yet known, some experimental results published during the last few

* Membrane potential changes were measured externally by sucking the cells into the open end of the recording electrode.

years have led to a better understanding of at least some parts of this chain.

It is now well established that *C. reinhardtii* swims in the breast-stroke style with a rotational component so that the cell rotates one to two times per second around its long axis. Moreover, although the photo-receptor pigment is still unidentified, it is highly probable that it is located in the areas of either the plasmalemma or the outer chloroplast envelope (or both) overlying the stigma. Another possibility is that the intramembrane particles of the outer chloroplast envelope in this area are the photoreceptor molecules, whereas the particles in the corres-ponding area of the plasmalemma are ion gates. In any case, both the rotation of the cell and the asymmetrical position of the photoreceptor structure with respect to the long axis of the cell favour the two-instant mechanism rather than the one-instant mechanism (Feinleib, 1977; Boscov & Feinleib, 1979). If the light comes from the side, the photoreceptor is periodically irradiated if it faces the light, but is screened by the stigma and the chloroplast for most of the time during one rotation.

It is now established that calcium ions are required not only for photophobic responses and photoaccumulation, but also for phototactic orientation. Phototaxis is saturated at an external Ca^{2+} concentration of 10^{-4} mol 1^{-1}. If we assume that the intramembrane particles in the stigma area of the plasmalemma are Ca^{2+} gates, the absorption of photons by the photoreceptor would result in a calcium influx, since the intracellular Ca^{2+} concentration is probably less than 10^{-6} mol 1^{-1}. This calcium influx would cause a membrane depolarisation.

The induction of membrane depolarisation by light has been demons-trated only for *Haematococcus pluvialis*, but is also probable for *Chlamydomonas* according to the experiments of Marbach & Mayer (1971). From the ultrastructural studies it cannot yet be decided whether or not the flagella can be regarded as a compartment separated from the cell lumen. The substitution of magnesium for the extracellular calcium causes the phototactic reactions to be decreased to about 20–50% of the control. However, the phototactic reactivity is further gradually decreased to almost zero with dilution of the intracellular calcium by cell division and leakage.

Based on these considerations the following hypothetical model of the phototactic reaction chain of *Chlamydomonas* is proposed (Fig. 11), which draws in part on the model of the mechanophobic sensory transduction chain of *Paramecium* as suggested by Hildebrand (cf Haupt, 1980). If a *Chlamydomonas* cell moves toward the light source

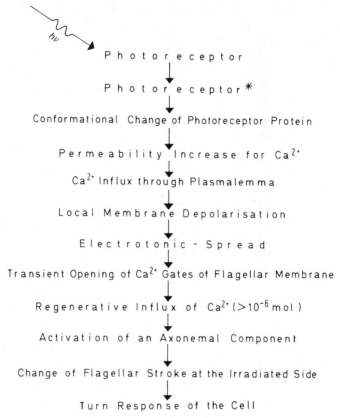

Fig. 11. Hypothetical scheme of the phototactic reaction chain of *Chlamydomonas*.

or away from it (demonstrating a positive or negative reaction respectively) the photoreceptor membrane overlying the stigma is continuously screened or equally irradiated during rotation. If the light direction is changed so that light impinges from the side, the photoreceptor passes through the light beam. Photons are absorbed by the photoreceptor pigments causing an excited state and, hence, a conformational change of the photoreceptor protein. The following possibilities then exist: the photoreceptor molecules are located in the plasmalemma and themselves serve as Ca^{2+} gates; they are located in the outer chloroplast membrane and cause the gates in the plasmalemma to open; or they are located in both membranes. Whichever is the case, the Ca^{2+} permeability of the plasmalemma increases transiently. Since the Ca^{2+} concentration in the medium is at least two orders of magnitude higher than in the cell, a calcium influx results. The membrane responds with a local depolarisation.

The signal transfer to the flagellar apparatus is unclear. In the scheme in Fig. 11 an electrotonic spread along the plasmalemma is assumed, whereas Melkonian (1978) considered that the root tubules linking the suggested photoreceptor site in the chloroplast envelope to the flagella might be used for the signal transfer. In response to this unknown signal, the Ca^{2+} gates open in the membrane of the flagellum at the stigma side causing a regenerative Ca^{2+} influx into the flagellum as suggested for the cilia of *Paramecium* (see Eckert & Machemer, 1975).

Many questions arise at this point of the reaction chain. For a positive response, the flagellum at the stigma side must beat more weakly. But how is the simultaneous reaction of the other flagellum prevented: by a block or by a 'short circuit'? Perhaps the distal connecting fibres play a part in the coordination of the beat of both flagella. If the Ca^{2+} influx into the flagellum facing the light weakens its beat, what is the response to a stronger stimulus which causes a negative reaction? According to the experiments of Hyams & Borisy (1978), a stronger Ca^{2+} influx should weaken the flagellar beat even more and thus elicit a stronger positive reaction rather than a negative one. Moreover, are the flagella really separate compartments with an ionic strength independent of the ion milieu inside the cell? If so, is an ion influx into the flagella also possible from the cell lumen? The results of the experiments in which the external calcium was removed suggest that it is, but do not indicate how it is controlled.

By the influx of Ca^{2+} ions an axonemal component is activated. This component is obviously not ATPase, since Bessen, Fay & Withman (1980) have observed that calcium changes the waveform of axonemes which lack a flagellar membrane and, hence do not have the Ca^{2+}-activated ATPase. They suggest that calcium binds directly to an axonemal component and so alters the flagellar waveform. Whether or not calmodulin, which according to Gitelman & Witman (1980) occurs in both the cell bodies and the flagella, is involved in this flagellar response cannot be decided as yet.

The phototactic reorientation is brought about by the change of the synchrony of the flagellar beat. Depending on the irradiance and on external factors such as pre-irradiation, CO_2, oxygen, pH and ion milieu (Nultsch & Throm, 1975; Nultsch, 1977, 1979) the cells turn either towards the light source or away from it. Much work needs to be done in order to elucidate the relationships of these factors to the sense of the response, and to analyse the flagellar beat pattern in positive and negative responses, respectively. Further high speed microcinematographic analyses would be helpful in resolving these mechanisms. Thus, this suggested reaction chain is extremely speculative and can serve only as a working hypothesis for future investigations.

I am indebted to Dr Ursula Rüffer, Dr Jürgen Pfau and Dr Hartwig Schuchart for good cooperation; to Dr M. E. Feinleib for giving us a stock culture of the mutant strain 622 E and to the Deutsche Forschungsgemeinschaft for financial support.

References

BESSEN, M., FAY, R. B. & WITMAN, G. B. (1980). Calcium control of waveform in isolated flagellar axonemes of *Chlamydomonas*. *J. Cell Biol.* **86**, 446–55.

BOSCOV, J. S. (1974). Responses of *Chlamydomonas* to single flashes of light. Thesis, Tufts University Medford.

BOSCOV, J. S. & FEINLEIB, M. E. (1979). Phototactic response of *Chlamydomonas* to flashes of light II. Response of individual cells. *Photochem. Photobiol.* **30**, 499–505.

BUDER, J. (1917). Zur Kenntnis der phototaktischen Richtungsbewegungen. *Jb. Wiss. Bot.* **58**, 105–220.

ECKERT, R. & MACHEMER, H. (1975). Regulation of ciliary beating frequency by the surface membrane. In *Molecules and Cell Movement*, ed. S. Inoue & R. E. Stephens, pp. 151–64. New York: Raven Press.

FEINLEIB, M. E. (1977). Photomovement in microorganisms: strategies of response. In *Research in Photobiology*, ed. A. Castellani, pp. 71–84. New York & London: Plenum Press.

FEINLEIB, M. E. (1980). Photomotile responses in flagellates. In *Photoreception and Sensory Transduction in Aneural Organisms*, ed. F. Lenci & G. Colombetti, pp. 45–68. New York & London: Plenum Press.

FEINLEIB, M. E. & CURRY, G. M. (1967). Methods for measuring phototaxis of cell populations and individual cells. *Physiol. Plant.* **20**, 1083–95.

FEINLEIB, M. E. & CURRY, G. M. (1971). The nature of the photoreceptor in phototaxis. In *Handbook of Sensory Physiology*, ed. W. R. Loewenstein, vol. I, pp. 365–95. Berlin, Heidelberg & New York: Springer-Verlag.

FOSTER, K. W. & SMYTH, R. D. (1980). Light antennas in phototactic algae. *Microbiol. Rev.* **44**, 572–630.

GITELMAN, ST E. & WITMAN, G. B. (1980). Purification of calmodulin from *Chlamydomonas*: calmodulin occurs in cell bodies and flagella. *J. Cell Biol.* **87**, 764–70.

HALLDAL, P. (1958). Action spectra of phototaxis and related problems in Volvocales, *Ulva*-gametes and Dinophyceae. *Physiol. Plant.* **11**, 118–53.

HAUPT, W. (1980). Panel discussion. Sensory transduction and photobehaviour: final considerations and emerging themes. In *Photoreception and Sensory Transduction in Aneural Organisms*, ed. F. Lenci & G. Colombetti pp. 397–404. New York: Plenum Press.

HYAMS, J. S. & BORISY, G. G. (1975*a*). Flagellar coordination in *Chlamydomonas reinhardtii*: isolation and reactivation of the flagellar apparatus. *Science*, **189**, 891–93.

HYAMS, J. S. & BORISY, G. G. (1975*b*). The dependence of the waveform and direction of beat of *Chlamydomonas* flagella on calcium ions. *J. Cell Biol.* **67**, 186a.

HYAMS, J. S. & BORISY, G. G. (1978). Isolated flagellar apparatus of *Chlamydomonas*: characterization of forward swimming and alteration of waveform and reversal of motion by calcium ions *in vitro*. *J. Cell. Sci.* **33**, 235–53.

KASAI, S., MIURA, R. & MATSUI, K. (1975). Chemical structure and some properties of roseoflavin. *Bull. Chem. Soc. Japan*, **48**, 2877–80.

LITVIN, F. F., SINESCHEKOV, O. A. & SINESCHEKOV, V. A. (1978). Photoreceptor electric potential in the phototaxis of the alga *Haematococcus pluvialis*. *Nature, London*, **271**, 476–78.

MARBACH, I. & MAYER, A. M. (1971). Effect of electric field on the phototactic response of *Chlamydomonas reinhardii*. *Israel J. Bot.* **20**, 96–100.

MELKONIAN, M. (1977). The flagellar root system of zoospores of the green alga *Chlorosarcinopsis* (Chlorosarcinales) as compared with *Chlamydomonas* (Volvocales). *Plant Syst. Evol.* **128**, 79–88.

MELKONIAN, M. (1978). Structure and significance of cruciate flagellar root systems in green algae: comparative investigations in species of *Chlorosarcinopsis* (Chlorosarcinales). *Plant Syst. Evol.* **130**, 265–92.

MELKONIAN, M. (1980). Ultrastructural aspects of basal body associated fibrous structures in green algae: a critical review. *Biosystems*, **12**, 85–104.

MELKONIAN, M. & ROBENEK, H. (1980). Eyespot membranes of *Chlamydomonas reinhardii*: a freeze-fracture study. *J. Ultrastruct. Res.* **72**, 90–102.

NULTSCH, W. (1974). Movements. In *Algal Physiology and Biochemistry*, ed. W. D. P. Stewart. Botanical Monographs, Vol. 10. Oxford, London, Edinburgh & Melbourne: Blackwell Scientific.

NULTSCH, W. (1975). Phototaxis and photokinesis. In *Primitive Sensory and Communication Systems: the Taxes and Tropisms of Microorganisms and Cells*, ed. M. J. Carlile, pp. 29–90. London, New York & San Francisco: Academic Press.

NULTSCH, W. (1977). Effect of external factors on phototaxis of *Chlamydomonas reinhardtii*. II. Carbon dioxide, oxygen and pH. *Arch. Microbiol.* **112**, 179–85.

NULTSCH, W. (1979). Effect of external factors on phototaxis of *Chlamydomonas reinhardtii*. III. Cations. *Arch. Microbiol.* **123**, 93–9.

NULTSCH, W. (1980). Effect of blue light on movement of microorganisms. In *The Blue Light Syndrome*, ed. H. Senger, pp. 38–49. Berlin, Heidelberg & New York: Springer-Verlag.

NULTSCH, W. & HÄDER, D.-P. (1979). Photomovement of motile microorganisms. *Photochem. Photobiol.* **29**, 423–37.

NULTSCH, W. & HÄDER, D.-P. (1980). Light perception and sensory transduction in photosynthetic prokaryotes. In *Structure and Bonding*, ed. J. D. Dunitz, J. B. Goodenough, P. Hemmerich, J. A. Ibers, C. K. Jorgensen, J. B. Neilands, D. Reinen & R. J. P. Williams, vol. 41, pp. 111–39. Berlin, Heidelberg & New York: Springer-Verlag.

NULTSCH, W. & THROM, G. (1975). Effect of external factors on phototaxis of *Chlamydomonas reinhardtii*. I. Light. *Arch. Microbiol.* **103**, 175–9.

NULTSCH, W., THROM, G. & VON RIMSCHA, I. (1971). Untersuchungen an *Chlamydomonas reinhardii* Dangeard in homokontinuierlicher Kultur. *Arch. Microbiol.* **80**, 351–69.

RINGO, D. L. (1967). Flagellar motion and fine structure of the flagellar apparatus in *Chlamydomonas*. *J. Cell Biol.* **33**, 543–71.

SCHMIDT, J. A. & ECKERT, R. (1976). Calcium couples flagellar reversal to photostimulation in *Chlamydomonas reinhardtii*. *Nature, London*, **262**, 713–15.

SCHÖTZ, F. (1972). Dreidimensionale, maßstabgetreue Rekonstruktion einer grünen Flagellatenzelle nach Elektronenmikroskopie von Serienschnitten. *Planta*, **102**, 152–9.

SLEIGH, M. A. (1974). Patterns of cilia and flagella. In *Cilia and Flagella*, ed. M. A. Sleigh, pp. 79–92. London: Academic Press.

SONG, P.-S., WALKER, E. B., VIERSTRA, R. D. & POFF, K. L. (1980). Roseoflavin as a blue light receptor analog: spectroscopic characterization. *Photochem. Photobiol.* **32**, 393–8.

STAVIS, R. L. (1974). Phototaxis in *Chlamydomonas*: A sensory receptor system. Thesis, Yeshiva University, New York.

STAVIS, R. L. & HIRSCHBERG, R. (1973). *Phototaxis in* Chlamydomonas reinhardtii. *J. Cell Biol.* **59**, 367–77.

WALNE, P. L. ARNOTT, H. J. (1967). The comparative ultrastructure and possible function of eyespots: *Euglena granulata* and *Chlamydomonas eugametos*. *Planta*, **77**, 325–53.

LEAFLET MOVEMENTS IN *SAMANEA**†

ARTHUR W. GALSTON

Department of Biology, Yale University, New Haven, Conn. 06520, USA

The main lesson offered to photobiologists by the study of leaf movements is that of *time dependency*. Because of the existence of internal oscillations within many photosensitive organisms, it matters very much exactly when a photon impinges on a photoreceptor pigment. Thus, a standard red light given early in the cycle of movement may reduce the rate of leaf opening; later it may have no effect or be conducive to closing. Similarly, a standard blue light may promote opening at one time, retard closure at a second time and have zero effect at a third time. When the effect of light is plotted against time, one frequently sees a rhythmic oscillation in the efficacy of its action, with a period of about 24 hours. Such a rhythm is referred to as *circadian* (Bünning, 1973). In general, internal rhythms exert a stronger control over many processes, including leaf movements, than does light; but light may reset ('rephase') the clock which regulates rhythmic behavior. In this way, light and rhythms can interact in the measurement of daylength changes as the seasons progress (Vince-Prue, 1975; Hillman, 1976). Such light–rhythm interaction in time measurement is known as *photoperiodism*.

The main pigments involved in rephasing rhythms in plants, and thus in time measurement, are phytochrome and the yet uncharacterized blue-light absorbing pigment. Following photoexcitation of these pigments, hormonal and other micrometabolic regulatory mechanisms are altered, leading to transformations of turgor, relative growth rates, chemistry and form. All of these are involved in morphogenetic changes, such as the formation of floral primordia in previously vegetative plants, while only some operate in simpler processes, such as

* Aided by grants from the National Science Foundation and the National Aeronautics and Space Administration. Much of this work is the result of close collaboration with Dr Ruth L. Satter, to whom I express deep thanks. All conclusions in this paper are, however, my own.

† The genus name has recently been changed to *Pithecolobium*. To avoid confusion with the older physiological literature, I will retain here the name *Samanea*.

Fig. 1. Plants of *Samanea* with leaves in the open (right) and closed (left) condition.

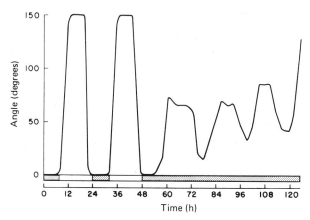

Fig. 2. Photocontrolled and rhythmic leaflet movement of *Samanea*. The abscissa bar indicates dark (stippled) and light treatments. Leaflets are closed in the dark and open in the light during 16 h light/8 h dark daily cycles. During the latter part of the light period, the leaves start to close rhythmically. When put into 'free-run' conditions in a long dark period, these leaves continue to oscillate with circadian periodicity, diminished amplitude, and a drift toward opening. *In vitro*, supplying sucrose increases the amplitude and halts the drift (see Fig. 5).

leaf movement, which do not involve growth. Thus, while leaf movements in *Samanea* involve entirely reversible turgor changes, they may be considered as paradigms for photoperiodic systems in which morphogenetic changes such as floral initiation occur.

We entered into the study of leaf movements because they seemed to us to offer a simple, rapid, convenient system for the analysis of light–rhythm interaction. Thanks to the careful foundation laid by such famous nineteenth century biologists as Darwin, Sachs, and Pfeffer, and in the twentieth century by Bünning and colleagues for *Phaseolus*, it has long been appreciated that leaf movements are indeed rhythmic, persisting with roughly circadian periodicity during a prolonged 'free run' period in total darkness (Bünning, 1973). Figure 1 shows *Samanea* plants in the 'asleep' or closed (left) and 'awake' or open (right) positions, and Fig. 2 shows opening and nyctinastic closure during alternating light and dark periods, followed by diminishing rhythmic oscillation in continuous darkness. During appropriate phases of the rhythm, these plants can respond to light of various wavelengths; light affects both the angle between petiole and stem ('hands of the clock') and the clockwork itself (shown by rephasing of the rhythm).

Movements of leguminous leaves like *Samanea, Albizzia, Phaseolus* and *Trifolium* occur because of turgor changes in key motor cells of the pulvinus (Satter, 1979; Satter & Galston, 1981; Fig. 3). This fleshy

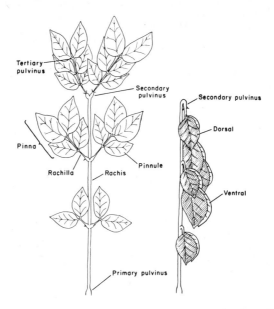

Fig. 3. Sketch of open and closed *Samanea* leaves. Our experiments utilised secondary pulvinules governing movements of *pinnae*. These tubular organs, straight in the open condition, bend like elbows during closure.

organ is found at the base of the petiole of the leaf or of the leaflet in a compound leaf. It consists of a central core of vascular tissue sur-rounded by a cortex, and at specific sectors of the periphery, by elastic, thick-walled, highly vacuolate motor cells. *Extensor* motor cells are those whose osmotically controlled change of volume leads to leaflet or leaf opening; *flexor* cells, in contrast, lose in volume during opening and both expand and change shape during closure (Fig. 4). In *Samanea*, the fleshy secondary pulvini, which control the movement of the *pinnae*, have been used in our investigations (Satter, Geballe, Applewhite & Galston, 1974).

Palmer & Asprey (1958*a, b*) introduced *Samanea* into experimental plant physiology almost 25 years ago. After showing that light caused pinnae to open, that darkness tended to cause closure, and that leaves continued to open and close rhythmically in extended dark periods, they slit the pulvinus longitudinally, leaving either the flexor or extensor intact. Pinnae with intact extensors showed pronounced rhythmic movement during prolonged darkness. Such plants also open in the light, but the maximum angle attained is no greater than that of rhythmic dark opening, which is somewhat less than the maximum

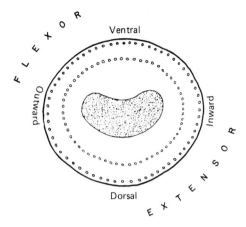

Fig. 4. Sketch of cross section of a *Samanea* pulvinule. The shaded area in the centre is the vasculature. The circles near the periphery indicate where microprobe measurements were made to yield the data of Fig. 7.

angles attained by intact open leaves in the light. Pinnae bearing only intact flexors show minor rhythmic movement, but are very responsive to changes in illumination.

Samanea pulvini can be conveniently studied *in vitro* as well as on the intact leaf (Satter *et al.*, 1974). A Y-shaped portion of the leaf containing paired *pinnae* attached to the petiole is stripped of its leaflets, then bisected longitudinally and each pulvinus inverted into a microvial containing 50–100 mM sucrose and any other required addenda (Fig. 5). Such skeletonized leaves open and close in response to light and dark signals, and continue to move rhythmically in extended dark periods. They thus contain the photoreceptors, clockwork and internal solutes with which to accomplish the turgor changes that cause movement. Furthermore, all components continue to function in an integrated manner as in the intact leaf, and except for a slight diminution in amplitude of oscillation, the movements shown by such excised pulvinules behave like the intact system.

We grew *Samanea* plants in controlled condition chambers with a 16 h light/8 h dark cycle (Simon *et al.*, 1976*a*, *b*). The pinnae are open during the first 12 hours of the light period, then close and remain closed until about two hours before the end of the dark period. Their slow opening at this point is greatly speeded by white light. Preliminary analyses by flame photometry had shown that the pulvinus is extraordinarily rich in K^+ ion; most pulvini contain about 400 mM of this ion on a total fresh

Fig. 5. Preparation of an *in vitro* pulvinule preparation. The terminal pinnae of a compound leaf are removed as a unit, and cut down to a small defoliated Y-shaped rachis–rachilla segment. This is bisected longitudinally to yield a pair of reactive pulvinules, which are then inverted into solutions in conical minivials.

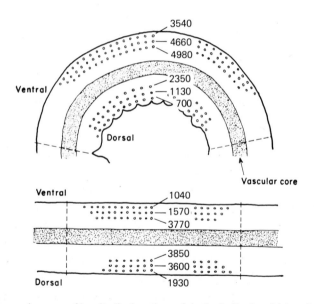

Fig. 6. Microprobe scans of longitudinal sections of rhythmically closed (above) and open (below) pulvinules. The shaded area is the vasculature; small circles are loci of microprobe measurements. There is essentially no longitudinal gradient. The figures represent K^+ scintillations on a roughly per cell basis. Note decrease in flexor (near ventral) and increase in extensor (near dorsal) during opening. Note also the decreasing gradient from vasculature toward peripheral motor cells.

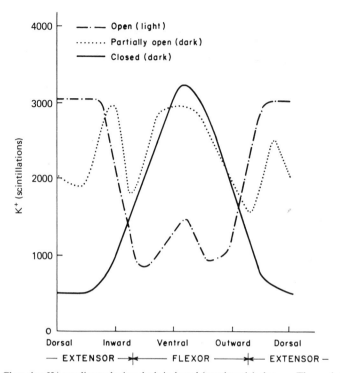

Fig. 7. Changing K^+ gradients during dark-induced (nyctinastic) closure. The periphery of the cross section of a pulvinule (see Fig. 4) has been opened up and represented as the abscissa. Thus, the regions labeled 'dorsal' at the left and right of the abscissa represent the same point of the opened-up circle. Note the sharp decrease of the high extensor plateau and increase in flexor peak during closure. From high plateau (opened) to nearest trough point represents the width of two cells.

weight basis, whether open or closed. When lyophilized longitudinal sections of frozen pulvini were examined in the electron microprobe, however, it became clear that K^+ is massively redistributed during movement, whether such movement is rhythmic or caused by light–dark transitions (Fig. 6). The K^+ content of the flexor (ventral) cells is almost five times that of the extensor (dorsal) cells in closed leaves, but less than half that of the extensors in open leaves. On a dry weight basis, extensor cells went from 16 to 4 mg g^{-1} dry weight during closure, while at the same time flexor cells changed from 19 to 26 mg g^{-1} dry weight. Midcortical and vascular tissues are higher in K^+ than are the more peripheral cells; we assume that these represent, respectively, the 'storehouse' and 'delivery pipeline' for pulvinal K^+. Microprobe examination of K^+ profiles in *transverse* lyophilized pulvinal sections (Fig. 7) revealed fairly sharp, rather than gradual transitions between

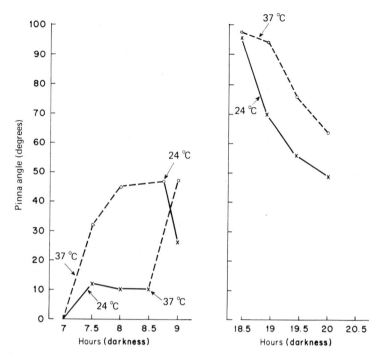

Fig. 8. Rhythmic opening (*left*) and closure (*right*) of *Samanea* pulvinules at two different temperatures. Raising the temperature reversibly increases the rate and steady state angle of opening (*left*) but decreases the rate of closing (*right*). This is compatible with a model indicating a temperature-sensitive active opening mechanism and a temperature-indifferent passive closing mechanism.

regions of high and low K^+, especially in the light. The space between the high plateau of the extensor (3000 K^+ scintillations) to the low region of the flexor (1000 K^+ scintillations) is occupied by only *two* cells! Later probes (Fig. 9) revealed even sharper patterns, and it now seems that sharp gradients may exist even across the walls of a single cell (R. L. Satter, unpublished). This indicates the operation of mechanisms other than simple diffusion. In dark rhythmic movement, the patterns are similar but of smaller amplitude.

The operation of metabolically powered systems during rhythmic movement is clearly seen by simple temperature experiments. Pinnae allowed to open at 37 °C open more rapidly and assume larger steady state angles (45° versus 10° opening) than similar pinnae at 24 °C (Fig. 8, *left*). When the pinnae at these two temperatures are moved to the opposite temperature, those originally at 37 °C close upon cooling and those originally at 24 °C open during warming. During closure, an apparently paradoxical result is obtained: the closing reaction is more rapid at 24 °C than at 37 °C (Fig. 8, *right*). A rational scheme to account for this would involve interaction of a metabolically-powered, temperature-sensitive process, and a temperature-insensitive system whose rate is determined by the phase of the rhythm. Let us assume, for example, an ion pump causing osmotic swelling and passive leakage causing shrinkage. When leakage is low, a pump operating more rapidly at higher temperature swells the cell faster (opening); when leakage is high, the faster pump at higher temperature slows emptying (closure). Thus, rhythms determine the broad pattern, environment (in this case, temperature) merely modulates the pattern. The Q_{10} for opening is approximately 2.5 and for closing about 0.8.

Phytochrome can be shown to regulate K^+ flux and thus pulvinal movement (Simon *et al.*, 1976*b*). When darkened pinnaexposed briefly to red light, their extensor K^+ value falls, their flexor K^+ value rises, and the pinnae close. Far-red light reverses these changes. An increase in temperature from 24 °C to 37 °C increases the differential effect of the two light treatments during opening, but not during closure. This implies that phytochrome controls an energetic process. Phytochrome also controls nyctinastic closure in previously illuminated leaves; far-red light tends to delay closing and increase the angle, while red light facilitates closing. As in rhythmic opening, the differential effect of light treatment increases with increasing temperature. From the experiments above, and from those of Palmer & Asprey (1958*a*, *b*), it seems clear that metabolic processes in the flexor cells lead to K^+ accumulation, higher turgor and leaflet closure. Rhythmic movements in the dark depend mainly on K^+-based turgor changes in the extensor cells, although the two systems, flexor and extensor, are tightly coupled in the intact pulvinus.

Detailed investigations with the electron microprobe have revealed that Cl^- accompanies K^+ during movement, partially preserving electroneutrality (Satter, Schrempf, Chaudhri & Galston, 1977). An H^+ for K^+ exchange probably accounts for the difference (Satter & Galston, 1981). The pattern of flow during opening is probably from flexor to extensor by way of the mid-ventral region. Upon reversal during closure, the flow seems to be back over the same route (Fig. 9). The mid-dorsal region, which has a lower background K^+, is assumed not to be involved in massive ion flux, although fluxes could be hidden in the high K^+ patterns of this tissue. In all cases, the mid-extensor: mid-flexor ratio of $K^+ + Cl^-$ is correlated with leaflet angle rather precisely

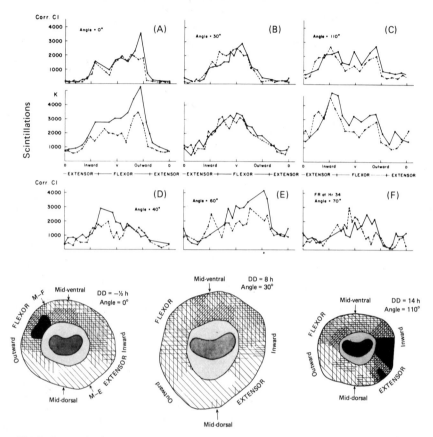

Fig. 9. Scans of K^+ and Cl^- profiles in cross sections of pulvinules at various stages of rhythmic movement. Upper scans show that K^+ and Cl^- profiles around the pulvinules change in the same way. Lower topographic sketches show distribution of K^+ scintillations from about 500 (single cross-hatched) progressively to over 3500 per 15 s (solid black) measurement period. Note movement of K^+ from flexor to extensor through the mid-ventral region (top) during opening. The reverse path is followed during closure.

(Fig. 10). Approximately 0.6 as much Cl^- as K^+ is present in the inner cortex storage area and motor tissue. Calculations of the ions migrating from flexor to extensor and reverse show a rough equivalence between K^+ and Cl^-, with a slight excess of the former. It appears that some other anion, possibly that of an organic acid, is involved in balancing the electrical charge during movement. An alternative possibility is the electrogenic extrusion of protons, creating an internal electronegativity that would favor the entrance of K^+ or other cations.

Direct evidence for rhythmic and light-regulated changes in the transmembrane potential in motor cells of *Samanea* was obtained by

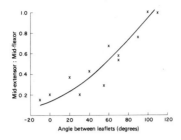

Fig. 10. The angle of opening is a simple function of the ratio of mid-extensor:mid-flexor $[K^+ + Cl^-]$.

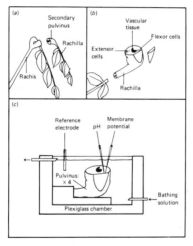

Fig. 11. Preparation of a pulvinule for bioelectric measurements. The pulvinule is cut transversely, then shaved longitudinally to expose the motor cells. It is then mounted in a lucite cell in a flowing medium, and microelectrodes are inserted with a micromanipulator while being viewed under a microscope.

impaling such cells with microelectrodes under a microscope, with the aid of a micromanipulator (Fig. 11; Racusen & Satter, 1975). During rhythmic movement in the dark, flexor cells hyperpolarize to $-100\,\mathrm{mV}$ during maximum opening and depolarize to $-35\,\mathrm{mV}$ during closure (Fig. 12). Extensor cells show a smaller amplitude rhythm, displaced in time; thus, maximum hyperpolarization (to $-85\,\mathrm{mV}$) occurs just as leaflet opening begins, and depolarization to about $-45\,\mathrm{mV}$ occurs just before closure is initiated. White fluorescent light of about 10 klux causes a prompt depolarization, progressing to a maximum of about $-20\,\mathrm{mV}$ after 50 min. Phytochrome is also involved in the regulation of

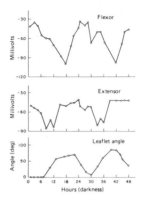

Fig. 12. As the pulvinus opens rhythmically (*below*), the flexor hyperpolarizes (becomes more negative, shown by declining curve). The extensor shows a smaller amplitude hyperpolarization, peaking just as opening begins. Both trends are reversed during closure, and show roughly circadian rhythmicity.

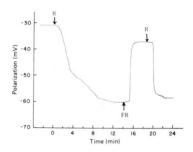

Fig. 13. Red light causes hyperpolarization of flexor motor cells and far-red light causes depolarization. The changes are sluggish during the first cycle, more rapid during subsequent cycles.

the transmembrane potential: red light causes hyperpolarization, and far-red causes depolarization in flexor cells. The first red flash produces a gradual effect over about 10 min; the subsequent far-red and red-induced changes show only a 90 s lag (Fig. 13). These studies tell us that rhythms and light signals, acting through membrane-generated bioelectric potentials, could provide the basis for ion fluxes that culminate in leaf movement.

The potentials generated in pulvinal motor cells, probably by H^+ secretion, may be partly annulled by an inward sucrose-proton co-transport (Racusen & Galston, 1977). Changing the pH around the impaled motor cells produced no marked effect on transmembrane potential until the addition of 10 mM sucrose. At that point, depolarization occurred over a 30 s period, the magnitude being proportional to

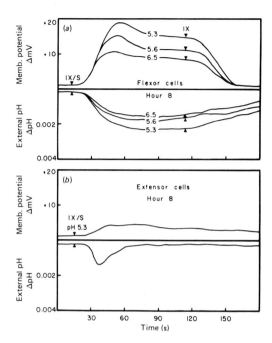

Fig. 14. When a pulvinus in basal salt–sucrose solution (1X/S) is exposed to a suddenly lowered pH (arrow), its motor cells (in this case flexor) depolarize (above) and the external medium becomes more alkaline (second level). The magnitude of the effect is proportional to the imposed pH gradient. When flexor cells react to the joint H$^+$–sucrose stimulus, the extensor cells (below) may not (see Fig. 15).

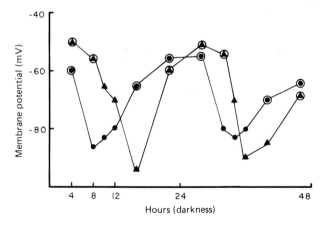

Fig. 15. When response of the motor cells to joint H$^+$–sucrose gradient is plotted against the endogenous circadian change in potential, an oscillating pattern of response is observed. In both extensor (●) and flexor (▲) a positive response, shown by the encircled data points occurs only when the potential is more positive than about −70 mV.

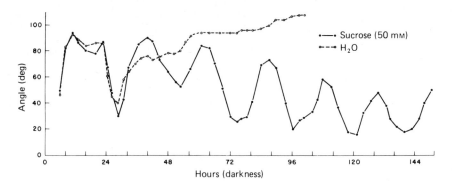

Fig. 16. *In vitro, Samanea* pulvinules require sucrose (ca. 50 mM) to maintain the rhythm for rhythmic perpetuation. In the absence of sucrose, the rhythm damps with the pulvinules in the open position (◯- - -◯).

the H^+ concentration (Fig. 14). At the same time, the external solution, monitored by a micro pH-electrode, becomes more alkaline. When flexor cells (above) respond electrically to such a joint presentation of sucrose and H^+, extensor cells (below) may or may not do so, and a plot of response versus time in darkness during a free-run rhythmic experiment shows that both flexor and extensor respond rhythmically, 4 h out of phase, and only when their potentials are less negative than about -70 mV (Fig. 15). The value for all sucrose – H^+-induced depolarizations was 16 ± 0.5 mV; it is not known whether this rhythmic ability to depolarize in the presence of sucrose – H^+ turns on and off all at once, or whether it is modulated. The effect of sucrose concentration on depolarization follows saturation kinetics, with a K_m in the range of 60–100 mM. Ions such as K^+, Na^+ and Ca^{2+} also induced concentration-dependent depolarizations of 4–12 mV. However, these effects were *not* affected by the addition of sucrose, nor were they accompanied by a change in the pH of the bathing medium.

The persistence of the rhythm in extended darkness in excised *Samanea* pulvini requires sucrose (50–100 mM) (Simon, Satter & Galston, 1976*a*, *b*). In the absence of sucrose, the rhythm damps with the pulvini in the open position, while supplying sucrose permits the oscillations to persist for several cycles (Fig. 16). In the related plant *Albizzia*, far-red light causes damping of the rhythm with the pulvini in the closed position (Satter, Applewhite, Chaudhri & Galston, 1976). Red light given every 24 h entrains the rhythm and prevents damping if given at the appropriate point in the cycle. At other points, red light rephases the rhythm, leading to an advance, a delay, or no effect, as in

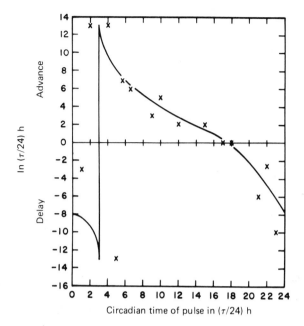

Fig. 17. A five-minute flash of red light given at various circadian times during a long dark period may advance, delay or produce no effect at all on the phase of the rhythm. Note that a 12 h delay is the same as a 12 h advance. This phase-alteration curve is typical for circadian systems.

other systems (Fig. 17). In all cases, the effect of red light is reversible by far-red, showing the involvement of phytochrome. Recent investigations reveal that blue light is also able to rephase the rhythm; more energy is required and the curve is shaped differently. In continuous white light, excised pulvini, incubated in 50 mM sucrose or water, show damping of the rhythm during the first cycle with the pulvini at an intermediate angle.

When K^+ and Cl^- move between flexor and extensor cells during leaflet movement, what is their path and how do they travel? Two possible pathways present themselves, the apoplastic and the symplastic. In the apoplastic mode, ions would move from cytoplasm through the membrane into the wall, thence either through extracellular wet wall space or into an adjacent cell by passage through its membrane. The elastic walls of motor cells would provide sequestration space for such ions, while the plasmodesmata would provide a symplastic pathway for the movement of ions from cell to cell without moving through any membrane (Morse & Satter, 1979).

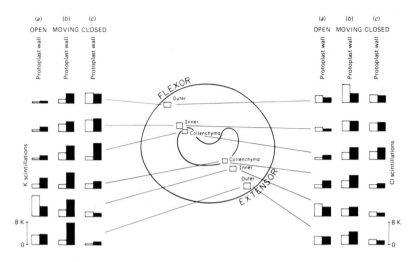

Fig. 18. Much of the K^+ of motor cells is found in the wall (solid bars) indicating a predominantly apoplastic pathway of movement. Changes of the ratio of K^+ in protoplast to K^+ in wall indicates entry and exit through the plasma membrane.

From observations on pulvinal cells of *Samanea, Mimosa, Albizzia* and related species (Satter, 1979), it is also clear that profound changes occur in vacuoles during the volume changes that cause movement. In their fully turgid state, motor cells tend to have a single large vacuole; when they lose volume, the vacuole breaks up into many smaller ones (Morse & Satter, 1979). Vacuolar tannins, abundant in the motor cells of *Mimosa* pulvini, also change form and distribution during leaflet movement. Although comparable studies have not been made with *Samanea*, these changes suggest some function which might be general in pulvini. The polyanionic membrane-bounded tannins are finely dispersed in the fully turgid cell, but are more clumped and precipitated after cell shrinkage. Concomitant with this decrease in volume and tannin aggregation is a release of adsorbed cations such as K^+, which then appear in the walls and extracellular volume. Substantiation of these early results, based on histochemistry, by high resolution energy-dispersive X-ray microanalysis of cryostated lyophilized material has recently been obtained (Fig. 18; Campbell, Satter & Garber, 1981). By focussing on regions of the wall lacking or containing plasmodesmata it was possible to ascertain that most of the K^+ leaving a shrinking extensor cell initially appears in the wall outside the membrane. The pathway involves intermediary passage through thick-walled collen-chyma *en route* from extensor to flexor. Thus, despite the presence of

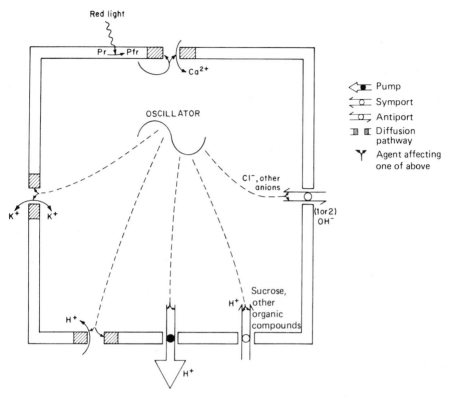

Fig. 19. A model of possible mechanisms of ion transport and control in a pulvinal motor cell. Phytochrome is envisioned as affecting only Ca^{2+} entry. This change could then alter other cellular processes, possibly through a calmodulin-based system. The oscillator may affect diffusion pathways for H^+, K^+, Cl^- and OH^-, as well as H^+ pumps, and H^+–sucrose symports, and Cl^-–OH antiports. The scheme is hypothetical; much remains to be demonstrated.

plasmodesmata in motor cells, the bulk of the evidence favors a predominantly apoplastic pathway of ion movement from extensor to flexor and return. Because of the nature of the ionic gradients in the apoplast of the *Samanea* pulvinus, we have recently postulated the intervention of transfer cells and at least one symplastic step in this otherwise apoplastic pathway (Satter & Galston, 1981).

A mechanism to explain these data has been recently proposed (Satter & Galston, 1981; Fig. 19). An oscillator is envisioned as influencing the state of the membrane such that passage of H^+, K^+, Cl^- and other anions is affected. H^+ could move either alone or as part of a H^+–K^+ exchange system; K^+ could also presumably diffuse freely in either direction, while Cl^-, on evidence from other plants, probably moves as part of a Cl⁻–OH⁻ antiport. The oscillator also affects the operation of a proton pump which secretes H⁺, creating the electro-chemical conditions for entry of other cations. The sensitivity of *Samanea* to oxygen (Sweet & Hillman, 1969) and the existence of a respiratory rhythm in O_2 uptake by *Samanea* pulvini (Satter, Hatch & Gill, 1979) suggest an energetic oscillation that would control the proton pump. Vanadate sensitivity of the energetic phase in the related plant *Albizzia* suggests participation of a membrane-localized ATPase (Saxe & Satter, 1979). Phytochrome is also envisioned as operating at a membranous locale, through modulation of a Ca^{2+}-diffusional pathway (Hale & Roux, 1980). The increase of cytoplasmic Ca^{2+} could then activate a calmodulin system, and thereby several transport-linked processes including membrane phosphorylation, membrane-localized ATPase and K⁺-diffusional pathways (see the article by Roux in this volume). The influence of phytochrome on the oscillator, shown by the rephasing results, could be exerted indirectly, through effects on the membrane, or directly, if the oscillator is located elsewhere.

As always, the leap from data to hypothesis is chancy. What we know for certain is that K⁺, Cl⁻ and possibly other ions are shuttled mainly apoplastically between flexor and extensor cells during leaflet move-ment, and that rhythms and light interact to control these processes. Rhythms continue to be obscure, although considerable evidence now localizes them, or at least some of their effects, in membranes (Njus, Sulzman & Hastings, 1974). Light operates through phytochrome and a blue-absorbing pigment, both through modulation of existing systems, and through phase-shifting. Where the oscillator and pigments are found is still predominantly a 'black box' with a few illuminated pinholes piercing the obscurity.

References

Bünning, E. (1973). *The Physiological Clock*. Berlin: Springer-Verlag.

Campbell, N. A., Satter, R. L. & Garber, R. C. (1981). Apoplastic transport of ions in the motor organ of *Samanea*. *Proc. Natl. Acad. Sci., USA*, **78**, 2981–84.

Hale, C. C. II & Roux, S. J. (1980). Photoreversible calcium fluxes induced by phytochrome in oat coleoptile cells. *Plant Physiol.*, **65**, 658–62.

Hillman, W. S. (1976). Biological rhythms and physiological timing. *Annu. Rev. Plant Physiol.*, **27**, 159–79.

Morse, M. L. & Satter, R. L. (1979). Relationships between motor cell ultrastructure and leaf movements in *Samanea saman*. *Physiol. Plant.*, **46**, 338–46.

Njus, D., Sulzman, F. & Hastings, J. W. (1974). Membrane model for the circadian clock. *Nature, London*, **248**, 116–1201.

Palmer, J. H. & Asprey, G. F. (1958a). Studies in the nyctinastic movement of the leaf pinnae of *Samanea saman* (Jacq) Merrill. I. A general description of the effect of light on the nyctinastic rhythm. *Planta*, **51**, 757–69.

PALMER, J. H. & ASPREY, G. F. (1958*b*). Studies in the nyctinastic movement of the leaf pinnae of *Samanea saman* (Jacq) Merrill. II. The behaviour of the upper and lower half pulvini. *Planta*, **51**, 770–85.

RACUSEN, R. H. & GALSTON, A. W. (1977). Electrical evidence for rhythmic changes in the cotransport of sucrose and hydrogen ions in *Samanea* pulvini. *Planta*, **135**, 57–62.

RACUSEN, R. H. & SATTER, R. L. (1975). Rhythmic and phytochrome-regulated changes in transmembrane potential in *Samanea* pulvini. *Nature, London*, **255**, 408–10.

SATTER, R. L. (1979). Leaf movements and tendril curling. In: *Encyclopedia of Plant Physiology, new series*, vol. 7, ed. W. Haupt & M. E. Feinleib. Berlin: Springer-Verlag.

SATTER, R. L., APPLEWHITE, P. B., CHAUDHRI, J. & GALSTON, A. W. (1976). P_{fr} phytochrome and sucrose requirement for rhythmic leaflet movement in *Albizzia*. *Photochem. Photobiol.*, **23**, 107–12.

SATTER, R. L. & GALSTON, A. W. (1981). Mechanisms of control of leaf movements. *Annu. Rev. Plant Physiol.*, **32**, 83–110.

SATTER, R. L., GEBALLE, G. T., APPLEWHITE, P. B. & GALSTON, A. W. (1974). Potassium flux and leaf movement in *Samanea saman*. I. Rhythmic movement. *J. Gen. Physiol.*, **64**, 413–30.

SATTER, R. L., HATCH, A. M. & GILL, M. K. (1979). A circadian rhythm in oxygen uptake by *Samanea* pulvini. *Plant Physiol.*, **64**, 379–81.

SATTER, R. L., SCHREMPF, M., CHAUDHRI, J. & GALSTON, A. W. (1977). Phytochrome and circadian clocks in *Samanea*: rhythmic redistribution of potassium and chloride within the pulvinus during long dark periods. *Plant Physiol.*, **59**, 231–5.

SAXE, H. & SATTER, R. L. (1979). Effect of vanadate on rhythmic leaflet movement in *Albizzia julibrissin*. *Plant Physiol.*, **64**, 905–7.

SIMON, E., SATTER, R. L. & GALSTON, A. W. (1976*a*). Circadian rhythmicity in excised *Samanea* pulvini. I. Sucrose–white light interactions. *Plant Physiol.*, **58**, 417–20.

SIMON, E., SATTER, R. L. & GALSTON, A. W. (1976*b*). Circadian rhythmicity in excised *Samanea* pulvini. II. Resetting the clock by phytochrome conversion. *Plant Physiol.*, **58**, 421–5.

SWEET, H. C. & HILLMAN, W. S. (1969). Phytochrome control of nyctinasty in *Samanea* as modified by oxygen, submergence, and chemicals. *Physiol. Plant.*, **22**, 776–86.

VINCE-PRUE, D. (1975). *Photoperiodism in Plants*. New York: McGraw-Hill.

A POSSIBLE ROLE FOR Ca^{2+} IN MEDIATING PHYTOCHROME RESPONSES

STANLEY J. ROUX

Department of Botany, The University of Texas at Austin, USA

Introduction

The idea that phytochrome could initiate morphogenic changes by promoting an increased concentration of free Ca^{2+} ions in the cytosol of cells has been present in the literature for over five years. Haupt & Weisenseel (1976) proposed this hypothesis mainly to stimulate discussion and to suggest new research approaches for studying the function of phytochrome *in vivo*. They defended the idea by pointing out that Ca^{2+} had already long been recognized as an important 'second messenger' in stimulus–response coupling in animals (Rasmussen, 1970) and that there were several known phytochrome responses that could be expected to be mediated by Ca^{2+}.

Today, seven years after Haupt & Weisenseel presented their ideas at the Easter School in Nottingham, their hypothesis is still very much alive. The data in direct support of that hypothesis are few, but the broader notion that changes in cytosolic Ca^{2+} concentration may initiate major changes in cellular functions in both plants and animals has received extensive support. It seems appropriate, then, to evaluate existing data on the phytochrome–Ca^{2+} connection and to propose ways of testing more rigorously how important that connection may be for understanding phytochrome responses *in vivo*. These will be the main goals of this chapter.

The control of cytosolic Ca^{2+} concentration by phytochrome can only be understood in the context of the more widely accepted hypothesis that, as an early step in the sequence of events leading to photomorphogenesis, phytochrome binds to cellular membranes and alters their functional properties. This topic has already been discussed in many review articles, most recently those of Pratt (1978, 1979), Rüdiger (1980), and Quail (1981). This discussion will only be updated here, reviewing mainly very recent papers on this topic, which were not discussed in the above-mentioned articles.

No coverage of Ca^{2+}-mediated cell functions would be complete without raising the question of the possible involvement of calmodulin (Cheung, 1980; Means & Dedman, 1980). This protein has been firmly established to be a regulator of many Ca^{2+}-dependent enzymes in animals, and of at least two in plants. Anderson & Cormier (1978) first proposed the likelihood that some phytochrome responses could be modulated through calmodulin. We will evaluate this hypothesis further here.

Phytochrome association with membranes

Many of the observed effects of phytochrome on Ca^{2+} fluxes are most easily rationalized by a model in which phytochrome has some direct interaction with one or more cellular membranes. There have been several different approaches to testing whether phytochrome has any significant affinity for membranes and, if so, whether that affinity results in an alteration of any membrane function. The approaches reviewed here were those taken by investigators who have published papers and research abstracts in the last two years on the topic of phytochrome in membranes. They include model membrane studies, studies on the phenomenon of phytochrome co-sedimentation with crude membrane preparations, assays of phytochrome binding to highly purified membrane preparations, and immunocytochemical localization studies.

Model membrane studies allow an evaluation of physical parameters which influence the association of phytochrome with lipid bilayers. Early studies (Roux & Yguerabide, 1973; Roux, 1974) provided evidence that phytochrome could alter the ion permeability of planar lipid bilayers and that this effect was dependent on both the conformation of phytochrome and the lipid composition of the membrane. The planar lipid bilayer system used in those early studies was not suitable for precise quantification of protein binding, nor did it readily permit physical analysis of the bound protein. For these purposes, liposomes are better model membranes to use. These spherical bilayers are technically simple to produce in large numbers and they can be manipulated in much the same way as isolated organelles or protoplasts. Two laboratories have recently reported initial findings on the reaction of phytochrome with liposomes.

Kim & Song (1981) used ^{125}I-labeled phytochrome to quantitate the binding of phytochrome to liposomes. They found that both Pr (red absorbing form of phytochrome) and Pfr (far-red absorbing form of phytochrome) bound to liposomes, but that as they raised the ionic strength of the reaction medium, Pfr bound in increasingly high amounts whereas the level of Pr binding remained relatively unchanged. They also observed that this preferential binding to liposomes of Pfr over Pr was promoted by having cholesterol in the bilayer, and that it was considerably magnified by raising the temperature to 300 K from 276 K. The authors argued that these and other data in their paper supported the idea that Pfr, but not Pr, bound to liposomes predominately through hydrophobic interactions (see Song, this volume).

Georgevich (1980), in our laboratory, recently completed a study on phytochrome–liposome interactions. He, too, used ^{125}I-labeled phytochrome to quantitate the binding of Pr and Pfr to liposomes; but focused more on comparing permeability changes induced in liposomes by these two forms of phytochrome. Under conditions of low salt in the reaction medium and no cholesterol in the bilayer, he found, as did Kim & Song (1981), that there was little difference in the quantity of Pr and Pfr bound to bilayer liposomes. But even under these conditions Pfr induced a rate of salt (KCl) efflux from the liposomes two to three times greater than did Pr, as monitored by conductivity measurements of the solution bathing the liposomes. This differential effect was evident at 298 K but not at 277 K. It did not result from gross changes in the 'leakiness' of the liposomes, as there was no differential leakage of ATP, glucose, or trypsin from liposomes after Pr and Pfr reacted with vesicles enclosing these substances.

Furuya, Freer, Ellis & Yamamoto (1981) quantified the binding of phytochrome to positively and negatively charged liposomes at 277 K by spectral methods. This method of quantification is limited by the presence of liposomes, which distort phytochrome spectra, and so the interpretation of their results was somewhat compromised. However, the evidence they obtained by this method supported the conclusion that at 277 K the initial binding of both Pr and Pfr to liposomes is determined primarily by electrostatic interactions. This conclusion does not address the nature of phytochrome binding at 298 K and, as such, does not contradict the conclusions of Kim & Song (1981).

The three model studies described above do not directly comment on the nature of phytochrome association with membranes *in vivo*. Their aim was to define more precisely some important physical parameters of the reaction of phytochrome with lipid bilayers. All of them found that both Pr and Pfr had the physical potential to bind to pure lipid bilayers. The two studies at 298 K found evidence that there was a qualitative difference in the binding of Pr and Pfr. In no study was this difference apparent at 277 K.

These model studies are like all others in that they inevitably over-simplify how proteins and membranes interact *in vivo*. This very over-simplification, however, can often reveal important aspects of the interaction which would have been difficult to discern in the complex setting of living cells (Racker, 1976), and thus can lead to new insights and the formulation of new hypotheses. One testable hypothesis suggested by all three model studies discussed above is that the binding of phytochrome to biological membranes may be determined more by the lipid composition of the membrane than by some protein 'receptor' (Roth-Bejerano & Kendrick, 1979).

One can also test the affinity of purified phytochrome for biological membranes. The results of such a study were recently published by our laboratory (Cedel & Roux, 1980a). We used the method of Georgevich, Cedel & Roux (1977) to quantify the binding of ^{125}I-labeled phytochrome to a well characterized membrane preparation which was highly enriched for intact, functioning mitochondria. We found that a significant fraction of the total ^{125}I-labeled Pfr that would bind to the mitochondrial preparation could be competitively inhibited from binding by including a 15-fold excess of unlabeled Pfr in the reaction mixture. We referred to this fraction as 'specifically bound' phytochrome because its binding could be competitively inhibited by phytochrome but not by other proteins such as bovine serum albumin, ovalbumin, or hemoglobin. We observed that the maximum specific binding of ^{125}I-labeled Pfr to mitochondria occurred within 30 s and was optimized in a reaction buffer containing 5 mM $MgCl_2$ at pH 6.8.

Scatchard plots of the binding data for Pfr indicated that there was a single high-affinity binding site with an extraordinarily high affinity constant of 1.79×10^{11} M. The binding data for Pr, in contrast, showed that there were only low-affinity sites for it. As predicted by the Scatchard plot, 'specifically-bound' ^{125}I-labeled Pfr was only slowly removed by multiple washes of the mitochondria, whereas the fraction that was non-specifically bound could be almost completely removed by a single wash. ^{125}I-labeled Pfr showed no specific binding to purified rat liver mitochondria.

The significance of *in-vitro* binding of phytochrome to mitochondria would be questionable, at best, unless the binding resulted in placing some mitochondrial function(s) under the photoreversible control of red and far-red light. Our laboratory has documented that when exogenously added phytochrome is reacted with mitochondria under conditions which maximize specific binding, at least two interrelated mitochondrial functions exhibit this photoreversible control: the rate of oxidation of exogenously added NADH by a Ca^{2+}-regulated inner-membrane dehydrogenase (Cedel & Roux, 1980*b*) and the net rate of uptake and release of Ca^{2+} (Roux, McEntire, Slocum, Cedel & Hale II, 1981).

The mitochondrial experiments were carried out to determine whether the association of phytochrome with mitochondria *in vitro* satisfied certain minimal criteria for the biologically relevant binding of a ligand to its receptor, as are described by Cuatrecasas (1974). There are too few data on the binding of Pfr to mitochondria and other intracellular sites to satisfy all of these criteria. However, the relatively low number of specific binding sites on mitochondria for Pfr and the high affinity of the binding are characteristics that would be expected of a ligand that is biologically effective at relatively low concentrations, as has been indicated for Pfr in some of its responses (Mandoli & Briggs, 1981). This, and the potentially important functional responses triggered by Pfr in mitochondria, encourage the further use of these organelles as model systems for studying the reaction of phytochrome with biological membranes.

The conversion of Pr to Pfr *in vivo* rapidly leads to a change in phytochrome which makes it more likely to co-sediment with cell particulates (become 'pelletable') during the centrifugation of a crude homogenate of the irradiated cells. The question of whether most of the Pfr associates with cell particulates before the disruption (*in vivo*) or afterwards (*in vitro*) has not been definitively settled, but Pratt (1980) and Quail & Briggs (1980) have both presented data that are more compatible with the former hypothesis. The latter two studies included high concentrations of Mg^{2+} in the homogenization buffer, which causes widespread aggregation of particulates in the crude extract and makes it extremely difficult to analyze whether Pfr is preferentially associated with membranes or with non-membranous particulates in the extract (Quail & Briggs, 1980). Watson & Smith (1981) have shown that enhanced pelletability does not require divalent cations. Their pelletable materials could be fractionated and examined for the possible occurrence of an association of Pfr with specific organelles.

When Mg^{2+} is not included in the extraction buffer, at least some of the *in-vivo*-generated Pfr ends up bound to mitochondria and other organelles. Slocum (1981) in our laboratory has demonstrated this in immunocytochemical localization studies using ferritin-labeled anti-phytochrome antibody. He reported that after labeling a preparation of membranes enriched for mitochondria (no exogenous phytochrome added) most of the label appeared on the mitochondria in the preparation. The amount of label on mitochondria was greater if the tissue source (etiolated oat seedlings) had been irradiated with red light prior to its extraction. Interestingly he found that the occasional etioplast and peroxisome present as contaminants in the mitochondrial preparation contained, on a per-organelle basis, more anti-phytochrome label than the mitochondria. Slocum also examined immunocytochemically the binding of exogenous phytochrome to mitochondrial preparations under conditions which Cedel & Roux (1980a) predicted would maximize specific binding. Again, most of the total label was found on the mitochondria in the preparation; and, again, on a per organelle basis, there seemed to be more label on the etioplasts and peroxisomes present.

Other preliminary reports on the localization of phytochrome on membranes have appeared recently. Verbelen, Butler & Pratt (1981) studied the localization *in situ* of phytochrome in oat coleoptile tips and bean hook epicotyl segments by immunofluorescence cytochemistry, using rhodamine-conjugated, goat anti-rabbit, immunoglobulin-G, antibodies. They found that phytochrome in dark-grown plants was uniformly distributed throughout the cell. The immunofluorescent staining of phytochrome in white-light grown plants was less intense than in dark-grown plants but still quite distinct. No clear assignment of phytochrome as localized on specific organelles was possible in these initial findings. However, the methods used, which included infiltrating the fixed tissue with sucrose and thin sectioning frozen sections on the coldstage of an ultramicrotome, have the potential to yield better resolution than was possible in previously used techniques.

Viestra, Tokuhisa, Newcomb, & Quail (1981) have described a new solid-phase antibody approach to identifying structures in crude extracts that contain phytochrome on their surfaces. Their preliminary data indicate that some of the pelletable phytochrome in crude extracts of irradiated oat seedlings is bound to membranous vesicles. Attempts to identify these vesicles are now in progress.

Taken together, the recent data reviewed in this section agree with earlier data in supporting the hypothesis that phytochrome develops a greater physical affinity for membranes following its activation by red light. They do not finally settle how or whether Pr associates with cellular membranes, or whether the greater affinity of Pfr for membranes is expressed by its binding to them both before and after, or only after, tissue extraction. Only studies of localization *in situ* now in progress will answer this question. In the absence of any evidence that the cellular environment is inimical to the association of Pfr with membranes, our working hypothesis is that at least some of the phytochrome in cells becomes tightly bound to one or more cellular

Table 1. *Selected literature reports on phytochrome-mediated effects on ion transport*

Ionic species	References[a]
H^+	Lurssen (1976); Pike & Richardson (1979)
K^{+b}	Satter, Marinoff & Galston (1970); Brownlee & Kendrick (1979); Pike & Richardson (1979)
Inorganic phosphate	Tezuka & Yamamoto (1975); Brownlee & Kendrick (1979)
Ca^{2+}	Dreyer & Weisenseel (1979); Hale & Roux (1980)

[a] Partial list only.
[b] Rubidium flux studies are included with K^+.

membranes soon after its photoconversion *in vivo*. We expect that such binding would have significant consequences for the function of the target membrane. In the next section we discuss the evidence that one such consequence is a change in the Ca^{2+} permeability of that membrane.

Phytochrome-induced changes in the permeability of membranes to Ca^{2+}

Given the close coupling of ion transport systems in plant membranes (Clarkson, 1977), it would be surprising if phytochrome could alter one of these without affecting others simultaneously. Table 1 calls attention to literature reports of phytochrome-promoted fluxes of H^+, K^+ and Ca^{2+}. The kinetic data and other details on these fluxes are not sufficiently complete to conclude whether they could all occur simultaneously or even whether they would all occur in the same plant. Nonetheless, the Ca^{2+} fluxes on which we will focus in the discussion below should not be assumed to occur as isolated or 'primary' events.

Weisenseel & Ruppert (1977) were the first to present convincing evidence that phytochrome could control the entry of Ca^{2+} into cells. They studied the effect of various external Ca^{2+} concentrations on the red-light-induced depolarization of the internodal cell membranes of *Nitella*. In all the cells tested the first response to red light was a depolarization. The magnitude of depolarization increased with the Ca^{2+} concentration of the medium up to an average maximum of $33\,\text{mV}$ in $5\,\text{mM}$ Ca^{2+}. In a Ca^{2+}-'free' medium (no EGTA added), the depolarization was only $1\,\text{mV}$. Repolarization would begin within several seconds after turning off the red light. Although far-red light alone had no effect on the resting potential of the membrane, if it was applied immediately after red light, it accelerated the rate of repolarization. The data did not rule out the possibility that photosynthetic pigments could also have participated in the responses to red light and far-red light, but they did implicate phytochrome as the major controlling pigment. The simplest interpretation of these results was that Pfr triggered an influx of Ca^{2+} into the cell, although the actual entry of Ca^{2+} into *Nitella* was not demonstrated.

That Pfr triggered an actual entry of Ca^{2+} into light-stimulated cells in threads of *Mougeotia* was suggested by the data of Dreyer & Weisenseel (1979). They used light microscope autoradiography to document that there was an increased accumulation of $^{45}Ca^{2+}$ in cells which had been irradiated with red light and that this effect was reversed by far-red light. In order to exclude photosynthesis as a cause of the increased accumulation of Ca^{2+}, Dreyer & Weisenseel incubated the irradiated cells in darkness in a non-radioactive medium for three to ten minutes before transferring them to a $^{45}Ca^{2+}$ medium. During this time, it was assumed, any light-induced changes in *Mougeotia* caused by the stimulation of photosynthetic pigments would have relaxed and returned to a pre-irradiation condition. Most of the phytochrome would still be in the Pfr state after this dark period (Wagner, 1974). The increased accumulation of Ca^{2+} promoted by Pfr could have resulted from a decreased efflux rate or an increased rate of uptake. The authors considered the latter interpretation more likely.

Our laboratory tested whether Pfr could alter Ca^{2+} fluxes in the coleoptile tip cells of *Avena sativa* (Hale & Roux, 1980). We used the Ca^{2+} indicator dye, murexide, to monitor changes in the Ca^{2+} concentration of the medium bathing the test cells: either apical segments of coleoptiles or protoplasts isolated from these sections. Our results indicated that red light promoted a net efflux of Ca^{2+} from the cells and that far-red light could reverse this change by promoting a Ca^{2+} re-entry into the cells. Using very different techniques, Newman (1981) has recently presented results of studies on electrical potential changes in oat coleoptiles, which indicate the possibility that red light induces a photoreversible Ca^{2+} efflux from these cells.

Our data agreed with those from Weisenseel's laboratory in favoring the hypothesis that phytochrome could control Ca^{2+} fluxes across the plasma membrane of plant cells; but they apparently disagreed on the direction of the Pfr-promoted Ca^{2+} flux. Their data indicated that Pfr promoted Ca^{2+} uptake–ours, Ca^{2+} efflux. Preliminary experiments in our laboratory suggest that this apparent discrepancy cannot be attributed solely to technique or organism differences: we have observed Pfr-promoted Ca^{2+} efflux from *Mougeotia* with the murexide technique and Pfr-promoted $^{45}Ca^{2+}$ efflux from oat coleoptile apical segments preloaded with the radiotracer (C. C. Hale II, unpublished observations).

A possible resolution of this discrepancy comes from a consideration of the fact that plasma-membrane Ca-ATPase pumps play a major role in maintaining low cytosolic Ca^{2+} concentrations in both animal and plant cells (Marmé, 1981). If there were a feed-back mechanism whereby the activity of these Ca-ATPases could be regulated by the level of Ca^{2+} in the cytosol, then a Pfr-induced uptake of Ca^{2+} into cells, as described for *Mougeotia*, could result in heightened activity of the plasma-membrane Ca-ATPases. If these pumps pushed Ca^{2+} out of the cell faster than Pfr promoted its entry, then the murexide method (Hale & Roux, 1980) would detect a net efflux of Ca^{2+} from cells, whereas the $^{45}Ca^{2+}$ method (Dreyer & Weisenseel, 1979) would still show that Pfr was promoting an uptake.

For this scenario to be plausible two assumptions have to be true (i) there has to be a plasma-membrane Ca-ATPase whose activity is significantly enhanced by increases in cytosolic Ca^{2+}; and (ii) Pfr has to release Ca^{2+} into the cytosol from one or more intracellular stores in addition to triggering Ca^{2+} uptake across the plasmalemma. Unless this second assumption were true, the light-activated rate of Ca^{2+} efflux could not long exceed its influx rate before the cytosolic Ca^{2+} levels would return to or drop below pre-irradiation levels.

As regards the first assumption, three different laboratories have obtained results supporting this conclusion (Dieter & Marmé, 1980*a*; Caldwell & Haug, 1981; C. C. Hale II, K. McEntire, R. L. Biro & S. J. Roux, unpublished). These papers will be reviewed in the next section of this chapter. Here we will only mention the most important inference of their results: any rise in the cytosolic Ca^{2+} concentration above 'resting' values in the submicromolar range would be expected to activate the calcium-binding protein, calmodulin, which in turn, should enhance the activity of calmodulin-dependent enzymes, including plasma membrane Ca-ATPases.

Apropos to this prediction, Weisenseel & Ruppert (1977) observed that the repolarization of irradiated *Nitella* cells could begin even while the red light was on, if the irradiation extended beyond two minutes. If one accepts the possibility that extrusion of Ca^{2+} ions would promote repolarization (just as their entry facilitated depolarization), then this observation could be considered consistent with the predictions of the calmodulin-Ca-ATPase hypothesis. A more detailed discussion of calmodulin and its possible role in mediating phytochrome responses will be the main focus of the next section.

We have published that the second assumption noted above may be true (Roux et al., 1981). We used murexide to monitor the uptake and release of Ca^{2+} by oat mitochondria. We found that Ca^{2+} fluxes in these organelles could be photoreversibly altered, red light diminishing the net uptake rate, and far-red light restoring this rate to its dark control level. To test whether the principal effect of Pfr was to reduce the activity of Ca^{2+} uptake pumps or to promote Ca^{2+} release from the mitochondria, we used the inhibitor ruthenium red, which specifically blocks the active uptake of Ca^{2+} by mitochondria. In the presence of this inhibitor, red light induced a net efflux of Ca^{2+} and subsequent far-red light reduced this efflux to nearly zero, the dark control level. The conclusion that Pfr can promote the release of Ca^{2+} from mitochondria seemed warranted. Although mitochondria are a major intracellular store of sequestered Ca^{2+}, they are certainly not the only organelles with high levels of Ca^{2+} (Slocum, 1981). If Pfr promoted the release of Ca^{2+} from any other organelle, this could also contribute to raising the cytosolic Ca^{2+} concentration.

Dieter & Marmé (1981*b*) recently reported that light irradiation of crude fractions of mitochondria isolated from etiolated *Zea* coleoptiles had no effect on either the Ca^{2+} uptake or efflux of these fractions. It is difficult to assess the significance of the discrepancy between their results and ours because the materials and methods they employed differed from ours in several major respects. Further studies are required to resolve this issue.

To summarize the main points discussed above, data from both our laboratory and Weisenseel's support the hypothesis that the activation of phytochrome by light would lead, at least transiently, to an increased concentration of Ca^{2+} in the cytosol of three species of plants. The experiments that led to those conclusions were extensive and technically complicated to execute. As a consequence, the much-needed confirmation of these findings in other organisms and a more detailed analysis of the original observations will both be major undertakings.

To elaborate briefly on this point, there is as yet no one simple method to obtain accurate and precise information on the rate of Ca^{2+} movement into and out of plant cells. If one uses intact tissue, the wall will act as a significant buffer zone for the free diffusion of Ca^{2+} into the medium bathing the tissue. If one uses protoplasts (at least, protoplasts from the apices of etiolated oat coleoptiles) obtaining sufficient numbers of stable cells in a preparation that is relatively free of debris and bacteria will be technically difficult (Hale & Roux, 1980). Interpreting the results of tracer experiments with $^{45}Ca^{2+}$ is more straightforward than interpreting murexide data (Ohnishi, 1978). But, as we discuss below, changes in the direction of Ca^{2+} fluxes could be expected to occur rapidly, and the time resolution potential of the murexide method (*c.* 1 s) is much more rapid than is usually possible by radiotracer methods. It seems to us that a combination of approaches using a variety of different subject materials will be needed to clarify how phytochrome is influencing Ca^{2+} fluxes in plant cells.

A potential role for calmodulin in mediating phytochrome responses

An appropriate starting point for this discussion is to review the evidence that calmodulin does occur in plants. Calmodulin is only one of a family of calcium-binding proteins which have regulatory functions in eukaryotic organisms, but it has quite distinctive properties (Jamieson, Bronson, Schachat & Vanaman, 1980). All calmodulins are acidic, heat-stable proteins with a monomer molecular weight near 17000. Because the structure of calmodulin has been highly conserved throughout evolution, antibodies against mammalian calmodulin cross-react well with calmodulin from any source. Interestingly, this criterion does not identify calmodulin as specifically as the criterion that calmodulin from any source will activate mammalian 3', 5'-cyclic nucleotide phosphodiesterase (Van Eldik, Piperno & Watterson, 1980). All calmodulins analyzed to-date have one trimethyllysine, one or two histidines, one or two tyrosines, one or no cysteines, and no tryptophan (Anderson, Charbonneau, Jones, McCann & Cormier, 1980). By these criteria, calmodulin has been unambiguously identified in algae, monocots, and dicots (Table 2). As judged by radioimmunoassay (Chafouleas, Dedman, Munjaal & Means, 1979), calmodulin represents about 0.4% of the total protein in a clarified crude extract of etiolated oat shoots (Roux, McEntire, Slocum, Cedel & Hale, 1981). On a molecular basis, the oat extract would have over an order of magnitude more calmodulin than phytochrome (S. J. Roux & K. McEntire, unpublished).

The first evidence that calmodulin served some function in plants came from Anderson & Cormier (1978). They discovered that calmodulin was the heat-stable protein activator of NAD kinase which had been partially characterized earlier by Muto & Miyachi (1977). They found that mammalian brain calmodulin could substitute for plant calmodulin

Table 2. *Plants in which calmodulin has been identified[a]*

Species	Reference
Pisum sativum	Anderson *et al.* (1980)
Arachis hypogea	Anderson *et al.* (1980)
Spinacea oleracea	Van Eldik, Grossman, Iverson & Watterson (1980)
Cucurbita pepo	Dieter & Marmé (1980*a*)
Hordeum vulgare	Caldwell & Haug (1981)
Avena sativa	Roux *et al* (1981)
Chlamydomonas reinhardtii	Gitelman & Witman (1980)

[a] Identification was by several criteria, including the ability to activate a mammalian Ca^{2+}-dependent cyclic nucleotide phosphodiesterase. This list includes references published through mid 1981 only. Current expectation is that calmodulin occurs ubiquitously among eukaryotes.

in activating plant NAD kinase. In this first report they indicated that the addition of Ca^{2+} and calmodulin stimulated the activity of relatively crude preparations of plant NAD kinase about five-fold. More recently, they have analyzed highly purified preparations of the enzyme and found they are activated over 100-fold by Ca^{2+} and calmodulin; that is, they are essentially inactive in the absence of Ca^{2+} and calmodulin (Jarrett, Charbonneau, Anderson, McCann & Cormier, 1980).

What made their observations particularly interesting to photobiologists was that Ogren & Krogman (1965) had shown that the conversion of NAD to NADP in green leaves was promoted by light; and Tezuka & Yamamoto (1972) had shown that this light response was photoreversible by red and far-red light, and thus probably under the control of phytochrome. Since calmodulin is present in unirradiated plant cells, the only additional requirement to activate, in sequence, calmodulin and NAD kinase would be an increase in the concentration of free Ca^{2+} in the cytosol. The inference that Pfr can mediate this increase seems warranted, assuming the validity of Tezuka & Yamamoto's findings (Frosch, Wagner & Mohr, 1974).

The data of Hale & Roux (1980) suggested that Pfr activates a plasma membrane calcium pump in oat cells. As indicated above, one way this finding could be reconciled with Dreyer & Weisenseel's conclusion that Pfr promotes Ca^{2+} uptake would be if there were a feedback system whereby increases in cytosolic free Ca^{2+} would lead to increased efflux of Ca^{2+} from the cell by the plasma membrane Ca-ATPases. That calmodulin regulated such a feedback system in red-blood cells was known before 1980 (Gopinath & Vincenzi, 1977; Jarrett & Penniston, 1978); but it was not until the papers of Dieter & Marmé (1980*a*, 1981*a*) that such a system was reported in plants. Dieter & Marmé used $^{45}Ca^{2+}$ to monitor the ATP-dependent uptake of Ca^{2+} into a crude microsomal fraction from *Cucurbita* (1980*a*) and *Zea* (1981*a*), and found that calmodulin stimulated this uptake in both cases. The membrane fraction was uncharacterized but was considered by the authors to be enriched for plasma membrane. The fraction of microsomes which utilized ATP to take up Ca^{2+} were assumed to represent largely inside-out vesicles of plasma membrane; so in their normal configuration in intact cells the Ca-ATPases on them would be pumping Ca^{2+} to the outside. Recently, Dieter & Marmé (1981*b*) have reported preliminary evidence that far-red light could inhibit the calmodulin stimulation of the corn microsomal Ca-ATPase. They proposed that far-red light could, by this mechanism, reduce the rate of Ca^{2+} extruded from cells and thus increase cytosolic Ca^{2+} levels and activate calmodulin.

Preliminary data from Caldwell & Haug (1981) indicate that gradient-purified, plasma membrane-enriched microsomes from barley roots also contain a Ca^{2+}-dependent ATPase that is stimulated by calmodulin. They found that the sensitivity of this Ca-ATPase to calmodulin stimulation was rapidly lost due, in part, to a loss of phosphatidylcholine from the membranes, which in turn was probably the result of phospho-lipase action on the membrane during the extraction. There is also a report of phospholipid requirements for the calmodulin stimulation of the erythrocyte Ca-ATPase (Gietzen, Tejcka & Wolf, 1980).

We have observed Ca^{2+} and calmodulin-dependent Ca-ATPase activ-ity on highly-purified preparations of plasma membrane from green oat leaf protoplasts (C. C. Hale II, K. McEntire, R. L. Biro & S. J. Roux, unpublished). These membranes were prepared by a novel method we have recently developed. When tiny plastic beads ($0.7 \mu m$ diameter) are phagocytized by protoplasts, they are enveloped by plasma membrane (Mayo & Cocking, 1969). These uniform 'phagosomes' may be rapidly purified away from other cellular membranes by a series of density-gradient centrifugations, during which they retain their membrane envelope. Another plasma membrane ATPase, the Mg^{2+}-dependent K^+-ATPase (Leonard & Hodges, 1973) is also present on the phago-somes, but is not stimulated by calmodulin.

If calmodulin were to control only the activities of NAD kinase and plasma membrane Ca-ATPases in plants, it would still be classified as a key regulator protein. There is some indirect evidence to suggest that

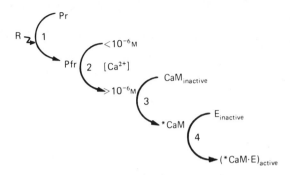

Fig. 1. Proposed model for the cascade of events initiated by the activation of phytochrome by light. 1. Phytochrome is activated by red-light irradiation. 2. Pfr mediates an increase in the concentration of free Ca^{2+} in the cytosol. 3. Calmodulin (CaM) is activated by the formation of a Ca^{2+}–CaM complex (*CaM). 4. The activation of *CaM-dependent enzymes by the binding of *CaM to them.

calmodulin may control even more enzymes than these two. Both Jarrett *et al.* (1980) and Dieter & Marmé (1980*b*) have reported that more than a half-dozen proteins from the clarified crude extracts of plants they studied bound to an immobilized-calmodulin affinity column in a Ca^{2+}-dependent fashion. Calcium controls a wide variety of plant functions, as reviewed by Marmé (1981), by mechanisms not yet well understood. Given the major role of calmodulin in controlling many Ca^{2+}-dependent functions in animal cells (Cheung, 1980), it seems plausible that more than two of the Ca^{2+}-dependent functions in plants may be regulated through calmodulin.

If the light-activation of phytochrome results in an increase in the cytosolic free Ca^{2+} levels of the irradiated cells, there should be some correspondence between those enzymes activated by calmodulin and those activated by phytochrome. Our model, illustrating this hypothesis (Fig. 1), is based on the evidence reviewed earlier that the correlation does hold for the first two calmodulin-controlled functions discovered in plants, NAD kinase and Ca-ATPase (Roux *et al.*, 1981). To emphasize a point made earlier, this model does not predict that the 'primary' function of phytochrome is uniquely to promote higher cytosolic concentrations of Ca^{2+}, nor does it exclusively require that all Ca^{2+} effects be mediated through calmodulin. The model rationalizes two enzyme responses already observed and suggests that there may be others.

The validity of the model can be examined by testing the accuracy of its predictions. One prediction is that some of the enzymes activated by phytochrome (Schopfer, 1977) will also be activated by calmodulin.

Strictly speaking, this prediction could be considered satisfied by the data reviewed above which indicate NAD kinase and Ca-ATPase are two such enzymes. But given the multiple functions of calmodulin in animal cells (Cheung, 1980), it would not be unreasonable to investigate whether one or more of the many other phytochrome-regulated enzymes in plants also give evidence of being modulated by calmodulin. Such enzymes would be expected to show a sensitivity to inactivation by EGTA (that is, show some Ca^{2+} dependency) and to have a Ca^{2+}-dependent affinity for immobilized calmodulin columns (Dieter & Marmé, 1980b; Jarrett *et al.*, 1980).

Stimulus induced increases in cytosolic Ca^{2+} concentrations tend to be transient: Ca^{2+}-sequestering and export mechanisms tend to lower Ca^{2+} levels back to pre-stimulus levels soon after the stimulus has ended (Baker, 1976). If this hypothesis is true also for phytochrome-stimulated Ca^{2+} fluxes, then enzymes which show more rapid activation kinetics following the photoconversion of phytochrome would be more likely candidates also to show some Ca^{2+}/calmodulin dependency.

A second prediction of the model is that phenothiazine drugs, which block the binding of Ca^{2+}-activated calmodulin to its target enzymes (Levin & Weiss, 1978), should interfere with some phytochrome responses. Such a prediction assumes that the drug will be taken up by cells in sufficient quantity to block a significant percentage of the calmodulin present. Although the actual binding of phenothiazines to calmodulin in an organism has not yet been demonstrated, several calmodulin-dependent processes in cells are known to be inhibited by these drugs, which is consistent with the notion that phenothiazines can complex with calmodulin *in vivo* (Weiss, Prozialeck, Cimino, Barnet & Wallace, 1980).

In preliminary studies conducted in our laboratory, Hale (1981) has shown that one phytochrome-dependent response in oats, the red-light inhibition of coleoptile elongation growth (DeLint, 1957), is significantly inhibited in plants that have been treated with low doses of the phenothiazine, chlorpromazine. The effective dosage of this drug (10^{-5} M) was low enough to permit the same rates of elongation as found in untreated plants. Labeling studies with ^{14}C-chlorpromazine indicated that under the experimental conditions used, significant quantities of the drug were being taken up by coleoptiles (R. L. Biro & S. J. Roux, unpublished observations). These data encourage further exploration of the phytochrome–calmodulin connection.

Elliot (1980) and Kordan (1980) have recently shown that relatively high doses of phenothiazines ($\geqslant 10^{-4}$ M) will disrupt certain plant growth functions. As Elliot pointed out, however, these results have to be interpreted cautiously because high doses of these drugs are known to have non-specific membrane effects (Seeman, 1972). Even at low concentrations of these drugs, observed phenothiazine effects only suggest a possible role for calmodulin in plants.

Hidaka has developed some highly specific anti-calmodulin drugs and some structural analogues which are as much as 100-fold less effective in binding to calmodulin and in blocking its activation of enzymes. Bolton, Chafouleas, Boyd & Means (1981) recently used one of these anti-calmodulin drugs, W13, and its dechlorinated (and inactive) analogue, W12, to demonstrate that calmodulin was an important regulator of the progression of cells from G1 to S phase in the cell cycle. The use of inactive analogues in parallel control experiments greatly facilitates the interpretation of how specifically an anti-calmodulin drug is blocking a calmodulin-regulated activity in cells. If these drugs are taken up effectively by plant cells, they may prove to be extremely valuable tools for probing calmodulin functions in plants.

Because current interest in investigating the role of calmodulin in plants is high, new information on this topic should begin to appear with increasing frequency in the near future. These data will inevitably comment, at least indirectly, on the plausibility of the proposed phytochrome–calmodulin model. The ultimate support for the model rests on the as yet only indirect data favoring the conclusion that phytochrome activation does indeed result in increases in the concentration of free Ca^{2+} in the cytosol. Because these increases are likely to be small (in an absolute sense) and transitory, direct confirmation of them will be a significant technical achievement. This effort, it may be confidently expected, will be aided by the new and improved methodologies (for example, Tsien, 1981) which have been and will continue to be inspired by the renewed interest in Ca^{2+} as a second messenger in stimulus–response coupling.

References

ANDERSON, J. M., CHARBONNEAU, H., JONES, H. P., McCANN, R. O. & CORMIER, M. J. (1980). Characterization of the plant nicotinamide dinucleotide kinase activator protein and its identification as calmodulin. *Biochemistry*, **19**, 3113–20.

ANDERSON, J. M. & CORMIER, M. J. (1978). Calcium-dependent regulator of NAD kinase in higher plants. *Biochem. Biophys. Res. Commun.* **84**, 595–602.

BAKER, P. F. (1976). The regulation of intracellular calcium. *Symp. Soc. Exp. Biol.* **30**, 67–88.

BOLTON, W. E., CHAFOULEAS, J. G., BOYD III, A. E. & MEANS, A. R. (1981). Anti-calmodulin drug inhibits G1/S transition in mammalian cells. *In Vitro*, **17**, 242.

BROWNLEE, C. & KENDRICK, R. E. (1979). Ion fluxes and phytochrome protons in mung bean hypocotyl segments. I. Fluxes of chloride, protons, and orthophosphate in apical and subhook segments. *Plant Physiol.* **64**, 211–13.

CALDWELL, C. R. & HAUG, A. (1981). Calmodulin stimulation of the barley root plasma membrane Ca^{2+}, Mg^{2+}-ATPase. *Plant Physiol.* **67**, (Supplement), 136.

CEDEL, T. E. & ROUX, S. J. (1980a). Further characterization of the *in vitro* binding of phytochrome to a membrane fraction enriched for mitochondria. *Plant Physiol.* **66**, 696–703.

CEDEL, T. E. & ROUX, S. J. (1980b). Modulation of a mitochondrial function by oat phytochrome *in vitro*. *Plant Physiol.* **66**, 704–9.

CHAFOULEAS, J. G., DEDMAN, J. R., MUNJAAL, R. P. & MEANS, A. R. (1979). Calmodulin. Development and application of a sensitive radioimmunoassay. *J. Biol. Chem.* **254**, 10262–7.

CHEUNG, W. Y. (1980). Calmodulin plays a pivotal role in cellular regulation. *Science*, **207**, 19–27.

CLARKSON, D. T. (1977). Membrane structure and transport. In *The Molecular Biology of Plant Cells*, ed. H. Smith, pp. 24–63. Berkeley: The University of California Press.

CUATRECASAS, P. (1974). Membrane receptors. *Annu. Rev. Biochem.* **43**, 169–214.

DE LINT, P. J. A. L. (1957). Double action of near infrared in length growth of the *Avena* coleoptile. *Meded. Landbouwhogesch. Wageningen*, **57**, 1–9.

DIETER, P. & MARMÉ, D. (1980a). Calmodulin activation of plant microsomal Ca^{2+} uptake. *Proc. Nat. Acad. Sci., USA*, **77**, 7311–14.

DIETER, P. & MARMÉ, D. (1980b). Partial purification of plant NAD kinase by calmodulin-sepharose affinity chromatography. *Cell Calcium*, **1**, 279–86.

DIETER, P. & MARMÉ, D. (1981a). Far-red light irradiation of intact corn seedlings affects mitochondrial and calmodulin-dependent microsomal Ca^{2+} transport. *Biochem. Biophys. Res. Commun.* **101**, 749–55.

DIETER, P. & MARMÉ, D. (1981b). A calmodulin-dependent, microsomal ATPase from corn (*Zea mays* L.). *FEBS Lett.* **125**, 245–8.

DREYER, E. M. & WEISENSEEL, M. H. (1979). Phytochrome-mediated uptake of calcium in *Mougeotia* cells. *Planta*, **146**, 31–9.

ELLIOT, D. C. (1980). Calmodulin inhibitor prevents plant hormone response. *Biochem. Int.* **1**, 290–4.

FROSCH, S., WAGNER, E. & MOHR, H. (1974). Control by phytochrome of the level of nicotinamide nucleotides in the cotyledons of the mustard seedling. *Z. Naturforsch.* **29c**, 392–8.

FURUYA, M., FREER, J. H., ELLIS, A. & YAMAMOTO, K. T. (1981). Electrostatic binding of proteins and phytochrome to differently charged liposomes. *Plant Cell Physiol.*, **22**, 135–44.

GEORGEVICH, G. (1980). Phytochrome interaction with lipid membranes. PhD Thesis, University of Pittsburgh.

GEORGEVICH, G., CEDEL, T. E. & ROUX, S. J. (1977). Use of ^{125}I-labeled phytochrome to quantitate phytochrome binding to membranes of *Avena sativa*. *Proc. Nat. Acad. Sci. USA*, **74**, 4439–43.

GIETZEN, K., TEJCKA, M. & WOLF, H. U. (1980). Calmodulin affinity chromatography yields a functional purified erythrocyte (Ca^{2+} + Mg^{2+})-dependent adenosine triphosphatase. *Biochem. J.* **189**, 81–8.

GITELMAN, S. E. & WITMAN, G. B. (1980). Purification of calmodulin from *Chlamydomonas*: calmodulin occurs in cell bodies and flagella. *J. Cell Biol.* **98**, 764–70.

GOPINATH, R. M. & VINCENZI, F. F. (1977). Phosphodiesterase protein activator mimics red blood cell cytoplasmic activator of (Ca^{2+}–Mg^{2+}) ATPase. *Biochem. Biophys. Res. Commun.* **77**, 1203–9.

HALE, C. C. (1981). Studies on the role of calcium in mediating phytochrome effects. PhD Thesis, University of Texas at Austin.

HALE, C. C. II & ROUX, S. J. (1980). Photoreversible calcium fluxes induced by phytochrome in oat coleoptile cells. *Plant Physiol.* **65**, 658–62.

HAUPT, W. & WEISENSEEL, M. H. (1976). Physiological evidence and some thoughts on localised responses, intracellular localisation and action of phytochrome. In *Light and Plant Development*, ed. H. Smith, pp. 63–74. London:Butterworths.

JAMIESON JR., G. A., BRONSON, D. D., SHACHAT, F. H. & VANAMAN, T. C. (1980). Structure and function relationships among calmodulins and troponin C-like proteins from divergent eukaryotic organisms. *Ann. New York Acad. Sci.* **356**, 1–13.

JARRETT, H. W., CHARBONNEAU, H., ANDERSON, J. M., McCANN, R. O. & CORMIER, M. J. (1980). Plant calmodulin and the regulation of NAD kinase. *Ann. New York Acad. Sci.* **356**, 119–29.

JARRETT, H. W. & PENNISTON, J. T. (1978). Purification of the Ca²⁺-stimulated ATPase activator from human erythrocytes. Its membership in the class of Ca²⁺-binding modulator proteins. *J. Biol. Chem.* **253**, 4676–82.

KIM, I.-S. & SONG, P.-S. (1981). Binding of phytochrome to liposomes and protoplasts. *Biochemistry*, **20**, 5482–9.

KORDAN, H. A. (1980). Largactil and hydroxylamine-induced geotropic disorientation in lettuce. *Z. Pflanzenphysiol.* **100**, 273–8.

LEONARD, R. T. & HODGES, T. K. (1973). Characterization of plasma-membrane associated adenosine triphosphatase activity of oat roots. *Plant Physiol.* **52**, 6–12.

LEVIN, R. M. & WEISS, B. (1978). Specificity of the binding of trifluoperazine to the calcium-dependent activator of phosphodiesterase and to a series of other calcium-binding proteins. *Biochim. Biophys. Acta*, **540**, 197–204.

LÜRSSEN, K. (1976). Counteraction of phytochrome to the IAA-induced hydrogen ion excretion in *Avena* coleoptile cylinders. *Plant Sci. Lett.* **6**, 389–99.

MANDOLI, D. F. & BRIGGS, W. R. (1981). Phytochrome control of two low-irradiance responses in etiolated oat seedlings. *Plant Physiol.* **67**, 733–9.

MARMÉ, D. (1983). Calcium transport and function. In *Encyclopedia of Plant Physiology*, New Series, vol. 15, *Inorganic Plant Nutrition*, ed. A. Laüchli & R. L. Bieleski. Berlin, Heidelberg & New York: Springer-Verlag, in press.

MAYO, M. A. & COCKING, E. C. (1969). Pinocytotic uptake of polystyrene latex particles by isolated tomato fruit protoplasts. *Protoplasma*, **68**, 223–30.

MEANS, A. R. & DEDMAN, J. R. (1980). Calmodulin–an intracellular calcium receptor. *Nature, London*, **285**, 73–7.

MUTO, S. & MIYACHI, S. (1977). Properties of a protein activator of NAD kinase from plants. *Plant Physiol.* **59**, 55–60.

NEWMAN, I. A. (1981). Rapid electric responses of oats to phytochrome show membrane processes unrelated to pelletability. *Plant Physiol.* **68**, 1494–99.

OGREN, W. L. & KROGMAN, D. W. (1965). Studies on pyridine nucleotides in photosynthetic tissue—concentrations, interconvertions and distribution. *J. Biol. Chem.*, **240**, 4603–8.

OHNISHI, S. T. (1978). Characterization of the murexide method: dual wavelength spectrophotometry of cations under physiological conditions. *Analyt. Biochem.* **85**, 165–79.

PIKE, C. S. & RICHARDSON, A. E. (1979). Red light and auxin effects on ⁸⁶Rubidium uptake by oat coleoptile and pea epicotyl segments. *Plant Physiol.* **63**, 139–41.

PRATT, L. H. (1978). Molecular properties of phytochrome. *Photochem. Photobiol.* **27**, 81–106.

PRATT, L. H. (1979). Phytochrome: function and properties. *Photochem. Photobiol. Rev.* **4**, 59–124.

PRATT, L. H. (1980). Phytochrome pelletability induced by irradiation *in vivo*. *Plant. Physiol.* **66**, 903–7.

QUAIL, P. H. (1981). Intracellular location of phytochrome. *Proceedings of the Eighth International Congress on Photobiology*, in press.

QUAIL, P. H. & BRIGGS, W. R. (1980). Phytochrome pelletability induced by irradiation *in vivo*. Mixing experiments. *Plant Physiol.* **66**, 908–10.

RACKER, E. (1976). *A New Look at Mechanisms in Bioenergetics*. New York: Academic Press.

RASMUSSEN, H. (1970). Cell communication, calcium ion, and cyclic adenosine monophosphate, *Science*, **170**, 404–12.

ROTH-BEJERANO, N. & KENDRICK, R. E. (1979). Effects of filipin and steroids on phytochrome pelletability. *Plant Physiol.* **63**, 503–6.

ROUX, S. J. (1974). Conductance changes induced by the glycoprotein, phytochrome, in model membranes. In *Proceedings of the Annual European Symposium on Plant Photomorphogenesis*, ed. J. A. De Greef, pp. 1–3. Antwerp: The University of Antwerp Press.

ROUX, S. J. & YGUERABIDE, J. (1973). Photoreversible conductance changes induced by phytochrome in model lipid membranes. *Proc. Nat. Acad. Sci., USA*, **70**, 762–4.

ROUX, S. J., MCENTIRE, K., SLOCUM, R. D., CEDEL, T. E. & HALE II, C. C. (1981). Phytochrome induces photoreversible calcium fluxes in a purified mitochondrial fraction from oats. *Proc. Nat. Acad. Sci., USA*, **78**, 283–7.

RÜDIGER, W. (1980). Phytochrome, a light receptor of plant photomorphogenesis. *Struct. Bond.* **41**, 101–41.

SATTER, R. MARINOFF, P. & GALSTON, A. W. (1970). Phytochrome controlled nyctinasty in *Albizzia julibrissin*. II. Potassium flux as a basis for leaflet movement. *Amer. J. Bot.* **57**, 916–26.

SCHOPFER, P. (1977). Phytochrome control of enzymes. *Annu. Rev. Plant Physiol.* **28**, 223–52.

SEEMAN, P. (1972). The membrane action of anesthetics and tranquilizers. *Pharmacol. Rev.* **24**, 583–655.

SLOCUM, R. D. (1981). Studies on the localization of phytochrome and calcium in light and gravity-stimulated plants. PhD Thesis, The University of Texas at Austin.

TEZUKA, T. & YAMAMOTO, Y. (1972). Photoregulation of nicotinamide adenine dinucleotide kinase activity in cell-free extracts. *Plant Physiol.* **50**, 458–62.

TEZUKA, T. & YAMAMOTO, Y. (1975). Control of ion absorption by phytochrome. *Planta*, **122**, 239–44.

TSIEN, R. Y. (1981). A non-disruptive technique for loading calcium buffers and indicators into cells. *Nature, London*, **290**, 527–8.

VAN ELDIK, L. J., GROSSMAN, A. R., IVERSON, D. B. & WATTERSON, D. M. (1980). Isolation and characterization of calmodulin from spinach leaves and *in vitro* translation mixtures. *Proc. Natl. Acad. Sci., USA*, **77**, 1912–16.

VAN ELDIK, L. J., PIPERNO, G. & WATTERSON, D. M. (1980). Similarities and dissimilarities between calmodulin and a *Chlamydomonas* flagellar protein. *Proc. Natl Acad. Sci., USA*, **77**, 4779–83.

VERBELEN, J.-P., BUTLER, W. L. & PRATT, L. H. (1981). Immunocytochemical localization of phytochrome in etiolated, illuminated and light-grown plants. In *Proceedings of the European Symposium on Light Mediated Plant Development*, Book of Abstracts, Abstract 8.19.

VIESTRA, R. D., TOKUHISA, J. G., NEWCOMBE, E. H. & QUAIL, P. H. (1981). A solid-phase antibody approach to identifying phytochrome-bearing structures in particulate fractions. *Plant Physiol.* **67** (Supplement), 130.

WAGNER, G. (1974). Some physiological properties of phytochrome in the alga,

Mougeotia, as studied by cytochalasin B and aminophylline. In *Proceedings of the Annual European Symposium of Plant Photomorphogenesis*, ed. J. A. De Greef, pp. 22–25. Antwerp: The University of Antwerp Press.

WATSON, P. J. & SMITH, H. (1981). In-vivo changes in etiolated oat seedling supernatant and pelletable phytochrome during a dark incubation after red irradiation. In *Proceedings of the European Symposium on Light Mediated Plant Development, Book of Abstracts*, Abstract 9.17.

WEISENSEEL, M. H. & RUPPERT H. K. (1977). Photochrome and calcium ions are involved in light-induced depolarization in *Nitella*. *Planta*, **137**, 225–9.

WEISS, B., PROZIALECK, W., CIMINO, M., BARNETT, M. S. & WALLACE, T. L. (1980). Pharmacological regulation of calmodulin. *Ann. New York Academy of Sciences*, **356**, 319–45.

EPILOGUE

By way of summing up the Symposium this epilogue attempts to draw attention to some of the common principles and differences defined during the meeting. Written from the viewpoint of a plant physiologist, it represents only one impression.

The starting point was to consider photoreceptor pigments and their physiological properties. The molecules adopted for photobiology have large extinction coefficients for efficient absorption and high quantum yields for photochemical reactions (Presti). There are perhaps no more than half a dozen kinds which act as major biological photoreceptors: the chlorophylls, flavins, carotenoids, rhodopsins and the biliproteins such as phycocyanin, phycoerythrin and phytochrome. Nevertheless, each of these few types of photoreceptor pigment, variously modified and organised into different structures, may supply an organism with information about the intensity, spectral composition, direction, duration and polarisation of the light in its environment, and may also be used to capture light energy for biological processes.

When considering photoreceptor physiology, the distinction between those photoreceptor systems which function to capture light energy for biological processes and those which give information about the radiation environment was clearly drawn. An important distinction can also be made between systems where information about the light environment is processed to modulate behaviour (vision, photomovements) and those where the information is processed to make long-lived changes in development (photoperiod-dependent changes). The effector chains are likely to be very different at least in the later stages of transduction, for the latter processes involve cellular differentiation. Whether the immediate events following photon capture are the same remains unanswered.

Biological photoreceptors probably evolved from molecules having other functions in more primitive organisms, for example: haem to chlorophyll; phytochrome and phycobilins and β-carotene to retinal

(Presti). Is it, however, logical to conclude that each signal-transducing photoreceptor will have only a single specific function from which its mechanism of action might be deduced (Smith)? Several of the examples given at the meeting argue against this concept. Consider how the same chromophore is used in functionally different pigments: in *Halobacteria* a retinal–protein complex is used for energy capture via a light-driven proton pump (Hildebrand), and yet retinal–protein complexes are also widely used in vision. The photoreceptor of *Stentor* controls two different mobile responses: one depending on the perception of an intensity change with time, and the other depending on the detection of a gradient in space (Song). Again, the UV photoreceptor in insects probably evolved in the functional context of exploiting skylight cues for orientation by detecting patterns of polarisation, but is also used for recognising floral patterns (Wehner). So, even in the same organism, more than one function can be attributed to a single photoreceptor, presumably involving differences in its organisation within the cell and possibly differences in its mechanism of action. Phytochrome also appears to have more than one function: to detect light quality (responses to shade); to detect direction (chloroplast orientation); and to discriminate between light and dark (photoperiodism). Thus the same photoreceptor has evolved to detect different properties of light and it is not clear which, if any, of these can be called the fundamental perceptual function from which the basic mechanism of action might be deduced. Perhaps more than one mechanism of action has also evolved.

The functional approach which emphasises the operation of a photoreceptor in the natural environment of the organism can, nevertheless, be valuable in elucidating photoreceptor physiology. For example, Burkhardt pointed out the necessity for carrying out behavioural studies to determine the communcation value of colour vision in various organisms. Under water the exploitation by red algae of the chlorophyll–carotenoid window (Barber), and the correlation between the preferred depth at which fish swim and their peaks of spectral sensitivity are interesting examples of adaptation to an ecological niche. Likewise visual perception at the far-red end of the spectrum by some tropical insects is probably an adaptation to a dense vegetation habitat (Burkhardt). The fascinating discussions which considered how different properties of light such as polarisation (Wehner), direction (Haupt, Colombetti & Lenci) and spectral composition (Burkhardt, Smith) can be perceived and made use of by a variety of plants and animals confirmed the value of this functional approach.

Photobiology can be considered as photochemistry in membrane systems, and the importance of the organisation of photoreceptors within membranes was emphasised again and again. In photosynthesis, where light is used as a source of energy, the arrangement and photochemical characteristics of the accessory pigments lead to high efficiency in light harvesting by facilitating energy transfer among molecules of the light-harvesting antennae and, finally, to the reaction centre pigments (Barber). A remarkably similar arrangement between a photostable pigment and rhodopsin was considered for the visual process in flies (Franceschini). Fluidity of the membrane and movements in the lipid bilayer allow sequential interactions by components, as for example in vision (Chabre) and photosynthesis (Barber, Mathis). A proton gradient across the membrane is essential for ATP formation in photosynthesis (Mathis) and several other photoresponses appear to depend on the transfer of ions across membranes (Galston, Hidebrand, Stieve, Kendrick). An ordered environment may also be important to populate the lowest excited singlet state of chlorophyll (Mathis) and is clearly essential for the perception of polarised light (Haupt, Wehner).

Is phytochrome an exception? Although it can associate with membranes in model systems (Song, Roux), it appears to be present mainly as a soluble protein *in vivo*. Evidence for a functional association with membranes comes mainly from physiological studies (Galston, Kendrick), the most telling of these being the polarised light studies of chloroplast movement where photoreceptor orientation must occur (Haupt). Present evidence does not suggest that phytochrome is a transmembrane protein like rhodopsin. Another unusual feature of phytochrome is that it appears to be active in the ground state (Kendrick), rather than in the excited state as in other biological systems (Chabre, Presti). A comparison of rhodopsin with phytochrome is interesting. Isomerisation of the retinal chromophore seems to be common to all visual systems (Chabre); in phytochrome, however, although the chromophore appears to undergo re-orientation with respect to the protein (Song), there is no gross photoisomerisation. In rhodopsin, isomerisation of the retinal chromophore leads to the formation of an active site in a part of the protein distant from the chromophore (Chabre), whereas in phytochrome the (possible) active site on the protein appears to lie close to the chromophore (Song). Thus, in these major light-sensing pigment systems of plants and animals, differences rather than similarities in the early events following photon capture were highlighted.

The possiblity that calcium is a transmitter in vertebrate vision was proposed in 1971, yet today the hypothesis remains neither proven nor disproven. It appears unlikely that calcium is a transmitter for either excitation or adaptation in vertebrates (Stieve), but there is evidence for such a role in the adaptation process in invertebrates (Hillman). It was also suggested some five years ago that calcium may be a transmitter/messenger in phytochrome-mediated processes; evidence to support this concept was presented (Roux), but there are few direct data and the theory is by no means firmly established (Kendrick). Phytochrome studies are perhaps in the same position as were vision studies a few years ago, and the future of the 'calcium as transmitter' hypothesis is uncertain. Calcium does, however, appear to have some function in transduction or associated processes in many systems. Examples discussed at the meeting included: phytochrome-dependent chloroplast orientation (Haupt); ciliary beating in *Stentor* (Song); calcium-dependent phosphorylation of a gating protein or calcium binding in vision (Stieve); and photokinesis in *Chlamydomonas* (Nultsch).

The role of ion movements in transduction processes was discussed by several speakers. The elegant studies of leaf movements described by Galston showed that a massive re-distribution of potassium is necessary for the opening and closing responses to light and is probably coupled to a proton pump. The mode of action of phytochrome in other systems may also involve an ion pump or the opening of ion channels (Kendrick) and, of course, the opening and closing of sodium channels is an essential component of transduction in vision (Chabre, Stieve). The importance of ion transfers across membranes was also discussed in relation to the capture of light energy: in light-driven proton and sodium pumps in *Halobacteria* (Hildebrand) and proton transfer coupled to electron flow in photosynthesis (Mathis).

The meeting undoubtedly provided a valuable forum for the exchange of ideas about photoreceptor physiology in plants and animals and for the exchange of information about techniques for studying various photobiological problems on different time scales and degrees of biological complexity. This exchange would have been facilitated had there been a common usage of terminology and units by all contributors!

Daphne Vince-Prue

AUTHOR INDEX

Numbers in italic type indicate pages in the reference lists

Author index

SUBJECT INDEX

Subject index